Aquaculture in Shallow Seas:
Progress in Shallow Sea Culture

EDITOR

Takeo Imai

TRANSLATED FROM JAPANESE

1982

A. A. BALKEMA / ROTTERDAM

Translation of:
Senkai Kanzen Yoshoki (Senkai Yoshoku no Shinpo).
Koseisha Koseiku Publishers, Tokyo, 1971

Translator: Miss M.G. Alamelu
General Editor: Dr. V.S. Kothekar

ISBN 90 6191 022 6

Printed in India at Pauls Press, New Delhi

Contributors

Akiyama

Ito Susumu

Iino Takashi

Imai Takeo

Iwasaki Hideo

Oizumi Shigekazu

Oba Toshio

Kittaka Jiro

Kanno Hisashi

Kikuchi Syougo

Kurata Hiroshi

Kurogi Munehira

Koganezawa Akimitsu

Saito Katsuo

Sakai Seiichi

Sato Ryuhei

Shibui Tadashi

Shiraishi Kagehide

Sugawara Yoshio

Nishikawa Nobuyoshi

Numachi Renichi

Fuji Akira

Hudinaga Motosaku

Mori Katsuyoshi

Yamamoto Gotaro

Hoshido Tadao

Foreword

Multiple resources for mankind exist in the sea. Japan, under the auspices of the International Biological Program, is primarily concerned with augmenting mankind's food supply. Basic policies regarding Japan's approach to developing food resources from the ocean were formulated last year by the Ocean Science and Technology Council (*Kaiyo Kagaku Gijutsu Shingikai*). These policies led to the establishment of an important project, the Research and Development Program, whose aim is to cultivate fisheries. The number of such international and cooperative research projects, all concerned with marine resources and their effective utilization, have increased significantly in recent years, although some are still proposals rather than actualities. In theoretical and practical research, and in technical *savoirfaire*, particularly in the development of techniques for shallow sea culture, Japan's contribution is unique; quite understandably therefore, she leads the world in these investigations. Other nations, however, are rapidly gaining interest in coastal aquaculture.

Despite the problems relating to industrialization and urbanization, the marine products industry has been interested in coastal aquacultural enterprises for some time. Adaptations to the changing social milieu will inevitably have to be considered by those concerned with furthering the development of coastal aquaculture, and it is heartening to see that some steps have already been taken in the past twenty years in the improvement of technological methods. Spectacular advancements have been made in the following aquaculture fields: seaweed culture such as laver (*Undaria pennatifida*) and tangle, bivalve suspension culture with oysters and scallops, and their related seeding technology, prawn (*Penaeus japonicus*) and abalone culture, and their related seed stock production, and finally, the culture of several species of fish. Taking an overall view, the government has initiated developmental and structural reforms in oceanic research which, while making qualitative adaptations necessary, has resulted in an acceleration of aquaculture studies. Now, not only Kyushu, Tohoku, and Hokkaido, but regions further up the Japanese coast, are concerned with the study of the nature of marine organisms. Furthermore, aquaculture is now collaborating with civil and mechanical engineers to further marine investigations. The propagation and culture of marine organisms, or the development of commercial farming, is impossible, of course, without basic research concerned with the organisms *per se*. We believe that biological production through cultural operations can only progress as a science and a viable industry to the extent that basic research and practical techniques are simultaneously employed.

Professor Takeo Imai (Professor Emeritus, Tohoku University) has devoted considerable energy to the study of shallow-water culture, and his achievements in this field are worthy of note. On the occasion of his retirement in March, 1967 from the College of Agriculture, Tohoku University, we summarized the biological researches to date and assessed the current aquaculture techniques applied to the following marine organisms: laver, oysters, scallops, abalones, and prawns. The progress in the culture of these species over the past few years is very significant. Our purpose in summarizing that progress was to focus attention on the course that future studies in the field of aquaculture should take. Obviously, some of our research is still not finalized, but nevertheless, we decided to publish the results of the data processed over the past three years.

We sincerely hope that the development of appropriate research methodology and practical techniques will eventually establish aquaculture as a separate science, and place "through culture" on a firm foundation. If this publication assists in the attainment of these objectives, we will be more than gratified.

BOARD OF EDITORS

Preface

Japanese aquaculture has utilized the natural resources of the sea to maximal advantage, and augmented those resources by artificial inducements very successfully. Achievements are particularly striking with regard to many species of fish and seaweed. Progress in these areas is primarily due to the development of techniques through sustained experimentation by scientists in fishery research. The epochal advancement in technology is, undoubtedly, successful seed stock production, as this is the most important single factor for effecting a breakthrough in coastal aquaculture, i.e., the production of quality seed by artificial stimulation.

Biological data based on research have been accumulating since the beginning of the Showa Period (1926). Studies concerning the life history of marine organisms important to aquaculture, such as their feeding habits, reproductive cycle, and environmental requirements, had to precede studies on the techniques for artificial breeding and culture in natural surroundings, in water tanks, and in ponds. The development of life-sustaining techniques brought about improvements in aquaculture fisheries, which led to additional areas of research. Information about feed cultivation, nutrition, environmental control, and how best to sustain life in various environments, is making it possible for man to control the entire reproductive cycle of some marine organisms, from the collection of seed and seed stock, through the various stages of life right up to full maturity. As a result, a broadening of the traditional, quantitatively oriented, biological production to a qualitatively oriented aquaculture is foreseeable.

The process of artificial and planned biological production, which begins with seed and seed stock and results in a mature adult, I have named "through culture" (*kanzen yōshoku*). Though the species presently under cultivation and hence subjects of "through culture" are few, I am firmly convinced that the expanding field of basic research and technical developments will bring many additional species of shallow-water, aquaculture, marine organisms within its purview.

In the present paper, we have focused attention on just five of the marine organisms currently under investigation through aquaculture. All these organisms are endowed with biological traits proper to through culture, though their suitability for culturing varies greatly. These organisms include laver, oysters, scallops (whose culturing has developed remarkably in recent years), abalones (an important seashore resource for which artificial breeding is now being established), and prawns (for which seeding techniques are now well-developed). The status of the basic research and the technical achievements for each of these organisms has been summarized, and we hope that this paper will prove useful in indicating what areas further research should explore.

This paper is by no means a definitive statement on aquaculture. Since the inception of our work three years ago however, we have been able to process data with regard to the biological characters of each of the organisms mentioned in the preceding paragraph, and the techniques used in culturing them. Though our findings are tentative, we nonetheless decided to make them public. Any additions or comments from colleagues engaged in the same field, or an allied one, which would make the present work more complete, are welcomed.

We sincerely thank the members of the editorial staff who worked tirelessly to bring this book to fruition. Professor Munehisa Kuroki of Hokkaido University, Professor Gotaro Yamamoto of the Tokyo University Oceanic Research Institute, Professor Motosaku Hudinaga, Managing Director of Hudinaga Prawn Research Institute, and Professor Ryuhei Sato, Assistant Professor at Tohoku University deserve special mention. We also thank our colleagues and others who contributed numerous manuscripts and data, and those who offered suggestions. We also acknowledge our indebtedness to those who edited, proofread, and handled other office tasks so patiently, particularly Mr. Takeo Sugano, Section Chief at Tohoku Regional Fisheries Research Laboratory, Mr. Yoshio Sugahara and Mr. Katsuyoshi Mori, Teaching Assistants at Tohoku University College of Agriculture.

We are deeply grateful to the members of the staff of Koseisha Koseikaku Publishers whose efforts resulted in the present publication.

December 1, 1970   TAKEO IMAI
The entire volume is in Japanese

Contents

Part I. The Evolution of Seaweed Culture

Part II. The Evolution of Oyster Culture

Part IV.   The Evolution of Abalone Culture

Part V.   The Evolution of Prawn Culture

PART I

The Evolution of Seaweed Culture

CHAPTER I

Biological Research on Seaweed

1. Introduction

The historical background of various aspects of seaweed research in Japan is given in nearly every section of this work; the following is therefore a summary of the work completed to date.

Although Suringar (1870) was the first to refer to nori, his only contribution was to register its scientific name. Kjellman (1897) was the first to initiate systematic research on Japanese seaweeds. Among the Japanese workers, Okamura (1900, 1901) studied the ecology and germination of *Porphyra tenera* (Asakusa nori). His valuable book entitled *Asakusa Nori* (1909), contains an elaborate account of the historical background of seaweed culture as well as the various methods of culture, ecology, and classification. During 1909 to 1920, Yendo published several papers on the classification of seaweeds. In 1930, Okamura published a book entitled *Suisan Shokubutsu* (*Marine Plants*) in which he described the growing beds, the life history, the bamboo pole installation (hibitate) the significance of net exposure, and the holdfasts of seaweeds as these pertained to *P. tenera*.

In 1932, Ueda carried out a systematic study, following the plan of Okamura's work, on the *Porphyra* of Japan. He examined not only the cultured laver, but also the classification of 18 types of *Porphyra*, including the so-called rock seaweed (iwa-nori), and added new details to the data already published. Ueda is said to have laid the foundation for *Porphyra* classification. Following the classification of *Porphyra*, a number of papers were published by Tanaka (1952) and several others; thus about 30 species of *Porphyra* have been identified and classified to date.

Kunieda (1939), Suto (1950, 1957), Arasaki *et al.* (1952), Arasaki (1957, 1957a) and Kurogi (1957, 1961) wrote papers on the variability or quality of culturing. *P. tenera* and *P. yezoensis* are the most widely cultured seaweeds, followed by *P. kuniedai*, *P. angusta*, and *P. pseudolinearis*; however, a few other species have also been cultured. Among the cultured seaweeds, efforts are being made to improve the quality of laver by hybridization experiments (Suto, 1963).

The study of the life history of seaweeds progressed along with their classification. Particular attention was given to the fate of carpospores which formed in the winter, and the reappearance of spores in the autumn; this aspect was studied by Yendo in 1919. Okamura and his colleagues (1901, 1909, 1920) expressed the view that carpospores either remained at the bottom of the sea or floated freely. In this regard, a "summer seaweed" hypothesis was formulated, following the discovery by Ueda (1929, 1937) and Kusakabe (1929) of the formation of neutral spores (monospores) by summer seaweeds. Kunieda (1939), on the other hand, published a "dormancy" hypothesis for carpospores on the basis of the results obtained from a different course of investigation. Validity was claimed with equal force for both hypotheses.

In 1949, Drew reported the presence of the Conchocelis phase in *P. umbilicalis*, and Kurogi (1952, 1953) confirmed it in *P. tenera* and a few other species of *Porphyra*. The life history of *Porphyra* was understood with somewhat greater clarity following the discovery in 1956 and 1961 that, in addition to the presence of Conchocelis in *P. kuniedae*, the plant also passes the summer in the form of summer seaweed, depending upon the period and the place. This knowledge was utilized in the research carried out after World War II, and artificial seedlings were tried out by several workers using the Conchocelis form. This brought about a great change in seaweed culture, and considerably advanced it.

Ueda's *Textbook on Seaweed Culture* (*Kaiso Yoshoku Dokuhon*, 1952, 1958) summarizes the research and culture techniques of seaweed up to the early part of the postwar period.

Cytological research, especially on nuclear development, was carried out by Ishikawa (1921) with reference to the life history of seaweed. The observations of Dangeard (1927), who also worked on this aspect, differed from those of Ishikawa however. Nevertheless, both agreed that *Porphyra* is haploid and that meiosis occurs at the time of carpospore formation, following fertilization. This was the view held until Magne (1952) created some doubts; then Toyama (1957), Yabu and Tokida (1963), and others in Japan, insisted that the meiotic division did not occur after fertilization. According to Migita (1967), who explained the nuclear phases occurring throughout the entire life cycle of *Porphyra*, the meiotic division occurs when spores form from Conchocelis.

As yet the ecology of seaweed has not been studied in detail. However, the spore-forming period, the holdfast, the vegetation period, and the habitat have been under observation for a long time (Okamura, 1909, 1930). The study of holdfasts led to the discovery of horizontal hibi (nets fixed to bamboo poles) made from soda-hibi (twigs used for seaweed culture), which were in use in the early Showa period (beginning from 1926) (Ueda, 1952; Kaneko, 1931 to 1937). The changeover from a soda-hibi to a fixed or floating type of horizontal bamboo network came about mainly from the study of holdfasts of seaweed and the period and location of spore germination.

After World War II, the germination, growth, disappearing period, types of reproduction, and the relationship between the vegetative period and asexual reproduction, as well as the seasonal changes in holdfasts and their distribution, were studied in detail. These studies were carried out by Kurogi (1957, 1961), Kurogi and Kawashima (1966), Fukuhara *et al.* (1963, 1968) on cultured *Porphyra* and rock seaweed (iwanori). Yoshida *et al.* (1964) studied the changes in the density of epiphytes on the net, changes in growth, and the relationship of both to the yield rate. Satomi *et al.* (1968) estimated the production of seaweed from photosynthesis and respiration rate. They studied the yield rate per unit, taking losses into account. Earlier he and his colleagues published their findings on the changes in the photosynthetic rate resulting from the aging of seaweed (Satomi *et al.*, 1967). There are also reports by Furuhata and Ogawa (1957), Katada and Furuhata (1962), Tokida *et al.* (1961), and others on the culture of iwanori.

In order to identify those environmental conditions suitable for vegetation, studies were carried out on habitats by Matsue (1936), Matsudaira and Iwasaki (1953), Iwasaki and Matsudaira (1954, 1954a), Matsudaira and Hamada (1966), Tanida and Okuda (1956), Matsumoto (1959), and others; they analyzed the relationship of salt-nutrients, currents, etc., to the growth, and quality, or nitrogen content of sea-weeds. It should be mentioned that a study of seaweed habitats also includes an analysis of questions relating to fertilizers.

The physiology of seaweed culture constituted the subject matter of papers published prior to World War II by Fujikawa (1932 to 1937) and Takayama (1937). While Fujikawa studied growth in relation to water temperature, salinity, current, and nitro-gen in sea water, and the relationship between growth and quality of light, and the effect of cold storage on seaweed net, Takayama's studies were concerned with growth in relation to salinity. However, these studies of the prewar period were not compre-hensive.

From the postwar period, the growth of seaweed has been studied in detail in rela-tion to various factors such as water temperature, salinity, water current, salt-nutrients, exposure, drying, freezing, brightness, quality of light, etc. The influence of minute quantities of minerals, vitamins, growth stimulants, sunshine, and alternating periods of brightness and darkness on the growth of seaweed, have also been studied. Matura-tion in relation to these factors was carefully scrutinized. Sano (1955, 1955a), Iwasaki alone, or in collaboration with Matsudaira (1956, 1957, 1958, 1961, 1965), Tsuruga and Arata (1957), Kinoshita and Teramoto (1958, 1960, 1962), Matsumoto (1959), Akiyama (1959), Maekawa *et al.* (1959), Suto (1960, 1961, 1961a), Nakatani *et al.* (1961, 1962), Shimo and Nakatani (1963, 1964), Kanazawa (1963), Migita (1964), Shinmura (1965), and others, have done detailed research on chromatophores.

These investigations have made it possible to devise an indoor culture method for seaweed, and have contributed to the field of seaweed physiology. They also laid the foundation for the development of a method of culturing seaweed on land (Suto, 1960, 1961). Studies on seaweed freezing resulted in the perfecting on a technique for freezing seed net (Kurakake, 1966).

Reports on the diseases of seaweed appeared as early as 1909 by Okamura, and in 1909, 1926, 1930 by Okamura *et al.*; they contained an analysis of the disease caused by diatoms, and frost damage. Arasaki studied red rot and white rot diseases (1947). Soon after World War II, seaweed diseases became a major subject of study; bud damages, hole rot, tumor, diseases caused by ascidian bacteria and *Leucotrix*, cold damage, etc., were investigated (Akasaka, 1956; Tamura, 1956; Suto and Ume-bayashi, 1954; Nozawa, 1954, 1957; Nozawa and Nozawa, 1957, 1957a; Toyama, 1957; Arasaki and Arasaki *et al.*, 1956, 1960). There are still a number of unknown facts relating to the pathology of seaweed which need to be studied.

Once the life cycle of seaweeds was understood, a number of investigators concen-trated on the study of the Conchocelis phase in relation to artificial seedlings. The starting point of the investigation was a study of the morphology and growth of Con-chocelis, and the course of maturation and spore release. Later investigations covered a wide field, including the relationship of growth and maturation of the Conchocelis and spore release to water temperature; the relationship of spore release to brightness; the relationship between growth and maturation to the intensity or quality of light, photoperiodism, and the quality of water, e.g., salinity, pH, and salt-nutrients; the influence of growth stimulants; the problems relative to drying and freezing; and finally, the relationship between spore attachment and brightness. In addition to these aspects, investigations were also carried out on the influence of salt-nutrients, vitamins, and minerals on the growth of seaweeds by Kurogi and his colleagues (1952 to 1967), Suto *et al.* (1954), Takeuchi (1954), Suto *et al.* (1954, 1954a, 1957), Ogata (1955, 1961), Saito (1956, 1956a), Toyama *et al.* (1956), Maekawa and Toyama (1958), Migita, Migita and Abe (1959, 1962, 1966, 1967), Iwasaki (1961, 1965, 1967), and Honda (1962, 1964).

When it became possible to use artificial seedlings, diseases occurred in the Conchocelis phase at times during the culture, the nature of which is being studied (Arasaki *et al.*, 1956; Nozawa and Nozawa, 1957). The actual method of using artificial seedlings has been dealt with in several handbooks such as *Handbook of Artificial Seedlings (Jinko Saibyo no Tebiki)* by Kobayashi (1960) and *Handbook of Indoor Seedlings (Shitsunai Saibyo no Tebiki)* by Honda (1964).

A review of the investigations carried out from the Meiji era (1859-1911) to the present time shows that very striking progress was made in the postwar period, especially during 1955-1965. The annual production at present is around 3,500,000,000 sheets and the number of enterprises between 62,000 to 63,000. The pace of development is seen in the expansion of the growing beds and the common use of horizontal bamboo networks soon after the end of the war. Artificial seedlings, floating culture, synthetic fiber nets, power ships, etc., have become common since 1955. Of these, floating culture plays an important role in the widening of growing beds, which are termed "open-sea culture." Together with an increase in the production of thermal drying equipment in 1960, and the production of a seaweed processing machine in 1962, the use of a seaweed harvesting machine became common. The cold storage of seed net was first tried in 1965.

Seaweed investigations are carried out and reported on at various Fishery Experiment Stations in cities, in prefectures, and in the Nation of Japan. Other sources also provide valuable reference material, such as:

Suisan Zōshoku Danwakai (1957), *Suisan Zōshoku,* Vol. 4, No. 4 (Special Issue on Seaweeds).

Setonaikai Suisan Kaihatsu Kyōgikai (1958, 1959), *Setonaikai Kaisō Kenkyū Shūhō* 1, No. 2.

Seikaiku Suisankenkyūsho-Ariake-kai Suisan Kenkyūkai (1958, 1960), *Nori yōshoku* 1, 2.

Tōhokuchiku Kaisō Byōgai Bōcho Taisaku Kyōgikai (1961).

Research Report on Prevention of Diseases in the Seaweeds of Tōhoku (1958 to 1960).

Suisancho Chōsakenkyūbu Kenkyū Nika (1965).

Research Report on Coastal Aquaculture Techniques with Special Reference to Locations and Types.

2. Classification and morphological and ecological characteristics

The ordinary seaweeds, popularly known as nori in Japan, come under the genus *Porphyra* belonging to Phylum Rhodophyta, Class Protoflorideae, Order Bangiales, and Family Bangia.

In Europe, records relating to *Porphyra* can be found in systematic studies made as early as the 18th century. In 1753, Linnè included *Porphyra* under *Ulva* (at present, this belongs to the genus *Ulva*). This inclusion continued for some years, and it was not until 1824 that the word *Porphyra* was first used by C. A. Agardh; since that time this name has persisted. *Porphyra* is widely distributed both in the northern and southern hemispheres. It has been well-studied on the Atlantic coasts of Europe and North America, and the Pacific coasts of North America, Japan, Australia, and New Zealand. The studies of *Porphyra* on other sea coasts are not always satis-factory. The structure of *Porphyra* is quite simple and hence it is very difficult to find any special characteristics for identification and classification. While it is possible that there may be some species as yet unidentified, a doubt exists that all the species reported thus far would be accepted without question. Nevertheless, there are about fifty species of *Porphyra* recognized throughout the world (includ-ing those of Japan), although the classification of *Porphyra* needs further investigation.

Porphyra vulgaris Ag. of Suringar (1870) is the first species recorded in the systematics of the *Porphyra* of Japan based on the collec-tion by Textor. Kjellman then recorded six species in 1897. Yendo (1909, 1913, 1915, 1916, 1920) added six more species, and Oka-mura (1916) another one. Thus by 1920, the number of recognized species of *Porphyra* was 13. Ueda later carried out a comprehensive study of Japanese *Porphyra* (1932), adding five new species and four varieties, thus laying the foundation for the classification of *Porphyra* in Japan. After Ueda, several papers on new varieties and new species, and a re-examination of old species, were published by a number of investigators (Kawabata, 1936; Nagai, 1941; Tanaka, 1952; Katada, 1952; Ohmi, 1954; Mikami, 1956; Kurogi, 1957, 1961, 1963; Miura, 1961; Noda, 1964; Fukuhara, 1968). At present, there are about 30 species of *Porphyra* in Japan, including some varieties not yet studied. Throughout the world, there are a number of species of *Porphyra*, of which three or four are common to both the Atlantic coast and the eastern Pacific.

Subgenus *Porphyra*

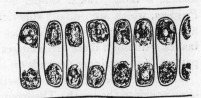

Subgenus *Diplastidia*

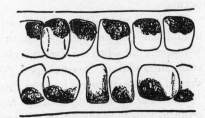

Subgenus *Diploderma*

Figure 1.1. Comparison of the three sub-genera *Porphyra*, *Diplastidia*, and *Diploder-ma*, on the basis of transverse sections.

2.1 Morphological characteristics

2.1.1 DIFFERENTIATION OF SUBGENERA

Porphyra has a leafy appearance with different shapes. It may have a single or double layer of cells (Figure 1.1). Kjellman (1883) distinguished the double-layered *Porphyra* from the single-layered, and placed it under a separate genus, *Diploderma*. But Rosenvinge (1893) classified the double-layered type as a sub-genus of *Porphyra*. He placed the single-layered form under the subgenus *Euporphyra*. Thereafter, Tokida (1935) discovered that the cells of some forms had a single chro-matophore while those of others had two. He included the *Porphyra* which have cells with two chromatophores in a new subgenus, *Diplastidia*. There are now three sub-genera under *Porphyra*. As the subgenus *Euporphyra* is included under the subgenus *Por-phyra*, according to the latest nomenclature and classification, the three subgenera are *Porphyra, Diplastidia,* and *Diploderma*. However, even in the subgenus *Diplastidia,* the cells in the young stage have only one chromatophore in some forms (*P. pseudocrassa,* Kurogi, unpublished), while in others the central cells have two chromatophores, and the marginal cells only one, at some stage of their life.

2.1.2 EXTERNAL FEATURES

The shape of the thallus was first considered a morphological characteristic for classifying *Porphyra* by species. Since each species has its own characteristic shape, it can be grouped before or during early maturation under one of three types. Some are circular, and some oval; some have elongated thalli which are ellipsoidal, linear, elongated obovoid, oblanceolate, or linear oblanceolate in which the thallus tapers at the upper and lower part, or the lower part only; and finally, some have elongated thalli which taper in the upper part, assuming an elongated, obovoid, linear, lanceo-late shape (Figure 1.2). When *Porphyra* mature and spermatia and carpospores are released, the thalli usually become short and wide. In other words, the thalli which were linear or lanceolate before maturation, gradually become irregular to later assume a broad, linear shape, or to become ellipsoidal or even spherical. Some *Porphyra,* which are initially spherical, become kidney-shaped or funnel-shaped and, at times, rapid growth on the sides usually results in an overlapping of the rounded bodies. Although there are changes in shape, these do not occur at random and the pattern of the changes depends on the type of *Porphyra*.

The rim of the body and, in particular, the upper one, is irregular and small, and often rough to the touch. The irregularity, smallness, and roughness of the rim occur after the release of carpospores and spermatia. Sometimes the thallus is deeply caved in. Often, the plants become brittle. Some show many or few wrinkles on the marginal part. When the plants with many wrinkles put out leaves, a rupture in the wall occurs.

The basal part of the thallus varies in form, ranging from V-shaped, cordate, to umbiliform (Figure 1.2). When the thallus is linear, the basal part is usually V-shaped; but when the thallus is elongated, ellipsoidal, or lanceolate, the basal part is semi-ellipsoidal or heart-shaped. When the thallus is oval, spherical, or kidney-shaped, the basal part is heart-shaped or umbiliform in most cases.

The color of the thallus differs from species to species. It may be completely red, reddish-brown, bordered by a patch of blue, etc. In many cases, the body color changes with aging and a color differentiation exists between the mature and immature parts. The thallus of *Porphyra* may be membranous and soft, or firm and coriaceous.

Porphyra are usually 10 to 30 cm in length, but may be shorter or longer.

Of the foregoing characters, the shape may differ according to the density of the epiphytes, and the color vary according to the depth of the vegetation and the availability of salt-nutrients in the sea water. Size may differ according to the depth of the vegetation and the location, i.e., whether the seaweed is situated inside a gulf

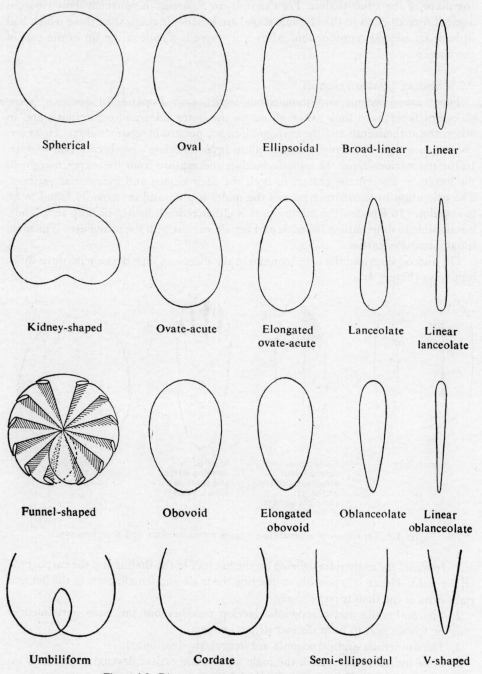

Spherical   Oval   Ellipsoidal   Broad-linear   Linear

Kidney-shaped   Ovate-acute   Elongated ovate-acute   Lanceolate   Linear lanceolate

Funnel-shaped   Obovoid   Elongated obovoid   Oblanceolate   Linear oblanceolate

Umbiliform   Cordate   Semi-ellipsoidal   V-shaped

Figure 1.2. Diagrammatic figures of the shape of *Porphyra*.

or in the open sea. In other words, the external features of seaweed seem to be quite dependent upon the environment. Moreover, in the case of the species of *Porphyra* which have asexual growth during the plumule stage, the asexual growth itself affects the shape of the adult thallus. For example, in *P. tenera,* those which show vigorous asexual reproduction in the plumule stage, are broader in shape than those which had little or no asexual reproduction. Moreover, there is a split at the tip in the case of the former.

2.1.3 SEXUAL CHARACTERISTICS

There are dioecious and monoecious varieties in *Porphyra*. In addition, some dioecious types have their antheridium on the outer side (androdioecious), and in others the antheridium and the carpogonium are present in separate areas. However, no dioecious type with only a carpogonium (gynodioecious) has been found to date. In the monoecious type, the gonads develop and mature from the upper margin to the lower, or toward the center, in both the androecious and gynoecious varieties. The coloration in the mature part of the males is poor and is normally found to be pale yellow. In females, the mature part is either reddish-brown or deep red. Thus, it is possible to differentiate the male and female organs with the naked eye. The male usually matures earlier.

The antheridium and the carpogonium in the dioecious type develop in three different ways (Figure 1.3).

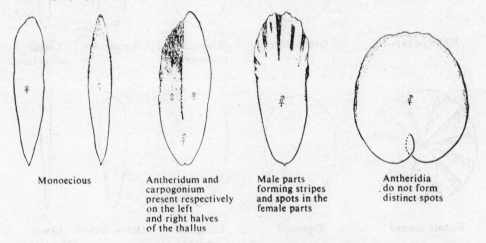

| Monoecious | Antheridum and carpogonium present respectively on the left and right halves of the thallus | Male parts forming stripes and spots in the female parts | Antheridia do not form distinct spots |

Figure 1.3. Formation of reproductive organs in monoecious and dioecious types.

1. In some, the antheridia develop on the left half of the thallus and the carpogonia on the right. Hence it is possible to describe the male and female parts as the left and right sides of the thallus respectively.

2. The antheridia and carpogonia develop together but the male part intrudes into the female part to form distinct stripes or spots.

3. The antheridia and carpogonia are irregularly distributed.

In those individuals in which the male and female organs develop respectively on the left and right side of the thallus, the male organs mature earlier. In some instances, the antheridia are carried away by the currents long before the full formation of the carpogonia and, as a result, the latter alone are present in the thallus. In types 2 and 3, the antheridia start developing from the upper margin, while the carpogonia form more toward the center. When the antheridia on the upper margin mature and the spermatia are released, this part of the body sloughs off, and the complete development of the carpogonia occurs on the inner side of the thallus, which thereby becomes exposed along the margin. This exposure leads to the release of carpospores. In some varieties, the antheridia develop either on the lower margin or toward the center; in the latter case, they are either in the form of stripes or spots inside the carpogonia or irregularly mix with the carpogonia. This is the difference between types 2 and 3. In the mature part of type 3, there are a number of minute bulges. Since in the dioecious types 2 and 3 antheridia only form at an early stage of maturation, these types may be mistaken as androecious unless very carefully examined.

In some *Porphyra* (both monoecious and dioecious) degenerated cells mix in with the female parts and cause bulging, spotted, or injured thalli.

When the thallus is thin, following the release of spermatia and carpospores, the gelatinous membrane on its surface sloughs off together with the membranous covering of the antheridia and carpogonia; when the thallus is thick the membranes remain intact.

2.1.4 STRUCTURE OF PORPHYRA

While some *Porphyra* have serrated margins in which each tooth contains 1 to 3 cells, in others the margin is smooth (Figure 1.4). The serrated margin is found only in those types which have a single cell layer and belong to the subgenus *Porphyra*. The serrations disappear in those places from which spermatia or carpospores are released, but persist in the immature part on the lower margin.

The vegetative cells on the surface may be triangular, rectangular, or polygonal in shape; no pattern is observed in the arrangement of these cells. The septa in some are very thin and the contents of the cells become closely massed, so much so that the cell layer looks like a miniature stone patio (*P. pseudolinearis*). The cells are small in the immature part and in areas where cell division occurs rapidly; they are larger in the central part. However, in the subgenus *Porphyra*, the cells become gradually larger toward the basal part. The cells of old plants are larger than those of younger ones. In the subgenera *Porphyra* and *Diplastidia*, both of which are single layered, the cells of the central and lower parts of the mature thallus differ conspicuously in being of standard size. The diameter of these cells is usually between 10 μ and 18 μ in the narrow part and 15 μ and 25 μ in the broader part. However, in the double-layered subgenus *Diploderma*, the cells are smaller in *P. bulbopes*, but slightly larger than in the single-layered variety in the other species.

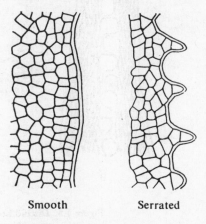

Smooth   Serrated

Figure 1.4. Thallus with smooth and serrated margin.

The cells near the basal part of the thallus are spherical, ellipsoidal, or obovoid in shape. Rhizoids grow from the lower part and attach themselves to the substratum. A number of rhizoids usually group together to form a small, discoid holdfast which then attaches to the substratum. However, the rhizoids of some *Porphyra* (*P. onoi;* Fukuhara, 1958) which adhere only to other seaweed, penetrate the thallus of the seaweed to which they remain attached (Figure 1.5). When the basal part of the thallus is closely observed with the naked eye, a peculiarly colored line is often noticed in the form of an arch at a height of 2 to 3 mm from the base. The rhizoids grow out of the cells of this part. Longitudinal sections of the plants of the subgenus *Diploderma* show that the rhizoids are either straight and directed downwards, or grow outwards, or turn toward the inner side, i.e., into the spaces between the cell layers of the two sides (Fukuhara, 1968).

It has been reported that in the two-chromatophored subgenus *Diplastidia*, the cell wall is double-layered in some parts (*P. onoi, P. punctata*), but this report requires further confirmation.

In *Porphyra* belonging to the single-layered variety, cell shape differs between that part in which the rate of cell division is high, and that part in which cell division is low. Some cells are elongated and some are almost square. Cells which are thicker, usually show longitudinal elongations.

The thickness of the thallus varies from one type of seaweed to another, and also varies within the thallus itself. The immature marginal part of the thallus is thin, and the central part thick, gradually becoming thicker toward the basal part. In the more mature part, the structure is either as thick as in the central section or even

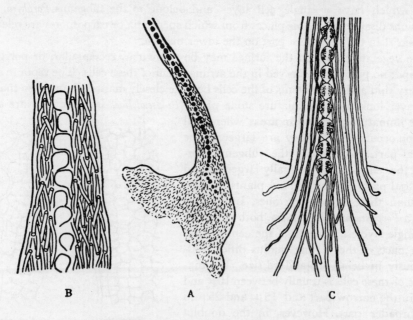

B   A   C

Figure 1.5. Discoid holdfast (A, B: *P. yezoensis*) and holdfast
penetrated by a group of rhizoids (C: *P. onoi*).
A and B—Kurogi, 1961; C—Fukuhara, 1968.

thicker. The immature thallus is thinner than the aged thallus. On the basis of the size of the central and lower parts of the mature thallus, the single-layered type can be divided into three groups: *viz.*, the thin variety of 30 to 40 μ in thickness, the medium variety of 40 to 60 μ, and the thick variety of 60 to 100 μ. In some cases the thickness exceeds even 100 μ.

2.1.5 REPRODUCTIVE ORGANS

The organs of sexual reproduction in *Porphyra* are the antheridium and the carpogonium, which are formed by the metamorphosis of the body cells of the upper marginal part.

When the cells of the antheridium multiply and mature, 32 to 128 nonmotile spermatia are formed. After fertilization, the cells of the carpogonium divide to form 8 to 32 carpospores, which in this condition, are called the cystocarp. The division of cells in the antheridium follows a specific pattern, namely, 8 to 16 divisions at the surface, and 16 to 32 in the cross section. Cell divisions in the carpogonium number 4 to 8 on the surface and 4 to 16 in the cross section. If the antheridium and carpogonium are imagined as rectangular objects, there would, at first, be several divisions passing through axis *a* and axis *b* on the surface, or there might be several parallel divisions on the surface passing through axis *c* (Figure 1.6). There are usually 32, 64, or 128 cells resulting from cell divisions in the antheridium, which follow this pattern: 32(a/4, b/2, c/4), 64(a/4, b/4, c/4 or a/4, b/2, c/8), and 128(a/4, b/4, c/8). In the cystocarp, cells are found in clusters of 8, 16, or 32, resulting from cell divisions which follow this pattern: 8(a/2, b/2, c/2), 16(a/2, b/2, c/4), and 32(a/4, b/2, c/4). A

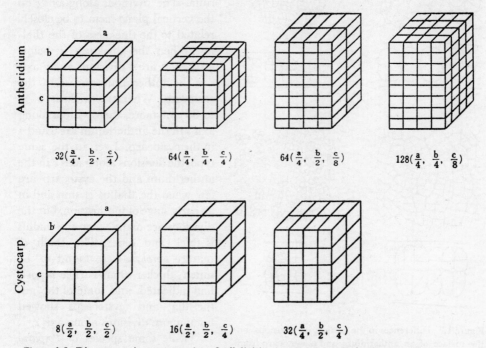

Figure 1.6. Diagrammatic representation of cell division in an antheridium and a cystocarp.

diagrammatic representation of cell division was first evolved by Hus (1902); later Ueda (1932) made certain changes in the patterns. According to Hus, Ueda, and Tanaka (1952), the mode of division in the different types is an important taxonomic criterion for *Porphyra*.

Usually the pattern of divisions in the antheridium and the cystocarp of various types is expressed in the relation of seaweed maturation to the maximum rate of cell divisions. Hence the divisions need not always be in line with the course shown in the diagrams. For example, in the case of a cystocarp in which the divisions occur in the form of 16(a/2, b/2, c/4), the carpospores are sometimes released even when the divisions result in 2(a/1, b/1, c/2), or 4(a/2, b/1, c/2), or 8 cells (a/2, b/2, c/2). Usually the division rate is low in the early stage of maturation, and in some growing beds the divisions are incomplete. These apply to the antheridium also. The early divisions in the antheridium and carpogonium are found to be along axis *c* on the surface. Even in the absence of divisions along axes *a* and *b* in the cystocarp, if the divisions occur along axis *c*, the carpospores released are capable of germinating. The absence of cell divisions along axis *c* in the cystocarp of *Porphyra* is very rare, but one such rare sample, termed c/1, was found in *P. tasa*.

Based on the pattern of their division, cystocarps can be divided into two types— one showing divisions into 4(a/2, b/2), and the other into 8(a/2, b/4) (Figure 1.7). It is difficult, however, to classify the types on the basis of the external appearance of antheridia. The divisions along axis *c* on the vertical plane seem to be closely related to the thickness of the thallus. When the thickness is below 40 μ, there are 4 divisions along axis *c* in the antheridium and 2 in the cystocarp. When the thickness is 40 to 60 μ or more, the divisions along axis *c* in the antheridium are 8 and 4 in the cystocarp. Even in the same type, the divisions along axis *c* in the antheridium and the cystocarp are few when the thallus is thin and in an early stage of maturation, but the divisions are many when the thallus is thick and older. The thalli of mature *Porphyra* were found to be much thicker than earlier reports had indicated, and some of the antheridia and cystocarps showed many more divisions along axis *c*.

While some species of *Porphyra* reproduce asexually by forming

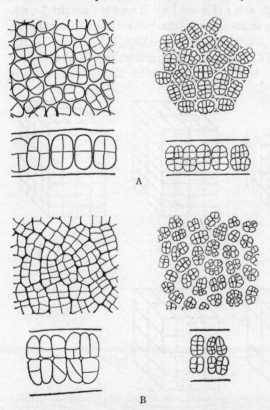

Figure 1.7. Differences in the pattern of divisions on the surface of an antheridium and a cystocarp from (A) *P. tenera* and (B) *P. pseudolinearis*.

neutral spores (monospores), these are absent in others. When the contents of the cells at the upper margin become dense, they give rise to neutral spores. Consequently, no cell division occurs, and the neutral spores are of the same shape and size as those of the cells at the margin of the thallus (*vide* Figure 1.11). The size of the asexually reproducing types may be large or small. The structure of the neutral sporangium is the same in all types.

2.1.6 EARLY CELL DIVISION IN THE LONGITUDINAL AXIS OF PLUMULES

The embryo of *Porphyra* is initially made up of a single row of cells which gradually assume the shape of leaves by longitudinal cell divisions. Kunieda (1939) was the first to give a detailed account of these early cell divisions. According to him, the number of cells (n) at the time of longitudinal cell division in the embryo, differs from species to species (Figure 1.8, Table 1.2). The number of cells also depends upon the conditions of growth. The number is small, for example, when the growth of the plumules is controlled. When n is small, the adult thallus is either spherical or broad, but when n is large, the adult thallus tends to be elongated. The number of cells, n, is usually 6, 5, or even less in *P. kuniedai, P. suborbiculata, P. seriata,* and *P. onoi,* all of which have a spherical shape in an adult stage.

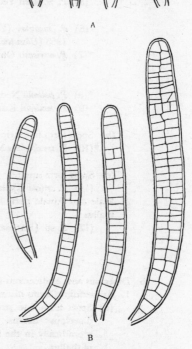

2.2 Ecological characteristics

2.2.1 DISTRIBUTION

The distribution of *Porphyra* is very wide, extending from the northern to the southern hemisphere, from the cold to the warm belt. There are also regional distributions according to types.

Of the *Porphyra* of Japan, some species are found in the frigid zone of Karafuto-Chishima and a part of Hokkaido (*P. ochotensis, P. amplissima, P. tasa,* and *P. bulbopes*), some in the subfrigid zone in the coastal regions of the Pacific Ocean along the southeastern part of Hokkaido and the northeastern part of Honshu (*P. onoi, P. pseudocrassa,* etc.), some in the temperate zone along the western coast of Hokkaido, Honshu, Shikoku, and Kyushu (*P. okamurai, P. dentata, P. kuniedai, P. seriata, P. tenera, P. angusta,* etc.), and some in the subtropical zone off the southern part of Kyushu (*P. crispata*) (Table 1.1). In addition to these, other species such as *P. umbilicalis* and *P. variegata* are found from the frigid to the subfrigid zone; *P. yezoensis, P.* sp. (Katada)

Figure 1.8. Plumules of (A) *P. kuniedai* and (B) *P. tenera*; the difference in the number of cells in plumules at the time of early longitudinal cell division is shown (Kurogi, 1961).

Table 1.1. Comparison of the main characteristics of various

----

1. THALLUS COMPOSED OF SINGLE-LAYER CELLS, CELLS OF SINGLE
   CHROMATOPHORES

1. Subgenus Porphyra

    A. Microscopic serrations present on margin   Body shape

      a. Dioecious

        1. Antheridium forms distinct spots

          (a) Spots form stripes on both sides at margin

            (1) *P. okamurai* Ueda, 1932 (*Kuronori*)   oval to oblanceolate

            (2) *P. suborbiculata* Kjellman, 1897 (*Maruba-amanori*)[1]   spherical to funnel-shaped

        2. Antheridium forms no distinct spots and mixes with carpo-
          gonium

           *(3) *P. crispata* Kjellman, 1897 (*Tsukushi-amanori*)[2]   spherical to kidney-
                                                          shaped, rounded

      b. Monoecious

           *(4) *P. dentata* Kjellman, 1897 (*Oni-amanori*)   broad-oblanceolate to
                                                             linear-oblanceolate

    B. Microscopic serrations absent on margin

      a. Dioecious

        1. Antheridium forms distinct spots

          (a) Spots form stripes

            (5) *P. yezoensis* Yeda, 1932 (*Susabinori*)   narrow-oblanceolate to
                                                             spherical

            (6) *P. kinositai* (Yamada and Tanaka) Fukuhara,   oblanceolate
                1968 (*Utasutsunori*)

           *(7) *P. moriensis* Ohmi, 1954 (*Kayabenori*)[3]   elongate-oval to broad-
                                                             lanceolate

           *(8) *P. palleola* Noda, 1964 (*Satsukinori*)[4]   oval to elongate-oval
            (9) *P. kuniedai* Kurogi, 1957 (*Maruba-asakusanori*)   oval to kidney-shaped

          (b) Spots form fine stripes on both sides at margins
           *(10) *P. tenuipedalis* Miura, 1961 (*Kaigara-amanori*)   elongate-ellipsoidal to
                                                             linear-oblanceolate

          (c) Spots form squares
            (11) *P. seriata* Kjellman, 1897 (*Ichimatsunori*)   spherical to kidney-shaped

        2. Male and female parts form on the left and right side of
          thallus

            (12) *P.* sp. (Katada, 1952) (*Somewake-amanori*)   elongate-oval, to spheri-
                                                             cal, comma-shaped

      b. Dioecious and Androecious (Androdioecious)

        1. Antheridium forms distinct spots

          (a) Spots form fine stripes on both sides at margin in
             dioecious thallus. Sometimes antheridia appear
             sporadically in the form of white spots on upper part
             of thallus.
            (13) *P. tenera* Kjellman, 1897 (*Asakusanori*)   ellipsoidal to linear-
                                                               oblanceolate

species of the genus *Porphyra* **found around Japan**

		Other Characteristic Features				
Thickness	Cell division in carpogonium (a × b/c)	Cell division in antheridium (a × b/c)	Vegetation period	Vegetation place	Distribution	Reference
ordinary	4/4	16/8	winter	intertidal zone	temperate zone	Ueda (1932)
ordinary	8/4	16/4	winter	intertidal zone	temperate, sub-temperate zones	Ueda (1932)
ordinary	4/8	16/8	winter	intertidal zone	subtemperate zone	Ueda (1932)
ordinary	4/4	16/8	winter	intertidal zone	temperate zone	Ueda (1932)
ordinary	4/4	16/8	winter	intertidal zone	temperate, subfrigid zones	Kurogi (1961)
ordinary	4/4	16/8	winter	subtidal zone	western coast Hokkaido	Tanaka (1952) Fukuhara (1968)
ordinary	4/4	16/8	winter	subtidal zone (adheres to other seaweed)	southern coast Hokkaido	Ohmi (1954) Fukuhara (1968)
slightly thin	4/2	8/4	winter	intertidal zone	Sado	Noda (1964)
thin	2 to 4/2	8 to 16/4	winter (to summer)	intertidal zone	temperate zone	Kurogi (1961)
slightly thin	4/4	16/8	winter	subtidal zone (adheres to molluskan shells)	Tokyo Bay	Miura (1961)
ordinary	4/4	16/8	winter	intertidal zone	temperate zone	Ueda (1932)
slightly thin	4/4	16/8	winter	below intertidal zone (mainly adheres to other seaweed)	temperate, subfrigid zones	Katada (1952) Fukuhara (1968)
thin	4/2	16/4	winter	intertidal zone	temperate zone	Kurogi (1961)

(Contd.)

Body shape

c. Dioecious
   1. Male and female parts form on left and right side of thallus in dioecious
      (14) *P. umbilicalis* (L.) Kutzing f. *(Chishimakuronori)*[5]   lanceolate to spherical

d. Monoecious
   (15) *P. angusta* Ueda, 1932 *(Kosujinori)*[6]   ellipsoidal to linear-oblanceolate

   (16) *P. pseudolinearis* Ueda, 1932 *(Uppuruinori)*   linear-oblanceolate to lanceolate

   (17) *P. ochotenesis* Nagai, 1941 *(Ana-amanori)*[7]   linear-lanceolate to spherical

   *(18) *P. irregularis* Fukuhara, 1968 *(Erimo-amanori)*   linear-lanceolate to spherical

   *(19) *P. crassa* Ueda, 1932 *(Atsuba-amanori)*[8]   oval to kidney-shaped

II. THALLUS COMPOSED OF SINGLE-LAYER CELLS, SOME OF WHICH HAVE TWO CHROMATOPHORES; SOME PARTS OF BODY WALL COMPOSED OF TWO LAYERS OF CELLS

2. **Subgenus Diplastidia**
   A. Microscopic serrations absent on margin
      a. Dioecious
         1. Antheridium does not form distinct spots and mixes with carpogonium
            (20) *P. onoi* Ueda, 1932 *(Ono-nori)*[9]   elongated oval to spherical
      b. Monoecious
         *(21) *P. punctata* Yamada and Mikami in Mikami (1956) *(Makure-amanori)*[10]   elongated oval to linear
         (22) *P. pseudocrassa* Yamada and Mikami in Mikami (1956) *(Makure-amanori)*   spherical to funnel-button-shaped (single-layered body wall)

III. THALLUS COMPOSED OF TWO LAYERS OF CELLS WHICH HAVE SINGLE CHROMATOPHORE

3. **Subgenus Diploderma**
   A. Microscopic serrations absent on margin
      a. Dioecious
         1. Antheridium does not form distinct spots and mixes with carpogonium
            (23) *P. amplissima* (Kjellman) Setchell and Hus, 1900 *(Benitasa)*[11]   oval to broad-linear

		Other Characteristic Features				
Thickness	Cell division in carpo- gonium $(a \times b/c)$	Cell division in antheri- dium $(a \times b/c)$	Vegetation period	Vegetation place	Distribution	Reference
ordinary	8/4	8 to 16/8	summer (to winter)	intertidal zone	frigid zone	Kirogi and Kawazu (1966)
ordinary	4/2	16/8	winter	intertidal zone	subfrigid zone	Kurogi (1961)
ordinary	8/4	8 to 16/8	winter	intertidal zone	temperate and subfrigid zones	Ueda (1932) Kurogi (unpub.)
thick	8/4	16/8	summer collection	intertidal zone	frigid zone	Nagai (1941)
ordinary	8/4	16/8	autumn	intertidal zone	Muroran to Hiro	Fukuhara (1968)
thick	4/4	16/8	winter	intertidal zone	western coast of Korea	Ueda (1932)
ordinary	4/2	16/4	winter	below intertidal zone	subfrigid zone	Ueda (1932)
slightly thick	16/4	16/8	collection in August	intertidal zone	Hidako Hokkaido	Mikami (1956)
thick	4/4	16/8	summer	intertidal zone	subfrigid zone	Mikami (1956)
fairly thick	4/1 to 2	8 to 16/ 4 to 8	summer	subtidal zone	frigid zone	Kjellman (1883) Kurogi (unpub.)

(Contd.)

Body shape

*(24) P. tasa (Yendo) Ueda, 1932 (Tasa)[12]   broad oval to kidney-shaped

*(25) P. abyssicola Ueda, 1932 (non-Kjellman) (Aka-nori)[13]   elongate-oval to kidney-shaped

2. Male and female parts appear on left and right side of thallus

(26) P. variegata (Kjellman) Ueda, 1932 (Fuiritasa)[14]   oval-elongate to comma-shaped

*(27) P. tenuitasa Fukuhara, 1968 (Usubatasa)   oval to spherical

*(28) P. bulbopes (Yendo) Okamura, 1916 (Fukurotasa)   oval, median cavity near base

b. Monoecious

*(29) P. occidentalis Setchell and Hus in Hus, 1900 (Kiirotasa)[15]   lanceolate (not clear in female thallus)

[1]Among the species mentioned above, the presence of Hiroha-maruba-amanori(F. latifolia Tanaka, 1952) has been reported from Ariake-kai and Yamaguchi Prefecture.

[2]According to Tanaka (1952), the cystocarp shows 8/4 fissions.

[3]Ohmi (1954) reported the species as dioecious or monoecious. but Fukuhara (1968) considered it as dioecious only.

[4]Fukuhara (1968) labeled this species as P. jezoensis.

[5]The presence of three strains of this species has been reported, viz., f. vulgaris (Ag.) Rosenvinge (Nagaba chishimakuronori), f. laciniata (Lightfoot) J. Ag. (Maruba chishimakuronori), and f. linearis (Greville) Rosenvinge (Hosoba chishimakuronori). The original place of vegetation was the European coastal region of the Atlantic Ocean, but the species has been variously located in Europe and, at present, there appears to be some controversy as to its exact position. P. umbilicalis (Chishimakuro) included in the Table, might conceivably be given the same position as P. purpurea (Roth) Ag. [syn. P. vulgaris Ag., P. umbilicalis f. vulgaris (Ag.) Rosenvinge, P. laciniata f. vulgaris (Ag.) Thuret.]

[6]From this species, f. sanrikuensis Kurogi 1963 (Nisekosuji nori) has been reported from the Iwate Prefecture.

[7]From this species f. lanceolata Tanaka 1952 (Nagaba-ana-amanori) has been reported from Chishima and Karafuto.

[8]Tanaka (1952) reported that this species, which may be either monoecious or androdioecious, with a cystocarp showing 4 or 8/4 fissions, was found in the coastal regions of Nihon-kai in Tohoku, Hokkaido, and Kanafuto. However, Fukuhara (1968) could not find the species in Hokkaido.

[9]According to Fukuhara (1968), the cystocarp and the antheridium show 4/4 and 16/8 fissions respectively.

[10]Fukuhara (1968) stated he could not find the species in Hidaka, reported as its original habitat.

[11]Three strains of this species have been reported, namely f. elliptica Nagai 1941 (Marubabenitasa), f. lanceolata Nagai 1941 (Nagababenitasa), and f. crassa Kawabata 1936 (Atsubabenitasa). Ueda (1932), Nagai (1941), and Fukuhara (1968) considered the species to be either dioecious or monoecious.

Thickness	Other Characteristic Features					
	Cell division in carpogonium (a×b/c)	Cell division in antheridium (a×b/c)	Vegetation period	Vegetation place	Distribution	Reference
very thick	4/1	16/8	summer collection	intertidal zone	frigid zone	Yendo (1913), Ueda (1932), Tanaka (1952)
ordinary	4/1	16/2		below intertidal zone	north of Hidaka, Hokkaido	Ueda (1932)
fairly thick	4/4	16/2 to 4	summer	subtidal zone (adheres to other seaweed)	subfrigid and frigid zones	Ueda (1932), Nakamura (1947)
ordinary	4/4	16/4	spring	below intertidal zone	western, southern coasts Hokkaido	Fukuhara (1968)
thick	8/2	16/4	summer collection	above subtidal zone	frigid zone	Yendo (1913) Tanaka (1944)
thick		16/4	summer collection	below intertidal zone	frigid zone	Nagai (1941)

According to Tanaka (1952) *P. occidentalis (Kiirotasa)* (Kawabata, 1936; Nagai, 1941) is another name for this species which is monoecious. Although it is not possible to infer this from Kjellman's original report in 1883 (which does not specifically exclude the existence of a monoecious form), nevertheless one can assume that most of the seaweeds belonging to this strain are dioecious. Specimens of dioecious forms collected by the author from the eastern part of Hokkaido conform to the description and diagrams given by Kjellman in 1883. Since a monoecious form of this species has not been discovered thus far, it must be regarded as dioecious as of now.

[12]Nagai (1941) considered the species to be monoecious, but Tanaka (1952) considered it dioecious and, often, even androdioecious.

[13]Kjellman (1883) classified *Porphyra* with a double-layered thallus under *Diploderma*, and the single-layered one as *Porphyra*. The monoecious forms were grouped under *Porphyra*. The specimens collected by Ueda were of the double-layered type and considered dioecious (or monoecious), and thus it is not possible to include them under *P. abyssicola* Kjellman. On the other hand, Tanaka (1952) and Fukuhara (1968) considered *P. abyssicola (Akanori)* (Ueda, 1932) to be only another name given to *P. onoi*. However, this species *(P. onoi)* shows a double-layered condition in some part of the body, but the vegetative part is usually single-layered. Hence, it would be improper to assert that *P. abyssicola (Akanori)*, which has a double-layered body wall, differs from *P. onoi (Ononori)* in name only. In view of the fact that the mode of formation of the antheridium and the cystocarp is the same in both the *P. abyssicola* (dioecious) and the *P. onoi* (collected by Ueda, the former may be regarded as coming under *P. onoi* pending further research.

[14]Fukuhara (1968) has renamed *P. variegata (Fuiritasa)* reported by Ueda (1932) and others, as a new species, *P. uedae* Fukuhara sp. nov. Nevertheless, the name *P. variegata* persists.

[15]Tanaka (1952) considered *P. occidentalis (Kiirotasa)* (Kawabata, 1936; Nagai, 1941) another name for *P. amplissima (Benitasa)*. It is not yet certain whether the opinions of Kawabata and Nagai are acceptable, as this plant may be related to the *P.* sp. of Kurogi (1966) and the *P.* sp. No. 2 of Fukuhara (1968).

(known in Japan as *Somewake-amanori*) and *P. pseudolinearis* are found from the sub-frigid to the temperate zone; and *P. kuniedai* ranges from the temperate to the sub-temperate zone. The distribution of other species such as *P. kinositai, P. moriensis, P. tenuipedalis, P. irregularis, P. crassa, P. punctata, P. tenuitasa,* etc., is not clearly defined.

It is interesting to note that the subgenera *Diploderma* and *Diplastidia* show a restricted distribution in the frigid and subfrigid zones.

2.2.2 HABITAT

Some *Porphyra* can withstand environmental changes such as drying, by living as epiphytes in the intertidal belts. Some live mostly under water in the subtidal zones. The majority of the species of the subgenus *Euporphyra* are found in the intertidal zones, while most of the species of the subgenus *Diploderma* are found in the subtidal zones (Table 1.1).

The substrata to which seaweeds adhere are rocks, stones, shells, other seaweeds, etc., but some species such as *P. moriensis, P.* sp. (Katada) (*Somewake-amanori*), *P. onoi,* and *P. variegata* attach only to other seaweeds (or grass).

Some forms, *P. pseudolinearis* for example, prefer places where there are tidal currents and waves, while others, *P. tenera* and *P. kuniedai*, grow in calm gulf regions. *P. yezoensis* grows in places where the conditions are neither too rough nor too calm.

2.2.3 VEGETATIVE PERIOD

Porphyra can be divided into two groups according to the period of luxuriant growth —those which grow from winter to spring, and those which grow from summer to autumn (Table 1.1). Within these two groups, however, species vary in their vegetative period. *P. kuniedai* grows luxuriantly from autumn to summer, and *P. umbilicalis* from summer to winter. Interestingly, seaweeds in the frigid zones have their most, luxuriant growth from summer to autumn.

2.3 Comparison of the major characteristics of different species of the genus Porphyra found in Japanese waters

Table 1.1 presents the more important characteristics of different species of the genus *Porphyra* which are found in Japanese waters. The genus is subdivided into subgenera which are easily distinguishable by the presence or absence of serrations, sexual characteristics, the development of the antheridium and the cystocarp, and other morphological and ecological characteristics. The description of the characteristics of some of the *Porphyra* reported from Japan was inadequate and, furthermore, the systemic position of some is not finally settled; these species are included in the Table. The number of cells (expressed as *n* of plumules) at the time of longitudinal fission in simple plumules with a single row of cells, is given in Table 1.2.

The following points should be kept in mind when reading Table 1.1:

1. Regarding the shape of antheridial spots, only special characteristics are mentioned;

2. Regarding the shape of the thallus, the degree of variation from the standard shape shown in Figure 1.2 is given;

3. The thickness of the thallus is classified as thin, normal, thick, and very thick.

Table 1.2. Number of cells at the time of cell division in the longitudinal axis in the plumule, with a single row of cells

Name of species	Number of cells	Reference
P. okamurai	8-12	Fukuhara (1968)
P. suborbiculata	3-6	Kurogi (unpublished)
P. crispata	?	
P. dentata	?	
P. yezoensis	6-8	Kurogi (1961)
	7-12	Migita (1960)
	5-7	Maekawa (1961)
P. kinositai	2-3	Fukuhara (1968)
P. moriensis	3-4	Fukuhara (1968)
P. patleola	?	
P. kuniedai	4-6	Kurogi (1961)
P. tenuipedalis	?	
P. seriata	4-7	Migita (1960)
	3-9	Kurogi (unpublished)
P. sp (Katada 1952)	2	Kurogi (1953), Fukuhara (1968)
	3-7	Kito (1966)
P. tenera	15-30	Kurogi (1961)
	16-27, 8-13	Migita (1960)
P. umbilicalis f.	2-9	Kurogi, Yoshida (1967, unpublished)
P. angusta	10-20	Kurogi (1961)
P. pseudolinearis	11-14	Maekawa (1961)
	8-16	Kurogi (unpublished)
P. ochotensis	?	
P. irregularis	7-13	Fukuhara (1968)
P. crassa	?	
P. onoi	(3)-5-8	Kurogi (1956, unpublished)
P. punctata	?	
P. pseudocrassa	?	
P. amplissima	?	
P. tasa	?	
P. abyssicola	?	
P. variegata	?	
P. tenuitasa	?	
P. bulbopes	?	
P. occidentalis	?	

A thallus is thin if it measures 40 μ or below at its central, lower, or mature part; it is classified as normal, thick, or very thick if it measures 40 to 60 μ, 60 to 80 to 100 μ, and over 100 μ respectively. The expressions "slightly thin" and "slightly thick" are relative to normal thickness, and "fairly thick" is relative to thick;

4. The fission in the carpogonium and the antheridium is expressed according to the number of fissions on the surface ($a \times b$) and the number of fissions on the vertical plane (number of fissions on c). Except in one instance, the number of fissions represents the maximum number;

5. The vegetative period is divided into winter when the seaweeds grow luxuriantly from winter to spring, and summer when they grow from summer to autumn. In the absence of a detailed report, the term "summer collection" is used to indicate the specimens collected in summer;

6. The habitats are divided into two categories, viz., the intertidal zone where the plants remain in places in which they become exposed at low tide, and the subtidal zone where the plants are not exposed. Places below the intertidal zone and above the subtidal zone are included in the subtidal zone;

7. Regarding distribution, Japanese waters have been divided into four categories—subtemperate, temperate, subfrigid, and frigid zones. The subtemperate zone includes the area south of the southern part of Kyushu (including capes on the southern parts of Shikoku

and Honshu). The temperate zone covers the other parts of Kyushu, Shikoku, Honshu, and the western coast of Hokkaido. The subfrigid zone covers the Pacific coast between the northern part of Honshu and the eastern part of Hokkaido, and the northern coast of Hokkaido. The frigid zone includes Chishima and Karafuto. The types which have a narrow distribution range are referred to by the names of their respective localities;

8. As already mentioned, there are still some problems in classifying some types of *Porphyra;* in order to avoid confusion, the classification given here is based on the main Japanese species and only those species which are fully known. However, some of the species described in the earlier reports from Japan are also included. Although there is no special reference to the details of the investigations carried out by the authors, the results of their investigations are mentioned at some places. The types which have not yet been studied by the authors are asterisked.

3. Reproduction and life cycle

3.1 Reproduction

Almost all the species of *Porphyra* reproduce sexually. However, there are a few which reproduce asexually.

The antheridium produces 32 to 128 small, nonmotile spermatia by repeated cell divisions along the horizontal and vertical planes of the thallus. The size of a spermatium of *P. tenera* is 3 to 5 μ in diameter (Kurogi, 1961).

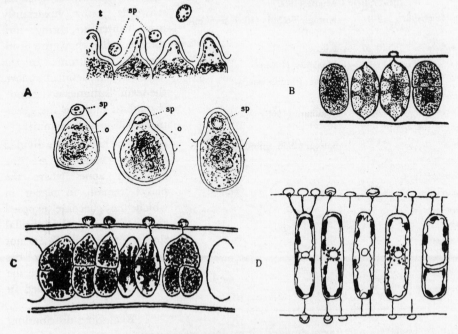

Figure 1.9. Carpogonium and fertilization.
A and B—*P. tenera* (A—Kunieda, 1939; B—Kurogi, 1961); C—*P. leucosticta*
(Berthold, 1882); D—*P. umbilicalis* f. *linearis* (Dangeard, 1927).

The carpogonium cannot be easily differentiated from other cells of the thallus by external appearance. Several authors have described the shape of the carpogonium in transverse section as well as the process of fertilization, but the descriptions given do not always agree. Furthermore, the shape and fertilization process seem to differ from species to species. Ishikawa (1921), Ueda (1932) and Kunieda (1939) observed in *P. tenera* that the carpogonium together with the exosporium, forms protuberances on both edges which act as trichogynes (Figure 1.9). According to Kunieda, spermatia adhering to a trichogyne are sucked in by the carpogonium.

The presence of fertilization tubes between the spermatia adhering to the surface of the thallus and the carpogonium, was reported in *P. leucosticta* by Berthold (1881, 1882), in *P. umbilicalis* by Rosenvinge (1909) and Grubb (1924), in *P. umbilicalis* f. *linearis* by Dangeard (1927), in *P. tenera* by Tseng and Chang (1955), and in *P. tenera, P. kuniedai, P. yezoensis* and *P. angusta* by Kurogi (1961). Berthold and Kurogi reported that the carpogonium, as seen from the sections, is spindle-shaped, and the two edges turn slightly inward (Figure 1.9). Kurogi's observations were carried out in vivo. Tseng and Chang reported that the carpogonium put out prototrichogynes and retracted them after fertilization. Grubb and Dangeard make no mention of the protuberances at the edges of the carpogonium (Figure 1.9). Yabu and Tokida (1963) reported the presence of prototrichogynes in *P. yezoensis*, but said

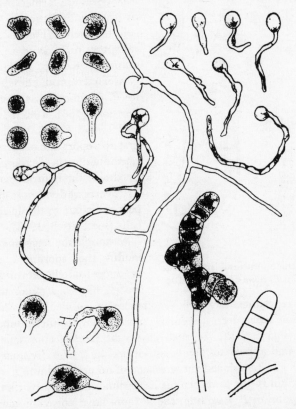

Figure 1.10. Carpospores and their germination in *P. tenera*. The thick branch shown at the bottom right of the figure is sporangial (Kurogi, 1953).

nothing about the fertilization tubes. Krishnamurthy (1959), however, reported that there was no connecting link between the carpogonium and spermatia in *P. umbilicalis* f. *laciniata*. In other words, there is no proof to support that there is fertilization in the species.

Suto (1963) reported that when spermatia were added to isolated cultures of the female thallus of *P. pseudolinearis*, *P. angusta* and *P. umbilicalis* (?), carpospores formed from the carpogonium which germinated and became filamentous structures. If spermatia were not introduced, the carpogonium did not produce any carpospore capable of growing into a filamentous structure. Self-fertilization occurs in dioecious *P. tenera* and *P. yezoensis*; Suto (1963) states that few carpospores capable of forming filamentous structures, are produced when there is only self-fertilization.

The carpogonium usually gives rise to carpospores in clusters of 8 to 32 by the division of cells on the surface, either parallelly or at a right angle to the surface. Released carpospores grow into germinative tubes and become filamentous structures (Figure 1.10). Hence, the morphology here is different from that of the plumule. The diameter of the carpospores in *P. tenera* is (1C) to 11 to 14 to (18) μ (Kurogi, 1961).

Asexual reproduction depends upon the neutral spores (monospores). The neutral spores are formed in the sporangium by the condensation of the vegetative cells themselves (Figure 1.11). The neutral spores in *P. tenera* are 12.5 to 15 μ in diameter (Kurogi, 1961).

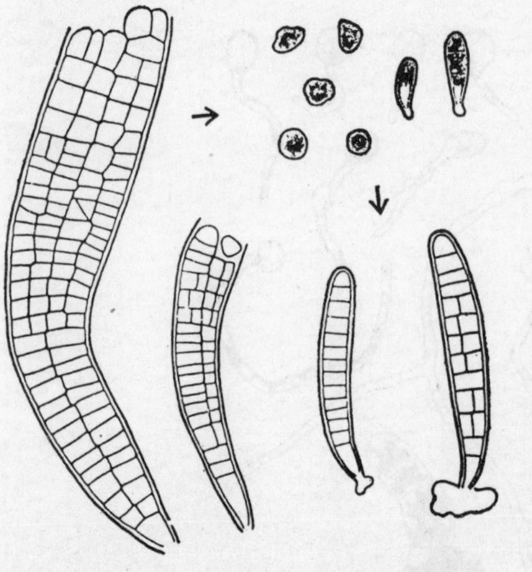

Asexual reproduction by neutral spores in *P. leucosticta* was first reported by Berthold (1881, 1882). In Japan, Ueda (1929) and Kusakabe (1929) were the first to report the occurrence of asexual reproduction in *P. tenera*. According to Berthold, the neutral spores are formed by a vertical division of the mother cell, perpendicular to the surface, into 2 to 4 cells (see Fritsch, 1945).

Although the vegetative cells directly modify into sporangia, each of which produces a single neutral spore, in some species like *P. yezoensis* and *P. kuniedai*,

Figure 1.11. Formation of neutral spores and their germination in *P. tenera* (Kurogi, 1961).

the thallus which reproduces asexually becomes sexual or vice versa. Hence the same thallus can, in some cases, form a neutral sporangium, antheridium, or cystocarp (Kurogi, 1961) (Figure 1.12). In such cases, the neutral sporangia are found in groups of 2 or 4, and appear to have been formed by 2 or 4 divisions of the mother cell. Berthold has also mentioned the presence of an antheridium, a cystocarp, and a neutral sporangium in the same thallus in specimens studied by him.

With the exception of *P. leucosticta*, the authors have not come across any report on asexual reproduction in *Porphyra* from other countries, in particular Europe or North America. Table 1.3 lists several species of *Porphyra* found in Japan. Of these,

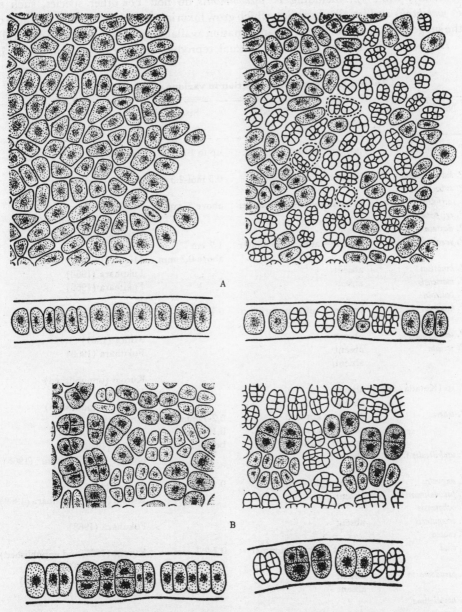

Figure 1.12. Neutral sporangium, antheridium, and cystocarp found together in *P. yezoensis*.

A—(left) Neutral sporangium forming part of a thallus in which both the neutral sporangium and the antheridium are present; (right) a part in which the antheridium and the neutral sporangium mix. B—(left) A part in which both the neutral sporangium and the cystocarp were found in a mixed condition; (right) a part in which the antheridium, the neutral sporangium, and the cystocarp were found in a mixed condition (Kurogi, 1961).

some exhibit asexual reproduction, some do not, and for some no information is available on this aspect. Eight species, including *P. tenera*, are found to reproduce asexually, while 10, including *P. pseudolinearis* do not. For other species, such a. *P. umbilicalis* and its allied species, which grow luxuriantly in the summer, and those of the subgenus *Diploderma*, there is no information available on the subject (see Table 1.3).

Almost all the species exhibiting asexual reproduction develop neutral spores in

Table 1.3. Asexual reproduction in various species of *Porphyra*

Name of species	Present/ absent	Period	Size	Reference
P. okamurai	present	autumn	up to 1 cm	Tanaka (1952)
	present	autumn		Fukuhara (1968)
P. suborbiculata	present	autumn, spring	0.5 mm-2.3 cm	Kurogi (unpublished)
P. suborbiculata f. latifolia	present	winter to summer	above 2 mm	Migita (1960)
P. crispata	present			Tanaka and Kofuji (1952)
P. dentata	absent			Fukuhara (1968)
P. yezoensis	present	autumn to winter	1-7 cm	Kurogi (1961)
			above 0.2 mm	Migita (1960)
P. kinositai	absent			Fukuhara (1968)
P. moriensis	absent			Fukuhara (1968)
P. palleola	?			
P. kuniedai	present	autumn, spring to summer	0.1-5.6 cm	Kurogi (1961)
P. tenuipedalis	absent			Miura (1961)
P. seriata	absent			Fukuhara (1968)
	absent/ ?			Kurogi (unpublished)
P. sp (Katada, 1952)	absent			Fukuhara (1968)
	absent			Kurogi (unpublished)
P. tenera	present	autumn	0.2-1 mm	Ueda (1937)
	present	autumn	0.15-1 mm	Fukuhara (1961)
	present	autumn	below 1 mm	Migita (1960)
P. umbilicalis f.	absent			Kurogi and Kawazu (1966), Fukuhara (1968)
P. angusta	present	autumn	0.1-1 mm	Kurogi (1961)
P. pseudolinearis	absent			Kurogi (1959), Fukuhara (1968)
P. ochotensis	?			
P. irregularis	absent			Fukuhara (1968)
P. crassa	?			
P. onoi	present	winter, summer	0.7-2.7 mm	Kurogi (1956, and unpublished)
	present	winter	1-3 mm	Fukuhara (1968)
P. pseudocrassa	absent			Kurogi and Kawazu (1966)
	absent			Fukuhara (1968)
P. amplissima	?			
P. tasa	?			
P. abyssicola	?			
P. variegata	?			
P. tenuitasa	?			
P. bulbopes	?			
P. occidentalis	?			

the autumn when they are young. However, in *P. yezoensis* neutral spores form up to February or March and *P. kuniedai*, *P. suborbiculata* and *P. onoi* form neutral spores from spring to summer. Further, in *P. tenera*, *P. kosuji*, etc., the neutral spores form when the thallus is microscopic in size or just large enough to be seen with the naked eye. But in *P. yezoensis*, *P. kuniedai*, etc., neutral spores form even when the thallus is as long as 5 to 6 cm or longer. In *P. tenera*, *P. angusta*, etc., there is no mixing of the neutral sporangium, the antheridium, and the cystocarp, but in *P. yezoensis*, *P. kuniedai*, etc., a mixed condition can often be seen in thalli of considerable sizes. In early spring, or just before, and with changes in environment, the sexually mature thalli of *P. kuniedai* become asexual and begin to produce neutral spores.

The *P. suborbiculata* f. *latifolia* Tanaka *(Hirohamaruba-amanori)* found in Ariake-kai, reportedly forms neutral spores only between autumn and summer and does not develop sexual reproductive organs (Migita, 1960).

Data with regard to the conditions under which asexual reproduction occurs in *Porphyra* are not adequate. However, the following conditions may be assumed as essential:

First, they should be young. The formation of neutral spores stops even during the period of asexual reproduction, if the size of *P. tenera* exceeds 1 mm. In *P. yezoensis*, sexual reproductive organs develop when the thallus is old. On the other hand, the sexually mature thallus of *P. kuniedai* gradually becomes asexual. Hence, for asexual reproduction, it is not always essential that the thallus should be young.

Second, the water should be of the right temperature. It is reported that there is a relationship between the water temperature and the formation of neutral spores in *P. suborbiculata* (Kusakabe, 1929), *P. kuniedai* (Kinoshita and Shibuya, 1942), *P. yezoensis* and *P. tenera* (Kurogi and Sato, unpublished paper). In *P. yezoensis*, in which the neutral spores form during the period of autumn to winter and the carpospores only in the post-spring period, it seems that the duration of daytime is also an important factor (Kurogi, unpublished).

Unlike carpospores, when neutral spores germinate the plumule grows straight from the substratum giving rise to a single row of cells which result from transverse fission. This is followed by further cell divisions in a longitudinal plane, as a result of which the structure looks like the leaf of an ordinary *Porphyra* (Figure 1.11).

3.2 Life cycle

With the discovery of certain facts (Drew, 1949, 1954; Kurogi, 1953), the life cycle of *Porphyra* could be detailed. For a long time the fate of the carpospores and their mode of germination was not clearly known. The carpospores were thought to remain dormant throughout the summer, giving rise to fresh crops of plants in the autumn. Since the germination of carpospores into filaments is unusual, this "dormancy theory" was held by Berthold (1881, 1882), Kunieda (1939), Suto (1948, 1950a, b), and others. In the so-called "summer seaweed theory" of Ueda (1929, 1937), Kusakabe (1929), Wada (1941), Kinoshita and Shibuya (1941, 1942), Kinoshita (1943, 1949), and others, the carpospores do not hibernate but germinate into plumules. They pass the summer in the form of minute thalli. Filamentous plumules of carpospores are not usually seen. A third theory holds that spores formed in the filaments germinate from carpospores (Kylin, 1922; Rees, 1940, 1940a, and others). Dangeard (1931) proposed that the buds formed on the filaments germinate from carpospores.

The dormancy theory favored by Kunieda and others is based primarily on studies of *P. tenera*, while Kylin's view is based on the study of *P. laciniata*, Rees' on the study of *P. umbilicalis*, and Dangeard's on the study of *P. leucosticta*, *P. umbilicalis* and *P. miniata*. Very little experimental proof exists to support any of these theories.

In 1949, Drew reported that the filaments germinating from the carpospores of *P. umbilicalis* are similar to those of *Conchocelis rosea* Batters (1892), a type of shell-boring algae. Like Batters, Drew also found that structures similar to sporangia formed. Kurogi (1953) reported the presence of a Conchocelis phase similar to the one described by Drew, in *P. tenera, P. yezoensis* (described as the *Maruba* type of *P. tenera*, i.e. *P. kuniedai*), *P. suborbiculata, P.* sp. (Katada) (described as *P. umbilicalis*), and *P. pseudolinearis*. Drew further reported that the spores are released from the sporangium and develop into embryos as happens in the natural Conchocelis, as observed in *P. tenera* and other species of Japan by Shinsaki (1954), Suto (1954) and others. The presence of the Conchocelis phase has also been confirmed in *P. tenera* by Tseng and Chang (1954, 1955), in *P. capensis* by Graves (1955), in *P. perforata* by Hollenberg (1958), and in *P. leucosticta, P. linearis, P. umbilicalis, P. purpurea* and *P.* sp [*amethystea* Kütz (?)] by Kornmann (1961, 1961a).

Thus during the period in which the *Porphyra* disappears, it is said to be in the Conchocelis phase. While it is now possible to grow the filaments germinating from carpospores in experimental cultures without making them bore into shells (Iwasaki, 1961), such filaments have not been observed in nature.

3.3 Nuclear phase

Ishikawa (1921), Dangeard (1927), Magne (1952), Toyama (1957), Krishnamurthy (1959), Yabu and Tokida (1963), Kito (1966), Kito *et al.* (1967), and Migita (1967) have studied the nuclear phase and nuclear divisions in *Porphyra*.

According to Ishikawa (1921), the nuclear division in *P. tenera* occurs between amitosis and mitosis; other authors feel that its occurrence is the same as in higher plants.

The chromosomes in *Porphyra* are usually small and few in number (Table 1.4).

Two different views exist with regard to the meiotic division of the nucleus: one holds that meiosis occurs when the carpogonium divides after fertilization (carpospores haploid) (Ishikawa, Dangeard, and Tseng and Chang), and the other maintains that the carpogonium divides in the absence of meiosis (carpospores diploid) (Magne, Toyama, Yabu and Tokida, Kito and Kito *et al.* For the latter, there is no indication of the exact stage at which the meiotic division occurs.

Migita and Abe (1966) found that the sporangium of the filaments in *P. yezoensis* and *P. tenera* divide once or twice before releasing the spores (see Figure 1.14). On this basis, Migita (1967) studied the nuclear phase of the leafy thallus and filaments of *P. yezoensis* and reported that the somatic cells of leafy thallus, neutral spores, and spermatia were haploid ($n=3$), and that the fertilized carpogonium, carpospores, somatic cells of filaments, and the sporangium were diploid ($2n=6$). The meiotic

Table 1.4. Number of chromosomes in the somatic cells of *P. tenera*

Name of species	Chromosomes	Reference
(Japanese species)		
P. tenera	3	Ishikawa, 1921
,,	4	Toyama, 1957
,, (Chinese species)	5	Tseng and Chang, 1955
P. yezoensis	3	Yabu and Tokida, 1963
,,	3	Migita, 1967
P. moriensis	4	Kito et al., 1967
P. sp. (Katada, 1952)	4	Kito, 1966
P. pseudolinearis	4	Kito et al., 1967
P. onoi	3	Yabu and Tokida, 1963
P. pseudocrassa	3	Kito et al., 1967
P. umbilicalis f. linearis[1]	4	Kito et al., 1967
P. amplissima	3	Kito et al., 1967
(European species)		
P. umbilicalis f. linearis[2]	2	Dangeard, 1927
P. linearis[3]	4	Magne, 1952
P. umbilicalis f. laciniata	5	Krishnamurthy, 1959

[1]This plant might be mistaken for a European species judging from the collection period.
[2,3]These two species are considered identical in Europe.

division of the sporangia is present in the filaments, and thus the spores become haploid.

Assuming this to be common among the various types of *Porphyra*, the nuclear phase of the plant can be explained by the facing diagram.

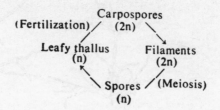

3.4 Filaments

When a slide is used as a substratum, the filaments germinating from carpospores branch out in the form of creepers and form clusters of 2 to 3 mm in size (Figure 1.10). The filaments consist of a single row of cells. The branches are lateral or sometimes opposed, and separate from one another; every cell puts out a branch. The filaments grow apically and the cells are columnar with bulges here and there. The cells in *P. tenera* are 4 to 5 μ in diameter and their length is 5 to 10, or more, times greater than their width. The dimension of the cells are more or less the same in other species of *Porphyra*. The chromatophores are adjacent to the outer walls and are either in the form of bands or discs.

The sporangial branches arise either laterally or at the tips of the branches. They are broader than the vegetative branches and may be either single or bifurcated. The sporangial branch has a single row of short cells (the sporangium). The sporangium in *P. tenera* is 11.5 to 18 μ in diameter and as long as it is broad. These dimensions are more or less the same as in other *Porphyra*. At the center lies a single stellate or massive chromatophore. The sporangium divides 2 or 4 times to form spores in *P. yezoensis* and *P. tenera* (Migita and Abe, 1966).

The Conchocelis filaments, boring into the shell, are slightly different from those seen on a slide (Figure 1.13). The filament is made up of a single row of cells which grow in three directions or in a radial manner, especially those close to the holdfast. Viewed from above, the holdfast looks like a disc. Branches grow from the

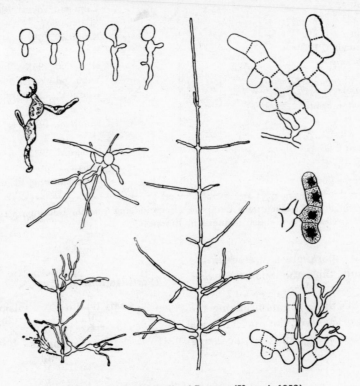

Figure 1.13. Conchocelis of *P. tenera* (Kurogi, 1953).

Plumules in early stage (upper left), branches on the marginal part of
a mature thallus (middle), and a sporangial branch (right).

upper part of each cell either in opposite directions or in groups of 3 or 4. The branches divide several times, and a number of main branches with lateral branches form. The upper part of the main branch is made up of columnar cells; it has long, slender, lateral branches widely separated from each other. The cells of the central part of the filament are relatively short, and the part from which the branches extrude, bulges slightly. The branches are very close to one another, appearing clustered. In *P. tenera* the cells in the narrow part of the filament are 3 to 5 μ, and in the bulging part, 7.5 to 12.5 μ in diameter; these dimensions are more or less common for all types of *Porphyra*. As observed on slides, the chromatophores are adjacent to the outer walls, and are either in the form of bands or discs. Growth is apical.

The sporangial branches form laterally and are broader than the vegetative ones.

In large plants, a branch divides repeatedly to give rise to 20 or more branches, each of which consists of a single row of short cells (sporangium). At the center of each cell, there is a single stellate or lumpy chromatophore. In *P. tenera*, the sporangium is 10 to 12.5 μ in diameter and more or less the same in length.

The sporangia, as observed by Migita and Abe (1966) in *P. yezoensis* and *P. tenera*, form spores either directly without fission, or by dividing into two (Figure 1.14). The spores in *P. tenera* are (9) to 10 to 13 to (14) μ in diameter. Like the neutral spores, these also germinate and become plumules (Figure 1.15).

The color of Conchocelis differs in different species of *Porphyra*. In *P. tenera*, *P. yezoensis* and *P. angusta*, it is dark purple, but in *P. kuniedai*, *P. suborbiculata*, *P. pseudolinearis*, *P.* sp. (Katada), and other species, it is reddish. However, color varies under cultural conditions. When the temperature or the degree of brightness is high

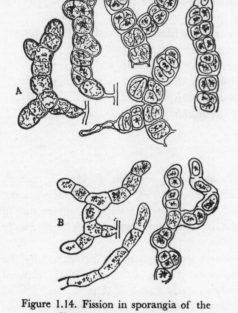

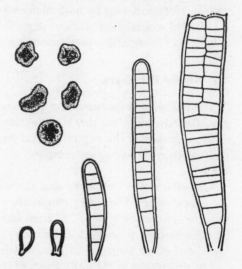

Figure 1.14. Fission in sporangia of the filaments in *P. yezoensis*.

A—free filaments; B—Conchocelis (right) and undivided sporangia (left) (Migita and Abe, 1966).

Figure 1.15. Spores of Conchocelis and their germination in *P. tenera* (Kurogi, 1953, 1961).

and the day long, the purple colored filaments turn reddish, and the red filaments become peach colored. At the same time, the branches grow close to one another under higher temperature, greater brightness, and longer days (Kurogi and Sato, 1962, 1962a; Kurogi and Akiyama, 1966). The size of the disc giving rise to filaments also varies. In *P. pseudolinearis*, it is large, and in *P. kuniedai*, small; in other words, one grows rapidly and the other grows slowly (Kurogi, 1961; Kurogi and Akiyama, 1966). Finally, a detailed examination of the thickness of sporangial branches shows that the filaments differ from species to species.

The filaments found in natural conditions bore into various types of shells at the bottom of the sea. While the leafy thalli of *Porphyra* are mostly found in intertidal zones, the filaments grow in subtidal zones. This is because the filaments cannot withstand desiccation and strong light, but are able to grow even when the light is very weak (Kurogi and Hirano, 1955; Takeuchi et al., 1956).

In experimental cultures of the filaments, when the same were allowed to mature and release spores, the branches usually withered. However, if the branches did survive, they continued to live for more than a year. The physiological characters of the filaments will be discussed later.

4. Life history of major species

4.1 Outline

Apart from the useful varieties of *Porphyra*, the details of the life history of many forms are not clear. In general, the life history of *Porphyra* can be divided into the following categories:

I. Luxuriant growth of leafy thalli in the winter, of filaments mostly in the summer.
   A. Secondary growth occurs due to asexual reproduction.
      a. Summering in the form of filaments (e.g., *P. tenera*).
      b. Summering by both filaments and leafy thalli (e.g., *P. kuniedai*).
   B. No asexual reproduction (e.g., *P. pseudolinearis*).
      a. No asexual reproduction (e.g., *P. umbilicalis*).

4.2 Winter Porphyra

The following is an account of *P. tenera, P. angusta, P. yezoensis, P. kuniedai*, and *P. pseudolinearis*, based on the investigations of Kurogi (1959, 1961). Although some differences exist in the vegetative and reproductive periods in various regions due to differences in air and water temperature, the life history of these species is essentially the same.

Matsushima Bay is shallow and the water temperature is easily influenced by the air temperature. The mean temperature observed during a five-year study showed that the temperature was 25°C from late July to early and mid-August, and 4°C at its minimum in January and February (Figure 1.16). The mean density also was low, i.e. 1.02 or even less in July, which is the rainy season in Japan (Figure 1.16).

The differences in the daily levels of the tide are significant. While in the summer and winter such differences are significant for full tides, in the spring and autumn, they are significant for ebb tides. The mean tidal level for syzygy tides is 145 cm, and the mean difference is 130 to 170 cm (Figure 1.17). The monthly mean tidal level is minimal in April and maximal in September (Figure 1.18).

4.2.1 Life History Theories

1. **Porphyra tenera** (Asakusanori). The shape of *P. tenera* usually ranges from elongated oval to linear oblanceolate (Figure 1.19). It is the most popularly grown species in Japan. In natural conditions, *P. tenera* is found attached to rocks and other substrata, and seldom adhering to other seaweeds. Although this species grows abundantly inside or at the mouth of a gulf, it thrives just as well outside the gulf region.

To grow *Porphyra*, bamboo poles combined with net and twigs are fixed in mid-

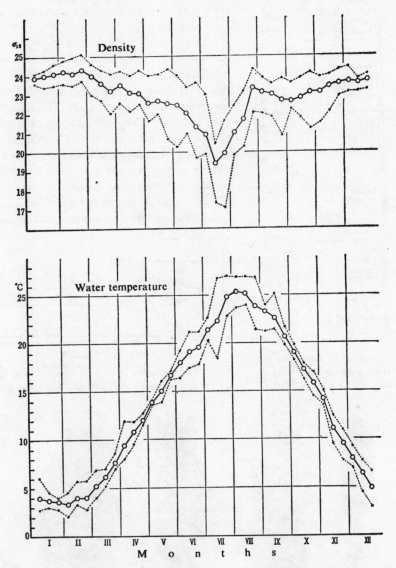

Figure 1.16. Temperature and density of water in Matsushima Bay (Kurogi, 1961).

or late September; plumules of 1 mm size are seen only in mid- or late October. The seaweed grows up to 15 to 20 cm or more by mid- or late November, and becomes abundant between December and March. It decreases in April and by May disappears (Figures 1.21 and 1.22).

From late September to early November, neutral spores form minute plumules of 150 μ and 1 mm in length, followed by secondary growth. With further plumule growth, the formation of neutral spores stops. By late October, the plumules attain a length of 3 to 5 cm or more, and begin to develop sexual organs. From early November carpospores are released, and the production and release of carpospores

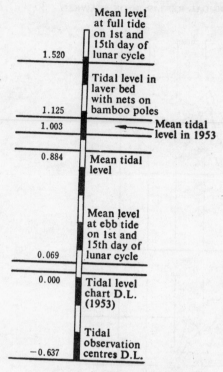

Figure 1.17. Condition of tidal level in the vicinity of Matsushima Bay (R. Ayu) (Kurogi, 1961; data partly modified).

continues up to the final vegetative period of the plant, or until the plant is carried away by the currents.

The plumules derived from the spores which germinate by early October give rise to neutral spores and then to reproductive organs. But most of the plumules derived from the spores germinating in early November, give rise to sexual reproductive organs without passing through the stage of producing neutral spores. It takes 7 to 10 days or even more for the spores to settle and grow into plumules capable of developing neutral spores. The tips of the neutral spores which form plumules, split. This condition remains until the plumules

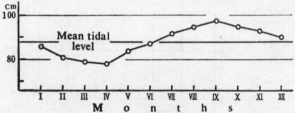

Figure 1.18. Monthly variations in mean tidal levels in the vicinity of Matsushima Bay (R. Ayu) (Kurogi, 1961).

Figure 1.19. *Porphyra tenera* (*Asakusanori*).

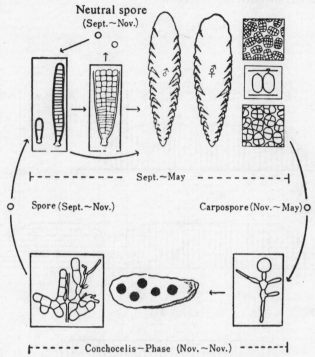

Figure 1.20. Life cycle of *P. tenera* (Kurogi, 1961).

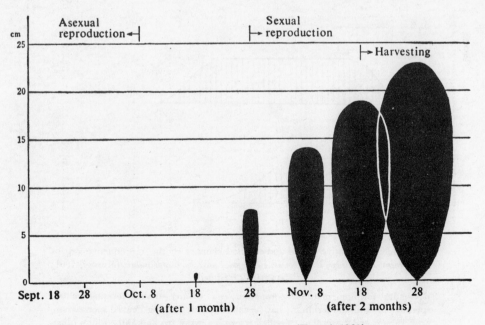

Figure 1.21. Growth chart of *P. tenera* (Kurogi, 1961).

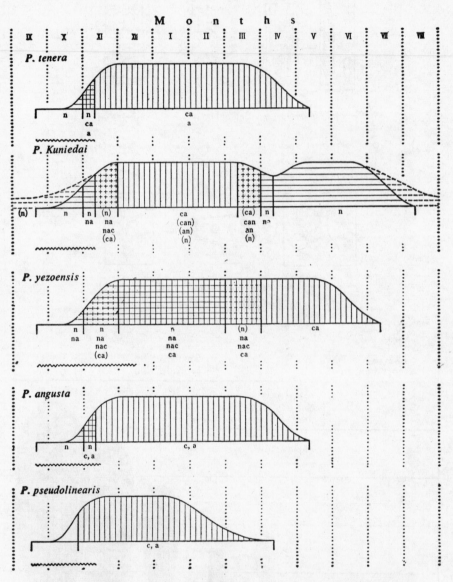

Figure 1.22. Vegetative period and seasonal changes in the reproductive organs of *P. tenera*, *P. kuniedai*, *P. yezoensis*, *P. angusta*, and *P. pseudolinearis* (Kurogi, 1961; data partly modified).

n—thallus forming neutral sporangium; na (=an)—thallus forming neutral sporangium and antheridium; nac (=can)—thallus forming neutral sporangium, antheridium and cystocarp; ca—thallus forming cystocarp and antheridium (dioecious); c—thallus forming cystocarp (gynoecious); a—thallus forming antheridium (androecious) 〰 = spore releasing period of Conchocelis.

attain 4 to 6 cm in length, at which time the liberation of the carpospores or spermatia begins. When the neutral spores form from late September to early November, the water temperature is 20 and 21° to 14 and 15°C. In in-vitro experiments,
the neutral spores formed at a water temperature of 15° and 20°C, but not at 10°C
(Kurogi and Sato, unpublished).

The carpospores adhere to various types of shells and grow into Conchocelis. Under
culture conditions the Conchocelis form spots of 1 cm in diameter by August or
September. From July onwards sporangia begin to develop, and from late September
to early November a large number of spores are released (Kurogi and Hirano, 1956)
(Figure 1.23). The water temperature during this period is 23° to 14.5°C. This is

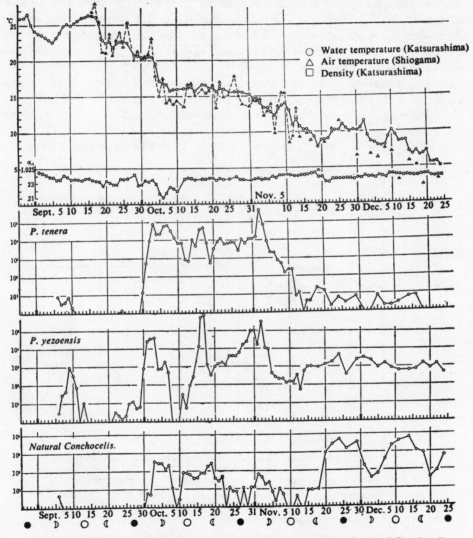

Figure 1.23. Conchocelis of *P. tenera* and *P. yezoensis*; spore release of natural Conchocelis
[experiments carried out in the sea (Kurogi and Hirano, 1956; data partly modified)].

more or less equal to the water temperature at which the plumules of *Porphyra* produce neutral spores. This temperature is much the same as the water temperature of 24.5 to 15°C when the Conchocelis form sporangia and release the spores (Kurogi and Hirano, 1956a; Kurogi and Akiyama, 1966). This period of spore release corresponds to that of *P. tenera* in which the Conchocelis give rise to sporangia within a few days (Kurogi, 1959; Kurogi and Sato, 1962; Honda, 1962).

The filaments exhibit periodicity in the release of spores. According to Takeuchi *et al.* (1954) and Honda (1962), the spore release increases with every tide. Kurogi (1963), Kurogi and Hirano (1956, 1956a), who studied *P. tenera* and *P. yezoensis*, reported the spore releasing periods as covering 12 to 21, 14 to 18, 5 to 8, 12 to 15, 20, 8 to 16, or 12 to 14 days. The variations in the period of spore release are very clear when the spore release is in full swing, or when the temperature is relatively high. The period is also related to the number of days required for maturation under various conditions. On the other hand, it is also influenced by the fluctuations in water temperature and density (Kurogi and Hirano, 1956a; Toyama *et al.*, 1956; Yamazaki *et al.*, 1957; Maekawa and Toyama, 1958, 1959; and Honda, 1962).

Even during the course of a day, spore release shows some difference at various hours. Soon after dawn spores are released in large numbers for 3 to 4 hours (Yamazaki, 1954; Suto *et al.*, 1954; Kurogi and Hirano, 1955a, 1956) (Figure 1.24).

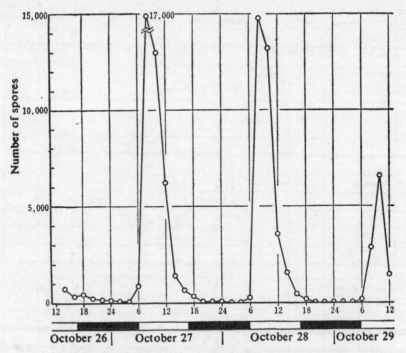

Figure 1.24. Cycles of spore release in Conchocelis of *P. tenera* (Kurogi and Hirano, 1955) on various days.

2. **P. angusta** (Kosujinori). *P. angusta* resembles *P. tenera* in shape but is monoecious (Figure 1.25). It is one of the major types cultivated in the Tohoku region and

grows well near the mouth of gulfs. It was mainly studied in Kesen-numa Bay. In natural conditions, it is found adhering to rocks and mussel.

The vegetative period, reproduction, life cycle, and maturation period of the Conchocelis of this type are almost identical to the same in *P. tenera*. Moreover, the water temperature and maturing conditions for the Conchocelis of this species resemble those for *P. tenera*. However, the temperature suited for spore release differs for *P. angusta* (Kurogi and Sato, 1962a; Kurogi and Akiyama, 1966).

3. **P. yezoensis** (Susabinori). *P. yezoensis* is broader than *P. tenera* and varies from oval to oblanceolate (Figure 1.26). It is a type of *Porphyra* which is widely grown, second only to *P. tenera*. In natural conditions, it is found slightly above the intertidal zone, adhering to rocks, barnacles, mussels, and other seaweeds. It grows well at the mouth of a bay as well as outside it.

The life cycle of this type resembles that of *P. tenera*. While the period of existence is not different from that of *P. tenera*, the period of disappearance is delayed and thus the vegetative period is longer. One of the special features of this species is its long period of asexual reproduction.

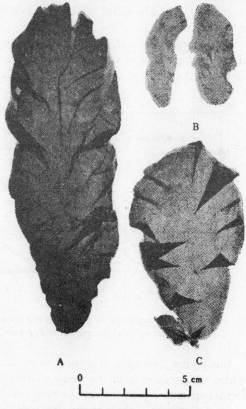

Figure 1.26. *P. yezoensis* (Susabinori).
A—sexual thallus; B—asexual thallus forming neutral spores; C—thallus forming neutral spores and spermatia.

Figure 1.25. *P. angusta* (Kosujinori).

Plumules of *P. yezoensis* are visible to the naked eye from early or late October. They become fairly large by late November or early December and exhibit luxuriant growth during December to May. From June onwards their growth begins to decrease and they almost disappear in July. Under natural conditions they show maximum growth from March to May.

Plants that grow to some size in February or before March begin producing neutral spores; with further growth there are neutral sporangia as well as antheridia in the same thallus. Later on the thallus develops a cystocarp as well. In the final stage however, instead of the neutral spores, only antheridia and cystocarps are produced, i.e. after April, the plants directly develop antheridia and cystocarps without the intervening formation of neutral spores. Hence, from October to March a number of plumules grow as a result of active secondary reproduction from neutral spores. On the other hand, after November and up to the disappearing period, they give rise to carpospores only. Thus, in October only neutral spores form and in the post-April period only carpospores.

The size of the plant producing neutral spores is 1 to 7 cm, although even smaller plants less than 1 cm sometimes produce neutral spores. Sometimes plants may be as large as 11 cm. The plant which forms only sexual reproductive organs after April, is 3 to 4 cm or more.

As mentioned earlier, the plumules develop up to March. However, the growth of plumules appearing in the autumn is rapid, that of those appearing in the winter is slow, and that of those appearing in the spring is rapid once again (Figure 1.27).

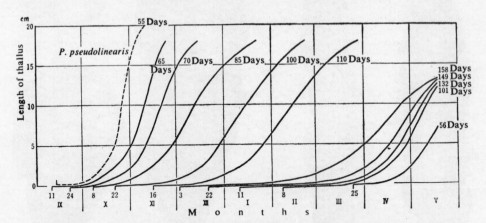

Figure 1.27. Comparison of growth on the basis of the spore-adhering period of *P. yezoensis* (chart includes the growth of *P. pseudolinearis*) (from Kurogi, 1959, 1961).

During the formation of neutral spores in *P. yezoensis* from October to March, the water temperature is in the range of 18 to 4 to 7.8°C, and during the carpospore formation after April, it is in the range of 14 to 4 to 20°C. *P. angusta* changes from asexual to sexual reproduction, although the asexual reproduction does not seem to be influenced by the water temperature as in *P. tenera*. Kurogi (unpublished) has reported the occurrence of asexual reproduction in *P. yezoensis* when the days are short. On the other hand, the carpospores which are released, pass the summer in the form of Conchocelis attached to shells at the bottom of the sea. The Conchocelis in turn produces a large number of spores from September to December, which ultimately grow into the thalli of the next generation (see Figure 1.23).

The period of spore release in Conchocelis is longer than that in *P. tenera* (Kurogi, 1953 described as *P. kuniedai;* Maekawa and Toyama, 1959a). This is supported by the experimental fact that the optimum temperature for the maturation of Conchocelis is 10°C or below (Kurogi and Akiyama, 1966). As in *P. tenera,* maturation is accelerated when the days are short (Kurogi and Sato, 1962a).

In any case, the long period of vegetative growth in *P. yezoensis* is due to the long period of spore release in Conchocelis and the neutral spore formation in the leafy thalli.

4. **P. kuniedai** (Maruba asakusanori). This plant is usually round in shape (Figure 1.28). Although the plant sometimes grows independently, it usually mixes with *P. tenera* and *P. yezoensis*. In natural conditions, it is found adhering to rocks, pebbles,

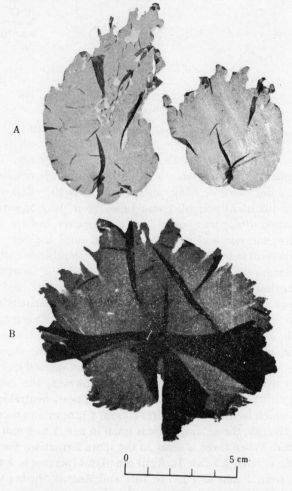

Figure 1.28. *P. kuniedai*: A—asexual thallus forming neutral spores; B—sexual thallus.

and sometimes to *Gloiopeltis* and agarophytes. It grows well both in a bay and at its mouth. It also grows abundantly in the vicinity of estuaries and outside a bay.

It passes the summer as summer seaweed reproducing asexually (Figure 1.29). It is large enough to be seen with the naked eye in late September and early October. The microscopic plumules continue to form until early or mid-November. The early

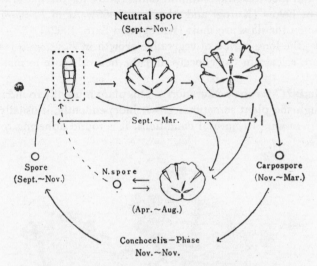

Figure 1.29. Life cycle of *P. kuniedai* (Kurogi, 1961).

plumules attain a length of 10 to 15 cm or more in December. The plants start to gradually disappear from February or March but growth remains abundant until April or May. After April or May, microscopic or slightly larger plumules appear again and grow luxuriantly in May and June. From July onwards, they gradually decrease and almost disappear in early August. However, in some places, a small part of the vegetation survives and continues to grow until the next vegetative period.

Neutral spores form in large numbers from late September or early October to early November onwards; sexual reproductive organs gradually develop and by mid- or late November, thalli with neutral sporangia and antheridia, and some with even a cystocarp, can be seen. Although the thalli produce both neutral spores and carpospores, the thalli have only sexual reproductive organs from December to late February or early March, and hence produce carpospores only. In mid- and late March, the reproductive organs gradually become asexual and thus one encounters antheridia, or cystocarps and neutral sporangia, or antheridia and neutral sporangia. The plants now produce carpospores and neutral spores. However, the capacity to produce carpospores is gradually lost, but until the latter disappear, neutral spores are released.

The thallus which forms neutral spores is from 1 mm to 5 to 6 cm or even more in size. Initially, though, the neutral spore is small in size, 1 to 4 mm in late September or early October. When there is delay in the spore formation, the size increases and by late October, reaches 1.5 cm. In April the plants increase to 2 to 4 cm, but again become small, from 1 mm to 1 cm in July and August, during which period they produce neutral spores.

On the basis of the germination period and its life history, *Porphyra* can be arranged into four groups:

1. Plants which are visible in their early stages during late September and early or mid-October, reproduce asexually in autumn and sexually in winter, and disappear by March or April, thus passing through asexual and sexual reproductive stages;

2. Plants which are visible from late October to early or mid-November, do not reproduce asexually in autumn, but reproduce sexually in winter; some disappear in March or April, the remainder reproducing asexually after April, thus passing through an asexual reproductive stage only, or sexual and asexual reproductive stages;

3. Plants which are visible after late November but do not grow well in winter and do not develop reproductive organs, but grow after March or April and reproduce asexually, thus having only an asexual reproductive stage;

4. Plants which appear after April and reproduce asexually until they disappear.

Asexual reproduction is in full swing from late September or early October to early November when the water temperature is 22 to 14°C, and for some time after April when the temperature is 9 to 25°C. On the other hand, when the water temperature is 8 to 3°C from December to February, the plants reproduce sexually. Kinoshita and Shibuya (1942) reported that the temperature for the formation of neutral spores was 15 to 20°C in *P. yezoensis* (Susabinori).

The carpospores which are formed and released in the winter pass the summer in the form of Conchocelis. From May onwards they develop sporangia, and in November spores are gradually released. The spores germinate and become the plumules of *Porphyra* which, in turn, give rise to a luxuriant growth in the next crop. When *Porphyra* of the spring-summer period live until August or September, the neutral spores formed by them and the spores formed by the Conchocelis, together produce the next luxuriant crop of *Porphyra*.

As in *P. tenera*, the period of spore release in Conchocelis continues up to November (Kurogi and Akiyama, 1966). The appearance of visible plumules is slightly earlier than in *P. tenera*. It thus appears that either the plumules originate from summer seaweeds, or that spore release from Conchocelis takes place early. In *P. kuniedai* the sporangia develop well even when the days are long, with 15 hours of daylight (9 hours of darkness). However, the spores are released when the days are short (Kurogi and Sato, 1962a) and their release is poor when the light is strong (Kurogi and Akiyama, 1964). There are reports on the release of Conchocelis spores having been observed in nature in June (Kurogi, 1961), although it is not yet clear whether these spores take part in the formation of summer seaweed.

Summer seaweed is seen when the summer temperature is low, or the temperature of the growing beds is low, and also in the years and situations in which the plants are abundant just before spring. The plants adhere firmly to rocks which dry easily and are well-exposed to light, but in such a habitat they disappear rapidly. They survive on substrata which do not dry easily.

5. **P. pseudolinearis** (Uppuruinori). This seaweed is shaped like a curved lance. Longer specimens are about 30 to 50 cm or more (Figure 1.30). It was considered a rock laver in the past, but since 1955 it has been cultured as a rapidly growing laver. It grows best outside bays, not doing well inside them.

This plant is characterized by the absence of asexual reproduction (Figure 1.31).

P. pseudolinearis appears earlier than *P. tenera*. By early October, it has reached 2 to 3 mm and sometimes even 1 cm in size. By early or mid-November, it attains 20 to 30 cm or more, and is most luxuriant from mid- and late November to mid- and late December. In January, the thallus gradually decreases in size and most of the plants disappear in February or March. The vegetative period is shorter than *P. tenera* (see Figures 22 and 27).

Figure 1.30. *P. pseudolinearis*
(*Uppuruinori*).

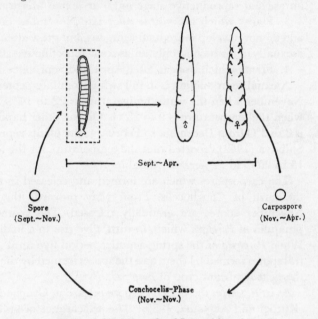

Figure 1.31. Life cycle of *P. pseudolinearis*.

Carpospores are released from mid-November until the plants disappear. No secondary reproduction due to neutral spores is found during the period of growth from minute plumules to adult structures.

The carpospores which are released pass the summer in the form of Conchocelis. The sporangia begin to appear from June onwards, and spores are released from September, giving rise to a prolific growth of laver in the next season.

It seems that the spore release of the Conchocelis continues for a long period as in *P. tenera*, or even slightly longer than in *P. tenera*, extending up to early or mid-November (Kurogi and Akiyama, 1966). While the Conchocelis develop sporangia even on long days with 14 hours of daylight, the spores are rapidly released only on short days (Kurogi and Sato, 1962a). The release of spores is also poor when the light is strong (Kurogi and Akiyama, 1965).

4.2.2 ADHERING ZONE

The following is an account of the spreading of a laver net at various water levels, during the period of laver culture in winter when it grows profusely.

1. **Spreading laver net at various water levels.** During the seeding period in late September to early October, the water level is taken into consideration when spreading the net. The most suitable water level may differ from year to year and vary with the changes in the growing season.

The water level best suited for net spreading, according to data collected during 1953 to 1954 in Matsushima Bay, was 12.2 cm above the annual mean tidal level (Kurogi, 1961). This corresponds to 7.9 and 2.7 cm above the monthly mean tidal levels in September and October respectively. When this is compared with the tidal levels during the period 1951 to 1956, the water level for net spreading was 10.7 to 12.5 cm above the annual mean tidal level, and was more or less the same as the monthly mean tidal levels of September and October. Thus it was 8 cm below and 5.8 cm above the monthly mean tidal levels (see Figure 1.17).

This water level is found either above the water surface or within the water, depending upon the high and low ebb tides which occur daily. The exposure time at this water level differs according to the month and the season. Surveys carried out between 1953 and 1954 indicate that the seasonal variations in exposure time can be expressed in terms of the mean tidal levels, and the division of a full day into daytime and nighttime on the basis of the time of sunrise and sunset (Figure 1.32). Between late September and early October, the seeding period, the mean exposure time in a day is 10 hours and 17 minutes (maximum, 14 hours and 30 minutes; minimum, 5 hours and

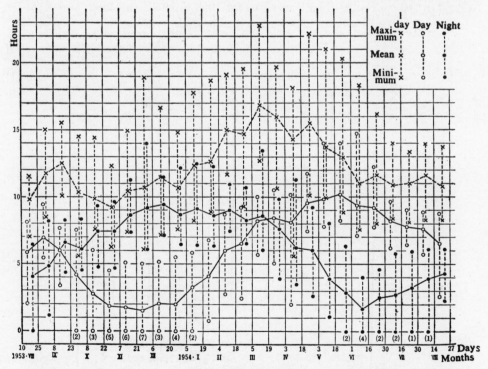

Figure 1.32. Seasonal variations of exposure time of laver net level (Kurogi, 1961). Mean values for the day for each tide are shown. Figures given in parentheses below 0 show the number of unexposed days.

30 minutes), during daytime *per se*, 4 hours and 12 minutes (maximum, 7 hours and 30 minutes; minimum, 0), and during nighttime, 6 hours and 5 minutes (maximum, 8 hours and 17 minutes; minimum, 4 hours and 30 minutes). If the days are not divided into daytime and nighttime, the mean exposure time studied in relation to low tides is 5 hours and 52 minutes (maximum, 14 hours and 30 minutes; minimum, 0).

When the daily exposure time of the nets at a particular water level is studied in relation to seasons, one finds that it increases from January onwards, reaching 15 hours in February to April, the final period of cultivation. During this period, the monthly mean tidal level is low. The exposure time during the day gradually decreases to only two hours in November or December and again increases in March to eight hours. On the other hand, exposure time at night increases after the period of seeding from 8 to 9 hours from November to March.

The water level in which the laver net is spread can be changed within the limits of 10 cm above, and 20 cm below, the usual water level; changes are made after carefully watching the variations in the water temperature and the vegetative conditions of the laver. However, when the laver culture is to be continued after February or March, the aforementioned water levels for the nets are not appropriate as they result in longer periods of exposure than would be good for the plants.

2. **Adhering zones in natural conditions.** A study of the poles supporting the laver net for the cultivation of *P. tenera* established that the range of the adhering level was 80 to 110 cm on the tidal level chart for 1951 to 1956. In March and April the plants were carried away by the currents at this level.

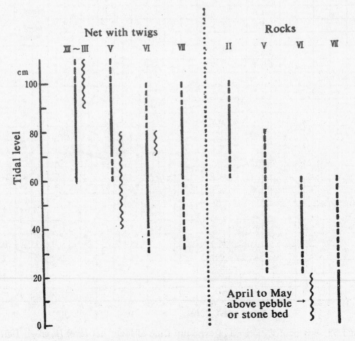

Figure 1.33. Seasonal variations in the adhering zone of *P. kuniedai* (Kurogi. 1961).
〜〜〜 indicates the adhering of plumules.

The growth of *P. yezoensis* is not so rich under natural conditions up to December, but the plant adheres to rocks in the same tidal level as that of *P. tenera*. After this period, it adheres to plants like *Gloiopeltis* growing at a lower level, and after March, attaches to *Chondria, Cystophyllum,* and *Eckloris* a little above or below the bottom tidal line (0 cm).

P. kuniedai is rarely found attached to other seaweeds up to March or April, but is found on pebbles and rocks at a tidal level of 60 to 110 cm. By May or June they occupy a lower level of 40 to 80 cm, but in July they occupy a slightly higher position, i.e. 70 to 80 cm (Figure 1.33). They are also found growing profusely after April or May near the low tide line when a new substratum is artificially introduced.

Seasonal changes of adhering zones are almost absent in seaweeds such as *P. tenera*, which have a short spore germination period and a short vegetative period. Such changes take place strikingly in *P. yezoensis, P. kuniedai,* etc. which have long germination and vegetative periods. The changes in the adhering zone are either a higher level in winter, a lower level in autumn, and a slightly higher level in summer. The changes are related to the life cycle of the laver itself as well as to the changes in the substratum. In other words, the mean tide level in spring is low, and on those days when the tides are high, the period of ebb tide at noon is prolonged; this results in a longer period of plant exposure in the nets at a higher level. Diatoms and other plants detach from the rocks and disappear, thereby facilitating the adherence of *Porphyra*. As the other types of algae degenerate by aging, it becomes easier and easier for the spores of *Porphyra* to attach and settle down.

4.2.3 CHARACTERISTICS OF CULTURES

The following is an account of the ecological features pertaining to the culture of the aforementioned five species of *Porphyra* (see "Life History of Major Species," p. 51). The regions suitable for the vegetation of these five species have already been discussed.

Table 1.5 presents the harvesting period for various types of *Porphyra*. If the vegetative period and the harvesting period of *P. tenera* and *P. angusta* are taken as

Table 1.5. Harvesting period for various species of *Porphyra*

Name of species	Harvesting period
P. tenera	late November to February (March)
P. yezoensis	late November to April (May)
P. kuniedai	early December to April (May)
P. angusta	late November to February (March)
P. pseudolinearis	early November to December

normal, the harvesting period of *P. yezoensis* and *P. kuniedai* is long, and that of *P. pseudolinearis* is short. The fact that the first harvesting of *P. kuniedai* is delayed, as compared to other types, is probably due to growth stagnation resulting from the formation of neutral spores in a large number in late October. However, when *P. kuniedai* passes the summer in the form of summer seaweed, it can be harvested much earlier.

Among the five species of *Porphyra* discussed here, *P. tenera* and *P. kuniedai* are found in and at the mouth of bays, *P. angusta* at the mouth of bays, and *P. yezoensis* and *P. pseudolinearis* outside bays. When more than two species of *Porphyra* germinate at the same place, the type which attaches to the laver net varies according to when the net was spread (Table 1.6). When the net is spread very early, the order of dominance

Table 1.6. Net-spreading period and types of *Porphyra* which adhere to the net

Net-spreading period	Dominant species	Species in mixed growth
Early to mid-September	*P. Pseudolinearis (P. kuniedai)* *	*P. tenera, P. yezoensis,* or *P. kuniedai*
Late September to early October	*P. tenera (P. kuniedai)* *	*P. pseudolinearis, P. yezoensis,* or *P. kuniedai*
Mid-October to early November	*P. yezoensis (P. kuniedai)* *	*P. tenera* or *P. kuniedai*
Mid- to late November	*P. yezoensis*	*P. kuniedai*
December to March	*P. yezoensis*	
April to July	*P. kuniedai*	

*When *P. kuniedai* passes the summer as summer seaweed and grows luxuriantly.

is: *P. pseudolinearis, P. tenera,* and *P. yezoensis.* When the seedlings grow in spring-to-summer, *P. kuniedai* dominates the net. When this species passes the summer and becomes abundant, it is still the dominant species from early September to early November. In places where *P. angusta* germinates, it is possible to substitute *P. tenera* for *P. angusta.* Mixed cultures of *P. tenera* and *P. angusta* are almost nonexistent in the Tohoku region. When *P. pseudolinearis* grows mixed with *P. tenera,* the growth of the former is poor while that of the latter is luxuriant. In a mixed growth of *P. tenera, P. yezoensis,* and *P. kuniedai,* the growth of *P. tenera* is poor and that of *P. yezoensis* or *P. kuniedai* abundant. In Japan, this is known as "mekawari" which means a "bud exchange" phenomenon.

Apart from the differences in the germinating regions, the differences in the type of *Porphyra* in relation to a particular net-spreading period also influences the spore release period of Conchocelis, the asexual reproduction period, and early or delayed growth. Among these, although the details of the period of spore germination of the Conchocelis in *P. pseudolinearis* and *P. kuniedai* are not complete, those of *P. tenera* and *P. yezoensis* have already been explained, and the details of their asexual reproduction given. Regarding the latter, the number of days required for the formation of neutral spores from the day of spore adherence is also important. In *P. tenera* and *P. angusta,* this is 7 to 10 days, in *P. kuniedai,* 20 to 30 days or more, and in *P. yezoensis,* 30 to 40 days. If the number of spores adhering is more or less the same in these species, that which produces neutral spores earlier and develops secondary buds, becomes the dominant species.

4.3 Summer Porphyra

The following is an account of *P. umbilicalis* and *P. pseudocrassa* based on the studies carried out by Kurogi and Kawashima (1966-1968) in the Nemuro region of Hokkaido. The water temperature of the Nemuro region in Japan is low; in January and

February it is lower than −1 to −2°C. The temperature does not increase much even in the summer so that in August it is still only around 18°C (Figure 1.34). The

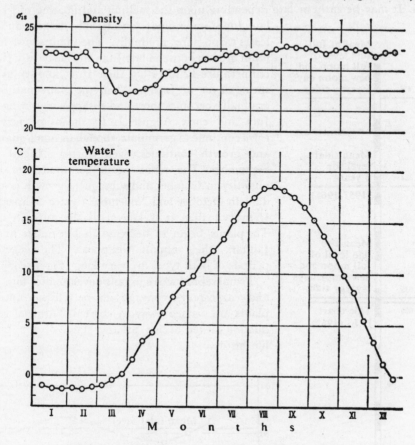

Figure 1.34. Temperature and density of water in Nemuro.

density of water is influenced by the melting and floating of ice, and tends to decrease in March and April. Using those at Hanasaki as the standard, tidal conditions have been studied. The difference in the ebb tide following high tide is 145 cm (annual mean value) (Figure 1.35). This differs very little from that in Matsushima Bay. In other regions, freezing is seen from late December to late March, and floating ice is seen in the coastal regions from early February to late March.

4.3.1 LIFE HISTORY OF VARIOUS SPECIES

1. **P. umbilicalis** (Chishimakuronori). This species varies greatly in shape and may be lanceolate, elongated-oval, or spherical. It has a length of 5 to 20 cm (Figure 1.37) and may be dioecious or monoecious. It is found in Karafuto-Chishima and along the eastern coast of Hokkaido. It grows as a rock laver and is cultured in some parts of Japan. It remains attached to rocks in the intertidal zone.

The life cycle is similar to that of *P. pseudolinearis*, but its conditions of growth in the winter and the summer are reversed.

In mid-May (water temperature 7 to 8°C), the size of the seaweed ranges from 1 to 2 mm. It may be early or late depending upon the melting and floating of ice. By late June, it becomes considerably larger and releases carpospores gradually. The plant shows maximum growth between late June and early July (water temperature 12 to 14°C); then it is known as the spring laver in Japan. When the plant matures, it gradually becomes short and disappears during late July and early August. In the meantime, new *Porphyra* continue to germinate, though in small quantity, and growth continues even during August and September. The plant again germinates in large quantity in October and subsequent growth is abundant in October and November (water temperature 15 to 5°C); then it is known as the autumn laver. The plants begin to decrease in December and by January they almost disappear. The vegetative periods of both types are very long (Figure 1.38).

Sexual reproductive organs develop after late June and carpospores continue to be released until the plants are carried away by currents. Asexual reproduction never occurs at any stage of the plant's life history.

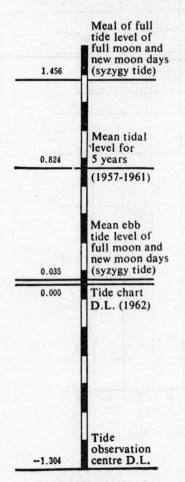

Meal of full tide level of full moon and new moon days (syzygy tide)

1.456

Mean tidal level for 5 years
(1957-1961)

0.824

Mean ebb tide level of full moon and new moon days (syzygy tide)

0.035

0.000   Tide chart D.L. (1962)

Tide observation centre D.L.

--1.304

Figure 1.35. Tidal levels in Hanasaki (1962).

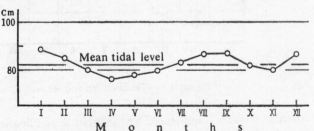

Figure 1.36. Changes in the mean tidal level in Hanasaki (1951 to 1961).

Carpospores grow into Conchocelis in the winter. In the spring, spores are released by the Conchocelis and the plants once again grow luxuriantly. In vitro experiments show that the Conchocelis give rise to sporangia from December onwards, and by March or April they release a large number of spores; in the open seas, however, spore release continues up to early October. The water temperature during this period is 0°C or 1 to 18 to 14.5°C. The Conchocelis of *P. umbilicalis* give rise to sporangia when the water temperature is in the range of 1 to 20°C, but the release of spores is more rapid when the temperature is below 15°C, and sporangia develop well when the days are long (Kurogi and Sato, 1967; Kurogi *et al.*, 1967).

The periodicity of spore release from Conchocelis is the same as in *P. tenera*. The

Figure 1.37. *P. umbilicalis.*

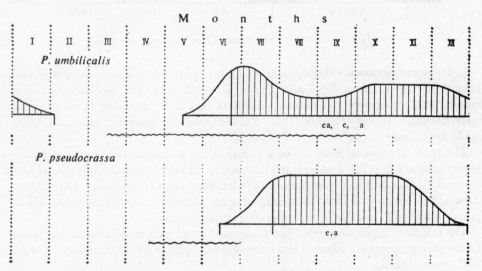

Figure 1.38. Vegetative period and development of reproductive organs
in *P. umbilicalis* and *P. pseudocrassa* (Kurogi and Kawazu, 1966).

Legend same as in Figure 1.22.

time of day for spore release differs, however, from *P. tenera,* the most rapid release occurring around noon (Figure 1.39).

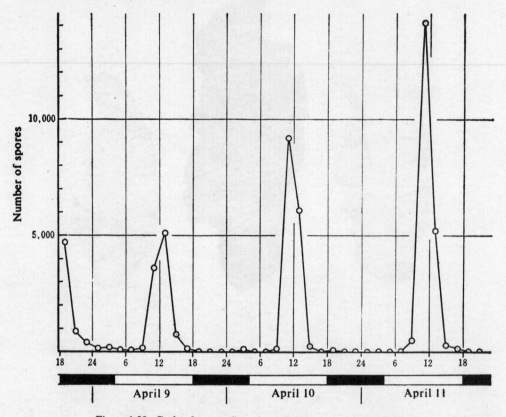

Figure 1.39. Cycle of spore release from Conchocelis of *P. umbilicalis.*

2. **P. pseudocrassa** (Makureamanori). This species is spherical in shape and belongs to the subgenus *Diplastidia* (Figure 1.40). It is usually monoecious and is found along the eastern and northern coasts of Hokkaido, adhering to rocks in the intertidal zone. In some parts of Japan, it is used as a rock laver and efforts are being made to cultivate it.

The plant appears later by about a month, than *P. umbilicalis,* i.e. in mid-June (water temperature, 12°C) when it is 1 mm in size. It attains considerable size by late July, and is abundant between August and October (water temperature, 18 to 12°C). Asexual reproduction is found at no stage during this period (see Figure 1.38). Compared to that of *P. umbilicalis,* the vegetative period is short.

The released carpospores grow into Conchocelis and pass the winter in that form. In the spring, spores are released and once again the plants grow luxuriantly (Kurogi *et al.,* unpublished).

The characteristics of Conchocelis are not completely known. However, it seems that the sporangia form from February or March and spores are released from April onwards. In the open seas, the release of spores continues up to late June (water temperature, 13°C). The release of spores is very active when the water temperature is 10 to 15°C or less, and the days long (Kurogi and Yoshida, unpublished).

4.3.2 ADHERING ZONE

Compared to species which grow abundantly in the winter, the vegetative zone of *P. umbilicalis* and *P. pseudocrassa*, which grow luxuriantly in the summer, is somewhat

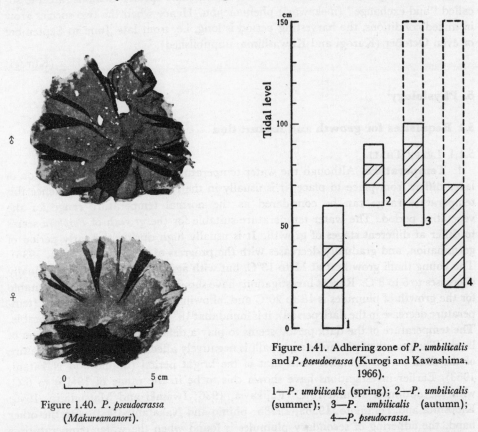

Figure 1.40. *P. pseudocrassa* (*Makureamanori*).

Figure 1.41. Adhering zone of *P. umbilicalis* and *P. pseudocrassa* (Kurogi and Kawashima, 1966).

1—*P. umbilicalis* (spring); 2—*P. umbilicalis* (summer); 3—*P. umbilicalis* (autumn); 4—*P. pseudocrassa*.

different. In the Nemuro region, the tidal conditions differ very slightly from those in Matsushima Bay. Freezing conditions and floating ice clean the beaches, especially the intertidal zone.

The spring laver of *P. umbilicalis* (a plant which grows luxuriantly from June to July) is found at 0 to 60 cm on the tidal chart, but more often 20 to 40 cm. But the summer laver (August to September) is found at 60 to 90 cm, but more often at 70 to 80 cm on the chart. The autumn laver, found after October, shows a wider distribution at 50 to 100 cm, especially at 60 to 90 cm, and sometimes even at 150 cm (Figure 1.41). The changes in the adhering zone of this plant resemble those of *P. kuniedai*.

At times *P. pseudocrassa* is even found at a height of 150 cm on the tidal chart, in areas where frequent spindrifts occur, but normally it lives at levels between 20 to 70 cm. As the germinating period of spores and the vegetative period are not as long as in *P. umbilicalis*, seasonal variations in the adhering zone are not striking.

4.3.3 Special Features in Cultivation

P. umbilicalis can be cultivated in April either by natural or artificial seeding. The harvesting period is short, from late June to late July.

In natural seeding, *P. pseudocrassa* is often mixed with *P. umbilicalis*. *P. umbilicalis* grows abundantly until August when the growth declines; *P. pseudocrassa* grows abundantly from August to October. In other words, these species demonstrate the so-called "bud exchange" (mekawari) phenomenon. Hence when the two species grow in mixed conditions, the harvesting period is long, i.e. from late June to September or even October (Kurogi and Kawashima, unpublished).

(Kurogi)

5. Physiology[1]

5.1 Requisites for growth and maturation

5.1.1 Leafy Thalli

1. **Temperature.** Although the water temperature suitable for the cultivation of laver differs from place to place, it is usually in the range of 2 to 22°C. Hence this temperature range can be considered as the normal temperature range for the vegetative period. The water temperature suitable for the growth of *Porphyra* seems to differ at different stages of growth. It is usually high during the early period of germination, and gradually decreases with the progress of growth (Fujikawa, 1936). The young thalli grow best at 11 to 13°C, but with aging the temperature gradually decreases to 6 to 8°C. Recent investigations have shown that the temperature suitable for the growth of plumules is 18 to 20°C and, allowing a difference of 4 to 8°C (temperature decrease in the dark period), it is found that 18°–14°C or 18°–10°C is favorable. The temperature of the dark period seems to play a definite role in the vegetation of laver. The growth of plumules and thalli is negatively affected when the temperature of the dark period is higher than that of the bright period (Shimo and Nakatani, 1963). Earlier investigations have shown this to be in the range of 15(16) to 6°C, with the optimum around 10°C (Fujikawa, 1936; Iwasaki and Matsudaira, 1956; Kinoshita and Teramoto, 1958a, 1958b; Shimo and Nakatani, 1963). On the other hand, the adhering of secondary plumules is found when the water temperature is above 17 to 18°C (Suto, 1961a; Shimo and Nakatani, 1963). The characteristic feature of *Porphyra* thalli is its remarkable ability to withstand freezing conditions. When the thalli are frozen under semi-drying conditions (−18 to −20°C), most of them are able to survive for even four months (Migita, 1964).

Since metabolic activity increases at high temperatures, factors related to the supply of nutrition such as water current, disturbances in deep water, etc., play important roles. Furthermore, bacteria, diatoms, etc., which cause damage to *Porphyra*, multiply rapidly at high temperature. Hence the influence of temperature should not be considered *per se*.

2. **Light.** Systematic data on the influence of the quantum and quality of light in relation to the vegetation of *Porphyra* are meager.

---

[1]Unless otherwise stated, the seaweed referred to in this section is *P. tenera*. However, it is possible that some very closely allied species may come under this section.

The saturation point of light during the photosynthesis of *Porphyra* is said to be 10,000 (Tsuruga and Arata, 1957) or 20,000 lux (Kinoshita and Teramoto, 1958). However, *in vitro* experiments carried out under a fluorescent lamp for 15 days showed that the plants grew well at 6,000 lux, and that the color of the plants changed to dark brown when the light was below 2,000 lux. When the strength of light is increased, the black coloration fades and a red color appears and gradually deepens. When the light is above 8,000 lux, the color of the entire plant pales (Kinoshita and Teramoto,

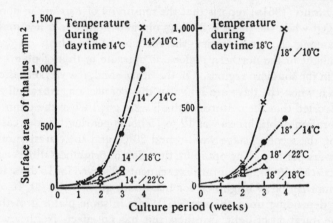

Figure 1.42. Influence of temperature on the vegetation of *Porphyra* thalli during short days (10 hours of brightness/day) (Shimo and Nakatani, 1966).

1958a). On the other hand, experiments carried out for 16 days under fluorescent and ultraviolet lights have shown that growth improves when the light is increased up to 10,000 lux (Matsumoto, 1959). According to Nakatani *et al.* (1961), the vegetation of *Porphyra* (cultured for 5 days in an incandescent light) is good when the light is above 5,000 lux. Beyond this level, there is not much difference in the growth of thalli. The crepe-like texture of the plant, known as chirimen in Japan, decreases with an increase in light intensity. No adverse effect of light is seen, however, even at 15,000 lux. The amount of solar radiation suitable for the vegetation of *Porphyra* is in the range of 150 to 500 cal/cm^2 per day (Shinmura, 1965). However, Shinmura's findings are not applicable under all conditions, because the culture fluid, light source, etc., are not the same in all investigations.

The red and yellow parts of sunlight are much more effective than the green and blue regions in increasing the size and weight of laver, as well as the rate of nitrogen assimilation. The effect of yellow rays in increasing the surface area of *Porphyra* is particularly impressive. The red and yellow colors of the plants fade away and are replaced by black and green. Similarly, when the plants receive red rays, the red color fades and the plants become green in color (Fujikawa, 1936). Light of short wave lengths (400 mμ, 460 mμ) is favorable for the vegetation of *Porphyra* thalli; the vegetation deteriorates when the wave length increases to 700 mμ, and the color alters toward red (Nakatani *et al.*, 1961). However, these results are not conclusive.

Recently it has been shown that a fixed period of darkness is necessary for the growth of *Porphyra*, and the period of brightness should be 8 to 9 hours per day (Iwasaki and Matsudaira, 1958). According to Iwasaki (1961), the maturation of *Porphyra* thalli is accelerated by allowing the plants to remain under optimum temperature conditions when the days are long. He further claims that it is possible to make *Porphyra* complete its life cycle in about five months and enter the next cycle. It seems that the maturation of *Porphyra* thalli is accelerated when the days are long and bright for more than 12 hours; the plants mature in 3 to 4 weeks after germination. However, if daylight is available for only 9 hours or less, the plants do not mature (Suto, 1961a; Iwasaki, 1965). Suto (1961a) reports that the sensitivity of *Porphyra* to photoperiodic action is observed when the plant is 20 days old and from 2 to a few millimeters in length. Even among the same species, the reaction of the plants on days of long duration of sunlight in the northern regions, is opposite to that of plants on days of short duration in the southern regions. On the other hand, few workers assert that the plants grow well when the days are long. In this connection, Nakatani and Shimo (1962) have reported that the cultures of thalli were good when the temperature was low and the duration of brightness was 12 to 15 hours per day. In this case, the temperature during the seedling stage was around 20°C, and the duration of brightness was 15 to 18 hours per day. The optimum duration of brightness differs according to species (Tanaka *et al.*, 1965). In culture experiments observed for 15 days, it was found that the optimum period of brightness for maximum growth was 20, 16, or 12 to 16 hours per day depending upon the species. While in some plants growth is sharply affected by the duration of light, in others this has no effect. Neither variations in the degree of brightness nor duration of light have any effect on the growth of plumules and thalli, and the extent of the growth of these parts can be estimated according to the amount of light radiation (degree of brightness multiplied by duration of brightness) (Teramoto and Kinoshita, 1962). It is said that for healthy and rapid growth the amount of light radiation should be 50,000 lux hours/day (i.e., 10,000 lux for 5 hours/day) for plumules, and 70,000 lux hours/day for thalli. These results are somewhat contradictory, but the difference is mainly due to the differing conditions and period of culture (the culture experiments required a long duration). The difference is secondarily due to the biological aspect and the quantitative increase. Such differences

Table 1.7. Influence of photoperiodic action on the thalli of *P. tenera* **(Iwasaki, 1965)**

Duration of brightness (hours/day)	Formation of reproductive cells	Formation of carpospores and their release	Growth after 60 days (maximum length in cm)
	After germination		
24	21 to 25 days	After 36 days	1.4
14	21 to 25 days	After 36 days	1.6
12	23 to 27 days	After 40 days	2.3
10	55 days	After 80 days	11.0
9	60 days	Not clear	6.2
8	60 days	Nil (90 days)	4.8

Note: Culture conditions = 4,000 lux, 13 to 15°C, culture medium SWI.

emphasize the necessity for further research on the problems related to the degree of brightness (Table 1.7).

3. **Chlorine concentration and desiccation.** In nature, *P. tenera* occurs under a wide range of chlorine concentrations. The density of sea water suitable for the cultivation of laver is said to be 1.015 to 1.023 (Cl: 11.4 to 17.2‰) (Takayama, 1937). With reference to photosynthetic activity, Cl 12.0 to 18.0‰ was found to be suitable for *Porphyra* vegetation (Iwasaki and Matsudaira, 1956).

Usually when the thalli of *Porphyra* are exposed daily to air for a fixed period, growth is delayed, but the plants are found to be healthier. The yield rate is at its maximum when the laver nets are exposed daily for two hours at the mean level of high and ebb tides (Kaneko, 1935). The yield rate is very poor if the exposure period is over 4 hours. The same applies to the photosynthetic activity, which is increased by about 20% when the laver nets are exposed for 2 hours only per day. When the plants are exposed for more than 4 hours, this activity is less than that of those remaining in the water (Iwasaki, 1965). While the factors involved in this physiological effect of exposure are not understood, it seems that changes in the osmotic pressure during exposure have an influence. In addition, the ability of *Porphyra* to resist desiccation is perhaps a deterrent to harmful and/or competing organisms.

5.1.2 CONCHOCELIS

1. **Temperature.** In nature, the carpospores are usually formed and released from early November onwards (water temperature, about 15°C) (Kurogi, 1961). Honda (1962) reported that 10 to 15°C is suitable for the carpospores to bore into shells. The growth of Conchocelis (especially the growth of the main branch) is good at 15 to 24°C, and there is hardly any difference in the growth rate within this temperature range (Kurogi and Hirano, 1965a). Growth becomes restricted at 27°C, and the branches are very close to one another. At 15 to 27°C, sporangia form but their formation is much more rapid at 21 to 27°C and slightly slower at 18°C. It is further delayed when the temperature is still lower (15°C). The size of the sporangia is maximal at 24°C, followed by 21°C, and 27°C. The size, however, decreases with a fall in temperature. The spores seem to be released at 15 to 24°C soon after the formation of sporangia. The release of spores can also be seen at 9 to 24°C, but the optimum temperature for the release of spores is between 17 to 20°C (in the case of *P. angusta* it is around 17°C) (Suto *et al.,* 1954). It seems that changes in temperature accelerate the release of spores. Kurogi and Hirano (1956a) reported that a maximal number of spores were released at 18 to 21°C, that the release was somewhat restricted at 24°C, and very poor at 15°C. When the Conchocelis in which the sporangia form are cultured at various temperatures, it is found that the release of spores is accelerated by lowering the temperature. On the other hand, the release becomes slower and slower as the temperature is increased. The restriction was very significant, especially at 24°C, and the influence of temperature variations could be seen in 2 to 6 days.

2. **Light intensity.** For the growth of Conchocelis, 2,000 lux is said to be optimum (Ogata, 1961). Although growth improves with an increase in light intensity (3,000 lux) (Honda, 1962), it is adversely affected when the sun falls directly on the plant (Kurogi and Hirano, 1955). On the other hand, in the case of free-living Conchocelis, maximum growth was reported to have been observed at 3,500 lux (Iwasaki, 1961).

The release of spores is either accelerated or restricted, according to the increase or decrease in light intensity. In *P. tenera*, the release of spores is unaffected when the culture is grown in the range of 4,000 to 2,000 lux. However, in *P. kuniedai* and *P. pseudolinearis*, the release of spores is poor in the culture under high light intensity (4,000 lux), even though the sporangia are well-formed. The release of spores in

Table 1.8. Influence of the photoperiod (short and long days) on the Conchocelis of *P. tenera* (Iwasaki, 1961)

Temperature	Photoperiod (bright hrs/day)	Light intensity (lux)	Formation of sporangia*	Appearance of thalli*	Comments
13–15°C	8	1500–2500	I after 19 days	<46 days	well-grown thalli (1 to 1.5 cm)
			II < 22 days	22–31 days	< 96 days 3 cm
		300–500	I after 27 days	nil (48 days)	
			II < 48 days	96 to 184 days	small thalli
18–20°C	11	1500–2500	I after 19 days	56 to 84 days	well-grown thalli which die out very soon
			II after 23 days	after 31 days	
		300–500	I after 19 days S1	nil (240 days)	
			II 23–31 days S1	nil (184 days)	Colony of large Conchocelis
13–15°C	continuous	1500–2500	I after 35 days S1–S2	nil (240 days)	**
			II after 31 days S1–S2	nil (184 days)	
		300–500	I after 84 days S1	nil (240 days)	
			II 72–96 days S1	nil (184 days)	
20–26°C	continuous	600–1000	I after 35 days S1	nil (240 days)	
			II after 31 days S1	nil (184 days)	
		100–200	I after 56 days S1	nil (240 days)	
			II	S1	nil (184 days)

Culture Medium: Experiment I—ASP; Experiment II—SWI.

  * Number of days after inoculation (Conchocelis).

  ** Conchocelis inoculated into the new culture medium after 50 days and transferred to short-day conditions (8 hours light per day). Spores and plumules were seen within a month.

S1: Enlarged cells (immature sporangia ?); S2: Tissues formed from enlarged cells.

these species improves when the light intensity is low (1250)-710-200 lux (Kurogi and Akiyama, 1965). According to Honda (1962), this restrictive effect on spore release is seen even at such a low light intensity as 20 to 40 lux.

3. **Photoperiodicity.** The existence of day-long action in the formation of monosporangia on short days (Kurogi, 1959a) prompted a study on the photoperiodic action of Conchocelis. When free-living Conchocelis were cultured on short days (8 to 11 hours of daylight) with 1,500 to 2,500 lux light intensity, monosporangia formed in 2 to 3 weeks. The plumules were reported to have appeared in 3 to 8 weeks. Furthermore, no changes in growth were seen within the range of 13 to 20°C. However, if the light intensity was in the range of 300 to 500 lux, the development of plumules was delayed (after 96 to 184 days) when the photoperiod covered 8 hours per day. When the photoperiod covered 11 hours per day, no thalli were seen in 180 to 240 days. Neither spores nor plumules apparent with continuous lighting, even when the light intensity and temperature varied (Iwasaki, 1961; Table 1.8). In Conchocelis found in shells, the day-long activity affected the maturation of the plants even at low light intensity (maximum intensity in a day, 50 to 100 lux) (Kurogi et al., 1962; Kurogi and Sato, 1962, 1962a; Kurogi and Akiyama, 1965). Weak light on long days does not bring about a short-day effect; if the light intensity is weak, maturation takes a correspondingly longer period. When the Conchocelis which mature on short days were cultured during the period of long days, however, no effect on the release of spores was evident. According to Honda (1962), the release of monospores is accelerated on short days and controlled on long days. The effect of photoperiod on the growth and maturation of Conchocelis, as observed by Kurogi and Sato (1962), is shown in Table 1.9.

It would seem therefore that before making a decision on the dark period suitable for growth, the formation of monosporangia, the release of spores, the water temperature, and the light intensity, other factors should be considered. The dark period is assumed to be in the range of 22 to 23 hours to 12 to 13 hours in a 24-hour cycle. Among *Porphyra*, photoperiodic reaction differs from species to species. The dark period which would suit growth, the formation of monosporangia, and the release

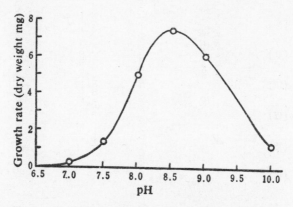

Figure 1.43. Influence of pH on the growth of free-living filaments (Iwasaki, unpublished).

of spores is assumed to be more than 14 hours in *P. tenera,* the same or a little less in *P. angusta* and *P. yezoensis,* 11 to 12 hours or more in *P. kuniedai,* and 14 to 16 hours or more in *P. pseudolinearis* (Kurogi and Sato, 1962a).

4. **Other conditions.** Shell boring by carpospores occurs smoothly when pH is 7.0 to 9.0, but is adversely affected when pH exceeds 9.0 (Honda, 1962). While the growth of filaments does not show much variation in sea water with pH of 5.6 to 10.6 (Ogata, 1961), the free-living filaments are reported to show maximum growth at pH 8.5 (Iwasaki, unpublished) (Figure 1.43).

The filaments are weak in withstanding desiccation (Kurogi and Hirano, 1955; Takeuchi *et al.,* 1956). They also cannot withstand water with a low chlorine concentration. According to one report, filaments treated with fresh water stopped growing but survived for 13 days, and when transferred back to sea water, resumed their growth (Kurogi and Hirano, 1955). The specific gravity of sea water suitable for the boring of shells by the Conchocelis, and for its growth, is 1.02 to 1.03 (Cl: 15.1 to 22.3‰). When the specific gravity is less than 1.01 (Cl: 7.8‰), the shell-boring phase is not seen (Honda, 1962).

Table 1.9. Effect of photoperiod on the growth and maturation of
P. tenera **Conchocelis (Kurogi and Sato, 1962)**

Photoperiod		Growth		Formation of sporangia		Release of spores
Dark Period	(Bright Period)	Growth	Branch color	Size	Location	
23	(1)					
22	(2)					
21	(3)	↑	↑ sparse with green patch	↑	from periphery to center	few
20	(4)					
18	(6)	growth continues good	close (becomes dark) with red patch	formation of sporangia good } best	inside central part	large in number maximum
16	(8)					
14	(10)					
12	(12)					
11	(13)	↓	↓	↓		nil
10	(14)					
9	(15)					

5.2 Nutritional requirements

5.2.1 LEAFY THALLI

1. **Nitrogen.** The normal, minimal, and absolute amount of nitrogen required for laver are 5.5 to 7.0, 4.0 to 5.0, and 1.2 to 1.3% respectively (Iwasaki and Matsudaira, 1956). Nitrates are best suited (Iwasaki and Matsudaira, 1958; Matsumoto, 1959); although the absorption of nitrogen from ammonia is rapid, it is inferior to nitrates. This is because the growth is adversely affected when the concentration of ammonia is 20 mg/l (Iwasaki and Matsudaira, 1956). According to Iwasaki (1956), when the nitrogen content (as nitrates) in sea water exceeds 3 mg/l, the growth is better in still culture. Absorption is maximum at 7 mg/l, and hence when the concentration is less than 0.7 mg/l, both the nitrogen and phosphorus contents of the laver decrease.

2. **Phosphorus.** *Porphyra* shows a wide range of phosphorus content and neither the normal nor the minimal requirement for laver is definitely known. However, the absolute amount required for growth is assumed to be 0.07 to 0.08% in terms of dry weight. Phosphorus absorption is high when the activity of the laver is maximal. The absorption increases with increasing concentrations of phosphorus in sea water, up to 0.3 mg/l. That absorption is also related to the degree of catalytic activity in sea water has also been established. Like the unicellular algae, these plants also seem to have a capacity to store phosphoric acid (Iwasaki and Matsudaira, 1956). There seems to be no difference between the organic and inorganic types, but the organic source seems to be slightly superior because of solubility and retentivity factors (Iwasaki, 1965).

3. **Growth promoting substances.** Investigations carried out thus far on the leafy thalli of *Porphyra* have been based on culture experiments in which other micro-organisms were also mixed. Further work is needed on this aspect.

Photosynthesis in *Porphyra* thalli is greatly accelerated when Fe, Mn, Co, Cu are provided as EDTA chelates (Iwasaki and Matsudaira, 1957). In addition to these, when an improved form of P 1 mineral solution, in which Zn and B chelate with EDTA, is added to artificial or natural sea water, the growth in the indoor culture is as good as that under natural conditions (Sato, 1960). Photosynthesis in *Porphyra* is also accelerated by vitamin B_{12}, glutamate, sodium sulphide, etc. (Iwasaki and Matsudaira, 1957). Vitamin B_{12} is also considered essential for the vegetation of *Porphyra* (Kanasawa, 1963). Regarding plant hormones, gibberellin (1 μg/l) is said to accelerate the growth of leafy thalli (Kinoshita and Teramoto, 1958b, 1958), and kinetin (0.1 ppm) the growth of plumules (Shimo and Nakatani, 1964). Amino acids such as lysine, B-alanine, serine, and asparagine, and purines such as guanine, adenine, and uric acid also accelerate the growth of *Porphyra* (Teramoto and Kinoshita, 1960).

5.2.2 CONCHOCELIS

There are very few investigations of the nutritional physiology of the Conchocelis and its relationship to growth. This is probably due to the fact that Conchocelis grow after boring into shells. Iwasaki (1965, 1967) studied the nutritional requirements of an aseptic culture of free-living Conchocelis and found that the growth was better when urea, L-asparagine, DL-lysine, and sodium nitrate were provided as nitrogen sources. However, the effective concentration of urea and ammonium salt as nitrogen sources is relatively low, i.e. 50 mg/l or less.

Phosphate is utilized at a low concentration (P 10 mg/l or below) in both organic and inorganic forms. It seems that sodium glycerophosphate is an excellent source of phosphoric acid. While the yield rate is high, the optimum concentration range is also wide (1 to 50 mg/l).

Calcium is necessary for the growth of Conchocelis and the optimum concentration is in the range of 0.1 to 1 g/l. Minerals such as Fe, B, Mn, Zn, Sr, Rb, Li, and I are also essential for growth; iodine particularly has an accelerating effect on growth, even though its effective concentration range is narrow (Iwasaki, 1967).

Although vitamin B_{12} is essential to the growth of Conchocelis, it can be replaced by several substances containing benzimidazol and resembling B_{12}. In fact, substances like 2-methyl mercapto-adenine or 5-methyl benzimidazol cobalamine, are said to be better than B_{12} itself. Plant hormones such as kinetin, adenine, indole acetic acid and valeric acid accelerate the growth of Conchocelis. Valeric acid in particular increased

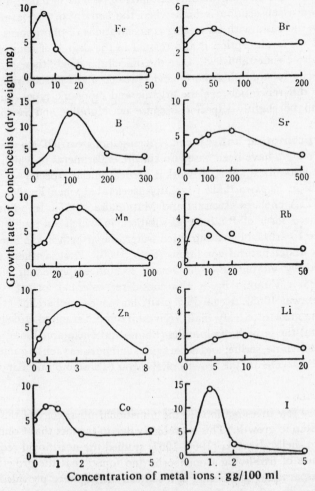

Figure 1.44. Growth of Conchocelis and the effect of minerals (Iwasaki, 1967).

the growth of the control plants by almost 500%. In the presence of vitamin B_{12}, even purines such as xanthine, hypoxanthine, guanine, etc., and pyrimidines such as uracil, methylcytosine, thymine, etc., have a growth-accelerating effect (Iwasaki, 1965).

5.2.3 PHYSIOLOGY AND CULTIVATION BEDS

The vegetation of *Porphyra* is greatly influenced by the physical and chemical properties of sea water. Water temperature, salt-nutrients, chlorine concentration, and water movements are particularly influential and, in some places, the degree of transparency of the sea water may also affect growth. In the past, laver seemed to grow abundantly only in favorable environments. But artificial means have since greatly widened the range of cultivation; laver can now be grown under even adverse conditions.

The water temperature required for an optimum cultivation period of laver is 10 to 8°C. Sudden changes in the water temperature (sudden rise) adversely affect the vegetation of laver, and hence in shallow places which are easily influenced by atmospheric temperature, water movement plays a very important role. Growth is restricted when nitrogen is less than 0.2 mg/l, and phosphorus less than 20 µg/l (Matsudaira and Iwasaki, 1953). Phosphorus is stored in the thalli and hence, unlike nitrogen, there seems to be no problem with this element. Since it is not possible to hope for an adequate supply of salt-nutrients from open seas, estuaries where land water rich in such salt-nutrients flows in, are best suited for the growth of laver. A supply of salt-nutrients can also be expected from the bottom of bays where putrifying organic matter exists. The mingling of the upper and lower layers of water by seasonal winds in the winter is also a favorable factor. As already mentioned, the dispersion of water by water currents is very important, especially when the water temperature is high and the region is poor in nutritional elements. The water current suitable for the vegetation of laver is usually 20 cm/sec for sea water (Matsumoto, 1959). However, when the sea water is rich in salt-nutrients, the water current could be 10 cm/sec, but if it is poor in such, the current should be 30 cm/sec. The water current has been found to be rapid when the temperature is low. Up to a certain degree of water temperature, it is assumed that the growth of laver could be normalized by adjusting the duration of exposure through the adjustment of laver nets, but with regard to nutrition, a careful study of the water current is essential, and an effective method of supplying fertilizers should be devised.

(Iwasaki)

CHAPTER II
Techniques of Seaweed Culture

1. Introduction

To trace the development of laver cultivation from a collector's interest to an organized industry using bamboo poles and nets is not an easy task. Records show that *Porphyra* was cultivated on poles in some parts of Tokyo Bay as early as 1675 to 1680 (Okamura, 1909), but doubtless pole culturing began much earlier there since laver culturing was pursued in Hiroshima Bay during 1596 to 1614 (Seaweed Report, 1967). The pole and net device was originally installed to catch fish, but when laver were frequently found to adhere to these nets, its culture by this method began. In Tokyo Bay the pole and net method of laver culture was initially practiced in a restricted area around Omori. During the 19th century this method gradually spread to other areas: Hamana-ko (1820), Culture and Education period (Bunsei Nenkon) (1818-1829), the Chiba Prefecture and Mikawa Bay (1844-1847), Matsukawa Ura-Kesen'numa (1854-1859), and the Aichi Prefecture (1861-1863). With the advent of the Meiji era (1867-1912), laver culture by the pole and net device became more and more widespread. But it was not until after 1955, when most of the areas from Hokkaido to Kagoshima were found suitable, that the expansion of laver culture became significant. By 1902, however, its culture had reached industrial proportions and laver had become an important marine product. In other words, laver culturing had ceased to be a private pursuit and became a business owned and managed by cooperatives and companies.

In the early stages of laver cultivation, the bamboo poles holding the hibi in place were fixed, and both seeding and culturing were carried out in the same bed. It soon became apparent, however, that the same place was not always suitable for both seeding and culturing. In 1884 it was discovered that the hibi with the seedlings could be transplanted from the seeding bed to another more suitable for culturing. This method of transplanting clearly differentiated between seeding beds (taneba) where the spores settle down, and the culture beds in which the seedlings developed. However the areas suitable for seeding beds were few, especially in Ariake-kai, thus limiting the production and collection of seedlings, and preventing the spread of laver culture to wider areas.

Subsequent to the work of Drew in 1949 on the life history of *P. umbilicalis,* the life cycle of cultured *Porphyra*, especially *P. tenera,* was finally understood; based on this understanding, a method for the artificial seeding of *Porphyra* was developed. The areas adopting this artificial seeding method began to multiply in 1955, and now natural seeding persists in very few places in Japan. Gradually, artificial seeding on an extensive scale became possible and, as can be seen from Figure 1.45, this technique has greatly increased the laver yield.

In the early Showa era (1925-1930), a horizontal method of culturing which involved the use of hemp nets or floating hibi (bamboo blinds), was practiced. The use of the net hibi spread from Tokyo Bay to other parts of Japan, and the floating hibi made of bamboo blinds acting as floating nets, came into use mainly in western Japan. The net hibi, i.e. hemp nets stretched on floating bamboo poles, is now widely

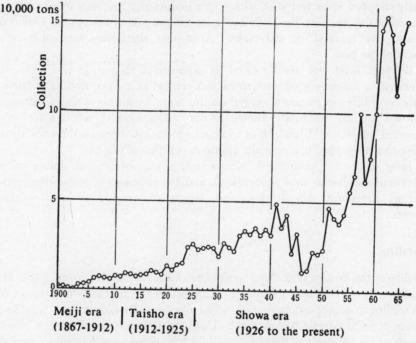

Figure 1.45. Transition in laver yield (Seaweed Report, 1967).

used, while the floating hibi made of bamboo blinds, twig nets, etc., is still in use in Kumamoto, Hiroshima, and a few other places. The changes in the type of hibi made during the period 1961 to 1966 are shown in Table 1.10. Bamboo blinds were replaced by natural fibers such as palm and hemp. Synthetic fibers like "Salan" were gradually introduced from 1955.

Table 1.10. Types and amount of hibi used for laver culture
(Data from the Ministry of Agriculture and Forestry)

	1961	1966
Pole nets	2,699,000	3,950,000
Bamboo blinds	385,000	113,000
Twig nets	14,848,000 m²	2,387,000 m²

The production rate after the Meiji era (1912) is shown in Figure 1.45. From 1955 production has shown a steady and significant increase.

About 36,000 enterprises existed in 1947 (Kimura, 1957), which increased to about 68,000 in 1960 to 1961; of the latter, 62,000 to 63,000 are still functioning (Tohoku Fisheries Experiment Station, 1965).

In the early stages of laver culture, the harvested laver was dried by spreading it on straw mats. Later the paper-making method of drying was imitated and laver was spread in thin sheets to dry. To make these thin sheets of laver, either wooden frames were placed on the mats and measured quantities of laver mixed with fresh water poured in an even sheet within the frame on the mat ("casting method" or nagetsuke), or finely chopped laver was kept with water in vats and the mats fitted with frames immersed in this mixture, the laver being mixed and allowed to spread uniformly on the mats ("wet method" or mizutsuke). At present, the former method is the more common of the two.

In the past, laver was usually dried by exposure to the sun in two ways. In the first method, a fence covered with straw was erected at a slant, and straw mats with thin sheets of laver in frames arranged on this fence by means of small bamboos. In the second method, popularly known as the Osaka method, wooden frames were constructed which could hold 18 or more seaweed mats. Because the Osaka method is easier to manipulate, it is now the common practice of the two.

As laver cultivation progressed, mechanization was introduced. Laver collecting vessels (ships and boats) were modernized, and the processes of harvesting, washing, pounding, spreading, dehydration, and drying are now done with mechanical assistance.

2. Seeding

Seeding is the process of making spores from Conchocelis or neutral spores formed in thalli settle on hibi and germinate. This process is also known as spore fixing. Since seeding is an important step in laver cultivation, its stabilization was the main objective of culture technique research. Until the Conchocelis phase was discovered as a part of the life cycle of *Porphyra,* the hibi were fixed after selecting a period and place considered (by experience) as best suited for spore adherence. The settling of spores was left to nature. Now shell-boring Conchocelis are cultured and the collected spores affixed to the hibi. Two methods of seeding are currently known—natural and artificial.

2.1 Natural seeding

In preparing seedlings, the places for pole fixation are selected on the basis of long experience with the fishing grounds. These places are called taneba (tracts of sea). In western Japan the taneba areas are inlets where the short-necked clam (*Tapes* sp.) lives (Fujikawa, 1957). Regarding the conditions of the sea and other topographical factors, the common features of taneba are not yet clear.

Just when the poles with nets for growing laver should be fixed also relies on knowledge gathered through experience, since the fixation period varies according to the type of *Porphyra* and the growing region. In the Tohoku region, the seeding period for *P. pseudolinearis* is from early to mid-September, for *P. tenera* and *P. angusta* from mid-September to early October, and for *P. yezoensis* in October. The seeding period in regions south of Chubu in Honshu Island is slightly delayed, late September to October being formally recognized as the seeding period; in these regions *P. tenera* germinates profusely. According to experiments carried out on the Conchocelis of *P. tenera* (Kurogi and Hirano, 1956a), the release of spores from Conchocelis is high when the water temperature is in the range of 15 to 24°C. Spore release is particularly maximal between 18 to 21°C. In natural conditions, the period in which the temperature of the sea water is below 23°C is considered suitable for seeding, which agrees with the results of the experiments on the Conchocelis phase.

The period of seeding, dependent upon the temperature conditions of the water, is also closely related to tidal waves. Kurogi (1961) carried out several experiments with regard to the period of fixing bamboo poles, and found that spores settled better when they were fixed on the 3rd or 4th day of the full moon or new moon (Figure 1.46). But experience had established this fact long before Kurogi's experiments (Ueda, 1958).

Spores do not adhere well when the salinity of the water is diluted by rain.

The best period for fixing the poles is when the water temperature, the tidal waves, and the salinity are simultaneously suitable.

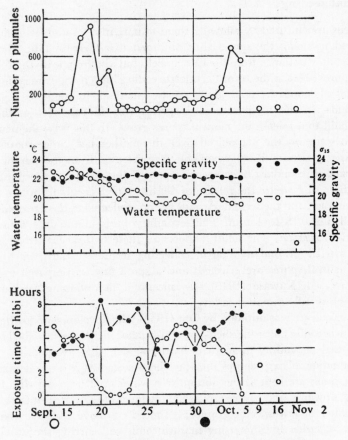

Figure 1.46. Number of plumules of *P. tenera* adhering after one month of planting poles on different days (Kurogi, 1961).

○—full moon; ●—new moon.

The level at which the poles for a horizontal hibi are fixed is important for seeding. When the level is too high and the exposure time too long, the resistance to desiccation in the early embryos is weak, despite the fact that the spores may adhere well (Akiyama, 1959). Hence the plants soon die. Too low a level is also not good because it encourages the prolific growth of green algae and diatoms. The water level must be determined with a view to providing, on the average, four hours of exposure in the daytime (from sunrise to sunset). This level is very near the mean water surface during September and October, or the seeding period. However, this is only a general view and the water level differs according to the water-holding capacity of the materials used in the hibi. Thus a water level suitable for hemp and coir yarn nets, which do not dry out easily, must be lowered for nets made of synthetic fibers such as Salan, etc., which dry out quickly. Again, the exposure time is shortened in places near the open sea due to the presence of swells and high waves. Hence, in these places the water level should be much higher than the standard level.

2.2 Artificial seeding

Conchocelis are cultured by allowing them to bore into a suitable substratum such as shells, and inducing the spores thus obtained to adhere to hibi. However, the spores are not only obtained from the Conchocelis but also from leafy thalli in the form of neutral spores. Hence the term "artificial seeding" has a wider meaning.

When neutral spores are used, there are two methods in vogue for collecting the secondary buds. In one, a separate net is spread over the pole nets (mother nets) on which the thalli that release the neutral spores grow. In this way, the neutral spores are made to settle on the net placed over the mother net. Since spores lose their adhering capacity (Suto, 1950) when they are released from the parent body, the adherence rate of secondary buds becomes poor if the distance from the mother net is considerable. In *P. tenera* the release of neutral spores occurs from late September to early November, and the secondary buds are not collected as the size of the plumules is very small, i.e. 0.15 to 1 mm. The secondary buds are mainly collected in *P. yezoensis* and *P. kuniedai*. The neutral spores in these two species are released from large thalli. In the second method of collecting secondary buds, the thalli which release the neutral spores are crushed, and a spore fluid is prepared in which the hibi are immersed. Kawase (1940) was successful in obtaining seedlings by this method, which is still employed today.

Conchocelis culture was initiated by Ota (1953) who developed an artificial seeding method by successfully culturing shell-boring Conchocelis. Substrata shells of bivalves are usually used depending upon their hardness, size, and availability.

The introduction of carpospores into the shells is known as the carpospore fixing process. The shells are kept at the bottom of a water tank or vessel with their inner surface facing upwards, and covered by a shallow layer of sea water (it is not advisable to use sea water of low specific gravity; it should be above 1.02). Carpospores are then introduced either by immersing mature thalli and stirring the water now and then, or by adding the carpospore fluid prepared by crushing carpospores in sea water. The boring rate of carpospores into shells depends upon various conditions of sea water such as pH, specific gravity, temperature, brightness, etc. The boring rate is high when the pH is 7 to 8, specific gravity about 1.025, temperature 10 to 15°C, and brightness 3,000 lux (Honda, 1964). It is better to have 10 to 20 filaments of Conchocelis in a 1 cm² area, because if the density exceeds a certain limit, diseases occur easily and the spore release is poor.

In culture experiments in which the seeding process is carried on up to autumn by using the germinated Conchocelis on the shells, the culture is carried out either by arranging the shells horizontally without disturbing the Conchocelis, or by passing a thread through holes made in the shells, and hanging this string of shells vertically in water tanks. The first procedure is known as plane culture, and the second as vertical. When culturing is done on a small scale, plane culture in a small vessel (capable of holding 50 shells) is suitable, but in large-scale culturing, vertical culture seems to be more profitable because of the larger unit area and the lower cost (Figure 1.47).

Figure 1.47. Conchocelis culture. Top: vertical culture; Bottom: plane culture.

The water temperature during the early period of culturing is low, the brightness is relatively increased (about 3,000 lux), and if the temperature is high during the summer, the place is darkened. If the water temperature is above 25°C, the luminous intensity is kept in the range of 500 to 800 lux or below. During this period, a number of adjustments have to be made such as changing the culture water, adjusting the specific gravity, adding salt-nutrients, etc., so that the sporangia will form in large numbers during the autumn or seeding period.

Various types of diseases occur during the culturing of Conchocelis such as yellow spot, red rot, green rot, etc. (Shinsaki *et al.*, 1959). Causal factors and preventive measures have not yet been established for every disease. Yellow spot causes the maximum damage. It often occurs in the Conchocelis which are ash-black in color, a phenomenon that takes place particularly when the water temperature is between 20 to 27°C. Yellow spots appear and spread, the number of spots rapidly increasing until the entire colony of Conchocelis is infected, and the shell surface completely covered by a yellowish-white coating. The causal organism (bacteria) was first discovered by Nozawa and Nozawa (1957). They suggested treatment for a few days with either boric acid (500 to 1,000 ppm), or with sea water with a low salinity (specific gravity, 1.005) as control measures. As a preventive measure, dihydrostreptomycin (1 unit/cc) should be administered during the culture development.

Artificial seeding with Conchocelis culture (Figure 1.48) may be carried out in open fields or indoors.

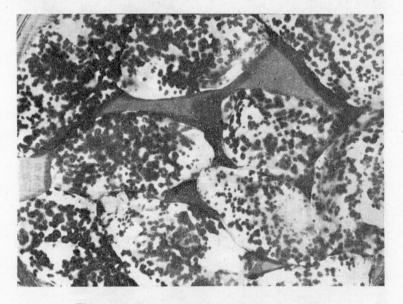

Figure 1.48. Conchocelis of *P. tenera* (in oyster shells).

Different methods are employed in open-field cultures. The simplest (and now obsolete) is to scatter the shells in which the Conchocelis have grown luxuriantly. Another common method is to hang the shells with Conchocelis vertically under the hibi (Figure 1.49). When net hibi are used, 10 to 20 nets are spread one above the other, and the shells hung vertically below the nets at the rate of 100 to 200 shells per 1.2 × 18 m. Since Conchocelis are sensitive to exposure, they are lifted out in polyethylene bags or troughs and then transferred to the nets. Hibi can either be fixed from the start, or allowed to float for some time and then fixed.

Figure 1.49. Artificial seeding in open fields (Matsushima Bay).

The major difficulty in open-field seeding occurs during the period when the spores are released from the Conchocelis. The shells holding the cultured Conchocelis have to be kept immersed in the sea for a few days before the spores become ready for release (Honda, 1964). The spore releasing period of Conchocelis can be ascertained more or less accurately on the basis of the research carried out by Migita and Abe (1966).

In the indoor seeding method, spore release from the Conchocelis on shells occurs inside the tank; bamboo poles with nets are then dipped into the tank and the spores allowed to adhere. Since it is possible to observe the release of spores and their adher-

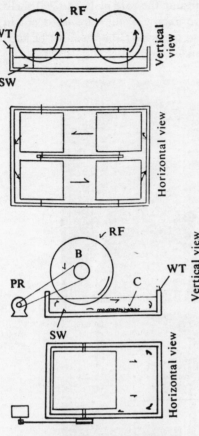

Figure 1.50. Diagram of a rotation apparatus for artificial seeding (Honda, 1964).

RF—rotating frame; WT—water tank; SW—sea water; PR—power and reduction gear; B—belt; C—conchocelis.

ing capacity, seeding can be carried out very systematically. Since conditions vary from place to place, other methods of seeding have been devised, for example, the rotation method (Figures 1.50 and 1.51), the air-stirring method (Figures 1.52 and 1.53), etc.

Even in the course of a day the spore release from Conchocelis follows a definite pattern, and in *P. tenera* the maximum release is found during the forenoon. Since the capacity of the spores to settle is lost soon after their release, the duration of the seeding period is limited. However, the spore release period can be manipulated during artificial seeding, since it is possible to control the conditions of brightness and darkness.

Spores are very sensitive to exposure just after adhering, as are their embryos (Akiyama, 1959). Hence when the spores have firmly settled, they are kept for one day in sea water before being transferred to the growing beds.

The Conchocelis of *P. tenera* and allied species are released in large numbers when the days are short (Kurogi, 1959a). In view of this, and also the fact that spores are not released at high temperatures (around 27°C), spore release from Conchocelis can be suitably controlled.

On the other hand, the capacity of the spores to adhere is retained at a low temperature; hence spores can be preserved for two weeks in sea water mixed with glucose at −5 to −15°C. This is of considerable significance in seeding (Honda, 1964).

Even though the water content of the leafy thalli decreases due to drying, plants do not die even when preserved at low temperatures; on the contrary, they resume their growth, given suitable conditions, when returned to the growing beds. This fact has made it possible to develop a new technique of laver preservation in nets, i.e. net poles which have a growth of laver thalli are preserved at about −20°C, after which the thalli are released in sea water. This method is now used for preserving various types of laver nets (Kurakake, 1966).

3. Culture

Before the development of net poles, bamboo blinds, etc. (known as horizontal hibi), dry-twig bundles and bamboo were used. The preparation of dry-twig bundles is very simple (Figure 1.54). Although it is not necessary to make provisions for upward and downward movements, little or no loss occurs due to storm or high waves, and the quality of laver obtained is good (Ueda, 1958), nevertheless, these advantages are offset in that the manipulation of the twig bundles is rather complicated and the laver yield is poor. Since it is not possible to exercise any control, no measures against

Figure 1.51. Artificial seeding by a rotation type apparatus.

damage can be taken; damage is often caused by obstructions in the flow of water. Consequently, this device has been discontinued (Figure 1.55).

At present the horizontal type of hibi is commonly used in laver culture. Of the various types of horizontal culture methods, that in which the nets are held in place by fixed bamboo poles (known as *ami*-hibi) is the most common. This method was first developed in 1925. In the past, the net material was either hemp or coir yarn, but now the nets are generally made of synthetic fibers such as Salan, etc.

Several methods have been suggested for affixing the net poles, arising from the conditions of the areas in which the growing beds are laid. In places where the tidal difference is small, the fixed type of net pole is adopted (Figure 1.54 B and C); where the difference is large, the floating type is used to compensate the insufficient solar radiation during full tide (Figure 1.54 D and E). Where water is too deep to affix the poles, the raft type is used (Figure 1.54 F). In all these methods, it is possible to adjust the water level and, on the basis of this adjustment, control exposure time. Finally, in places where the water is very deep and the waves relatively rough, the

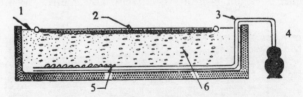

Figure 1.52. Diagram of bubbling apparatus for artificial seeding.

1—water surface; 2—net hibi; 3—vinyl tube; 4—air compressor; 5—shells with Conchocelis; 6—air bubbles.

Figure 1.53. Artificial seeding by a bubbling apparatus.

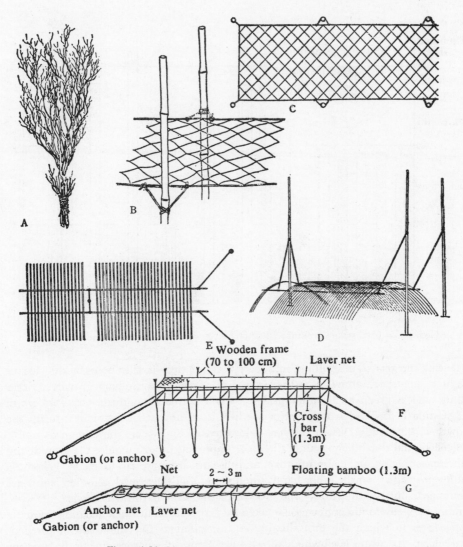

Figure 1.54. Various types of hibi and culture methods.
A—twig bundle (*soda*-hibi); B and C—nets fixed to bamboo poles (*ami*-hibi); D and E—bamboo blinds (floating type) (*sudara*-hibi); F—raft type; G—floating type (A—Okamura, 1909; B and C—Ueda, 1958; D and E—Kaneko, 1931; F and G—Kurogi, 1958).

floating culture (also known as the beta current, Figure 1.54 G) in which the nets are never exposed, is often practiced to further the growth of leafy thalli.

In many cases, the place where the seeding is done, differs from that in which the culture is grown. When the cultivation depends solely on natural seeding, the seeding tracts (*taneba*) are automatically limited; hence, as mentioned earlier, hibi are often transplanted. When dried, partly grown leafy thalli can withstand long-distance transportation.

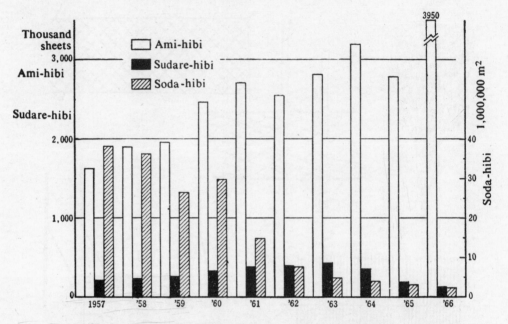

Figure 1.55. Changes in the number of different types of hibi used during various periods.

During the growth of seedlings into leafy thalli on nets fixed to poles by the various methods mentioned above, care is exercised to prevent the rampant growth of other plants such as green algae, diatoms, etc. which normally compete with *Porphyra* in the seeding beds. The water level of the hibi is an important factor in insuring the growth of laver only. When the net is lowered in the water, the exposure time is shortened, and thus the growth of the laver improves; unfortunately, the growth and adherence of other algae also improve, and diseases are more apt to occur. When the net level is raised, on the other hand, the diseases and the growth of other plants are minimized, but simultaneously the growth of the laver is stunted. Hence, all these conditions of environment have to be taken into consideration when deciding upon the water level to which the hibi should be transplanted. On the basis of previous experience, that water level which allows 4 to 5 hours of exposure during the daytime (from dawn to dusk) has been found as standard. Since the time and level of tides vary daily, the water level which would allow 4 to 5 hours of exposure differs daily; hence the nets must be lowered and raised several times to guarantee the essential amount of exposure. Exposure time calculated on the basis of a particular period of tidal waves seems to meet this demand.

When such adjustments are made, the cultured plants grow to 15 to 20 cm in length —enough for harvesting—about 50 days after seeding, i.e. from autumn to winter, during which period the temperature gradually decreases (see Figure 1.21).

In the course of culturing, the density of leafy thalli increases rapidly during the early period of growth in *P. tenera*, *P. yezoensis*, etc. and these species produce neutral spores in the initial period of germination. These neutral spores grow into secondary buds.

The following is an account of the course of vegetation as studied in *P. tenera* (Yoshida *et al.*, 1964). The density of thalli is expressed in terms of the number of thalli for 10 cm length of net thread. In the case of artificial seeding, it is possible to have a fairly high density from the very beginning, and when about 50 spores settle and germinate in late September, they grow to 120-160 μ in 10 days. Filaments with a double layer of cells grow and start releasing neutral spores. The release of neutral spores continues up to early November (Kurogi, 1961). In this way the density of leafy thalli on the nets rapidly increases, reaching a maximum in late November to early December. Usually 1,000 to 2,000 plants are seen on the hibi. After December density decreases (Figure 1.56), and when the culture period finishes in March, only about 100-300 plants remain.

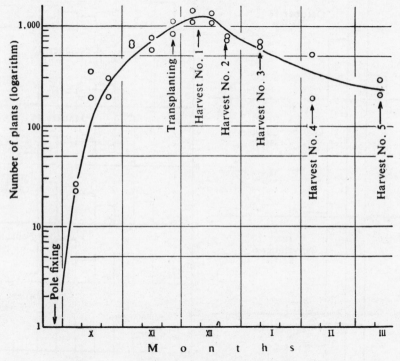

Figure 1.56. Changes in the density of thalli during culture
(Yoshida *et al.*, 1964).

As shown in Figure 1.57 in which the growth sections are converted to a log scale, the changes in the length of thalli up to early November (when they are visible to the naked eye), almost form a logarithmic regular curve. Later, some of the plants grow rapidly and can be differentiated as a group of 0.1-1 mm and another group of more than 10 mm. The larger group grows further from 10-100 mm, and continues as the main group up to the end of the harvesting period.

When thalli have grown to some extent, they are harvested. This does not mean that all the thalli on the nets are harvested. Only the larger ones are cut either at the basal or at the upper part. The harvesting of thalli is best done when they attain the

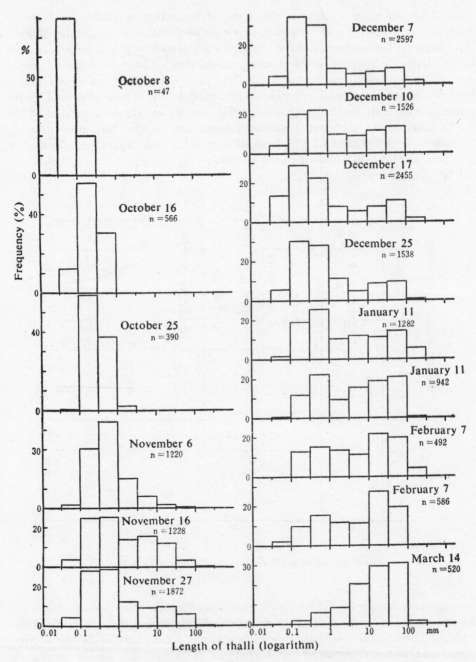

Figure 1.57. Seasonal variations in growth composition (Yoshida *et al.*, 1964).

size of 10-15 cm (the best grown ones, called tobi, are 20-30 cm) (Figure 1.58). In the past, the harvesting was done by hand. When the harvesting was done by hand, only the upper part was plucked, leaving 7-8 cm of the basal part in the nets

However, when the specimens plucked by hand were inspected, more than half of them were found to have been torn out at the basal part. At present, several types of machines are used in plucking the thalli. When the laver is harvested by using a rotat-

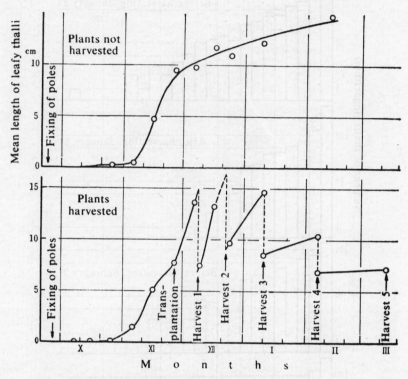

Figure 1.58. Growth of leafy thalli (Yoshida *et al.*, 1964).

ing knife, the number of plants cut at the upper part only is more than when the laver was hand-plucked. When the difference in the length before and after harvesting was studied, it was found that those of 7-8 cm length had decreased, and almost all the plants of 10 cm and longer had been harvested (Figure 1.59).

Since the remaining parts continue to grow, the harvesting is repeated several times in the same net until the end of the growing period. As 50 plants, more or less, are harvested for every 10 cm length of net thread, about 300-400 are harvested at the time of the maximum density of the plants adhering to the net (1,000-2,000 thalli).

There are very little reliable data available on the size of a collection from one net (*ami*-hibi). Table 1.11 shows a maximum of 2,500-3,000 for 1.2 × 18 m of the net. However, the mean value obtained at various culture beds would be much lower.

4. Improvement of breed

In laver culture, in addition to the main type of *Porphyra*, several other types are also found on the same pole net and, depending upon the period, different species dominate the net. According to the period of germination, they are divided into autumn, winter,

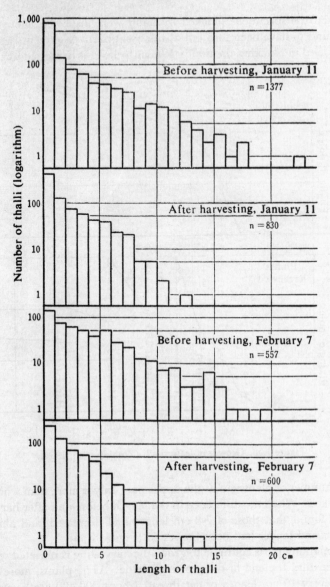

Figure 1.59. Changes in the length composition before and after harvesting
(Yoshida *et al.*, 1964).

cold, and useless or abnormal (*baka*). In addition to these, various types were named after the region, growth morphology, quality, and color pattern. Systematic studies have been carried out in detail with *P. tenera, P. kuniedai, P. angusta, P. yezoensis, P. pseudolinearis*, etc. (Kurogi, 1961). The early growth period and the period when the reproductive cells are formed, vary according to the species (see Figure 1.22). By taking into account the early and late vegetation period, the period of spore release, and some other features, combined or blended cultivation of several species of *Porphyra* in the same bed seems feasible.

Table 1.11. Production of dry laver from one net (*ami*-hibi) (1.2 × 18 m)

Species of *Porphyra*	Name of prefecture	Culture method	Year of study	Harvest					Total
				I	II	III	IV	V	
P. tenera	Miyagi	fixed	1962 to 1963	424	813	700	670	350	2958
P. tenera	Fukushima	fixed	1954 to 1955	1257	760	550	—	—	2567
P. tenera	Kumamoto	fixed	1964 to 1965	1500	822	63	—	—	2466
P. angusta	Kumamoto	fixed	1964 to 1965	1305	966	363	—	—	2858
P. pseudolinearis	Miyagi	floating	1958 to 1959	700	1200	800	—	—	2700

As knowledge on the culture of laver in natural conditions is increasing, and more and more of the characteristics of *Porphyra* are understood, it is now possible to culture a single species continuously by utilizing Conchocelis. Based on this fact, selective experiments to isolate the breed of *Porphyra* excelling in yield rate, quality, resistance to disease, etc. are being carried out in Fishery Experiment Stations in various Prefectures of Japan. For this purpose, various types of *Porphyra* were collected from different regions (report from the Fisheries Department, Government of Japan, 1965).

Even those species of *Porphyra* which have not been used in culture, are presently under experiment. For example *P. umbilicalis*, growing in the Hokkaido region, has a vegetative period from spring to summer. This property was utilized in culture experiments and efforts were made to prolong the growing period of the species. The results were encouraging (Kurogi and Yoshida, 1966). However, it is not easy to make a successful artificial seedling of *P. umbilicalis* as it is a rock seaweed; hence it is very difficult to improve the breed.

In the culture experiments of the *Porphyra* species, not only the endemic Japanese varieties, but many exotic species from other countries are likewise under investigation.

Due to the development of new culturing media and methods, it became possible to grow seaweeds from seed to full maturity in laboratory conditions (Suto, 1960). Moreover, since the products in the form of Conchocelis could be preserved, it became possible to carry out hybridizing experiments between the various types of *Porphyra* (Suto, 1963).

Typical varieties from various regions were selected and cultured indoors. The plants thus grown were separated while immature and cultured separately. When carpogonia matured, a few small pieces of the thalli were cut and kept in separate vessels and spermatia added. Since in the case of the monoecious form no Conchocelis grows from the carpogonium, in the absence of any added spermatia the presence of Conchocelis will confirm the occurrence of hybridization. On the other hand, while it is not possible to prevent self-fertilization in dioecious *Porphyra*, the percentage of such fertilization is very low. Hence, when the rate of Conchocelis formation is very high after the addition of spermatia, hybridization is assumed to have occurred. The results of several hybridization experiments are given in Table 1.12; they show that hybridization occurs easily among some species. The resulting hybrid varieties have characteristics intermediate to those of the parent ones. Among these hybrids, there are some which are heteroploid or polyploid, for example, P × Tp.

Table 1.12. Results of hybridization experiments (Suto, 1963)

♀	♂	P	U	A	T	Tk	Tg	Tp	Y
P. pseudolinearis (Kesen'numa)	P		31	22	21	21		23*	1
P. umbilicalis ? (Yokohama)	U	21		23					
P. angusta (Chiba)	A	22			21	21		22*	22*
P. tenera (Chiba)	T			23*	22				
P. tenera (Matsukawa-ura)	Tk			23*	23	22			
P. tenera (Mangoku-ura)	Tg								
P. tenera (Kesen'numa)	Tp								
P. yezoensis	Y								

Note: The data in the table are given in two digits. The number in the unit place expresses the formation and growth of Conchocelis as well as the shape of spores and that in the tens, the growth of thalli of next generation: 1—poor, 2—normal, 3—good. Asterisk denotes appearance of unusual thalli; underlined digits indicate that similar results were obtained in more than two replicates.

Regarding U × A, Tg × T, and T × Tp, experiments were also carried out in the sea, and the resultant growth was not at all inferior to the normal *Porphyra*.

Cytological investigations (Yabu and Tokida, 1963; Migita, 1967) have shown that *Porphyra* thalli are haploid, the Conchocelis phase is diploid, and the number of chromosomes is 3-4, which suggests that only a limited number of hybrids can be expected. Conditions such as the occurrence of fission in the next generation thalli, are not favorable to hybridization, and the possibilities for obtaining good breed through hybridization are meager. Moreover, even when hybrids are obtained, there are various problems involved in the selection of culture methods; to date no one has succeeded in growing the best breed of *Porphyra* by hybridization.

Studies on the structure of polyploids, the structure of unusual strains obtained from sudden mutations caused by gamma-ray treatment, and on the selection of the best strain for the culture beds, etc. are presently underway.

5. Culturing places

Laver cultivation is carried out in the regions of the Pacific coast from north to south; but for the most part, there is no cultivation along the coastal regions of the Sea of Japan. The growing places are found inland in various regions and a description is therefore inappropriate. However, all these growing places have one feature in common, namely, each is economically suitable for laver cultivation in line with modern techniques. The conditions of growing places with regard to weather, sea conditions, water quality, topography, etc. will be discussed separately.

5.1 Weather

Except rarely, weather conditions are not inimical to growing beds for laver cultivation in any part of the coastal regions of Japan. In fact, the northwest wind from autumn to winter should be favorable to laver as it abets the mixing of sea waters. However, these places are not suitable since culture equipment may be broken by strong winds at the time of low pressure, or by seasonal winds.

5.2 Water temperature

If water temperature were considered alone, the entire Japanese coast would be suitable for laver cultivation. In the Nemuro region on the eastern coast of Hokkaido (Figure 1.60), the maximum water temperature is 18°C and the minimum −1.0°C. But since floating ice reaches the coast in January to March, it is not possible to install culture equipment during this period and hence it is impossible to carry out

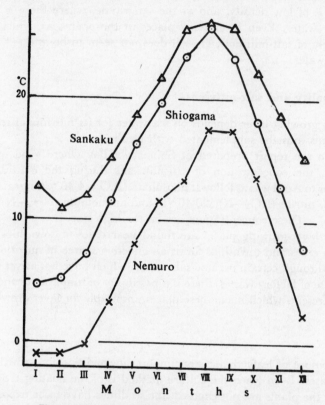

Figure 1.60. Seasonal variations in water temperature at
Nemuro, Shiogama, and Kumamoto (Sankaku).

cultures in the winter. Although Kurogi and Kawashima (1966) did succeed in culturing certain selected species of *Porphyra* between spring and summer and between autumn and winter, this is considered to be an exceptional case. In the Tohoku region, the minimum water temperature is about 5°C, and the maximum about 25°C (Culture Section, Tohoku Fisheries Experiment Station, 1965). The minimum water temperature gradually increases to 10°C toward the southern regions (Figure 1.60).

5.3 Specific gravity

Most of the growing places of laver are found in the river estuaries of the Bay, and the density of sea water is usually lower than the water of the open seas. Physiological activities such as respiration, photosynthesis, etc. decrease in sea water of low specific gravity (Ogata, 1964). When large quantities of fresh water flow into the growing places and cover the surface, there is often an inimical effect on the growth of laver. Sometimes there are crop failures resulting from the overflow of fresh water into the growing places near the estuaries. Since salt-nutrients are mostly obtained from land water, the growing places are selected mostly in the regions where the sea water is of low density, and at the same time, where there is an effective inflow of river water. Even in growing places in the open seas, when there is an adequate supply of salt-nutrients, growth does not seem to be affected by the high density of sea water.

5.4 Water quality and salt-nutrients

As naturally growing laver depends on sea water for its nutritional requirements, the salt-nutrients must be supplemented with fertilizers.

According to the report prepared at Shinagawa Bay where good quality laver was produced, the concentration of salt-nutrients during the cultivation period from November to April was as follows: phosphate (P_2O_5), 4-40 γ/l, nitrate (NO_3—N) about 300 γ/l, nitrite (NO_2—N), 20-40 γ/l, and ammonium (NH_3—N) 200-400 γ/l (Matsue, 1936). These concentrations are on the high side.

Most of the laver-growing places are found near cities or townships. While the drain water in cities and townships forms an effective source of nutrition, it has an adverse effect through certain organic substances, which have not as yet been identified. In the case of factory wastes, there is a possibility of tumor and cancer inducing factors being present, which make these places unsuitable for laver growing.

5.5 Tide

Periodic exposure of leafy thalli is an important condition for growth, except in some special cases. Even when the culture method is of the floating type (beta current) in which the plants are not exposed, the seedlings have to be exposed between the time of collection and transplantation to the floating hibi. Hence it is difficult to grow the laver at places with little tidal difference. Usually it is necessary to choose a place where the tidal difference at the time of full tide is 1-1.5 m or more.

5.6 Water current

While tidal current is an important factor in the movement of water at growing places, factors like wind current are also important. Since the movement of water plays an important role in preventing an increase in pH, caused by the consumption of carbon dioxide and the (supplementary) supply of salt-nutrients, places where the water flow is poor (especially when these are poor in salt-nutrients) are not suitable for laver culture. As stated earlier, the results of indoor cultures carried out by Matsumoto (1959) show that a current of about 20 cm/sec is suitable for culturing in ordinary sea water. However, in places rich in salt-nutrients, the current should be 10 cm/sec, but in those deficient in the same, 30 cm/sec.

As a simple method of measuring the movement of water in growing places, Matsudaira and Hamada (1966) placed iron plates vertically in the sea and, by measuring the decrease in the weight of the plates less the amount of rust collected from them, were able to judge water movements. Sheets of iron measuring 5 × 10 cm (0.8 mm thickness) were polished and weighed before being placed vertically in the water. At fixed intervals the plates were removed, the rust scraped off, the sheets weighed again, and the difference between the first and second weight determined. It was learned that in water regions where salt-nutrients are not rich, the culture of laver could not be carried out if the weight decrease of the plates, minus rust, fell below 80 mg/100 cm²/day.

Water movement in the areas of cultivation is definitely affected by the installation of bamboo poles. Thus poor results in close planting can be partly attributed to the obstruction of free water flow.

5.7 Topography

Bays are used to cultivate laver, mainly because they are easy and convenient, and secondly, not exposed to strong winds or high waves. Tokyo Bay, Ise Bay and Ariake-kai are some of the main regions in Japan where cultivation is carried out. When dry twigs and bamboo poles were solely used, only those places between the mean water surface and near the lowest ebb tide, and those with sand or mud, were selected for cultivation. When horizontal hibi came in, cultivation extended to deeper areas, where long, branched twigs were used. Then the floating hibi and rafts were introduced and cultivation spread to regions of greater water depth. Beds were no longer restricted to floors of sand and mud, as even rocky beds were found to be suitable. With the development of additional techniques, the water basins of many more areas are expected to be utilized in the future.

5.8 Fertilizers

The necessary nutrients which are not sufficiently available in natural sea water, are replenished artificially through a supply of fertilizers, in order to improve the production of laver in places where it has been poor. Since sea water is frequently found to be deficient in nitrogen, this should be the main ingredient in the added fertilizers.

Various investigations have been done with artificial manuring, but these still require detailed study (Second Annual Report of the Fisheries Department, Government of Japan, 1965, Suisanchoken-Ni).

5.9 Preparation of growing places

Several investigations are underway to increase the area of laver cultivation. In places where the water current and water mixing are not good, improvement through artificially created springs is being tested at Matsushima Bay near the Miyagi Prefecture (Watanabe et al., 1966).

Wiers and breakwaters have been erected in places where culture implements could be damaged by high waves, and reports to date show that these measures are proving effective.

6. Diseases: prevention and remedy

The yield and quality of laver may be affected by various diseases. The following is an account of the diseases identified to date, and their causal factors.

6.1 Red rot (Arasaki, 1947)

This disease is found in all the growing places of Japan; its damage peak occurs between November and December and in March. The disease rarely occurs in winter when the temperature is low. It is prevalent during the rainy season when the chloride content is low, and the water temperature relatively high. It abounds in thalli which are bright in color, and is comparatively infrequent in those which have lost their color or begun to turn bright yellow. The first symptom of the disease is the appearance of red-rust spots whose centers change to yellow-green as the spots enlarge, until only the margin retains the red-rust color. Individual spots are circular in shape, but when the disease is severe, spots fuse and give rise to irregularly shaped patches. The spots generally start in the central part and spread toward the basal part. When the disease reaches an advanced stage, the thalli degenerate in the center, become brittle, and finally break into small pieces which slough off.

A microscopic study of the disease shows that colorless hyphae, 1.5-3 μ in diameter, pass through the body cells and run along the longitudinal axis of the body at the center. Infected cells die and the color changes to red-violet (Figure 1.61).

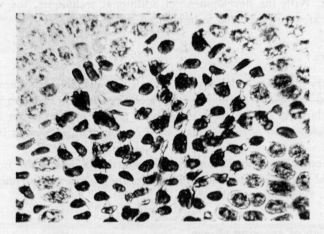

Figure 1.61. Red rot. Hyphae passing through *Porphyra*.

Reports indicate that the disease is caused by the *Pythium* sp. belonging to the class Phycomycetes. Usually the infection is caused by zoospores. During a period of low temperature, they remain dormant (passive) and develop ova (Figure 1.62).

No drug which would effectively eradicate red rot has been discovered to date. When infection occurs and seems likely to continue for a long time, the level of the poles is raised to increase exposure time, and the plants are harvested earlier than usual. This is the only control measure known and adopted at present.

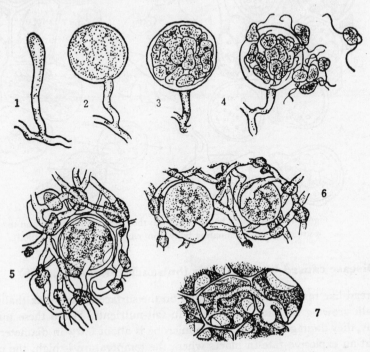

Figure 1.62. Red rot.
1 to 4—formation of zoospores; 5 and 6—sexual reproduction;
7—germination of zoospore (Arasaki, 1947).

6.2 Ascidian bacteria disease (Arasaki, 1960)

This disease is caused by the parasitic *Olpidiopsis* sp. belonging to Phycomycetes. This is widely known in various regions. The parasite penetrates only one cell of the thallus and does not spread to the neighboring cells as does the *Pythium* sp. Consequently, although the thalli may be infected by this parasite, it is very difficult to differentiate infected plants from healthy ones with the naked eye. Observations under a low power microscope show the infected part as a slightly shiny, white spot (Figure 1.63).

Very often this disease occurs simultaneously with another. It is assumed that the damage it causes is slight. However, if the disease occurs when the thalli are small, it may cause bud damage. Since the parasite also infiltrates the carpospores, the culture grown from artificial seeding done with Conchocelis will be badly affected.

No drug or control measure has been developed to date for the ascidian bacteria disease.

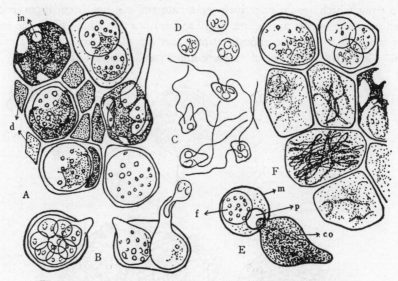

Figure 1.63. Ascidian bacteria disease: the parasite in *Porphyra* and the zoospores (E—germinating body from carpospore) (Arasaki, 1960).

6.3 Disease caused by Leucothrix (Suisancho Dai-ni-ka, 1965)

Thread-like microbes are often found on the surface of the weak thalli of *Porphyra*, or thalli growing in water basins rich in salt-nutrients. When these microbes grow densely, they damage the thalli. The microbe is about 0.6 μ in diameter, and multiplies at an explosive rate in places where the temperature is high, the chloride content low, the water current stagnant, and the salt-nutrients abundant.

Arasaki (1947) suspected the thread-like microbes infecting *Porphyra* thalli to be either *Actinomyces* or *Myxophyceae*. Recently they were identified as *Leucothrix mucor* (Fujita and Zenitani, 1967).

6.4 Disease caused by benthic diatoms (Okamura, 1909)

Benthic diatoms grow profusely on the surface of *Porphyra* thallus. When dried, the thallus appears covered with a bluish-white powder. On removing the diatoms, the thallus is found to have no luster, an effect which greatly diminishes the value of the product (Figure 1.64). The diatoms found in *Porphyra* belong to *Synedra*, *Licmophora* and *Achnanthes* (Kato and Kawamura, 1956). While *Synedra* attach to the main thallus directly by some adhesive secretion and remain affixed even in strong winds and high waves, *Licmophora* and *Achnanthes* attach by long adhesive manubria and are found mostly inside bays.

Benthic diatoms do not infect healthy *Porphyra* thalli but those which have been weakened for some reason or the other.

The diatoms multiply rapidly when the density of the sea water is low and the exposure time insufficient. As a preventive measure, the bamboo poles should be raised and the exposure time increased.

6.5 White rot disease (Arasaki, 1947)

This disease occurs frequently in thalli of considerable growth during the period from early November to December. It was first reported from Tokyo Bay, Mikawa Bay, and Ise Bay, and later from various culture centers in the Tohoku region (Association for the Prevention of Pests, Byogai Taisakukyo, 1961).

The disease occurs on rapidly growing thalli of 2-3 cm in length or more. The disease usually starts as red patches on the tip of the thallus, the color gradually changing to yellow and then white. The basal parts of the thalli are not so easily affected, which distinguishes this disease from others (Figure 1.65).

It seems that the disease is not caused by any parasitic organism but is due to a physiological disturbance in the *Porphyra* body itself, resulting from adverse environmental conditions.

The causes for this disease are now said to be many. It may occur, for example, when the formation of warm water basins in Okiai (open sea, offshore) seem imminent, when the temperature

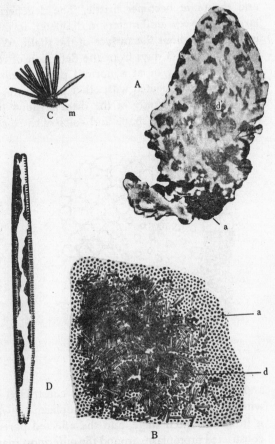

Figure 1.64. Disease caused by benthic diatoms.
A and B—thallus infected by diatoms: a—*Porphyra* thallus; b—diatoms; C and D—adhering *Synedra*: m—adhesive manubria (Okamura, 1909).

and chloride content of water suddenly increase, when there is a sharp decrease of *Skeletonema* due to unusually rapid currents, when the minimum temperature is increased by a continuous S-SW wind, when the climate is continuously calm and windless, etc. When the disease is expected to occur, the level of the nets should be raised to increase the exposure time. If the disease is in a fairly advanced state, it is better to lower the level of the nets or loosen them so as to allow some movements in the waves. This will reduce the damage caused by the disease.

Since the causal factor of white rot is not a parasite, detailed investigations in various regions need to be done to determine its origin.

6.6 Hole rot disease (Suto and Umebayashi, 1954)

This disease is first observed when minute holes appear on the upper part of the plant. The periphery of the hole is initially blue, but the color gradually disappears and the plant becomes brittle. The degeneration spreads to other regions and a large hole, a few millimeters in diameter, forms at the center. Ultimately the disease is seen throughout the surface of the thalli. When progress is rapid, the thalli break up within 2 or 3 days from the detection of their diseased condition. If the disease progresses slowly or at a normal pace, within 5 to 7 days the affected plants are simultaneously infected with other parasites.

A microscopic study of the disease shows that initially 2 or 3 cells are infected and, as a result, the shape and color of these cells change. They become constricted

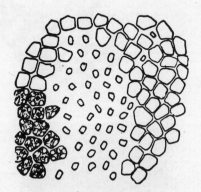

Figure 1.65. White rot disease
(Nozawa, 1957).

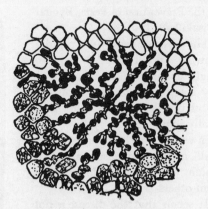

Figure 1.66. Hole rot disease
(Nozawa, 1957).

and turn red-violet. A part of the constricted protoplasm is found in close contact with other healthy cells, the protoplasm of which also constrict when the disease is in an advanced stage, and the affected parts present a radial pattern. When the constricted protoplasm around the outermost margin fuses with the protoplasm of other healthy cells (Figure 1.66), degeneration begins at the central part and a hole occurs.

If a mechanical stimulation is applied to the cells of *Porphyra,* and these are then placed in sea water with a low chloride content, it is possible to simulate the type of cell changes mentioned above (Nozawa and Nozawa, 1957). Hence hole rot occurring in natural conditions, may be due to injuries caused by sand particles brought in by river water during and soon after heavy rains, and the concomitant reduction in chlorine the additional water causes. The disease notwithstanding, these two conditions have an adverse affect on the cultivation of *Porphyra.*

Exposing the acts and sprinkling or spraying chemicals containing Mn, Ca, etc., which strengthen the protoplasmic membranes, are suggested as countermeasures (Arasaki, 1956).

6.7 Tumor (Toyama, 1957)

Diseases which commonly occur in water basins polluted by drainage water from factories on townships, are atrophy (chijimi), crenullation (pamanento), and distrophy (fuka).

Protuberances, which can be discerned with the naked eye, form on both surfaces of the thalli; when these increase in number, the thalli begin to shrink. Subsequently, their growth is greatly affected and they lose their luster; as a result, the value of the product decreases. When the disease is very serious, the color of the affected colonies of *Porphyra* also fades and the plants look like minute areas of overgrown thick grass. In this condition, the crop is unfit for marketing.

When sections of the plants infected with the disease are studied under a microscope, abnormal cell divisions at various places, and some layers of relatively round cells which form lumps, are seen. The circumference of these cells eventually shrinks. Thus in an advanced stage of the disease, the whole plant shows an irregular accumulation of cells in 5 to 10 layers (Figure 1.67). Folds which have a crepe-like texture are sometimes found at the end of the culturing season when salt-nutrients are deficient. But the latter wrinkled condition differs from the diseased one in that no abnormal cell division occurs in this instance.

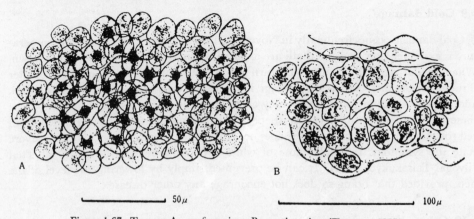

Figure 1.67. Tumor: A—surface view; B—section view (Toyama. 1957).

Since it is possible to cause abnormal cell division by culturing thalli in dilute solutions of nitromin, KCN, phenol, etc. and also by X-ray irradiation, it seems that tumorous diseases are caused by minute quantities of certain materials present in the drainage water from factories or townships. Since there is no specific countermeasure, one has to acknowledge that such areas are undesirable for the cultivation of *Porphyra*.

6.8 Bud damage (Akasaka, 1956; Tamura, 1956)

This damage occurs in *Porphyra* during September to October when they grow from microscopic plumules to thalli of 2-3 cm in length, and the affected plants either die or are lost in the water current.

A loss of color in buds which can be seen with the naked eye, is the first indication of this particular disease. Shortly thereafter, the thalli turn red, the petals turn white, and degeneration begins. During this period, the thalli bend sharply and look like hooks. Sometimes the edges of the thalli become denticulated and ragged.

Microscopic investigations show that the stellate chromatophores found in the cells of the thalli have blanched. The thalli shrink and their color changes to violet. Ultimately they lose their chromatophores and turn white. When the damage is severe, the entire thallus falls off.

The causal factors of this damage are not yet known. The conditions which frequently give rise to it are said to be high water and air temperature, continuous calm and windless climate, high density of water, stagnant water of low density due to rains, unusually high tides, etc. It is also thought that very close cultivation of plants contributes to water stagnation, another condition favorable to this particular damage.

Exposing the healthy buds by raising the level of the nets, is a good preventive measure when bud damage is expected. But when it has already taken hold, lowering the level of the nets, or loosening them until they float freely in the water, seems a better countermeasure.

6.9 Cold damage

Cold damage occurs frequently in November to December when the night temperature is extremely low and the thalli are exposed during the night or just before dawn. When the damage is light, the thalli turn red in color but recover if kept in sea water throughout the day. When the damage is severe the cells of the affected part shrink and become violet in color. Later they change to white with visible red spots. Since these red spots have an irregular shape, cold damage is easily distinguished from red rot. Usually, the damage *per se* does not cause much harm, except that in a severe state the symptoms resemble those of white rot and growth is adversely affected (Byogai Taisakukyo, 1961). It can be prevented simply by lowering the level of the nets, provided that doing so does not encourage any other disease.

6.10 Other diseases

Besides the above, diseases variously known as hamagusare, migake, etc. are reported from Japanese fishing villages, but the nature of them has not yet been determined. Even the diseases discussed hitherto require further investigation with regard to causal factors and reliable countermeasures.

7. Harvesting and processing

All operations relating to laver, from the various stages of culture to the preparation of dry laver sheets, are preplanned and systematically carried out in all the fishing villages of Japan.

Harvesting, previously done by hand, is now done with implements expressly designed for the various methods of culturing with bamboo poles.

The harvested laver is collected and carried to the processing units in bamboo baskets, in which the laver is kept until the water has drained off. The plants are then chopped into small pieces. Previously, chopping was done with knives, but now machines in which the thickness of the chopped section can be increased or decreased according to the hardness of the thalli, are used. Adjusting the size of the cut sections of thalli ensures a uniform thickness in the laver sheets.

Of the different methods of molding the laver into sheets, "casting" (Nagetsuke) has been commonly adopted. In this method, a suitable quantity of water is added to the chopped plants (about 13-14 l of fresh water for 1 kg of plants) and mixed thoroughly. Laver drying mats are spread out and sheeting frames set up on them. The mixture of laver and water is poured onto the framed sections where it spreads, and the water runs out through the small eyelets in the bamboo mats. Thus, only that laver of a definite thickness and shape remains on the mats. As this method of sheeting requires technical training and is slow, a seaweed sheeting machine has now been developed which offers the following advantages: minimal technical skill is involved in operating the machine, the product is of uniform quality, and the efficiency rate is higher.

The next step after sheeting is drying. There are two methods of drying, $viz.$ open-air, sun-drying and drying indoors with heaters. In the first method, common throughout Japan, mats are fitted to frames for easy transportation. However, in the Kanto region, sun-drying is carried out by fixing the mats on slanted fences covered with straw. First the backside of the mats (the side without the laver sheet) is exposed to sunlight; when this side has dried thoroughly, the frontside is exposed. The sun-drying method is greatly influenced by climatic conditions and also requires a wide working area. Hence the indoor drying method is becoming more and more popular, especially in the Tohoku regions where the climate in winter is very uncertain. The structure and design of the drying rooms are of different types. The fuel used for drying is usually heavy oil. The mats with the laver sheets are first sent to the dehydration chambers where the water content is greatly reduced by centrifugal force; then the mats are removed to the drying rooms.

When the laver has dried thoroughly, the sheets are stacked in bundles of 10 each.

Thus, from collection to drying, machinery and other equipment are presently used in laver manufacture; nevertheless, human labor plays a major role, and laver manufacturing is undeniably a small-scale industry.

8. Tank culture

Conchocelis culturing is relatively easy; the difficulty lies in obtaining mature thalli from spores inside the laboratory. Although the growth rate for the first 10 days in indoor cultures (in sea water enriched with N, P, and Fe, or simple, artificial sea water), is more or less the same as in the outdoor cultures, the growth rate in the former slows down thereafter (Suto, 1961). Suto (1960) successfully cultured $Porphyra$ inside the laboratory by following the seaweed culturing method of Provasoli $et\ al.$ (1954). In this case, they used natural or artificial sea water enriched with salt-nutrients and EDTA chelated Fe, Mn and other minerals in minute quantities. The conditions for culture were further improved by supplementary supplies of CO_2 and the growth rate was reported to be almost the same as that in natural conditions. There are also reports from other investigators on the successful culture of $Porphyra$ on a large scale (Nakatani, 1964; the research laboratories of several companies). According to Suto (1961), 2-4 kg of $Porphyra$ can be grown in one ton of culture fluid, provided the conditions are suitable and the rate of production is calculated to be 0.4-0.8 kg/day. In order to make a success of tank culture on land, there are a number of problems to be solved such as the selection of the right culture fluid for plant growth at various stages, the selection of the right plant strain, methods for improving growth rate and preventing the growth of diatoms, etc. If these problems are fully solved, it should be possible in the near future to culture the plant on a large scale in tanks on land. It appears that that day is not far off !

(Yoshida and Akiyama)

Bibliography

1. AGARDH, C. A., 1824. *Systema Algarum*. Lund, 312.
2. AKASAKA, Y., 1956. Miyagi-ken kesen-numa-wan oyobi mangokuura ni okeru norino me-itami ni tsuite (On Bud Damage of *Porphyra* of Kisen-numa Bay and Mangoku-Ura, Miyagi Prefecture), *Suisan Zoshoku (Aquaculture)*, **4** (2), 41-43.
3. AKIYAMA, K., 1959. Asakusanori yoga no taikansei ni tsuite (On the Resistance of *P. tenera* to Desiccation), *Nihon Suisan Gakkai Tohoku Shibu Kaiho (Bulletin of Tohoku Branch of the Japanese Society of Scientific Fisheries)*, **10** (1, 2), 55-56.
4. ——, 1960. Amanori-rui no Itojotai no eiyo hanshoku to sono zoshoku (Reproduction and Culture of *Porphyra* Filaments), *Tohoku Suisan Kenkyujo Kenkyu Hokoku (Bulletin of Tohoku Regional Fisheries Research Laboratory)*, (17), 45-49.
5. ARASAKI, S., 1947. Asakusanori no fuhaibyo ni kansuru kenkyu (On Diseases of *P. tenera*), *Nihon Suisan Gakkaishi (Bulletin of the Japanese Society of Scientific Fisheries)*, **13** (3), 74-90.
6. ——, 1950. Asakusanori no yoshoku (Culture of *P. tenera*), *Suisan Jijo (Monthly Research Report of Fisheries)*, 11.
7. ——, 1954. Asakusanori no kagakuteki saibai (Scientific Culture of *P. tenera*), *Kagaku (Science)*, **24** (5), 218-223.
8. ——, 1956. Asakusanori no byogai to sono taisaku (Diseases of *P. tenera* and Their Countermeasures), *Shokubutsu Boeki (Plant Protection)*, **10** (6), 243-246.
9. ——, 1957. Asakusanori no ikushugakuteki Kenkyu. Hinshu no shikibetsuho to isemikawa-wan-san no nori hinshu (Breeding of *P. tenera*: I. Quality Difference and the *Porphyra* Breed Grown in Isemigawa Bay), *Suisangaku Shusei*, 805-818.
10. ——, 1957a. Asakusanori no hinshubetsu to ikushu (Breeding and Quality Difference of *P. tenera*), *Suisan Zoshoku (Aquaculture)*, (4), 32-38.
11. ——, 1960. Amanori-rui kiseisuru tsubojokin ni tsuite (Ascidian Bacteria Infecting *Porphyra*), *Nihon Suisan Gakkaishi (Bulletin of the Japanese Society of Scientific Fisheries)*, **26** (6), 543-548.
12. ——, T. FUJIYAMA, S. SUTO, Y. KITAGAWA, Y. MATSUE, and F. MATSUURA, 1952. Asakusanori no ikushu gijutsu (Breeding Techniques of *P. tenera*), Brochure, 9 p.
13. ——, ——, and Y. SAITO, 1956. Asakusanori hoshi no fuchaku to kaikyo ni tsuite (Conditions of the Sea and the Settling of *P. tenera* Spores: II), *Nihon Suisan Gakkaishi (Bulletin of the Japanese Society of Scientific Fisheries)*, **22** (3), 167-171.
14. ——, K. NOZAWA, and Y. NOZAWA, 1959. Asakusanori itojotai no byogai ni tsuite (Diseases of *tenera* Filaments), Brochure, 16 p.

15. ARASAKI, S., A. INOUE, and Y. KOUCHI, 1960. Nori no byogai, tokuni 1959-nen ryoki Tokyo-Wan okubu de miraret aganshubyo, tsubojokin-byo ni tsuite (A Study of the Diseases of *Porphyra*, Particularly of the Tumors and Ascidian Bacteria Found on *Porphyra* in Tokyo Bay in 1959), *Nihon Suisan Gakkaishi (Bulletin of the Japanese Society of Scientific Fisheries)*, **26** (11), 1074-1081.

16. BATTERS, E. A. L., 1892. On *Conchocelis*, a New Genus of Perforating Algae, in *Phyc. Mem.* by Murray Georges, I, 25-28, pl. VIII.

17. BERTHOLD, G., 1881. Zur Kenntnis der Siphoneen und Bangiaceen, *Mitt. Zool. Stat., Neapel*, (1), 72-82.

18. ——, 1882. Die Bangiaceen des Golfes von Neapel und der angrenzenden Meeresabschnitte, *Fauna und Flora des Golfes von Neapel*. Monographie, 8, 1-28.

19. DANGEARD, P., 1927. Recherches sur les *Bangia* et les *Porphyra*, *Botaniste*, 18, 183-244.

20. ——, 1931. Sur le développement des spores chez quelques *Porphyra*, *Trav. Cryptogam. déd. à L. Mangin, Paris*, 85-96.

21. DREW, K. M., 1949. *Conchocelis* Phase in the Life History of *Porphyra umbilicalis* (L.) Kütz., *Nature*, 164, 748-749.

22. ——, 1954. Studies in the Bangioideae: III. The Life History of *Porphyra umbilicalis* (L.) Kütz. var. *laciniata* (Lightf.), *J. Ag. Ann. Bot.*, N. S., **18** (70), 183-211.

23. FRITSCH, F. E., 1945. *The Structure and Reproduction of the Algae*. The University Press: Cambridge, II, 939.

24. FUJIKAWA, K., 1932. Chosen-nori no seiri ni kansuru Kenkyu (Physiology of Korean Seaweeds) (Report No. 3), *Annual Report of Chosen Sotokufu Fisheries Experiment Station* (1930), 32-125.

25. ——, 1936. *Ibid.* (Report No. 4). *Ibid.* (1932), 1-135.

26. ——, 1937. *Ibid.* (Report No. 5). *Ibid.* (1933), 1-131.

27. ——, 1957. Nori no jinko saibyo to tennen saibyo (Natural and Artificial Seedlings of *Porphyra*), *Suisan Zoshoku (Aquaculture)*, **4** (4), 10-14.

28. FUJITA, Y., and B. ZENITANI, 1967. Asakusanori no yotai ni chakusei suru itojosaikin *Leucothrix mucor*: I. Ippan biseibutsugakuteki seijo narabini hatsuiku kankyo yoin ni tsuite (On *Leucothrix mucor* Infecting the Thalli of *P. tenera*: I. Common Microbiological Characteristics and Environmental Factors Causing the Outbreak), *Nagasakidai Suisangakubu Kenkyu Hokoku (Bulletin of the Faculty of Fisheries, Nagasaki University)*, (22), 81-89.

29. FUJIYAMA, T., 1957. Asakusanori ganshubyo no saiboka gakuteki Kenkyu (Histological Studies of the Tumors of *P. tenera*), *Suisangaku Shusei*, 829-840.

30. ——, 1957a. Nori no ganshubyo (Tumors of *Porphyra*), *Suisan Zoshoku (Aquaculture)*, **4** (4), 69-73.

31. FUKUHARA, E., 1958. Amanori no fuchaku kikan ni tsuite (Yoho) [Holdfasts of *Porphyra* (Preliminary Report)], *Hokkaido Suisan Shikenjo Geppo (Monthly Report from the Hokkaido Fisheries Experiment Station)*, 15 (7), 314-318.

32. ——, 1958a. Utasutsunori ni tsuite (Study of *P. kinoshitai*), *Ibid.*, 15 (5), 218-222.

33. ——, 1963. Hokkaidosan amanori ni tsuite (Study of *Porphyra* of Hokkaido),

Ibid., **20** (2), 41-57.

34. FUJIYAMA, T. 1968. Hokkaido kinkaisan amanori-zoku no bunruigakuteki narabi ni seitaigakuteki Kenkyu (Classification and Ecology of the Genus *Porphyra* of Hokkaido and Its Neighborhood), *Hokkaido Suisan Kenkyujo Kenkyu Hokoku (Bulletin of Hokkaido Regional Fisheries Research Laboratory),* (34), 40-99.

35. GRAVES, J. M., 1955. Life Cycle of *Porphyra capensis* Kütz., *Nature,* 175 (4452), 393-394.

36. GRUBB, V. M., 1924. Observations on the Ecology and Reproduction of *Porphyra umbilicalis* (L.), *J. Ag. Rev. Algol.,* 1, 223-234.

37. HOLLENBERG, G. J., 1958. Culture Studies of Marin Algae: III. *Porphyra perforata, Amer. J. Bot.,* **45** (9), 653-656.

38. ——, and I. A. ABBOTT, 1966. Supplement to Smith's Marine Algae of the Monterey Peninsula. Stanford Univ. Press: California, 130 p.

39. HONDA, N., 1962. Asakusanori-rui no yoshoku ni okeru jinko saibyo ni Kansuru Kenkyu (Study on Artificial Seedlings Used for the Culture of *P. tenera*), *Okayama-Ken Suisan Shikenjo Rinji Hokoku (Bulletin of Okayama Prefecture Fisheries Experiment Station)* (Special number), 67 p.

40. ——, 1964. *Shitsunai saibyo no tebiki (Handbook on Indoor Seedlings).* Zennoriren: Tokyo, 65 p.

41. HUS, H. T. A., 1902. An Account of the Species of *Porphyra* Found on the Pacific Coast of North America, *Proc. Calif. Acad. Sci.,* III: *Bot.,* **2** (6), 173-240.

42. ISHIKAWA, M., 1921. Cytological Studies on *Porphyra tenera* Kjellman I, *Bot. Mag. Tokyo,* 35, 206-218.

43. IWASAKI, H., 1961. The Life Cycle of *Porphyra tenera* in vitro, *Biol. Bull.,* **121** (1), 173-187.

44. ——, 1965. Asakusanori no seiri, seitai ni kansuru Kenkyu (Physiology and Ecology of *P. tenera*), *Hiroshima Daigaku Suisan Chikusan Gakubu Kiyo (Journal of the Faculty of Fisheries and Animal Husbandry: Hiroshima University),* **6** (1), 133-211.

45. ——, 1965a. Nutritional Studies of the Edible Seaweed *Porphyra tenera:* I. The Influence of Different B_{12} Analogues, Plant Hormones, Purines and Pyrimidines on the Growth of *Conchocelis, Plant and Cell Physiol.,* 6, 325-336.

46. ——, 1967. Nutritional Studies of the Edible Seaweed *Porphyra tenera:* II. Nutrition of *Conchocelis, J. Phycol.,* **3** (1), 30-34.

47. ——, and C. MATSUDAIRA, 1954. Matsukawaura asakusanori yoshokujo no Kenkyu: I. Asakusanori no chisso, rin ganyuryo ni eikyo suru kankyo yoin ni tsuite (Culture Bed of *P. tenera* in Matsukawa-Ura: I. Environmental Factors Influencing the Nitrogen and Phosphorus Content of *P. tenera*), *Nihon Suisan Gakkaishi (Bulletin of the Japanese Society of Scientific Fisheries),* **20** (2), 112-119.

48. ——, ——, 1954a. *Ibid.,* II. Kanoseisanryoku ni tsuite (Productivity), *Ibid.,* **20** (5), 380-385.

49. ——, ——, 1956. Studies on the Physiology of a Laver, *Porphyra tenera* Kjellm., *Tohoku J. Agr. Res.,* **7** (1), 65-83.

50. ——, ——, 1957. Studies on the Physiology of a Laver, *Porphyra tenera* Kjellm., III. Chemical Factors Influencing the Photosynthesis, *Ibid.,* **8** (1), 47-54.

51. IWASAKI, H., and C. MATSUDAIRA, 1958. Asakusanori no baiyo: I. Baiyojoken ni kansuru yobijikken (Culture of *P. tenera*: I. Preliminary Experiments Related to Culture Conditions), *Nihon Suisan Gakkaishi (Bulletin of the Japanese Society of Scientific Fisheries)*, **26** (6-7), 398-401.

52. KANAZAWA, A., 1963. Sorui no vitamin (Vitamins of Algae), *Ibid.*, **29** (7), 713-731.

53. KANEKO, M., 1931. Norifuko ni kansuru Kenkyu (Study on Seaweeds) (Report No. 1), *Zenra Nando suishi Hokoku*, (4), 1-20.

54. ———, 1932. *Ibid.* (Report No. 2), *Ibid.* (5), 1-18.

55. ———, 1934. *Ibid.* (Report No. 3), *Ibid.* (7), 1-27.

56. ———, 1934a. *Ibid.* (Report No. 4), *Ibid.* (7), 28-52.

57. ———, 1935. *Ibid.* (Report No. 5), *Ibid.* (8), 1-46.

58. ———, 1936. *Ibid.* (Report No. 6), *Ibid.* (9), 1-59.

59. ———, 1937. *Ibid.* (Report No. 7), *Ibid.* (10), 1-61.

60. KATADA, M., 1952. Nihon-kai nanbu ni miidasareta amanori-zoku no isshu ni tsuite (Yoho) [Species of *Porphyra* Found in the Southern Part of the Sea of Japan (Preliminary Report)], *Nissuiken Soritsu San (3)—Shunen Kinen Ronbunshu*, 85-86.

61. ———, and K. KOHATA, 1962. Konkurito-men zosei ni yoru iwanori no seisankoka (Kaiso zoshoku jigyo no seisankoka ni kansuru Kenkyu) [*Iwaranori* Grown on a Concrete Surface (Study on Seaweed Culture Techniques, Report No. 3)], *Suisan Zoshoku (Aquaculture)*, **10** (3), 27-36.

62. KATO, T., and M. KAWAMURA, 1956. Fuchaku keiso ni kansuru kenkyu Report No. 2 (Study on Diatoms: Report No. 2. On the Distribution of Diatoms Settling on *P. tenera*), *Nihon Seitai Gakkaishi (Japanese Journal of Ecology)*, **6** (1), 6-7.

63. KAWABATA, S., 1936. A List of Marine Algae from the Island of Shikotan, *Sci. Jap. Inst. Algol. Res., Fac. Sci., Hokkaido Imp. Univ.*, **1** (2), 199-212.

64. KAWASE, K., 1940. Nori jinko fuchaku ni tsuite (Artificial Settling of Seaweeds), *Zenra Nando suishi Hokoku* (13), 29-43.

65. KIMURA, G., 1957. Yakushin suru nori yoshokugyo no tenbo (Future Prospects of Laver Culture), *Suisan Zoshoku (Aquaculture)*, **4** (4), 1-3.

66. KINOSHITA, S., and K. TERAMOTO, 1958. Asakusanori no Kogosei ni kansuru ni, sanno chiken (Photosynthesis of *P. tenera*), *Sorui (Bulletin of the Japanese Society of Phycology)*, **6** (1), 11-16.

67. ———, ———, 1958a. Asakusanori no seicho ni taisuru hikari oyobi suion no eikyo (Influence of Light and Water Temperature on the Growth of *P. tenera*), *Nihon Suisan Gakkaishi (Bulletin of the Japanese Society of Scientific Fisheries)*, **24** (5), 326-329.

68. ———, ———, 1958b. Asakusanori no seicho ni taisuru shibererin no koka (Effect of Gibberellin on the Growth of *P. tenera*), *Sorui*, **6** (3), 85-88.

69. KINOSHITA, T., 1943. Hokkaidosan Amanori no seikatsushi ni kansuru Kenkyu. Susabinori no Kenkyu. Susabinori no natsu me no yoshoku [Study on the Life History of *Porphyra yezoensis* (Section 3): Culture of Summer Plumules of *P. yezoensis*], *Nihon Suisan Gakkaishi (Bulletin of the Japanese Society of Scientific Fisheries)*, **12** (1), 18-20.

70. KINOSHITA, T., 1949. Nori, tengusa, funori oyobi ginnanso no zoshoku ni kansuru Kenkyu (Study on the Culture of *Porphyra*, Gelidium, and Gloiopeltis), *Suisan Kagaku Gyosho*, II, Hoppo Publishers: Sapporo, 109.

71. ——, and S. SHIBUYA, 1941. Hokkaido-san amanori no seikatsushi ni kansuru Kenkyu. Susabinori no Kenkyu (Study on the Life History of *Porphyra* of Hokkaido: Report No. 1. *Porphyra yezoensis*), *Nihon Suisan Gakkaishi (Bulletin of the Japanese Society of Scientific Fisheries)*, 9 (6), 237-245.

72. ——, ——, 1942. *Ibid.* Report No. 2 (Section 2), *Ibid.*, 11 (2), 47-56.

73. KITO, H., 1956. Amanori-zoku sushu no saibogakuteki Kenkyu. Utsu romukade ni chakusei suru amanori-zoku no isshu ni tsuite (Cytological Studies on Some Species of *Porphyra*: I. A Species of *Porphyra* Adhering to Anthropods), *Hokkaido Daigaku Suisan Gakubu Kenkyu Iho*, 16 (4), 206-208.

74. ——, N. H. YABU, and J. TOKIDA, 1967. The Number of Chromosomes in Some Species of *Porphyra*, *Bull. Fac. Fish.*, *Hokkaido Univ.*, 18 (2), 59-62.

75. KJELLMAN, F. R., 1883. The Algae of the Arctic Sea: A Survey of the Species, Together with an Exposition of the General Characters and the Development of the Flora, *Vet. Akadem. Handl.*, 20 (5), 1-350.

76. ——, 1889. Om Beringhafvets algen-flora, *Ibid.*, 23 (8), 1-58.

77. ——, 1897. Japanska arter of slagtet *Porphyra*, *Bihang. Vet.-Akad., Handl. Afd.* III, 23 (4), 1-34.

78. KOBAYASHI, T., 1960. *Jinko saibyo no tebiki (Handbook of Artificial Seedlings).* Zennoriren, Tokyo, 64 p.

79. KOHATA, K., and Y. OGAWA, 1957. Iwanori no zoshoku ni tsuite (Culture of Iwanori), *Suisan Zoshoku (Aquaculture)*, 4 (4), 149-160.

80. KORNMANN, P., 1961. Die Entwicklung von *Porphyra leucosticta* im Kulturversuch, *Helgoland, wiss. Meeresunt.*, 8 (1), 167-175.

81. ——, 1961a. Zur Kenntnis der *Porphyra* Arten von Helgoland, *Ibid.*, 8 (1), 176-182.

82. KRISHNAMURTHY, V., 1959. Cytological Investigations on *Porphyra umbilicalis* var. *laciniata*, *Ann. Bot.*, N. S. 23, 147-176.

83. KUNIEDA, H., 1939. On the Life History of *Porphyra tenera* Kjellman, *J. Coll. Agr.*, *Tokyo Imp. Univ.*, 14 (5), 377-405.

84. KURAKAKE, T., 1966. *Noriamireizo no tebiki (Handbook on Cold Storage of Seaweed Net).* Zennoriren: Tokyo, 72 p.

85. KUROGI, M., 1952. Kaigara ni senko seru sorui ni tsuite (On shell-boring algae: Proceedings of the 17th Conference of the Botanical Association of Japan), p. 57, *Bulletin of Tohoku Regional Fisheries Research Laboratory* (3), 10-12.

86. ——, 1953. Amanori-rui no seikatsushi no Kenkyu. Daiippo. Kahoshino hatsuga to seicho (On the Life History of *Porphyra*: Report No. 1. Germination of Carpospores and Growth), *Tohoku Suisan Kenkyujo Kenkyu Hokoku (Bulletin of Tohoku Regional Fisheries Research Laboratory)* (2), 67-103.

87. ——, 1953a. Asakusanori no itojotai no tanhoshi hoshutsu ni tsuite (On the Liberation of Monospores from the Filaments of *P. tenera*), *Ibid.* (2), 104-108.

88. ——, 1956. Amanori-rui no seikatsushi, tokuni iwayuru natsunori ni tsuite (marubagata asakusanori no seikatsushi) [Life History of *Porphyra* with Special Reference to the Life History of Regional Summer Seaweed (Maruba-

type *P. tenera*) (Preliminary Report)], *Sorui (Bulletin of the Japanese Society of Phycology)*, **4** (1), 13-18.

89. KUROGI, M., 1956a. Amanori-rui no seikatsushi no Kenkyu. Oononori no hanshokukikan ni tsuite [On the Life History of *Porphyra* (in continuation of the previous report): Reproductive Organs of *P. onoi* (Oononori)], Proceedings of the 21st Congress of Botanical Association of Japan, p. 3.

90. ——, 1957. Yoshoku nori no shurui (Types of Cultured Laver), *Suisan Zoshoku (Aquaculture)*, **4** (4), 21-28.

91. ——, 1959. Uppurui-nori no yoshoku to ukashigata [Cultivation of *P. pseudo-linearis* (Uppurui nori)], *Ibid.*, **7** (1), 40-45.

92. ——, 1959a. Amanori-rui itojotai no seicho seijuku to kojoken: I. Tanhoshino keisei oyobi tanhoshi hoshutsu to nicchosayo (1) [Growth of *Conchocelis* of *Porphyra*—Maturation and Condition of Light: I. Daylight Action in Relation to the Formation of Monosporangium and the Release of Monospores (1)], *Tohoku Suisan Kenkyujo Kenkyu Hokoku (Bulletin of Tohoku Regional Fisheries Research Laboratory)* (15), 33-42.

93. ——, 1959b. Muroran-san no susabinori ni tsuite (On the Study of *P. yezoensis* Growing in Muroran), *Ibid.*, (15) 43-56.

94. ——, 1961. Yoshoku amanori no shurui to sono seikatsushi (amanori-rui no seikatsushi no Kenkyu) [Types of Cultured *Porphyra* and Their Life History (Study on the Life History of *Porphyra*, Report No. 2)], *Ibid.* (18), 1-115.

95. ——, 1963. Yamada-wan funakoshi-wan no yoshoku amanori no shurui to kosujinori no ichishinhinshu ni tsuite (Types of Cultured *Porphyra* in Yamada Bay and Funakishi Bay, Including a New Variety of *P. angusta*), *Ibid.* (23), 117-140.

96. ——, and K. HIRANO, 1952. Amanori-zoku no kahoshi no hatsuga oyobi seicho (Germination and Growth of Carpospores of *Porphyra*), Proceedings of the 1952 Congress of the Japanese Society of Scientific Fisheries, p. 8, *Bulletin of Tohoku Regional Fisheries Research Laboratory* (2), 17-18.

97. ——, ——, 1955. Kanso, Kaisui enbun, kosen ga amanori-rui no itojotai (Conchocelis-ki) ni oyobosu eikyo (Influence of Desiccation, Salinity, and Light on the *Conchocelis* Phase of *Porphyra*), *Tohoku Suisan Kenkyujo Kenkyu Hokoku (Bulletin of Tohoku Regional Fisheries Research Laboratory)* (4), 262-278.

98. ——, ——, 1955a. Asakusanori no itojotai no tanhoshi hoshutsu ni tsuite (2) hoshutsu nonis shuki [On the Liberation of Monospores from the Filaments of *P. tenera* (2): Daily Cycle of Spore Release], *Ibid.* (4), 279-282.

99. ——, and K. AKIYAMA, 1965. Amanori-rui no itojotai no seicho, seijuku to kojoken. IV. Tanhoshi no hoshutsu to okarusa (Growth of Filaments, Maturation, and Conditions of Light of *Porphyra*: IV. Relationship of the Liberation of Monospores to Brightness), *ibid.* (25), 171-177.

100. ——, and K. HIRANO, 1956. Amanori-rui no itojotai no tanhoshi hoshutsu ni tsuite (Umideno jikken) [On the Liberation of Monospores from the Filaments of *Porphyra* (Experiments Carried out in the Field)], *Ibid.* (8), 27-44.

101. ——, ——, 1956a. Asakusanori no itojotai no seicho, tanhoshino keisei, tanhoshi hoshutsu to suion to no Kankei (Relationship of Water Temperature and Growth of Filaments, Formation of Monosporangium, and the Libera-

tion of Monospores in *P. tenera*), *Ibid.* (8), 45-61.

102. KUROGI, M., and S. SATO, 1962. Amanori-rui no itojotai no seicho, seijuku to kojoken: II. Asakusanori no itojotai no seicho, seijuku to Niccho (Growth of Filaments, Maturation, and Conditions of Light of *Porphyra:* II. Growth and Maturation of Filaments of *P. tenera* in Relation to Daylight), *Tohoku Suisan Kenkyujo Kenkyu Hokoku (Bulletin of Tohoku Regional Fisheries Research Laboratory)* (20), 127-137.

103. ——, ——, 1962a. Shu ni yoru Nicchosayo no sai (Growth of Filaments, Maturation, and Conditions of Light of *Porphyra:* III. Differences in Daylight Action in Various Species), *Ibid.* (20), 138-156.

104. ——, K. AKIYAMA, and S. SATO, 1962. Amanori-rui no itojotai no seicho, seijuku to Kojoken: I. Tanhoshino keisei oyobi tanhoshi hoshutsu to Niccho-sayo (2) [Growth of Filaments, Maturation, and Conditions of Light of *Porphyra:* I. Daylight Action in Relation to the Formation of Monosporangium and the Liberation of Monospores (2)], *Ibid.* (20), 121-126.

105. ——, ——, 1966. Sushu no amanori no itojotai no seicho, seijuku to suion (Growth of Filaments, Maturation, and Water Temperature in a Few Species of *Porphyra*), *Ibid.* (26), 77-89.

106. ——, and S. KAWASHIMA, 1966. Dotochiho no haru-nori yoshoku (1) harunori yoshoku ni shiyo suru nori no shurui to sono tokucho [Spring Seaweed Culture in Hokkaido and Tohoku Regions (1): Types of *Porphyra* Used in Spring Seaweed Cultivation and Their Characteristics], *Hokkaido Suisan Shikenjo Geppo (Monthly Report of the Hokkaido Fisheries Experiment Station)*, **23** (5), 243-257.

107. ——, and T. YOSHIDA, 1966. Tohokuchiho ni okeru chishimakuronori no yoshoku (Culture of *P. umbilicalis* Found in the Tohoku Regions), *Tohoku Suisan Kenkyujo Kenkyu Hokoku* (26), 91-107.

108. ——, and S. KAWASHIMA, 1967. Itojotai no baishoku (Report No. 2 of the Spring Seaweed Culture in Hokkaido and Tohoku Regions: Culture of Filaments), *Hokkaido Suisan Shikenjo Geppo*, **24** (2), 51-62.

109. ——, and S. SATO, 1967. Chishimakuronori to makure-amanori no itojotai no seicho, seijuku to Niccho (Growth and Maturation of Filaments of *P. umbilicalis* and *P. pseudocrassa* in Relation to Daylight), *Tohoku Suison Kenkyujo Kenkyu Hokoku (Bulletin of Tohoku Regional Fisheries Research Laboratory)* (27), 111-130.

110. ——, ——, and T. YOSHIDA, 1967. Chishimakuronori no itojotai no tanhoshi hoshutsu to suion (Relationship between the Release of Monospores from the Filaments and Water Temperature of *P. umbilicalis*), *Ibid.* (27), 131-139.

111. ——, and S. KAWASHIMA, 1968. Saibyo (Report No. 3 of the Spring Seaweed Culture in Hokkaido and Tohoku Regions: Seedlings), *Hokkaido Suisan Shikenjo Geppo (Monthly Report of the Hokkaido Fisheries Experiment Station)*, **25** (5), 238-247.

112. ——, and S. SATO, n. d. Asakusanori no chuseihoshi keisei to suion (Relationship between the Formation of Neutral Sporangium and Water Temperature in *P. tenera*). (Mimeographed).

113. KUSAKABE, D., 1929. Asakusanori no yoshoku ni tsuite (On the Culture of

P. tenera), *Suiko Kenkyu Hokoku* (*Journal of Imperial Fisheries Institute, Tokyo*) **25** (2), 17-26.

114. KYLIN, H., 1922. Ueber die Entwicklungsgeschichte der Bangiaceen, *Arkiv. f. bot.,* **17** (5), 1-12.

115. LINNAEUS, C., 1753. *Species plantarum.* Stockholm.

116. MAEKAWA, K., 1961. Setonaikai, tokuni Yamaguchi ken engan ni okeru gyogyo no chosei kanri to shigenbaiyo ni kansuru Kenkyu (On the Control of Fishery Industries in Setonaikai, Particularly in the Coastal Region of Yamaguchi Prefecture), *Yamaguchi Naikai Suisan Shikenjo Chosa Kenkyu Gyoseki,* **11** (1), 1-483.

117. ——, and A. TOMIYAMA, 1958. Suion Chosetsu ni yoru asakusanori itojotai kara no hoshi hoshutsu no jiniseigyo ni tsuite (Artificial Control of Spore Release from the Filaments of *P. tenera* through the Adjustment of Water Temperature), *Suisan Zoshoku (Aquaculture),* **5** (4), 56-59.

118. ——, ——, 1959. Asakusanori no jinko tanetsuke ni kansuru Kenkyu. Suion chosetsu ni yoru amanori-rui itojotai kara no hoshi hoshutsu no jinitekiseigyo ni tsuite (On the Artificial Spore Fixation in *P. tenera*: Report No. 6. Artificial Control of Spore Release from the Filaments of *Porphyra* through the Adjustment of Water Temperature), *Yamaguchi Naikai Suisan Shikenjo Chosa Kenkyu Gyoseki,* **10** (1), 17-25.

119. ——, ——, 1959a. *Ibid.,* Report No. 7: Amanori-rui no itojotai kara no tanhoshi hoshutsu ni tsuite (On the Release of Monospores from the Filaments of *Porphyra*), *ibid.,* **10** (1), 27-31.

120. ——, N. AKIYAMA and A. TOMIYAMA, 1959. Asakusanori no jinko tanetsukeni kansuru Kenkyu. Yogano taikansei ni tsuite (On the Artificial Spore Fixing of *P. tenera*: Report No. 8. On the Resistance of Plumules to Descication), *Ibid.,* **10** (1), 33-38.

121. MAGNE, F., 1952. La structure de noyau et le cycle nucleaire chez le *Porphyra linearis* Greville, *C. R. Acad. Sc.,* 234, 986-988.

122. MATSUDAIRA, C., and H. IWASAKI, 1953. On the Enviromental Characteristics of Cultural Grounds of a Laver, *Porphyra tenera* Kjellman, in Matsushima Bay, *Tohoku J. Agr. Res.,* **3** (2), 277-291.

123. ——, and A. HAMADA, 1966. Teppan fushoku genryo ni yoru kaisui no ryudo no kani sokuteiho (A Simple Method of Measuring the Flow of Sea Water on the Basis of Decrease in the Amount of Rust on an Iron Plate), *Umi (Laver),* **4** (1), 8-13.

124. MATSUE, Y., 1936. Shinagawa Wan Asakusa-nori yoshokujo no kaiyokagakuteki seijo (Chemical Characteristics of Growing Beds of *P. tenera* at Shinagawa Bay), *Bulletin of the Japanese Society of Scientific Fisheries,* 7, 35-62.

125. MATSUMOTO, F., 1959. Nori seiiku ni taisuru kankyo. Tokuni suiryu no eikyoni kansuru Kenkyu (Environmental Conditions, Particularly the Influence of Water Current on the Vegetation of Laver), *Hiroshima Daigaku Suisan Chikusan Gakubu Kiyo* (*Journal of the Faculty of Fisheries and Animal Husbandry*), **2** (2), 249-333.

126. MIGITA, S., 1959. Nori itojotai no seicho ni oyobosu kankyo joken to baiyoeki no pH henka ni tsuite (Environmental Conditions for the Growth of *Por-*

phyra Filaments and the Influence of Changes in the pH of the Culture Fluid), *Nagasaki Daigaku Suisan Gakubu Kenyku Hokuku (Bulletin of the Faculty of Fisheries, Nagasaki University)* (8), 207-215.

127. MIGITA, S., 1960. Ariake-kai ni okeru yoshoku nori no shurui (Types of Cultured *Porphyra* in Ariake-kai), *Seikaiku Suiken (Laver Culture in Ariake-kai)*, 2, 75-82.

128. ——, 1962. Studies on Plantlets of *Conchocelis* Phase of *Porphyra*, *Rec. Oceanogr. Works Japan*, Special No. (6), 147-152.

129. ——, 1964. Amanori yotai no seitai toketsu hozon: I. Kaisuichu oyobi hankansojotai de toketsu hozon suita asakusanori yotai no seizon noryoku ni tsuite (Preservation of *Porphyra* Thalli by Freezing: I. On the Preservation of *P. tenera* Thalli by Freezing in a Semi-dry Condition), *Nagasaki Daigaku Suisan Gakubu Kenkyu Hokoku (Bulletin of the Faculty of Fisheries, Nagasaki University)* (17), 44-54.

130. ——, 1967. Toketsu asakusanori itojotai no seizon to karahoshi hoshutsu (Life of Filaments and Carpospore Release of *P. tenera* Preserved by Freezing), *Ibid.* (22), 33-43.

131. ——, 1967. Cytological Studies on *Porphyra yezoensis* Ueda, *Bull. Fac. Fish., Nagasaki University* (24), 55-64.

132. ——, and N. ABE, 1966. Amanori itojotai no karahoshi keisei ni tsuite (Formation of Carpospores in *Porphyra* Filaments), *Nagasaki Daigaku Suisan Kenkyu Hokoku (Bulletin of the Faculty of Fisheries, Nagasaki University)* (20), 1-13.

133. MIKAMI, H., 1956. Two New Species of *Porphyra* and Their Subgeneric Relationship, *Bot. Mag. Tokyo*, **69** (819), 340-345.

134. MIURA, A., 1961. A New Species of *Porphyra* and Its *Conchocelis* Phase in Nature, *J. Tokyo Univ. Fish.*, **47** (2), 305-311.

135. NAGAI, M., 1941. Marine Algae of the Kurile Islands, II, *J. Fac. Agr., Hokkaido Imp. Univ.*, **46** (2), 139-310.

136. NAKAMURA, Y., 1947. Fuiritasa, tokuni sono yuseitai ni tsuite (On the Study of *P. variegata* with Special Reference to Androecious Plants), *Shokubutsu Zasshi (The Botanical Magazine)*, **60** (703-714), 39-43.

137. NAKATANI, S., 1964. Ecological Studies on *Porphyra tenera*, Especially on Its Leafy Thalli, *J. Agr. Lab.* (5), 1-50.

138. ——, S. SHIMO, and Y. MINOHARA, 1961. Asakusanori jinkoyoshoku ni kansuru Kenkyu: II. Sushu hacho-kosen to yotai seiiku ni tsuite (On the Study of Artificial Culture of *P. tenera*: II. Effect of Various Wavelengths of Light Rays on the Growth of Thalli), *Nodenkenshoho (Journal of Agricultural Laboratory)* (1), 115-122.

139. ——, ——, 1962. Asakusanori-rui no seicho ni oyobosu Niccho, koryo oyobi suion no eikyo (Influence of Daytime, Luminous Intensity, and Water Temperature on the Growth of *P. tenera*), *Nodenken Hokoku*, 62004, 1-24.

140. Nihon Nori Shokuhin Shimbunsha (editorial) (1967), *Nori Nenkan*, Tokyo, 560.

141. NODA, M., 1964. On the *Porphyra* from Sado Island in the Japan Sea, *Sci. Rep. Niigata Univ.*, Ser. D. (Biol.), 1, 1-13.

142. NOZAWA, K., 1954. Asakusanori no (kusare) to chikei, Kaikyo to no Kankei (Relationship between Decomposition of *P. tenera* to Topography and Sea

Conditions), *Suisan Zoshoku (Aquaculture)*, **1** (3-4), 9-14.

143. Nozawa, K., 1957. Nori no hikisei-seibyo ni tsuite (Non-parasitic Diseases of *Porphyra*), *Ibid.*, **4** (4), 65-68.

144. ——, and Y. Nozawa, 1957. Kaiso no genkeishitsu ni kansuru Kenkyu: II. Asakusanori no (Anagusarebyo) ni tsuite (Study on the Protoplasm of Seaweeds: II. On the Hole Rot Disease of *P. tenera*), *Nihon Suisan Gakkaishi (Bulletin of the Japanese Society of Scientific Fisheries)*, **22** (11), 694-700.

145. ——, ——, 1957. Asakusanori itojotai no byogai ni kansuru Kenkyu [Study on the Diseases of *P. tenera* Filaments (Preliminary Report)], *Nihon Suisan Gakkaishi (Bulletin of the Japanese Society of Scientific Fisheries)*, **23** (7/8), 427-429.

146. Ogata, E., 1955. Perforating Growth of *Conchocelis* in Calcareous Matrices, *Bot. Mag. Tokyo*, **68** (810), 371-372.

147. ——, 1961. Nori itojotai no seicho ni kansuru Kenkyu (On the Growth of *Porphyra* Filaments), *Suiko Kenkyu Hokoku (Journal of Imperial Fisheries Institute, Tokyo)*, **10** (3), 423-500.

148. ——, 1964. Asakusanori no seiri to byori (Physiology and Pathology of *P. tenera*), *Shokubutsu Seiri (Physiology of Plants)*, **4** (3), 171-182.

149. Ohmi, H., 1954. New Species of *Porphyra*, Epiphytic on *Chorda filum* from Hokkaido, *Bull. Fac. Fish., Hokkaido Univ.*, **5** (3), 231-239.

150. Okamura, K., 1900. *Kaisogaku hanron (Introduction to the Study of Algae)*. Keigyosha: Tokyo, 269 p.

151. ——, 1901. Asakusanori ni Kansuru kenkyu (On the Study of *P. tenera*), *Suiko Kenkyu Hokoku (Journal of Imperial Fisheries Institute, Tokyo)*, **2** (1), 128-139.

152. ——, 1909. *Asakusanori (On P. tenera)*. Hakubunkan: Tokyo, 374 p.

153. ——, 1916. *Nihon Sorui meii* (Dai-ni-han) (*List of Japanese Algae*, 2nd ed.). Seibido: Tokyo, 362 p.

154. ——, 1930. *Suisanshokubutsu (Marine Plants)*. Iwanami: Tokyo, 32 p.

155. ——, K. Onda, and M. Higashi, 1920. Preliminary Notes on the Development of the Carpospores of *Porphyra tenera* Kjellm., *Bot. Mag. Tokyo*, 34, 131-135.

156. ——, S. Masuda, and Y. Miyake, 1926. Asakusanori no Kasumi-gaihiken (On the Cold Damage of *P. tenera*), *Suiko Kenkyu Hokoku (Journal of Imperial Fisheries Institute, Tokyo)*, **21** (6), 230-238.

157. Ota, F., 1960. Nori no jinkosaibyo to yoshoku ni kansuru Kenkyu (Artificial Seedlings and Culture of *Porphyra*), Kumamoto Suisan Shikenjo Jigyo Hokoku (Annual Report of the Kumamoto Prefecture Fisheries Experiment Station), 1953-1958, 123-135.

158. Provasoli, L., J. J. A. McLaughlin, and M. R. Droop, 1957. The Development of Artificial Media for Marine Algae, *Archiv. f. Microbiol.*, 25, 392-428.

159. Rees, T. K., 1940. Algal Colonization at Munbles Head, *J. Ecol.*, 28, 403-437.

160. ——, 1940a. A Preliminary Account of the Life History of *Porphyra umbilicalis* (L.), *Ag. Ann. Bot.*, N. S., **4** (15), 669-671.

161. Rosenvinge, L. K., 1893. Gronlands Havalger, *Meddel. om Gronland* III, 765-981, Kjøenhavn.

162. ROSENVINGE, L. K., 1909. The Marine Algae of Denmark: Contributions to Their Natural History. Part I. Introduction: *Rhodophyceae I (Bangiales and Nemalionales), Danske Vidensk. Selsk Skrifter., 7. Raekke, Naturv. og Mathem. Afd.* **7** (1), 1-151.

163. SAITO, Y., 1956. Kaichudeno asakusanori itojotai kara no hoshihoshutsu ni tsuite [On the Liberation of Spores from the Filaments of *P. tenera* (Preliminary Report)], *Nihon Suisan Gakkaishi (Bulletin of the Japanese Society of Scientific Fisheries)*, **21** (12), 1215-1218.

164. ———, 1956a. Asakusanori itojotai no seicho-seijuku ni oyobosu ni, san no yoinno eikyo ni tsuite (Influence of a Few Factors on the Growth and Maturation of *P. tenera* Filaments), *Ibid.*, **22** (1), 21-29.

165. SANO, T., 1955. Yoshoku nori no shikitakuhenka ni kansuru Kenkyu. Daiippo. Suiyoseishikiso no henka (On the Color Changes of Cultured *Porphyra*: Report No. 1. Changes in the Water Soluble Chromatophores), *Tohoku Suisan Kenkyujo Kenkyu Hokoku (Bulletin of Tohoku Regional Fisheries Research Laboratory)*, (4), 243-261.

166. ———, 1955a. Dai-niho. Shiyosei shikiso oyobi suiyosei shikiso no henka ni tsuite (On the Color Changes of Cultured *Porphyra*: Report No. 2. Changes in the Fat Soluble and Water Soluble Chromatophores), *Ibid.* (5), 64-78.

167. SATOMI, M., S. MATSUI, and M. KATADA, 1967. Net Production and Increment in Stock of the *Porphyra* Community in the Culture Ground, *Bull. Jap. Soc. Sci. Fish.*, **33** (3), 167-175.

168. ———, Y. ARUGA, and K. IWAMOTO, 1968. Effect of Aging on the Seasonal Change in the Photosynthetic Activity of *Porphyra yezoensis* Grown in the Culture Ground, *Ibid.*, **34** (1), 17-22.

169. SHIMO, S., and S. NAKATANI, 1963. Asakusanori-rui no seicho ni oyobosu niseionshuhanno ni tsuite (Influence of Solar Temperature Cycle Reaction on the Growth of *P. tenera*), Nodenkenpo, 63006, 1-13.

170. ———, ———, 1964. Asakusanori ni taisuru kainechin no koka ni tsuite (Effect of Kinetin on *P. tenera*), *Nodenken shoho* (5), 55-60.

171. SHINMURA, I., 1965. Nori no seiiku ni oyobosu kojoken ni tsuite (Effect of Light on the Vegetation of *Porphyra*), *Kagoshima-Ken Suisan Shikenjo Jigyo Hokoku (Monthly Research Report of Kagoshima Prefecture Fisheries Experiment Station)*, 309-321.

172. ———, H. SHIIHARA, and T. TANAKA, 1967. Ichimatsunori no itojotai no karahoshi hoshutsu ni oyobosu niccho joken [Effect of Daylight on the Liberation of Spores from the Filaments of *P. seraita* (Ichimatsunori)], *Sorui (Bulletin of the Japanese Society of Phycology)*, **15** (3), 123-126.

173. ———, and T. TANAKA, 1968. Shitsunai baiyo ni okeru ichimatsunnori yoga no seicho to suion (Relationship between the Growth of *P. seraita* Plumules and Water Temperature during Indoor Culture), *Ibid.*, **16** (1), 4-6.

174. Suisancho Chosa Kenkyubu Kenkyu Dai-ni-ka, 1965. Tekichi tekishu senkai zoshoku gijutsu kenkyu (nori) kekka hokoku (Results of the Investigations Carried out on the Shallow-water Culture of *Porphyra* to Discover a Suitable Region and a Suitable Species), 1-183 p.

175. SURINGAR, W.F.R., 1870. *Algae Japonicae musei botanici Lugduno-Batavi.* Harlemi, 39 p.

176. Suto, S., 1948. Kaisuichu kara eta asakusanori no akine oyobi tojigahoshi no seijoto teiryoho (kaiso hoshizuke no Kenkyu, dai-niho) [Characteristics of the Autumn Buds and Winter Spores of *P. tenera* (in the Study of Spore Fixation of Marine Algae, Report No. 2)], *Nihon Suisan Gakkaishi (Bulletin of the Japanese Society of Scientific Fisheries)*, **13** (5), 193-194.

177. ——, 1950. Tokyo-Bay san asakusanori no shurui (Yoho) [*P. tenera* Found in Tokyo Bay (Preliminary Report)], *Ibid.*, **15** (11), 649-652.

178. ——, 1950a. Asakusanori no hoshi no hoshutsu, fuyu oyobi chakusei (Kaiso-hoshitsuke no-Kenkyu, dai-Kyu-ho) [Liberation of Spores in *P. tenera* (in the Study of Spore Fixation of Marine Algae, Report No. 9)], *Ibid.*, **16** (4), 137-140.

179. ——, 1950b. Asakusanori no seikatsushi ni tsuite tokuni akini tachikonda hibi ni saisho ni tsuku hoshi no seishitsu (Kaisohoshizuke no Kenkyu) [On the Life History of the *P. tenera* with Special Reference to the Characteristics of Early Spores Settling on "hibi" (in the Study of Spore Fixation of Marine Algae, Report No. 10)], *Ibid.*, **16** (5), 171-174.

180. ——, 1954. *Ibid.*, Part III, *Ibid.*, **20** (6), 494-496.

181. ——, 1957. Tokyo-Wan o Chushintoshita nori no shurui (Types of *Porphyra* Found in Tokyo Bay and Other Places), *Suisan Zoshoku (Aquaculture)*, **4** (4), 28-32.

182. ——, 1960. Asakusanori no shitsunai baiyo no hoho ni tsuite (On the Indoor Culture Method of *P. tenera*), *Ibid.*, **7** (3), 7-11.

183. ——, 1961. Asakusanori no tairyobaibyo ni tsuite (Large-scale Culture of *P. tenera*), *Nosan Kako Gijutsu Kenkyu Kaishi*, **8** (1), 52-59.

184. ——, 1961a. Asakusanori no seicho, seijuku nado ni taisuru niccho koka (Effect of Daylight on the Growth and Maturation of *P. tenera*), Proceedings of the 1961 Congress of the Botanical Association of Japan.

185. ——, 1963. Intergeneric and Interspecific Crossings of the Lavers *(Porphyra)*, *Bull. Jap. Soc. Sci. Fish.*, **29** (8), 739-748.

186. ——, T. Maruyama, and O. Umebayashi, 1954. Asakusanori no conchocelis phase kara no hoshi hoshutsu ni tsuite (On the Release of Spores from the *Conchocelis* Phase of *P. tenera*), *Nihon Suisan Gakkaishi (Bulletin of Japanese Society of Scientific Fisheries)*, **20** (6), 490-493.

187. ——, and O. Umebayashi, 1954. Asakusanori no (Anagusarebyo) (Shinsho) ni tsuite (Hole Rot Disease of *P. tenera*), *Ibid.*, **19** (12), 1176-1178.

188. Takayama, H., 1937. Asakusanori no seicho narabini hinshitsu ni oyobosu hiju no hani ni tsuite (Effect of Density of Water on the Growth and Quality of *P. tenera*), *Suikenshi (Journal of Fisheries Research)*, 32, 66-69.

189. Takeuchi, T., T. Matsubara, M. Shimonaka, and S. Suto, 1954. Umi ni irata nori no Conchocelis phase kara no hoshi no hoshutsu to hibitate jiki (Release of Spores from the *Conchocelis* Phase of *P. tenera* and the Period of "hibi" Planting), *Nihon Suisan Gakkaishi (Bulletin of Japanese Society of Scientific Fisheries)*, **20** (6), 487-489.

190. ——, M. Shimonaka, A. Fukuhara, and H. Yamazaki, 1956. Asakusanori itojotai no seitai: III. Itojotai no chishijoken ni tsuite (Ecology of *Porphyra tenera* Kjellman Filaments: III. Factors Causing the Degeneration of Fila-

ments), *Ibid.*, **22** (1), 16-20.

191. TAMURA, S., 1956. Chiba-ken ni okeru nori no (meitami) ni tsuite (Bud Damage of *Porphyra* in Chiba Prefecture), *Suisan Zoshoku (Aquaculture)*, **4** (2), 44-54.

192. TANAKA, K., 1957. Uppuruinori no yoshoku ni kansuru Kenkyu (On the Culture of *P. pseudolinearis*), *Kagawaken Suisan Shikenjo Hokoku (Bulletin of the Kagawa Prefecture Fisheries Experiment Station)* (11), 10-29.

193. ——, T. TAKASUGI, and N. KANBARA, 1962. Amihibi ni yoru kuronori *P. okamurai* Ueda no seiiku jokyo ni tsuite (Growth of *P. okamurai* Ueda on Net "hibi"), Annual Report of the Year 1961 of the Kagawa Prefecture Fisheries Experiment Station (2), 53-56.

194. TANAKA, T., 1944. Nihonsan ushike nori tsonashokubutsu no bunruigakuteki Kenkyu (sono-ni) fukurotasa no kozo to seishokusaibo no bunretsu keishiki (Structure of *P. bulbopes* and Division of Reproductive Cells), *Shokken (Journal of Plant Research)*, **20** (5), 248-254.

195. ——, 1952. The Systematic Study of the Japanese Protoflorideae, *Mem. Fac. Fish., Kagoshima Univ.*, **2** (2), 1-92.

196. ——, and T. KOTO, 1952. Tsukushi amanori no seijo to natsunori ni tsuite (Characteristics of *P. crispata* with Special Reference to Summer Seaweed), Proceedings of the 1952 Congress of the Japanese Society of Scientific Fisheries, p. 8.

197. ——, I. SHINMURA, and M. KUBO, 1965. Amanori-rui hinshukan ni okeru seiiku ni oyobosu kojoken no kento (On the Study of Light Conditions Influencing the Growth of *Porphyra*), *Sorui (Bulletin of the Japanese Society of Phycology)*, **13** (3), 119-125.

198. TANITA, S., and T. OKUDA, 1956. Matsushima-Wan no suisan shigen ni kansuru Kenkyu. Dai-yonho. Seni suishitsu, teishitsu narabini teiseiseibutsu ni kisetsuteki seni (Study of the Marine Resources in Matsushima Bay: Report No. 4. Seasonal Variations in the Quality of Water and Sea Bottom as well as in the Living Organisms), *Tohoku Suisan Kenkyujo Kenkyu Hokoku (Bulletin of Tohoku Regional Fisheries Research Laboratory)* (6), 106-134.

199. TERAMOTO, K., and S. KINOSHITA, 1960. Asakusanori no seicho ni taisuru aminosan oyobi purin-rui no koka (Effect of Amino Acids and Purines on the Growth of *P. tenera*), *Sorui (Bulletin of the Japanese Society of Phycology)*, **8** (3), 90-95.

200. ——, ——, 1962. Asakusanori no shitsunai baiyo ni koteki shita kojoken no kento (Optimum Light Conditions for Indoor Culture of *P. tenera*), *Ibid.*, **10** (1), 12-17.

201. Tohoku Chiku Nori Byogai Bojo Taisaku Kyogikai, 1961. Tohoku ni okeru nori no byogaibojo ni kansuru shiken Kenkyu [On the Study of Preventive Steps against Seaweed Damage in Tohoku Regions (Covering the Period 1958 to 1960)]. 1-68 p.

202. Tohoku-Ku Suisan Kenkyusho Zoshokubu, 1965. Tohoku kaiiki ni okeru Senkaigyogyo (Shallow-water Fisheries in Tohoku). Collection of Papers (Nos. 1 and 2).

203. TOKIDA, J., 1935. Phycological Observations: II. On the Structure of *Porphyra onoi* Ueda, *Trans. Sappoo Nat. Hist. Soc.*, **14** (2), 111-114.

204. TOKIDA, J., 1954. The Marine Algae of Southern Saghalien, *Mem. Fac. Fish.,* *Hokkaido Univ.,* **2** (1), 1-264.

205. ——, 1966. Nihon kinkaisan Porphyra-zoku no shu no kensakuhyo (Study of Various Species of Japanese *Porphyra*), *Sorui (Bulletin of the Japanese Society of Phycology),* **14** (3), 146-149.

206. TOKIDA, J., and T. MASAOKI, and S. TSUBOKAWA, 1961. Konkurito nori sho no kanri ni tsuite (Control of Concrete Seaweed Tank), *Suisan Zoshoku (Aquaculture),* **9** (2), 79-86.

207. TOMIYAMA, A., K. YATSUYANAGI, and K. MAEKAWA, 1956. Asakusanori no jinko tanetsuke ni kansuru Kenkyu. Dai-niho. Teionshori ni yoru itojotai kara no hoshi hoshutsu ni tsuite (On the Artificial Seed Fixing of *P. tenera:* Report No. 2. Liberation of Spores from Filaments by Low Temperature Treatment), *Yamaguchi Naikai Suisan Shikenjo Chosa.* Kenkyu Gyoseki, **8** (1), 7-13.

208. TSENG, C. K., and T. J. CHANG, 1954. Study of *P. tenera:* I. Life History of *P. tenera, Shokubutsu Gakuho,* **3** (3), 287-302. (Chinese).

209. ——, ——, 1955. Study of *P. tenera:* II. Sexual Reproduction of *P. tenera, Ibid.,* **4** (2), 153-166. (Chinese).

210. ——, ——, 1955. Studies on the Life History of *Porphyra tenera* Kjellm., *Scientia Sinica,* **4** (3), 375-398.

211. TSURUGA, H., and T. NITTA, 1957. Kaiso no seiri-kagakuteki Kenkyu: I. Ondohenka, kanshutsu ga doka, kokyu ni oyobosu eikyo ni tsuite (Physiological Study of Marine Algae: I. Influence of Temperature Variations and Exposure on the Assimilation and Respiration of Marine Algae), *Naikai Suisan Kenkyujo Kenkyu Hokoku (Bulletin of Naikai Regional Fisheries Research Laboratory)* (10), 37-41.

212. UEDA, S., 1929. Asakusanori no seikatsushi ni tsuite (On the Life History of *P. tenera), Suiko Kenkyu Hokoku (Journal of Imperial Fisheries Institute, Tokyo),* **24** (5), 180-185.

213. ——, 1932. Nihonsan amanori-zoku no bunrui-gakuteki kenkyu (Classification of Japanese *Porphyra), Ibid.,* **28** (1), 1-45.

214. ——, 1937. Asakusanori no seikatsushi ni kansuru Kenkyu (On the Life History of *P. tenera), Nihon Suisan Gakkaishi (Bulletin of the Japanese Society of Scientific Fisheries),* **6** (2), 91-104.

215. ——, 1952. *Textbook of Seaweed Culture.* Zennori ren: Tokyo, 170 p.

216. ——, 1958. *Textbook of Seaweed Culture.* Zennori ren: Tokyo, 213 p.

217. WADA, H., 1941. Asakusanori no seikatsushi narabini hoshi, bunri saibo no fuchaku no kiko oyobi eiyosaibo no hatsuga ni tsuite (Life History of *P. tenera* with Special Reference to the Mechanism of Spore Fixation and Germination of Vegetative Cells), *Nihon Suisan Gakkaishi (Bulletin of the Japanese Society of Scientific Fisheries),* **10** (1), 47-59.

218. WATANABE, T., S. SATO, and K. SUZUKI, 1966. Matsushima-Wan no gyojo kaihatsujigyo ni tsuite (Development of Fishing Grounds in Matsushima Bay), *Suisan Doboku* (5), 9-15.

219. YABU, H., and J. TOKIDA, 1963. Mitosis in *Porphyra, Bull. Fac. Fish., Hokkaido Univ.,* **14** (3), 131-136.

220. YAMAZAKI, H., 1954. Asakusanori itojotai no seitai, I. (Ecology of *Porphyra tenera* Filaments, I), *Nihon Suisan Gakkaishi (Bulletin of the Japanese Society of Scientific Fisheries)*, **20** (6), 442-446.

221. ——, 1954a. *Ibid.*, II. Tokuni itojotai yori hoshutsu sareta hoshi ni tsuite (On the Study of Spores Released from *Porphyra* Filaments), *Ibid.*, **20** (6), 447-450.

222. ——, Y. FURUIDO, and S. ITO, 1956. Saikin no toka ni okeru nori no fusaku genin oyobi sono taisaku (Failure of Seaweed Crops, Some Reasons and Countermeasures).

223. ——, M. SHIMONAKA, and A. FUKUHARA, 1957. Asakusanori itojotai no seitai: IV. Kishaku kaisui no eikyo ni tsuite (Ecology of *Porphyra* Filaments: IV. Influence of Dilute Sea Water), *Nihon Suisan Gakkaishi (Bulletin of the Japanese Society of Scientific Fisheries)*, **23** (4) 195-198.

224. YATSUYANAGI, K., and H. YOSHINAGA, 1959. Iwanori-rui no zoshokugakuteki Kenkyu. Dai-ippo. Ami niyoru uppurui nori yoshoku shiken (Culture of *Iwanori*: Report No. 1. Experiments on *P. pseudolinearis* on Nets), *Yamaguchi Gaikai Suisan Shikenjo Kenkyu Hokoku*, **2** (1).

225. YENDO, K., 1909, 1915, 1916. Notes on Algae New to Japan: I, III, V, *Bot. Mag. Tokyo*, **23** (270), 117-133; **29** (343). 99-117; **30** (355), 243-263.

226. ——, 1913. Some New Algae from Japan, *Nytt Mag. f. Naturv.*, 51, 275-288.

227. ——, 1919. The Germination and Development of Some Marine Algae, *Bot. Mag. Tokyo*, **33** (388), 73-93.

228. ——, 1920. Novae algae japonicae, Decas I-III, *Ibid.*, **34** (397), 1-12.

229. YOSHIDA, T., Y. SAKURAI, and M. KUROGI, 1964. Yoshoku asakusanori no chakusei mitsudo, seicho to shuryo ni tsuite (Density, Growth, and Yield Rate of Cultured *P. tenera*), *Tohoku Suisan Kenkyujo Kenkyu Hokoku (Bulletin of Tohoku Regional Fisheries Research Laboratory)* (24), 89-101.

230. YOSHINAGA, H., 1961. Nihon-kai nanbu ni bunpu suru porphyra no shurui ni tsuite (Types of *Porphyra* Distributed in the Southern Part of the Sea of Japan), *Sorui (Bulletin of the Japanese Society of Phycology)*, **9** (2), 47-53.

231. ——, and K. YATSUYANAGI, 1959. Iwanori-rui no zoshokugakateki Kenkyu. Dai-niho. Yamaguchi-ken Nihonkaigan ni okeru kuronori no bunpu to keitai no henka ni tsuite [Culture of *Iwanori*: Report No. 2. Changes in the Distribution and Morphology of *P. okamurai* Ueda in the Coastal Regions of Yamaguchi Prefecture (Sea of Japan)], *Yamaguchi Gaikai Suisan Shikenjo Kenkyu Hokoku*, **2** (1).

232. ——, ——, 1960. Dai-sanpo. Yamaguchi-ken nihonkai engansan oniamanori to uppurui-nori ni tsuite [Culture of *Iwanori*: Report No. 3. On the Study of *P. dentata* and *P. pseudolinearis* in the Coastal Regions of Yamaguchi Prefecture (Sea of Japan)], *Suisan Zoshoku (Aquaculture)*, **7** (3), 12-17.

PART II
The Evolution of Oyster Culture

Several marine species are cultured on a mass scale; of these, the most widely and intensively cultured is the oyster. It was among the first, if not the first, of the marine species to be cultured and hence, comparatively speaking, oyster culture has the longest history. Existing literature on oysters and oyster culture is voluminous and rather unique in that it reflects an extraordinarily detailed and meticulous study of this species, from the various biological aspects to its culture techniques, including scientific details of academic interest and practical suggestions. Biological investigations on oysters have noticeably increased during the present century, which cover basic as well as ecological aspects; as a result of these discussions and studies, the scientific aspects of oyster culture have assumed great importance.

A significant advance in the culture methods adopted in Japan is represented by the changeover from bottom culture to hanging culture, a method not practiced in any other part of the world. This progress in oyster culturing resulted from the study of environmental conditions and technical advancement. Further technical modifications in the hanging culture method arose from the stability factors of different regions; these modifications are known as the simple hanging method, the raft hanging method, and the rope hanging method. Culture beds now spread from the end points of the inner bay to the coastal regions, and even to areas far out in the seas. Oyster production increases steadily from year to year along with the expanding culture beds.

Some of the more important aspects of past investigations, as well as those now in progress, are in the fields of physiology and ecology and, in particular, environmental and nutritional physiology with special reference to mass mortality. The findings of these investigations have contributed much to the progress of culture techniques. Producing artificial seedlings became feasible and efforts to improve the breed of oysters followed. As a matter of fact, the oyster was one of the first marine organisms to be used in breeding experiments. Based on biological research and technical progress, culture beds of improved quality are being tried, and strenuous efforts are underway to develop an oyster "nonpareil." Hence one can predict that oyster culture will continue to increase, attaining in the very near future the status of an entirely new industry.

Investigations made and published in Japan and elsewhere are reviewed in this section, which is divided into two chapters. Chapter I gives a résumé of the reports published thus far on the biological research of the fundamental characters of oysters; in Chapter II, the production of various types of oysters and their culture methods are discussed. The author hopes that the data compiled herein will be useful to investigators engaged in research work, to workers in the fisheries industry, and to those concerned with furthering culture techniques.

(Imai)

CHAPTER I

Biological Research on the Oyster

1. Introduction

More studies have been done on the oyster than on any other marine species. The literature now exceeds 1,000 published papers, which cover, among other features, the various aspects of biology, and the methods of oyster culture (Banghman, 1948; Ranson, 1952; Korringa, 1952).

The classification of oysters was initiated in the 18th century. The oyster known locally in Japan as *Magaki*, was the first to be given its scientific name, *viz.*, *Crassostrea gigas*, in 1793 by Thunberg. The biological and embryological aspects of oysters were studied as early as the 19th century. Regnault and Reiset (1849) measured the rate of respiration, while Bouchon-Brandely (1882a, b), Brooks (1879, 1880), and others, investigated the process of artificial fertilization and the development of oysters by this means.

Coming to the present century, Nelson (1918, 1921, 1926, 1927, 1928, 1936a, b, 1937, 1938) researched the mechanism of feeding and assimilation, and Yonge (1926) followed the same line of investigation. However, Nelson also studied the setting process, the ciliary movements, and the mechanism of spawning. Orton (1921, 1922, 1923, 1926, 1927, 1937) studied a variety of problems related to the characteristics of the European oyster including fertilization, liberation of larvae, changes in morphology resulting from the formation and growth of shell, and environmental conditions. He also studied various other aspects such as blood corpuscles and the causes of mass mortality in oysters. Amemiya (1926a, b, 1928a, b, 1929) studied hermaphroditism in oysters and the optimum conditions for their development. Hirase (1930), Wakiya (1929), and others, studied the classification of oysters and, in this connection, their conchological and malacological aspects. Investigators of the present era, such as Fujimori (1929), Tatsu (1933), Seki (1934), and others, have carried out extensive research on the classification of Japanese oysters such as *Magaki (C. gigas)*, *Suminoegaki (C. rivularis* Gould), *Iwagaki (C. nippona* Seki), etc.

During the period 1930 to 1940, Galtsoff (1930, 1931, 1932, 1934, 1935, 1938a, b, c, 1940) pursued both the physiological and chemical aspects of the mechanism inducing release of eggs and sperms. Cole (1931a, b, 1932a, b, c, 1934, 1936) investigated the phenomenon of maturation from the histological point of view. Cole *et al.* (1936, 1938a, b, 1939a, b) successfully reared larvae and studied their organization and setting. Imai *et al.* (1949, 1950) obtained larvae by artificial fertilization and reared them on a diet of cultured organisms. Thus successful breeding was achieved and a path opened for genetical study. Loosanoff *et al.* (1942, 1945, 1947, 1948, 1949, 1950, 1951) studied maturation, spawning, feeding, and the influence of environmental conditions on setting, using Virginia oysters as experimental specimens for this purpose, and thus simultaneously established the existence of physiologically different races of oysters. Ranson (1948) suggested the classifying of oysters on the basis of the shape of their shells, and succeeded in classifying a number of families whose position had been a subject of controversy for a long time.

The above covers only a part of the research done on oysters. It is fortunate that these past studies were well recorded. Takatsuki (1949) collected and published everything available on the subject up to 1946 in a work on oyster biology. Korringa (1952) collected 300 papers on oysters, both as a biological subject and as a marine product, written between 1940 and 1951, and published them in one volume. This was followed by the publications of Yonge (1960) and Galtsoff (1964) of general studies on the biological and fishery aspects of oysters.

Imai and Sakai (1961) carried out genetical research on *C. gigas* (*Magaki*) in Japan. The genetical characteristics of the regional strains of *C. gigas* (*C. gigas, C. laperousei, C. gigas* var. *sikamea*), were made clear by the authors. The classification of these had already been explained and finalized by earlier workers. The results not only paved the way for genetical research and the breeding of useful marine animals, but greatly helped the classification of oysters which, in the past, was mostly done on the basis of conchological and malacological features. Regarding subspecies, the presence of regional races, and the differentiation of characters, several papers have been written on ecology, physiology, antigenic properties, and chemical composition (Loosanoff, 1958; Numachi, 1959, 1962; Hillman, 1964; Li *et al.*, 1967). However, research has not yet fully established the genetical variations of individuals and, consequently, much work remains to be done in the areas of breeding and genetics.

Because large-scale hatcheries for shellfish were possible, the results of experiments to improve the breed and quality of oysters could be applied for commercial purposes (Loosanoff and Davis, 1963). Steady progress is being made in the study of the nutritional physiology of larvae, the nutritional value of feed organisms and their influence on larvae, and in constructing the culture equipment required for breeding, etc.

There are also a number of reports on the mass mortality of oysters, which often occurs at different regions (Andrews *et al.*, 1957; Haskin *et al.*, 1966; Hewatt *et al.*, 1956; Mackin, 1951, 1956; Mackin and Boswell, 1956; Menzel *et al.*, 1955; Burton, 1961; Takeuchi *et al.*, 1955, 1956, 1957; Ogasawara *et al.*, 1962; Kanno *et al.*, Imai *et al.*, Numachi *et al.*, Mori *et al.*, Tamate *et al.*, 1965; Imai *et al.*, 1968). By these studies, the pathological organisms or the pathology of fattening in oysters, could be more clearly understood and, at the same time, the normal metabolism and physiological functions of various tissues and cells of oysters, was elucidated. Regarding the organic pollutants present in the bays and the development of intensive culture methods, it is necessary to understand environmental conditions, including the condition of waste materials in the growing beds (Ito and Imai, 1955; Imai, 1960), as well as the management of the latter.

(Numachi)

2. Classification and distribution

The oyster belongs to Phylum Mollusca, Class Pelecypoda, Order Pseudolamelli-branchia, and Family Ostrea. Three genera and over 100 species have been identified to date. Some controversy over the classification of species still exists, so the present classification is not universal. However, the classification of oysters belonging to the Family Ostrea into three genera, *viz., Ostrea, Crassostrea,* and *Pycnodonta* by Ranson (1948a, b, c; 1950, 1960) and others, has been generally accepted (Gunter, 1950; Korringa, 1952). This classification is mainly based on the morphology and structure of the hinge plate of the larval shell, but it is also found to be consistent with the mode of reproduction, ecology, shells of adult animals, morphology of soft organs, the composition of antigen, etc.

The characteristics of each of the three genera are given below:

2.1 Classification

1. Genus *Ostrea* Linnaeus 1758

Mode of reproduction: hermaphrodite, larviparous, eggs (80 to 170 μ in diameter) and early larvae, fairly large in size. *Structure of hinge plate:* ligament (32 to 42 μ) present at the center of hinge plate (80 to 90 μ long). *Teeth:* present on both sides of the hinge plate, 2 each on the anterior and posterior parts of right valve and 3 (2) each on the anterior and posterior parts of left valve. (Ranson reported that both the right and left valves have 2 each but as seen in Figure 2.1 there is a third tooth. However, since the third tooth on the right valve is almost in a vestigial condition, it is shown as 2 on the right and 3 on the left valve. The teeth on the anterior side degenerate with the growth of larvae and, at the time of setting, they become vesti-gial.) *Ligament:* after setting, along with the formation of the calcareous shell, a new ligament forms inside the hinge plate (near the first anterior tooth) (Figure 2.2). *Shell:* almost circular, variations rare. The cavity of the lower valve (left valve) is shallow and does not extend to the lower part of the ligament. Prismatic layer deve-lops. Some species have minute teeth, soft parts, and no promyal chamber (space between the pallium and the epithelium covering the viscera such as the pericardium). *Adductor muscle* (white part of the smooth muscle): accounts for 50% of the muscles. *Habitat:* places ranging from the maximum tidal length, which become exposed during the spring, to as much as 15 m in depth, and water regions in far off seas with little turbidity.

2. Genus *Crassostrea* Sacoo, 1897

Ranson (1948c) classified these under the genus *Gryphaea,* but due to some diffi-culties in nomenclature, the term *Crassostrea* is now used (Gunter, 1950; Korringa, 1953; Ranson, 1960). *Mode of reproduction:* sexes separate (hermaphroditism rare), oviparous, eggs (40 to 60 μ) and early larvae small in size. *Structure of hinge plate:* compared to the genus *Ostrea,* hinge plate shorter (40 to 50 μ). There is a ligament at the center (15 to 25 μ in length) and also on either side (15 to 25 μ in length). *Teeth:* 2 in the right valve and 3 (2) in the left valve. Third tooth slightly larger on the left valve as in the case of *Ostrea* (Figure 2.3). *Ligament:* after setting, a strong

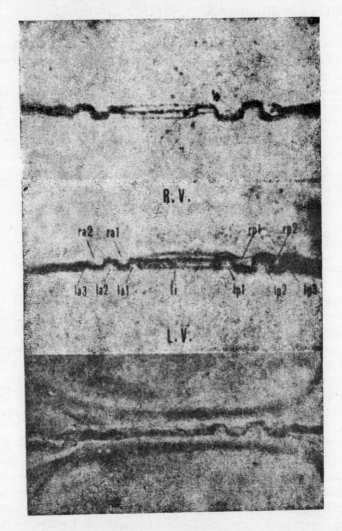

Figure 2.1. Genus *Ostrea*. Structure of hinge plate in larvae (Numachi, 1962).

Above—*Ostrea edulis* (shell height 189 μ); Middle—*O. lurida* (shell height 184 μ); Below—*O. edulis* (shell height 290 μ).

R. V.—right valve; L. V.—left valve; la 1, 2, 3—first, second, third anterior teeth on left valve; lp 1, 2, 3—first, second, third posterior teeth on left valve; ra 1, 2—first and second anterior teeth on right valve; rp 1, 2—first and second posterior teeth on right valve; li—ligament.

ligament forms at the anterior dorsal margin away from the hinge plate (Figure 2.2). *Shell:* usually elliptical or slightly elongated. Variations common: lower shell with deeper cavity, which extends up to the lower part of the ligament; some have small teeth. *Soft parts:* promyal chamber present, rectum does not pass through ventricle. *Adductor muscle:* comprises less than 30% of the meat. *Habitat:* usually found in bays and tidal belts; however, distribution also extends to open seas and coastal region, as this species can withstand a wide range of salinity and turbidity.

3. Genus *Pycnodonta* Fisher de Waldheim, 1807

Mode of reproduction: sexes separate, oviparous, eggs small. *Larval teeth*: on both the right and left valves 5 (6) teeth found on the surface of the hinge plate, and 10 at the anterior dorsal margin (Figure 2.4). *Shell*: lower valve comparatively deeper but the cavity below the ligament small; vacuole present at the inner part of shell; shells rough. *Soft parts*: promyal chamber present; rectum passes through ventricle. *Adductor muscle*: very small. *Habitat*: open seas; sometimes found at the bottom of the sea up to about 100 m (Thomson, 1954).

2.2 Species and their world distribution

Among the types of oysters distributed in different parts of the world, those given in Table 2.1 are very important because of their large distribution and commercial value. The table also shows the geographical distribution of these types.

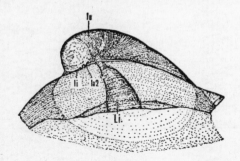

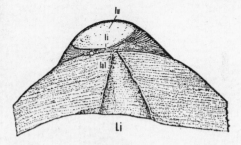

Figure 2.2. Ligament and its position after setting (Numachi, 1962).

Above—*C. angulata*; Below—*O. edulis*

In both cases the inner surface of the left valve is shown: Li—ligament; la 1—first anterior tooth on left valve; la 2—second tooth on left valve; li—ligament after setting; lu—umbo of left valve.

Table 2.1. Geographical distribution of oysters

Species	Distribution	
O. edulis (European flat oyster)	Europe (Mediterranean Sea, Scandinavian Peninsula, England)	Northern latitude 40-60°
O. lurida (Olympia oyster)	Pacific Coast of North America	Northern latitude 33-50°
C. angulata (Portuguese oyster)	Portugal, Spain, France	Northern latitude 37-45°
C. virginica (Virginia oyster)	Atlantic Coast of North America	Northern latitude 27-47°
C. gigas (*Magaki*)	Japan, Korea, China	Northern latitude 30-45°
C. commercialis (Sydney oyster)	Australia	Southern latitude 22-28°

1. *Ostrea edulis* Linne (European flat oyster, French oyster)

This species is distributed widely along the coastal regions of Europe off France, England, Denmark, and Norway. It is considered the type species of the genus *Ostrea*. The shell is circular and the cavity of the left valve shallow and almost flat. The shell length is sometimes about 10 cm (Figure 2.5). The larvae have a few

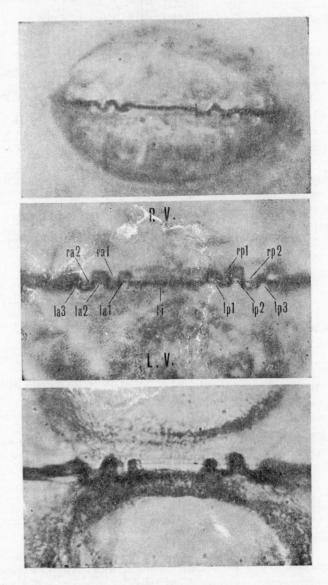

Figure 2.3. Genus *Crassostrea*. Structure of hinge plate in the larvae (Numachi).

Above—*C. angulata* (shell height 68 μ); Middle—*C. angulata* (shell height 110 μ);
Below—*C. gigas* (shell height 195 μ). Legend same as in Figure 2.1.

black spots and the shell and soft parts are pale yellow in color. This species is considered the most delicious of all oysters. Places like Arcachon near Bordeaux in France, Orei on the southern part of Gulitami, Essex Creek near Colchester in England, are well known for its cultivation. Recently stock fluctuations in various regions of Europe have posed a major problem for the oyster industries (Dalido, 1948; Spark, 1951).

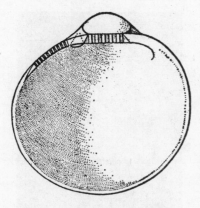

Figure 2.4. Shape of larval shell in
genus *Pycnodonta* (*P. hyotis* L.)
(Ranson, 1960).

Figure 2.5. *C. angulata*.

Imai *et al.* (1953) succeeded in large scale culture by transplanting oysters to Sanriku. From June to August (water temperature 16°C to 24°C) larvae of about 200 μ shell length were released in Sanriku. The larvae grew well at water temperatures of 20°C to 23°C and were able to settle down within 1 to 2 weeks.

This species is found below the ebb tide line, in cold places with high chlorine content, and in places in which clear sea water flows in from the ocean.

2. *Ostrea lurida* Carpenter (Olympia oyster)

This species is distributed along the Pacific coast of North America, and growing beds in cold places such as the Puget Sound, near Seattle, Washington, are well known. The shells are elliptical or oval in shape, but the lower shell has a greater convexity. The shell length is within 5 cm. The outer surface of the shell is black in color and the inner pale black to pale blue. The soft parts, particularly the mantle, are rich in black pigment. The animal has a peculiar sweet taste and is widely used in cocktails (Figure 2.6). Like *O. edulis*, larvae ranging up to about 175 μ in shell length are released in Sanriku from June. They are strong and can withstand high density of water.

3. *Crassostrea angulata* (Lamarck) (Portuguese oyster)

This species is mainly distributed in Portugal, England, and France. It is the type species of the genus *Crassostrea* and its shell gradually widens toward the posterior end. The cavity of the lower shell is deep. As seen in Figure 2.7, the color of the shell is like that of *C. gigas*. It is very difficult to differentiate the two types on the basis of their external features. *C. angulata* crosses well with *C. gigas* and both types show considerable resemblance in their characteristics even when crossed with a different species (Table 2.2). Ranson (1948a) asserts that the Portuguese oyster is a later generation of the *C. gigas* transplanted from Japan over 1,000 years ago. According to him both are one and the same species. Korringa does not agree with this view, reporting that the antigenic structure in the blood differs in the two species and, further-

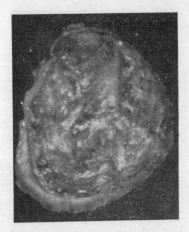

Figure 2.6. *O. lurida.*

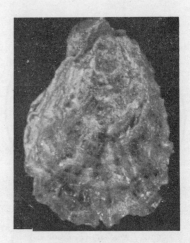

Figure 2.7. *O. edulis.*

more, the two differ even from the point of view of geographical segregation (Numachi, 1959, 1961). The water temperature suitable for early development and larval growth is slightly lower than that of the *C. gigas* of Hokkaido, and it is said to be a cold-water variety. In Europe, although the species is well known for its enormous vitality, as a food it is considered inferior to the European oyster.

Table 2.2. Degree of gamete segregation among the four species of
Crassostrea **(Numachi, 1962)**

Sperms	Eggs							
	C. gigas					Portuguese oyster	*C. rivularis*	Virginia oyster
	Hokkaido	Miyagi	Hiroshima	Ariake-Kai Type A	Type B			
C. gigas								
Hokkaido	—	—	—	—	—	—		
Miyagi	—	—	—	—	—	—	‡‡	+
Hiroshima	—	—	—	—	—	—		
Ariake-Kai								
Type A	‡‡	‡‡	‡‡	—	‡‡	‡‡	‡‡	
Type B	—	—	—	+	—	—	‡‡	
Protuguese oyster	—	—	—	—	—	—	‡‡	+
C. rivularis	‡‡	‡‡	‡‡	‡‡	‡‡	‡‡	—	—
Virginia oyster	+	+				+	—	—

— Rate of fertilization above 90%, no gamete segregation;
+ Rate of fertilization in concentrated sperm fluid (10⁻³ to 10⁻⁴ of dry sperm in sea water), 70 to 90%;
‡‡ Fertilization nil, complete segregation.

4. *Crassostrea virginica* (Gmelin) (American oyster)

This species is widely distributed along the Atlantic coast of North America from Prince Edward Island, near the southern part of Canada, to Texas, and is the most commercially important variety in the country. The temperature suitable for spawning differs according to regional distribution (Loosanoff and Nomejko, 1951; Stauber, 1950). Moreover, the oysters distributed near the estuaries in the southern regions grow and reproduce rapidly when a large amount of silt is present in the water. However, the mixing of silt does not seem to be effective in the northern regions (Loosanoff, 1958). It is said that the same species include strains which can be differentiated on the basis of physiological and ecological features. The species was transplanted from Prince Edward Island and other places, into the bay of Mangoku-Ura, Matsushima Bay, and Sanriku, but the growth, maturation, and meat content were not good.

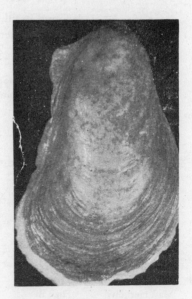

Figure 2.8. *C. virginica*.

The shells resemble those of *C. rivularis* of Japan. The cavity of the lower shell is small and the shell is hard with a smooth upper surface. The soft parts are pale yellow in color (Figure 2.8).

5. *Crassostrea commercialis* (Iredale and Roughley) (Sydney oyster, Sydney rock oyster)

This species is widely distributed from Vingan Bay of Victoria State on the eastern coast of Australia up to the shoal water bay of Queensland. It is cultivated in Queensland and New South Wales. While it is plentiful in the inner bay, it is also found attached to rocks in the open sea from the tidal belt to deep places (Thomson, 1954).

6. *Crassostrea cucullata* Born (Bombay oyster)

This species is found in large numbers near Bombay, India. The shell is thick and the cavity of the lower shell is deep. The shape of the shell varies.

Reports have also placed this species in the warm waters near Japan; however, Kuroda (1931) identified the Japanese type as *C. mordax* Gould.

2.3 Japanese species, breed and distribution

1. *Crassostrea gigas* (Thunberg) (Japanese oyster, Pacific oyster, Magaki)
   Ostrea gigas Thunberg, 1793; Lischke, 1871; Takatsuki, 1927; Amemiya, 1928; Hirase, 1930; Gunter, 1950.
   Ostrea laperousei Schrenk, 1867; Wakiya, 1929; Hirase, 1930.
   Ostrea talienwhanensis Crosse, 1862.
   Ostrea gigas var. *sikamea* Amemiya, 1928.
   This is a representative species of Japan both from an industrial and a distributional

point of view. It is found in all areas, from Hokkaido to Kyushu in Japan, and from China to Korea, and is considered the most important species in the Japanese oyster industry. As shown in Table 2.3, the size gradually increases as the distribution

Table 2.3. Comparison of morphological characteristics of *C. gigas*
of various regions in Japan (Sakai, 1965)

Morphological characteristics	Breed			
	Hokkaido	Miyagi	Hiroshima	Kumamoto
Growth	Very rapid	Rapid	Slow	Very slow
Size	Maximum	Large	Small	Minimum
Depth of shell	Shallow	Between that of Hokkaido and Hiroshima variety	Deep	Deep
Ratio of meat weight to total weight	Minimum	Low	Maximum	High
Smoothness of shell	Flat	Slightly wavy	Strikingly wavy	Strikingly wavy
Color of shell	Pale gray	Between that of Hokkaido and Hiroshima variety	Dark violet	Dark violet or brown
Mass mortality	High in southern part	High in southern part	High in northern part	Low in all regions
Period of spawning	Early	Later than Hokkaido variety	Later than Miyagi variety	Very early (eggs mature in winter)

spreads to a higher latitude. Shells found in Hokkaido are white in color and attain 40 cm in length. On the other hand, the species in a lower latitude is small in size and black in color. For example, in Kumamoto the shell length is within 5 to 8 cm. Variations in the morphology of shells are many, depending upon the environment. In the past, the *Magaki* of Tohoku and Hokkaido was called *C. gigas*, the *Magaki* of the region south of Kanto (Hiroshima) was called *C. laperousi*, and that of Ariake-Kai, *C. laperousei* var. *sikamea* (Figure 2.9). Thus each was assumed to be a different species or strain.

Imai and Sakai (1961) carried out breeding experiments on *C. gigas* by interchanging their regions. As shown in Table 2.3, there were genetical (hereditary) variations in growth, morphological characteristics of shell, weight of meat, mass mortality and maturation in cold versus warm water regions, period of egg and sperm release, etc. The differences clearly showed that these oysters could be distinctly divided into cold and warm varieties, the latter becoming more pronounced as the latitude changed. Nevertheless, these oysters were successfully crossed and produced offspring. Morphological, physiological, and ecological characteristics, as well as the composition of antigen in the serum, showed considerable difference according to the geographical distribution. When a comparison was made between the types found in rather distant regions, the morphological characteristics showed significant differences, but differentiating oysters of adjacent regions became difficult because the various characteristics are quite similar. Hence groups which are geographically nearer to one another are presumably interconnected and form a genetical continuity (Imai and Sakai, 1959; Numachi, 1962, 1965).

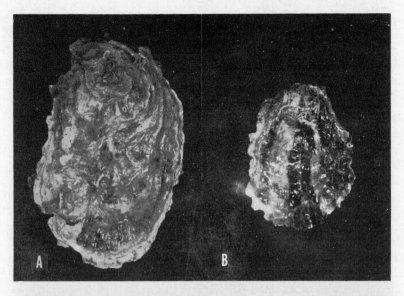

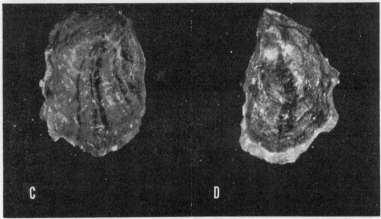

Figure 2.9. *C. gigas* found throughout Japan.

A—*C. gigas* found in Hokkaido; B—*C. gigas* found in Miyagi; C—*C. gigas* found in Hiroshima; D—*C. gigas* found in Kumamoto.

As shown in Table 2.2, type A oysters of Ariake-Kai did not cross with *C. gigas* of different regions (Numachi, 1958). The two types, A and B, form colonies alternately, and are found in the tidal belt, or they mix with other types grown in Ariake-Kai. It is not possible to differentiate the two types on the basis of their shell characteristics, but it is possible to differentiate them on the basis of egg size—those of type A are 44 to 48 μ in diameter, while those of type B are 53 to 56 μ in size.

2. *C. rivularis* (Gould) (Suminoegaki)

Ostrea rivularis Gould, 1861; Lischke, 1869; Wakiya, 1915; Amemiya, 1928.

Ostrea ariakensis Fujita, 1913; Wakiya, 1929; Lischke, 1871.

This species is distributed in Ariake-Kai. The shell is round or elliptical. The part near the hinge plate in the upper shell, is violet-brown in color. The lower shell is shallow and the umbo cavity below the hinge plate is very small. It differs from *C. gigas* in that a part of the rectum and anus are away from the soft parts. It is difficult to cross this species with *C. gigas* (Miyazaki, 1939; *vide* Table 2.2). Moreover, it shows peculiar features different from *C. gigas* when crossed with the Portuguese oyster and the Virginia oyster (Table 2.2). The antigen composition and the rate of embryonic development also differ from those of *C. gigas* (Numachi, 1959, 1962).

3. *C. nippona* Seki (Iwagaki)
   O. nippona Seki, 1934.
   O. multistriata Wakiya, 1929; *O. circumpicta* Hirase, 1930.

This species is found in all the regions of Japan. The shell is dark brown or violet in color. It is either elongated or oval in shape and measures 25 cm in length. In old specimens, the shell is thick and heavy and looks like a shoe horn. The species is widely used in biological research, particularly in the study of ciliary movement. Since *O. circumpicta* (*Kokegoromo*) was labeled as *Iwagaki*, it was often referred to as *O. circumpicta*. Furthermore, it was also assumed that *O. circumpicta* was an early stage of this species because of its small size (below 10 cm in shell length). However, *O. circumpicta* is a larval form belonging to the genus *Ostrea*, and the presence of a large number of teeth on both sides of the hinge plate clearly differentiates it from *C. nippona*.

4. *Crassostrea echinata* (Quoy and Gaimard) (Kegaki)
   O. echinata Quoy and Gaimard, 1835.
   O. spinosa Deshayes, 1836.

This species is a small variety usually found in the coastal regions where salinity is high. The shells show many variations. Slender, cylindrical spines radiate at more or less right angles from the surface of the upper (right) shell, which, however, are not seen in old specimens, and are sometimes absent in the young. Distribution is widespread, from various regions in Japan to regions between Indonesia and Australia. *O. spinosa* is said to be an earlier form of *O. echinata*.

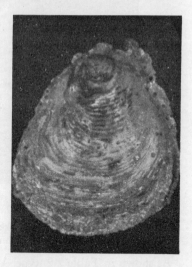

5. *O. denseramellosa* Lischke (Itabogaki)
   The shell is more or less circular in shape and flat. It resembles the European oyster (*O. edulis*).

Figure 2.10. *O. denseramellosa*.

A number of radiating strands are found on the surface of the lower shell. Compared to the European oyster, the color of the shell is darker (Figure 2.10). It is found widely distributed in the Pacific Ocean from Hakodate to Kyushu and in the coastal region of the Sea of Japan. However, the main habitat of this species is in the Setonai-Kai of Japan, and it remains in shallow water that is not exposed during ebb tide. At times it is found up to a depth of 40 cm. Cultivation of this oyster is possible, but since the yield is poor, the species is not important to commercial farming.   (Numachi)

3. Breeding and improvement of breed

Recently the phrase "toru gyogyo kara tsukuru gyogyo e" which means "from catching to cultivating" has become very popular in the fishery industries of Japan. Even in the field of marine biology, there are great expectations for improving the breed (Loosanoff, 1953, 1956; Korringa, 1952; Tamura, 1957). Steps have already been taken in this direction by researchers in agriculture and poultry, with fisheries a relative newcomer to the field.

The oyster industry was formed on the basis of culture which developed through studies on the improvement of rearing equipment, the structure of culture beds, the location of equipment, and the utility of growing beds. The industry is now able to control all phases of culture from seeding to harvesting, and even to control certain harmful organisms. However, the oyster industry differs from agriculture and poultry in that control over the selection of seeds has not yet been attained. Since seedlings are allowed to grow in the fishery beds, it cannot be said that the group grown are selected, and as yet there is no genetic control of the seeds either.

3.1 Present status of research: Hereditary differentiation in the breed of various regions and the characteristics of a hybridized breed

Nelson (1948b) pointed out the importance of genetical studies in the breeding of oysters, but this aspect was neglected by investigators until Imai and Sakai (1961) published their report. The collected *C. gigas* from various parts of Japan, and carried out artificial fertilization and rearing experiments for three generations, to study the purity and hybridization of breeds. After releasing the seeds in the rearing beds of various regions, they studied the genetical aspects. The results are shown in Figure 2.11 and Table 2.3, revealing that the growth, the morphological features of the shell, mass mortality, the spawning period, the glycogen content, and taste differ genetically.

The morphological features of shell, growth, glycogen content, and other features of F_1 obtained by crossing the regional strains, are given in Figures 2.12, 2.13, and 2.14, and Table 2.4. The intermingling of parental features is evident in every instance. The most striking feature is that the mortality of F_1 is lower than for its parents (Tables 2.5 and 2.6), which seems to show heterosis. Moreover, the changes in the morphological features of the shell and other characteristics of F_2 are more or less the same as those of F_1; F_1 also shows an intermingling of parental characteristics. It was interesting to note that the characters did not segregate.

Imai and his colleagues carried out an interspecies crossing, and were able to obtain the first crossed generation between a Portuguese oyster and a *C. gigas* (*Magaki*) from different regions. They were also able to rear the first generation up to maturation. There are several reports with regard to crossing different species (Seno and Hori, 1929; Galtsoff and Smith, 1932; Davis, 1950). However, all these investigations are confined to the study of the fertilization rate; no mention is made of rearing up to maturation. Hence Imai and his colleagues were the first to obtain a successful generation. In addition to an intermingling of the characters of their parents, F_1 of the cross between the Portuguese oyster and the *Magaki* also showed heterosis. Both

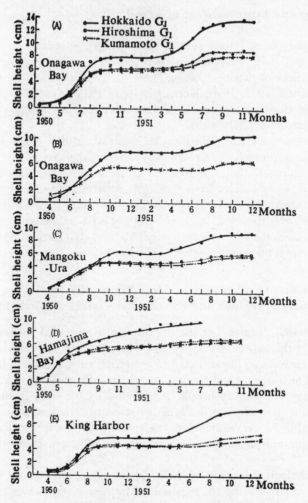

Figure 2.11. Composition of shell length in the first generation of
C. gigas grown in different regions (Imai and Sakai, 1961).

the males and females of the crossbreed matured, producing sperms and ova capable of normal fertilization. However, the second generation died out during the larval stage and could not be utilized as seedlings. However, the death of F_2 larvae, also known as the degeneration of speciation, is not yet confirmed, and investigations are presently underway to determine whether there were problems during rearing.

3.2 Commercial utilization of current information and future direction

In order to expand the fishing beds by effectively using the productivity in those places in which the best species available from the crossing experiments could be introduced, certain measures shown in Figure 2.15 were considered. Imai and his

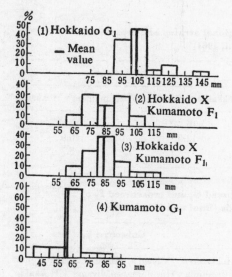

Figure 2.12. Shell length and distribution of the Hokkaido strain, Kumamoto strain, and the crossbreed of the two. Data collected at Onagawa Bay (Imai and Sakai, 1961).

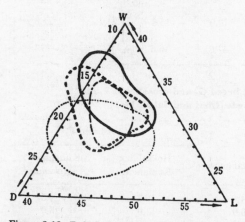

Figure 2.13. Relationship of shell length (L), shell width (W), and shell height (D) in Hokkaido strain (——), Kumamoto strain (....), the crossbreed of the two (Hokkaido and Kumamoto) (-----), and Kumamoto × Hokkaido (—.—.—.—) Onagawa Bay (Imai and Sakai, 1961).

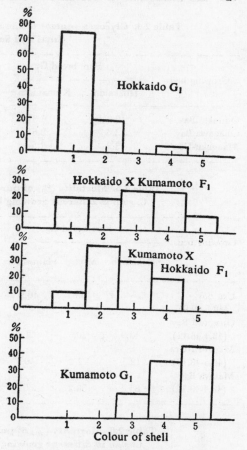

Figure 2.14. Color of the shells of the Hokkaido strain, the Kumamoto strain, and the crossbreed of the two (Imai and Sakai, 1961).

1—almost white; 2—partly black; 3—about half the body is black; 4—most of the parts are black; 5—entire body is black.

colleagues carried out genetical investigations of the various characteristics in stages *b* and *d* (Figure 2.15). They were able to determine the special characteristics of the species as well as the breed. However, there were many technical problems involved in obtaining the best breed from crossbreeding. Moreover, fundamental studies relating to a breed which could be introduced into the industries, are also necessary. In this section, the type and the transplantation of the type are discussed, and in considering the genetical characteristics of the wild variety, problems related to the improvement of a breed are discussed.

Table 2.4. Glycogen content of the regional strains and crossbreeds
(Imai and Sakai, 1961)

Growing beds	Pure breed G_1		Crossbreed F_1	
	Hokkaido	Kumamoto	Hokkaido × Kumamoto	Kumamoto × Hokkaido
Ominato Bay	16.3%	6.3%	10.7%	13.2%
Onagawa Bay	13.6	8.6	12.5	10.9
Mangoku-Ura	3.2	2.5	—	3.7

Table 2.5. Mortality (%) of pure breed G_3 and crossbreed F_1 of
C. gigas **in different growing beds (Imai and Sakai, 1961)**

Growing beds	Pure breed G_3			Crossbreed F_1			
	Hokkaido	Miyagi	Hiroshima	Hokkaido × Hiroshima	Miyagi × Hiroshima	Hiroshima × Miyagi	Hiroshima × Hokkaido
Usu Bay (48.5-49.2)	—	19.6	16.9	8.5	6.1	23.6	—
Onagawa Bay (48.4-48.12)	13.9	20.0	36.7	17.8	16.3	50.3	6.6
Mangoku-Ura (48.5-48.12)	12.5	22.4	40.7	10.0	15.3	24.1	21.8
Matoya Bay (48.4-48.12)	85.2	88.2	50.0	51.8	46.6	—	46.7

Table 2.6. Mortality (%) of pure breed G_3 and crossbreed F_1 of
C. gigas **in different growing beds (Imai and Sakai, 1961)**

Growing beds	Pure breed G_3		Crossbreed F_1	
	Hokkaido	Kumamoto	Hokkaido × Kumamoto	Kumamoto × Hokkaido
Ominato Bay	23.7	24.6	19.1	24.5
Onagawa Bay	54.7	38.0	42.0	26.0
Mangoku-Ura	48.8	40.8	—	31.8
Hamajima	100.0	65.2	95.4	94.3
Kagami Machi	98.0	71.0	95.3	98.0
Gig Harbor	38.4	44.8	17.2	38.0

3.3 Utilization of existing breeds

The classification of regional breeds of *Magaki* according to their genetical features, and the differences in their growth, meat content, mortality, etc. in various growing beds, have already been covered. The differences seen in the types available at various regions have been reported by Seki (1939), Mawara (1936), and investigators in North America. In North America, the Kumamoto strain of *Magaki*, which was

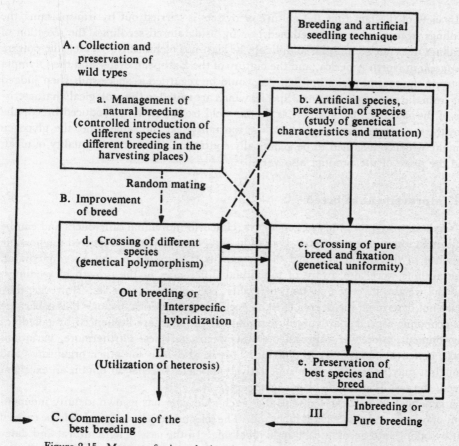

Figure 2.15. Measures for introducing the best breed for commercial purposes.

experimentally transplanted there, is said to have produced ova and sperms in winter. As already mentioned, this is one of the genetical characteristics of this breed. The fact that Kumamoto *Magaki* shows low mortality even in the cold regions of North America or Tohoku, and also the fact that it resembles the Olympia oyster in external features such as shell height and black color, was of great advantage in utilizing this variety in North America. However, the fact that of all the *Magaki* strains, the Kumamoto one has the lowest glycogen content in winter is a disadvantage. In a recent investigation, it was found that the Hiroshima *Magaki* has a lower degree of mortality than the Miyagi *Magaki* of Matsushima Bay (Imai *et al.*, 1965).

Among oysters, as already noted in *C. gigas* and the Virginia oyster, there are regional strains of the same species which can be differentiated on the basis of physiological and ecological characteristics, as well as economic importance. Hence, before attempting to improve the breed, the genetical characteristics of the existing wild variety should be carefully studied (Figure 2.15, I). At the same time, even in the case of transplanting the so-called best varieties such as the European oyster and the Olympia oyster, the presence of regional strains and their special features should be considered first.

In view of the fact that the culture of oysters is carried out by transplanting the seedlings from neighboring seed beds, or by using stored seedlings, the selection of seedlings does not seem to be so difficult, at least not technically. In fact, the culture of *Magaki* in North America, and the culture of the European oyster and the Olympia oyster in the coastal regions of Sanriku, could be regarded as successful when judged just from the rate of production. Species which are suited to the ecological characteristics of the growing places differ, as do the yield from a newly introduced breed, the growth, the increase in weight, and the mortality. At the same time, the glycogen content which is assumed to be genetically controlled, the taste, the quality of meat, and the price of the product also vary.

3.4 Improvement of breed

Needless to say, the study of genetics is related to individual differences and can be utilized in improving the breed. Loosanoff and Nomejko (1949) reported the possible presence of individual genetical differences in the growth of the Virginia oyster. The authors were also aware of individual differences in the growth of seedlings, and the weight, glycogen content, mortality, etc., of various species. However, they could not determine the degree of influence from the genetical factor. This is because the economic aspects, particularly the quantity of meat, are influenced by polydine; environment, moreover, also affects other characteristics. Furthermore, variations make genetical analysis very difficult. This is also true for other organisms, and though it may be possible for a superlatively efficient breeder to obtain an excellent breed, the task is still difficult for geneticists.

A genetical study clearly reveals that each character has its own identity independent of others, and that each is inherited. The selection of characters should be carried out by taking into account each individual one. In the case of agricultural and dairy products, it has been possible to accelerate the growth rate, or to fatten the body, or to make the organism more resistant to diseases and adverse environmental conditions. Imai *et al.* first considered the possibility of inducing spawning in water basins with very low temperatures in northern regions of the country, and thereby obtaining seedlings. They tried to produce a variety which would develop meat from autumn onwards. The authors are of the view that a model for the breeding should be prepared and, on the basis of this model, the necessary characters for breeding should be studied.

Fortunately, there are a number of strains and breeds of oysters which differ in genetical features. The authors selected such breeds as would suit model breeding, and from these some selected individuals were even subjected to pure breeding. In this way a number of strains were prepared (Figure 2.15 C). Hence, by studying the hereditary and independent characters and their continuity, it may be possible to obtain a group of very special and uniform characters. Crossing different strains, different breeds, and different species, will supposedly result in a breed or strain which combines the most desirable characteristics of the animals, and also utilizes the phenomenon of heterosis. For example, crossing the strong Kumamoto *Magaki*, the meat-rich Hiroshima *Magaki*, and the rapid-growing Hokkaido *Magaki* or Miyagi *Magaki*, should yield a strain containing all these desirable characters. Consequently, the pure breed could be further improved and the characters which are genetically controlled, better understood and better utilized.

In any case, in order to improve the breed, the crossing of species, strains, and individuals should be carried out over a long period and the resulting strains preserved (Figure 2.15 a, b, c). Again, in order to study the growth, spawning, meat, mortality, etc., the physiological continuity, and the breed, strain, and individual differences, it is necessary to carry out a large variety of experiments in the fishing grounds. Other important aspects are presently under study, namely, the preservation of old regional breeds, seedlings which are best for commercial purposes, and the incidental technical improvements necessary (Sakai, 1965; Numachi, 1965).

This type of hybridization to improve the breed involves large-scale experiments and expenditure, the results of which can only be assessed after a long and indefinite period of research. However, on reviewing the progress made so far in hybridization, it would appear that a number of new varieties could be obtained without much difficulty in the near future (Tanaka, 1948). Hybridization has resulted in the presence of excellent genetical features, which could be profitably used in commercial fields. Judging from the research data on oysters and the progress made so far in the rearing of them, it can be said that of all the marine animals the oyster is the first on which improvement of breed by hybridization has been attempted.

(Numachi)

4. Ecology

4.1 Geographical distribution

Oysters are widely distributed throughout the world except in the two polar regions (see Table 2.1). The low temperature of the polar seas is reportedly one of the major factors contributing to the absence of oysters in these areas (Takatsuki, 1949).

The migration of free-swimming oyster larvae occurs during a short period of 1 to 4 weeks, i.e. from the liberation of the eggs up to the setting of the larvae. Koganezawa *et al.* (1964) found that *Magaki* larvae, hatching in Matsushima Bay, are carried out to the open seas and settle at Ojika Hanto, which is 40 to 50 km away from the original place. Those species which are widely distributed in the bay, such as *Magaki*, most probably have a genetic continuity that is maintained in the open seas. Ranson (1951) reported that *C. margaritacea* of Madagascar settle on the coast of South Africa, which is 2,000 to 3,000 km away from their origin. There is some controversy over this report however (Korringa, 1952).

Among the species used for commercial purposes, there are some, for example, the *Magaki* of Japan, whose distribution has been artificially expanded by the transplantation of a large quantity to the Pacific coast of North America. In some species, the density and regions of distribution show a change of pattern throughout the year. The European oyster, which was found on a large scale in England, France, Holland, Norway, etc., has recently decreased, and is now virtually disappearing from almost all the fishing grounds (Millar, 1951; Hagmeier, 1941; Caspers, 1950; Dannevig, 1951). The reason for this decrease is not clear, although some investigators have suggested indiscriminate fishing (Dalido, 1948), low temperatures during ova and sperm release (Spark, 1950b, 1951), infection by pathological organisms (Cerruti, 1941), and retarding genetical factors (Gross and Smyth, 1946, which cause damage to the crop. On the other hand, Portuguese oysters have rapidly advanced to the northern regions, and efforts are being made to bring the European oyster to Arcachon (Lambert, 1946). It has been reported that the northern limit of the distribution of the Virginia oyster is now lower (Nelson, 1962).

4.2 Ecological distribution

There are various types of oysters. Some live in the coastal regions where salinity is high, and some live in bays where the flow of fresh water is regular. Water temperature and chlorine concentration are considered the two most important controlling factors in the horizontal distribution of oysters. Figure 2.16 shows the chlorine concentrations of the regions where various types of oysters live.

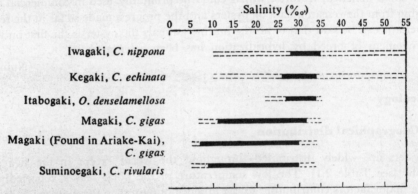

Figure 2.16. Salinity in areas in which Japanese oysters live (Amemiya, 1928).

(The original names of *O. circumpicta*, *O. spinosa*, and *O. sikamea* were changed to *C. nippona*, *C. echinata*, and *C. gigas*, which are popularly known as Ariake *Magaki* in Japan).

As mentioned before, the genus *Ostrea* is found in open sea basins while *Crassostrea* is found in places which vary greatly in salinity.

Amemiya explained the vertical distribution of the Japanese oyster by Figure 2.17.

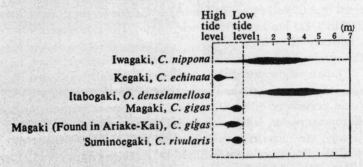

Figure 2.17. Vertical distribution of oyster species (Amemiya, 1928).

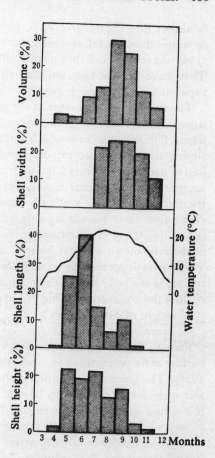

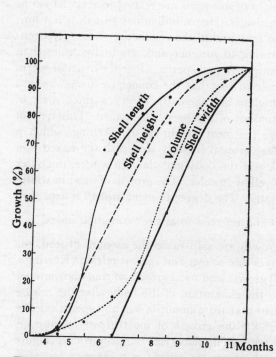

Figure 2.18. Virginia oyster: shell length, shell height, shell width, and formation of shell cavity. Percentage based on 100% as the total formation for one year (Loosanoff and Nomejko, 1949).

Figure 2.19. Virginia oyster: Growth of various parts of the shell versus water temperature (Loosanoff and Nomejko, 1949).

Usually the genera *Ostrea* and *Pycnodonta* are distributed in places which are not exposed, while *Crassostrea* is found in tidal belts.

4.3 Growth and meat

Oyster growth is usually explained in terms of increase in the shell size. Like other shells, the oyster's is mostly composed of calcium carbonate. It is hard and white conchiolin forms on some parts. Oysters directly absorb Ca-ions (0.4 g/1) from the sea water and form their shells with the help of a secretion from the slime glands present on the protuberance of the epithelium of the mantle. Hence the shell forms even when the animal is not fed. However, when the Ca concentration is less than 20% in the sea water, no shell forms (Belvelander and Benzer, 1948).

Shell growth is dependent upon water temperature. Growth is usually rapid between spring and summer when the temperature is high. In winter, the growth is almost nil (Figure 2.11 a, b, c, e). However, in places where the temperature in winter does not fall severely, there may be some growth. Loosanoff and Nomejko (1949) have reported that growth differs in various parts of the shell (Figure 2.18). They have further reported that the increase in the inner volume of the shell is rapid and that growth is closely related to water temperature (Figure 2.19).

The increase and decrease in the weight of soft parts are related to seasonal variations in the gonads or glycogen-storing tissues. Hence individual growth is not uniform. The follicular cells which remain scattered in the connective tissue (glycogen-storing tissue) increase rapidly from spring to summer and, gradually, more than half of the body becomes filled with mature ova and sperm. Once the ova and sperm are released, the gonads rapidly decrease in size, but the connective tissues around the follicular region, gradually accumulate glycogen. By winter a thick mass of connective tissue containing a large amount of glycogen has formed. This type of connective tissue development increases the meat weight or the swelling, which is commonly known as meat growth. Meat growth is easily seen with the naked eye. When growth occurs, the meat is thick and the body globular. In winter, the outer surface of the body is smooth and light yellow in color. The growth of meat in winter decides the commercial value of the oyster. The degree of meat growth is expressed as $\left(\dfrac{\text{weight of dry soft part}}{\text{inner volume of shell}} \right) \times 1{,}000$ and referred to as the "condition index."

Factors which influence the meat growth are said to be the amount of feed, the feeding activity, maturation, and changes due to ova and sperms release (Korringa, 1951). The temporary cessation of shell growth and meat growth during mid-summer and early autumn is said to be due to the exhaustion of the body after the release of ova and sperm. It has been confirmed that the amount of ova and sperm release, as well as the period of such release, affects the growth of meat greatly. However, this type of temporary growth arrest is not seen in places rich in salt-nutrients (Figure 2.20: Matsushima Bay culture). In such places the weight of meat even increases after maturation and the release of sperm and ova. It has been reported that the growth of shell and the growth of meat also, continue in places enriched with fertilizers (Lambert, 1950).

In places where the density of culture is high and the salt-nutrients poor, the shell growth and the meat growth after fertilization decline, due to a shortage of feed. Ito (1962) studied the problem of meat growth in fishing grounds in relation to the amount of feed available. The results showed that the rate of production of meat rapidly decreased in fishing grounds as the density of oyster population increased, and the flow of sea water slowed down. In Mangoku-Ura (Figure 2.20) where the flow of land water as well as open sea water is poor and the population high, not only is the shell growth poor, but there is hardly any growth of meat. The oyster found in this place contains a large amount of water in all four seasons, and the body is almost transparent; consequently, these animals are popularly known in Japan as "water oysters."

4.4 Maturation and fertilization

Maturation is controlled by water temperature. Even in winter it is possible to

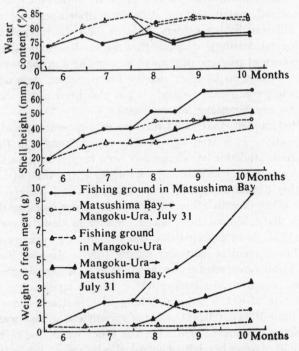

Figure 2.20. Growth of oysters in two fishing grounds of different nutritional salt content (Imai *et al.*, 1965). Matsushima Bay: rich in salt-nutrients. Mangoku-Ura: poor in salt-nutrients, spawning occurs in late August.

induce maturation by artificial temperature treatment. Loosanoff and his colleagues (Loosanoff and Engle, 1942; Stauber, 1950; Loosanoff and Nomejko, 1951; Loosanoff, 1956) carefully studied this problem, using the Virginia oyster. In Table 2.7 the number of days required for producing ova and sperms capable of fertilization, is shown for half the number of oysters reared at various temperatures in winter. Figures in parentheses indicate the number of days required for the release of ova and sperm. The results clearly show that in the maturation and release of sperm and ova, it is not the temperature at the time of fertilization, as believed earlier, but the tempera-

Table 2.7. Temperature and number of days required for maturation and production of ova and sperms in the Virginia oyster (Loosanoff, 1956)

Name of place	Rearing water temperature °C			
	12	21	24	27
Long Island Sound	68	15 (18)	5	5.5 days
New Georgia	—	55 (78)	32	22.5

( )=number of days required for release of ova and sperms; — =did not mature.

ture up to the stage of fertilization, which plays an important role. An interesting fact was simultaneously noted, i.e. the difference in temperature required by the same species in different fishing centers. Stauber calls these types, classified on the basis of physiological features, physiological races. Korringa (1952, 1957) also reported the presence of physiological races among the European oyster.

Fertilization in oysters, as in other marine invertebrates, does not always occur when the gonads are completely formed; it has also been noted that other factors besides maturation are necessary for fertilization.

Galtsoff reported in 1940 that rapid increase in temperature had a negative influence on fertilization. It is well known that rapid changes in the environment stimulate fertilization, and this knowledge has been helpful in inducing fertilization. That sperms and ova are released in large numbers at the time of high tide during full moon is another well-known fact (Orton, 1926; Korringa, 1947). But the release of sperm and ova is not directly related to the phase of the moon or the level of tide, but rather to the water temperature, water pressure or density, etc., at this time. Awati and Rai (1931) have reported that ova liberation in mature *O. cucullata* wa checked by the low current of sea water during July to August. Rapid changes in salinity seem to induce ova release just as changes in water temperature, etc. do. Other hypotheses regarding the factors which induce fertilization have been formulated. The release of ova is induced (along with a steady movement of shells) when spermatic fluid (0.03 to 0.02 ml of 1 g of sperms/100 ml sea water) is added in 30 liters of sea water (Galtsoff, 1938). On the other hand, the release of sperm is induced by ova fluid of other bivalves of chemical substances. While there are specific differences in spermatic fluid, no specific difference is evident in ova fluid. Regarding the difference in the interaction of sperm and ova, Takatsuki (1951) made a detailed study of the latter, in relation to the difference observed in shell-opening activity during fertilization.

Exactly what physiological changes are caused by physical or chemical stimuli is not yet clear. At present, this problem is being studied on marine invertebrates in relation to neural secretion (Kanatani, 1964; Chaet and McConnaughy, 1959, 1964).

4.5 Fertilization, development, and planktonic larval stage

Water temperature and salinity greatly influence fertilization, development, and larval growth. Fertilization occurs over a wide range of temperatures (Table 2.8). But water temperature suited to cell division and normal development up to the larval stage of type D is in a narrower range than the temperature range for fertilization (Table 2.8). At the lower limit of water temperature suited to the development of type D larvae, it is difficult for the larvae to attain the stage of setting. At 21°C, the Portuguese oyster and the Hokkaido *Magaki* grow and reach the stage of setting, but the umbo of Hiroshima and Kumamoto *Magaki* do not grow well, and find it difficult to attain a shell height beyond 200 μ. At low water temperature, not only is growth delayed, but also the setting of larvae because of an insufficient protrusion of umbo. Hence at the time of setting, the larvae are much larger than those grown at a higher temperature. In view of this, the planktonic period becomes

Table 2.8. Temperature range for the development of *Magaki* **and protuguese oysters time (y) required for the development at** $x°C$ **(Numachi, 1962)**

Types	Temperature range at which 80-90% fertiliza- tion rate is obtained	Temperature range at which more than 80-90% eggs deve- lop into D-type larvae	Time required for development (embryonic period)
Portuguese	14-29.5°C	18-23.5°C	$y=65.3/(x-8.9)$
Magaki			
Hokkaido	15-30	19-26	—
Miyagi	15-30.5	21-26	$y=63.7/(x-9.1)$
Hiroshima	15-30.5	21-26	—
Ariake-Kai	15.5-30.5	21-27	$y=62.9/(x-9.2)$
Suminoegaki			$y=67.1/(x-9.6)$

longer at a low water temperature. Table 2.9 shows the relationship between the length of the planktonic larval period and water temperature in various types of oysters. Davis has reported that Olympia oysters do not show changes in the body, even after 30 days, when the water temperature is 18°C, and growth did not continue beyond 220 μ of shell height when the temperature was 16°C. Experimental rearing of the European oyster, popularly known as the cold variety, showed that both the

Table 2.9. Planktonic larval period and water temperature

Types	Water temperature			Investigator	
	18°C	20°C	21°C		
Virginia oyster (found in Canada)	30	26	24 days	Needler	(1940)
European oyster	12	10	7	Korringa	(1941)
Olympia oyster	>30	16	—	Davis	(1949)

growth of larvae and its settlement followed a regular course when the water tempera- ture was above 20°C, but setting was delayed and the number of settled oysters was less when the water temperature was below 18°C.

The influence of water temperature is clearly seen during the early developing period (Figure 2.21 and Table 2.8). The time required for attaining the stage of type D larvae is found on a hyperbola as in the case of embryonic development, i.e. $y=294/(x-9.4)$ (Miyagi breed of *Magaki*, $y=$time required, $x=$temperature °C). The development is greatly delayed at low temperature.

Salinity also influence the rate and speed of growth. Amemiya (1928) and Seno *et al.* (1926) have studied this problem while Ranson (1948a) carried out ecological studies. The results obtained by these investigators show that salinity suited to deve-

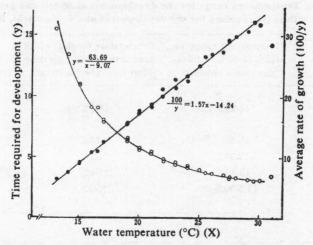

Figure 2.21. Relationship between the time required for the development of blastula and larvae, and water temperature in *Magaki* of Miyagi (Numachi, 1962).

lopment is not the same for even the same species (Table 2.10). However, when the rearing conditions of the parent oysters were changed, the salinity range suited to development could be determined on the basis of the salinity of the sea water to which the parent oysters had been exposed (Figure 2.22). Moreover, the salinity range suited to development changes according to the salinity to which the parent oysters are exposed (Numachi, 1965). This range is wide, between 14 and 37%,

Table 2.10. Salinity suited to development of *Magaki* and Portuguese oysters

Type	Suitable salinity	Investigator	
Magaki	20-26%o	Amemiya	(1928)
	23.3-28.5%o	Seno *et al.*	(1926)
	18-23%o	Ranson	(1948)
Portuguese	28-35%o	Amemiya	(1928)
	18-23%o	Ranson	(1948)

for both the *Magaki* of various regions, and for the Portuguese oyster. Hence, *Magaki* and Portuguese oysters are suited to the inner bay, whose waters undergo quick changes in salinity. Even in Mangoku-Ura where the salinity is 30 to 32%, the seedling collection is made on a commercial scale (Imai *et al.*, 1957) and the wide distribution of *Magaki* from the inner bay to the open sea might be thereby explained.

The larvae of the genus *Ostrea* differs from that of *Crassostrea* in that they usually do not grow well in low salinity. Davis (1962) carried out experiments on European oysters and the results are given in Figure 2.23. Regarding setting of larvae, the results shown in Table 2.11 were obtained. It is clear from these

results that the number of larvae settling is large when the salinity range is
22.5 to 25‰ and the same is reduced
when the concentration is below 20‰.

As shown in Figure 2.22, salinity
concentration greatly influences the speed
of growth. The growth speed is decided
on the basis of temperature and salinity.

In Table 2.12, a standard development
chart for Miyagi *Magaki* is shown. The
development of *Magaki* from other regions
and the Portuguese oyster, does not vary
much from this chart. Compared to the
results obtained in the past, the course of
development shown in this chart is rather
rapid, and the factor causing this is assumed to be the changes in salinity.

4.6 Setting

Larvae which are about to settle
develop byssal glands, and foot and eye
spots. A lens-like substance forms at the
opening part of the eye spot (Erdmann,
1934; Cole, 1938). They are passive to
light and color (Prytherch, 1934; Hopkins, 1937) and have a negative phototaxis (Nelson, 1924). The function of
these eye spots is not yet clear. The
larvae swim by the ciliary movement of

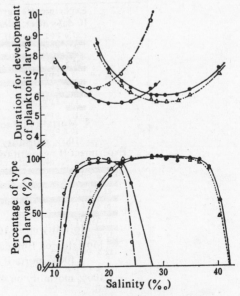

o Sea water with low salinity

△ Sea water with high salinity

⊙ Transplanted from high to
    low salinity

● Transplanted from
    low to high salinity

Figure 2.22. Differences in salinity suited to
development according to the water condition
in which parent oysters are reared, and the
temporary cum reversible variations
(Numachi. 1960).

the mantle in search of a suitable substratum to which they attach with the help
of a secretion from the byssal glands (Figure 2.24). The chemical nature of this
secretion is not known.

While showing the results in Figure 2.25, Prytherch (1934) has also reported
that the amount of secretion from the byssal glands is high when the salinity is between 15 to 20%. Takatsuki has explained the results of Prytherch by referring to
Igai in which the secretion of proteinous substance forms at a high rate when the
salinity is 20%. On the other hand, Davis *et al.* (1962) have found that the European
oyster settles easily during high salinity of sea water (Table 2.11). Prytherch further
studied the influence of various types of ions on the setting of larvae, reporting that
they settle only when the Cu ions are 0.05 to 0.6 mg/*l*. This concentration of Cu
ions is very much higher than in normal sea water.

Regarding the nature of the surface to which spat adhere, it is generally assumed
that they only prefer a smooth surface irrespective of the type of substratum (Nelson,
1924; Roughley, 1933). Some authors think that this depends upon the availability
of such a substratum, while others are of the view of that the spat prefer rocks which

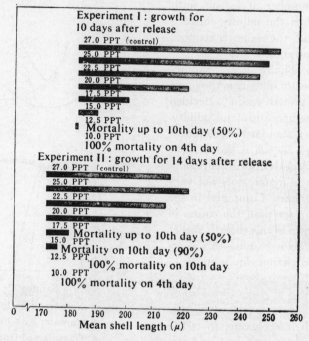

Figure 2.23. Growth of European oyster at various salinities (Davis *et al.*, 1962).

Table 2.11. Influence of low salinity on the setting of the larvae of the European oyster (9,000 larvae reared at 46 to 27‰ salinity up to the setting stage were dipped in 1 liter of sea water having various concentrations of salinity and the number of larvae setting were counted) (Davis *et al.*, 1962)

Salinity concentration	Number of larvae setting (mean)	
	Experiment I	Experiment II
26-27 ‰ (Control)	147	134
25.0	504	145
22.5	120	171
20.0	114	38
17.5	24	5
15.0	4	0.5
12.5	0	0
10.0	0	0
7.5	0	0

are cleaned by waves, rather than a muddy surface (Needler, 1941a; Korringa, 1951). Yet, in an experimental culture, it was noted that surfaces covered by a thin film of bacteria or fungi were preferred.

Table 2.12. Standard development chart of *Magaki* **(water temperature 20 to 21°C)**

Stages of development	Required time			
	hours/minutes		hours/minutes	
Release of 1st polar body	—	50	to	1	10
Release of 2nd polar body	1	—	to	1	20
1st cleavage	1	20	to	1	40
2nd cleavage	2	—	to	2	20
Morula	3	05	to	3	25
Blastula (beginning of rotating and movement, free living, floats)	5	20	to	5	50
	5	30	to	6	—
Gastrula	6	10	to	6	30
Trochophore, shell line appears	9	40	to	10	—
D-larvae					
1st stage of shell development completed	25	—	to	28	—

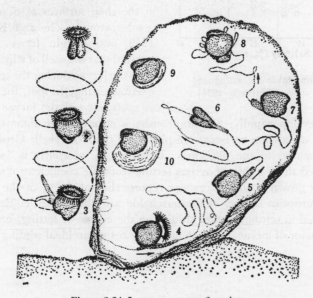

Figure 2.24. Larvae: course of setting.

1, 2—planktonic larvae; 3, 4—searching period; 5, 6, 7—creeping period;
8—setting; 9, 10—spat on the 19th and 20th days (Prytherch, 1934).

A difference of opinion exists with regard to the side of setting: the lower surface of a smooth substratum seems preferred (Wilson, 1941; Knight Jones, 1951; Schaffer, 1937), but Korringa found that more spat settled on the upper surface than on the lower, even though a higher mortality occurred there due to an abundance of mud and algae.

Regarding the pattern of distribution and the setting of larvae in a vertical direc-

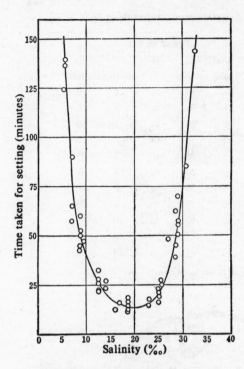

Figure 2.25. Time taken for setting and salinity of sea water (Prytherch, 1934).

tion, views once again differ (Cole and Knight Jones, 1949; Loosanoff and Engle, 1940; Mackin, 1946). All researchers agree, however, on the importance of water current. Korringa has reported that a large number of oysters settle in such places where the water current is low due to shielding materials. Thomson's view (1950) that a large number settle when the spaces between the setting plates are narrow, is also based on this factor, probably due to the formation of eddies by the shielding materials. On the other hand, it is also possible that a highly dense group of oysters form due to the behavior of the larvae. For example, one frequently finds a large number of young oysters adhering closely to one another on the shell surfaces of other oysters on a substratum. Cole and Knight Jones (1949) and Knight Jones (1951) have reported that the shell of a living oyster attracts the larvæ during the setting period.

After setting, the foot, the mantle, the eye spots, etc., of the larvae degenerate, but the gills develop rapidly. Simultaneously, a strong hinge ligament develops, and a new shell of calcium carbonate displaces the chitinous shell. Growth becomes rapid and, compared to the planktonic larval stage, mortality is low. Lambert (1946b) estimated the number of oysters setting from the coefficient of setting of the number of fully grown larvae. Korringa reports that only 1% of the fully grown larvae of the European oysters of Oosterschelde are able to settle. Regarding the problems involved in setting, further studies on the course of setting, environmental conditions, behavior of larvae, etc., are necessary before an ideal seeding method can be developed.

(Numachi)

5. Physiology

5.1 Feeding pattern

Oysters procure their food by filtering plankton from the sea water. The ciliary movement of the gills causes a water current into the mantle cavity as shown in Figure 2.26. The water thus drawn in passes through the branchial filaments and flows into the suprabranchial chamber. The plankton are thus caught by the lateral cilia put out by the filaments. The movement of the cilia carries the plankton around the gills and into the labial palp. The larger planktonic elements are prevented from entering the inhalant chamber and the heavier substances settle even before reaching the gills and being expelled. The substances thus carried to the labial palp are now sent to the mouth. When sand particles, etc. enter, the labial palp closes and they

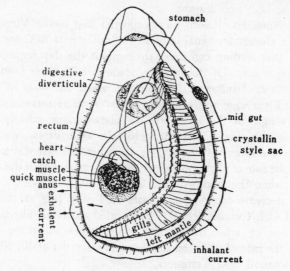

Figure 2.26. Structure of soft parts and direction of water current during feeding activity.

are moved into the pallial cavity. The labial palp selects the food material and experiments have shown that while it rejects yeast, it accepts a mixture of yeast and algae. In another experiment, when just a small amount of *Chromatium perty* (2 to 3 μ) was mixed with other algae, the mixture was rejected (Loosanoff, 1949a).

According to MacGinitie (1941, 1945) and Yonge (1946), the food is taken in not only by the gill filaments, but also by means of a mucous film secreted by the cells in a part of the gills. By this feeding mechanism, minute particles of 90 Å can be swallowed. Since the mucous secretion is more sensitive than the inhalant activity and both internal and external factors may stop the secretion, the rate of water inhaled does not always show the amount of food ingested. Korringa (1952) has pointed out that the space in between the filaments is 30 to 60 μ which is sufficient for the passage of flagellates. He has strongly supported the view expressed by MacGinitie in referring to the fact that the amount of food material composed of microorganisms in sea water to which *Chlorella* was added, did not tally with the amount of water filtered out and the concentration of microorganisms (Loosanoff and Engle, 1947a, b).

Substances like plankton, organic waste, fungi, flagellates, larvae of various invertebrates, sand, mud, spongy particles, etc., are usually found in the stomach of oysters. Since organic wastes are not assimilated, they do not constitute feed. On the basis of his study of the physiology of digestion, Takatsuki (1951) claims that oysters assimilate only the phytoplankton. He also asserts that organic waste is never assimilated. Animal food materials constitute only 10% or below of the total feed. Nelson (1947) did an analysis of the stomach contents of oysters and of the surrounding water. The results showed that *Skeletonema costatum,* etc., form an important part of oyster feed. The food material seems to be mainly composed of small organisms like single-celled algae and fungi. However, the organic contents differ according to the growing beds and seasons.

Loosanoff and Nomejko (1946b) have reported that in the Virginia oyster, food passes through the alimentary canal in 80 to 150 minutes at 20°C water temperature. They also found that feeding continues throughout the day regardless of the tidal conditions or time of day. Of the samples studied, 80% were found to have their stomach filled with food material. The rate of water filtration in these oysters is normally 5 to 25 liters/hour and sometimes even 31 to 34 liters/hour, which means 1,500 times the weight of the soft parts. In places where salinity is low, feeding often stops during ebb tide (Chestnut, 1946). In some cases, water temperature greatly influences the filtering activity. Again, when the sea water becomes acidic due to pollution, the rate of filtration continues to be normal until the acidity increases to pH 7.75, but when the acidity is pH 6.75 to 7.00, a temporary increase occurs; below pH 6.5 a decrease occurs, and at pH 4.14 only 10% of the normal rate is found (Loosanoff and Tommers, 1947). Suspended materials like silt also influence the rate of filtration, i.e. when the amount of silt suspended in the sea water is 0.1, 1.0, or 3 to 4 g/*l*, the rate of filtration decreases in the order of 40, 20, and 4% of the normal rate (Loosanoff and Tommers, 1948).

Loosanoff and Engle (1947a, b) have reported that when the concentration of algae is high, i.e. when the concentration of *Euglena* is about 3,000/ml, *Nitzshia closterium* about 75,000/ml, and *Chlorella* sp. more than 2,000,000/ml, the rate of filtration not only decreases but reaction to stimuli is also poor.

Since this effect was also seen in the filtered fluid of an algae culture medium, the presence of a poisonous action in the metabolic product of algae was deduced. This poisonous action did not weaken even when exposed to air. It was reported that the oysters recovered when transferred to normal sea water. When there is a decrease in the rate of filtration, the amount of apparent excreta is proportional, but the amount of excreta is inversely proportional to the density of the algae. *Gonyaulax,* etc., have also been reported as having the same type of poisonous effect on oysters (Loosanoff and Engle, 1947a; Kincaid, 1951).

Some substances which can be identified as hydrocarbons, and estimated by photographic methods with the help of N-ethyl-carbazole, apparently increase the filtering activity (Collier *et al.,* 1950). These are present in natural sea water (25 mg/ml for a corresponding amount of albinoze). When the concentration of these substances is very low, there is no feeding activity.

(Numachi)

5.2 Digestion and nutrition

The mode of digestion in oysters can be broadly divided into three types: intracellular digestion occurring in the fine tubules of the digestive diverticulum, extracellular digestion occurring in the stomach and intestine with the help of enzymes from the crystalline body, and digestion occurring due to the amoebocytes moving freely in the body tissues.

Figure 2.27 shows the digestive system of the Virginia oyster. The food which is selected by the gills and labial palps enters the mouth and esophagus and finally reaches the stomach. But food selection continues even in the stomach, and food particles which are relatively larger in size are passed into the crystalline style (Figure 2.28) to be broken down into smaller particles by the digestive enzymes (Table 2.13). This crystalline style is inside the crystalline sac, which lies parallel to the stomach and intestine. The cilia present on the inner surface of the sac move in a rotating fashion at a steady rate (70 to 80 rotations per minute). Simultaneously, a forward movement of the crystalline style causes the anterior edge to press against the stomach plate, eventually resulting in its total erosion (Figure 2.29). The movement of the cilia present in the duct pushes minute food particles in the stomach toward the fine tubules of the digestive diverticulum (Figure 2.30). Here the food particles are sucked in by the food vacuoles and then assimilated (Table 2.14). Digestion continues even in the intestine through the digestive enzymes of the crystalline body, and the digested material is absorbed. Amoebocytes are found in large numbers near the alimentary canal, which take in food material that flows inside

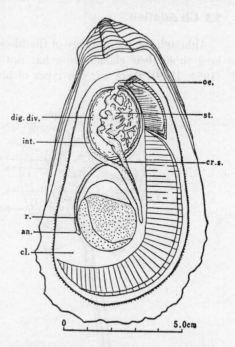

Figure 2.27. Digestive system of the Virginia oyster. Connective tissues cover labial palps of right side; stomach and intestine removed. an—anus; cl—exhalant current; cr. s—crystalline sac; dig. div—digestive diverticulum; int—intestine; oe—esophagus; r—rectum; st—stomach (Galtsoff, 1964).

and, aided by the digestive enzymes present in the cells (Table 2.15), the food is broken down and then assimilated. Amoebocytes are present not only near the alimentary canal, but also on the surface of the mantle and gills, where they perform the same function (Takatsuki, 1949; Galtsoff, 1964).

When the amount of food is more than what is required, the excess amount is stored in the body as either glycogen or fat. In most animals, regardless of the type of food, the excess is usually stored in the form of fat, but in the oyster the excess is usually stored in the form of glycogen. This is because oysters lead a sessile existence and hence to protect themselves from adverse environmental condition and enemies, they keep their shells closed for a long time; even when the shells open the respiration rate may not be sufficient. Hence, the energy required for the survival of this animal at such times is derived from the glycogen. Figure 2.31 shows the distribution of neutral fat and glycogen in the body of *Magaki*. It is clear from this figure that the distribution of these two stored materials shows a seasonal variation closely related to reproduction. The metabolic significance of this condition will be discussed under section 5.8.2 "Physiology of Reproduction."

5.3 Circulation

Although the morphology of the blood corpuscles of oysters has been studied for a long time, their classification has not yet been finalized. According to Takatsuki (1934, 1949), there are two types of blood corpuscles in the European oyster. One

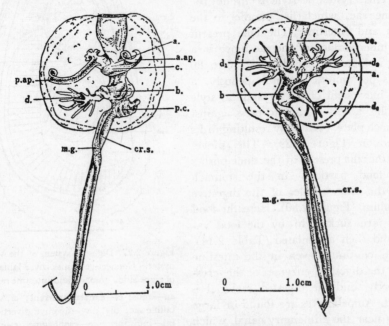

Figure 2.28. Stomach, crystalline style and esophagus of the Virginia oyster.

Figure on left shows the condition of the left side; figure on right shows the condition of the right side. a—anterior chamber of stomach; a. ap—anterior protrusion of cardiac sac; b—posterior chamber of stomach; c—cardiac sac; cr. s—crystalline sac; d, d_1, d_2, d_3—ducts leading into digestive diverticulum; m. g—mid gut; oe—esophagus; p. ap—posterior protrusion of cardiac sac; p. c—pyloric sac (Galtsoff, 1964).

Table 2.13. Digestive enzymes in the crystalline style of oysters

Types of enzymes	Quantity present	Investigators and types of oysters used in the experiment
Amylase	₩	Yonge (1926), *Ostrea*
Glycogenase	₩	*edulis*
Lipase	—	Yonge (1926), *O. edulis*
	+	George (1952), *Crassostrea virginica*
Protease	—	Yonge (1926), *O. edulis*

shows a number of granules inside the cells, and an irregular shape with many variations during the life of the oyster; the maximum length is 20 μ, and the nucleus is circular, about 3 to 4 μ, and does not change its shape. The other type of corpuscle has very few or even no granules inside the cells; the shape is mostly circular and varies little, being mostly confined to the formation of a broad pseudopodium; the size is 5 to 15 μ, and the round nucleus shows no change in shape. According to Tanaka (1964), the blood corpuscles in the Japanese oyster *(Magaki)* can be classified into 13 types, and the *Ostrea denselamellosa*, into 9 types. In large corpuscles

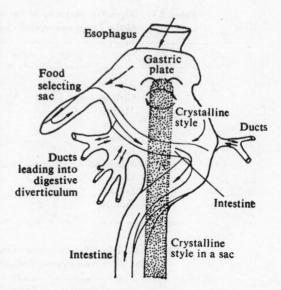

Figure 2.29. Stomach and crystalline style of the European oyster (Yonge, 1960).

with granules, the size is 60.5 to 64.8 μ with a nucleus of 23.0 to 27.0 μ in diameter. As shown in Figure 2.32, Mori *et al.* (1965b) found large Sudan III positive planocytes in *Magaki*. The size of a planocyte is 20 to 25 μ (8 to 10 μ for its nucleus). However, the authors had no opportunity to view the giant planocyte reported by Tanaka.

The source and the mechanics of blood-corpuscle formation in oysters are still unexplained, but the role of the corpuscle in food absorption and excretion (see section 5.6) is well known. The most significant feature of oyster blood is the absence of a respiratory pigment such as hemoglobin or hemocyanin, which easily combines with oxygen; the presence of other respiratory enzymes has not yet been confirmed.

The circulatory system is fairly well developed and in lieu of the capillaries found in vertebrates, a number of sinuses intervene between the veins and arteries. The vascular system is open (Figure 2.33). The heart, present in the pericardial cavity, has two auricles and a ventricle. Hopkins (1934) discovered two new Y-shaped organs, one on the right and one on the left side of the mantle near the exhalant current cavity. Although these organs show slow rhythmic move-

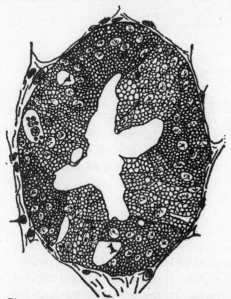

Figure 2.30. Fine ducts of digestive diverticulum in the European oyster. Minute particles of food have entered the cells (Yonge, 1960).

Table 2.14. Digestive enzymes in the digestive diverticulum of the European oyster (Yonge, 1926)

Types of enzymes	Quantity present
Amylase	‖‖
Glycogenase	‖‖
Maltase	+
Lactase	+
Sucrase	+
Raffinase	+
Salicinase	+
Lipase	+
Esterase	‖
Protease	‖

Table 2.15. Digestive enzymes in the amoebocytes of the European oyster (Takatsuki, 1949)

Types of enzymes	Quantity present
Amylase	‖
Glycogenase	‖
Maltase	+
Lactase	+
Salicinase	+
Sucrase	+
Esterase	+
Protease	+

ment, they pump the blood from the mantle ridge to the veins in the ctenidia. Again, according to Galtsoff (1964), the blood vessels in the mantle also function as an accessory heart. In view of the fact that the function of the heart could be checked by acetylcholine and cause an antibody reaction (Jullien, 1936), the heart is assumed to be myogenic. However, details of this aspect are not yet clear. The function of the so-called pacemaker is also not clear. The nerves controlling the function of the heart stem from the visceral nervous system. It is thought that the pacemaking system may also be influenced by these nerves (Krijgsman and Divaris, 1955). (Mori)

5.4 Respiration

Though the pallium plays some role in the respiration of bivalves, gills are the main organs of respiration. The constant movement of large number of cilia present in the gills causes a water current. This water current is useful for external respiration, feeding, and the removal of metabolic wastes. In the past, two methods for measuring the rate of respiration in oysters existed. One was concerned with the difference in the amount of oxygen dissolved in water before and after the experi-

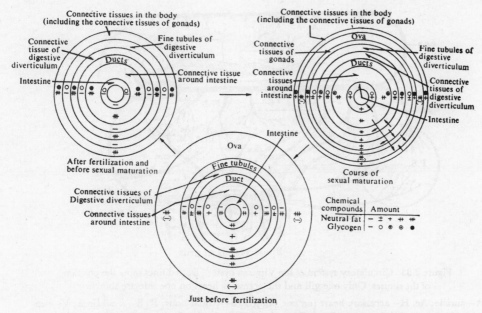

Figure 2.31. Seasonal variations in the distribution of glaycogen and neutral
fat in *Magaki* (Mori *et al.*, 1965; data slightly modified).

Arrows show the direction of shifting of stored fat.

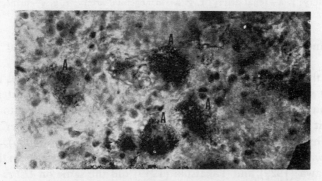

Figure 2.32. Large Sudan III positive planocytes found in *Magaki*
(A). Connective tissues around the intestine of two-years-old
oyster of Matsushima Bay during the spawning period. Gelatine
embedded sections (Mori *et al.*, 1965b).

ment. In this method, the influence of metabolic wastes on the results of the experiment was a major disadvantage; furthermore, it was not possible to know the actual
condition during the opening and closing of the shells. To avoid these disadvantages,
Galtsoff suggested the second method. Instead of stagnant water, he kept the oysters
in running water and tried to express the amount of oxygen dissolved during the
open and closed conditions of the shells in the form of a chemograph (Galtsoff, 1964).

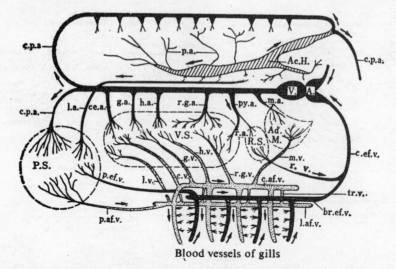

Blood vessels of gills

Figure 2.33. Circulatory system of the Virginia oyster. Dotted lines show the position of the sinuses. Only one gill and the accessory heart on one side are shown.

A—auricle; Ac. H—accessory heart (on one side); P. S—pallial sinus; R. S—renal sinus; V—ventricle; V. S—visceral sinus; br. ef. v—efferent branchial vein; c. af. v—common afferent vein; c. ef. v—common efferent vein; ce. a—cephalic artery; c. p. a—circum-pallial artery; c. v—cephalic vein; g. a—gastric artery; g. v—gastric vein; h. a—hepatic artery; h. v—hepatic vein; l. a—labial artery; l. af. v—lateral afferent vein; l. v—labial vein; m. a—adductor muscle artery; m. v—adductor muscle vein; p. a—pallial artery; p. af. v—afferent pallial artery; p. ef. v—efferent pallial artery; py. a—pyloric artery; r. a—renal artery; r. g. a—urinogenital artery; r. g. v—urinogenital vein; r. f—renal vein; tr. v—transverse vein (Galtsoff, 1964).

However, this method is highly complicated and very difficult to put into actual practice.

The study of the influence of oxygen tension on the exchange of gases reveals that the effect is not significant when the amount of oxygen in the sea water exceeds a certain limit, but when it is below this limit, the oxygen uptake falls to a very low level. This limit differs from type to type and in the Virginia oyster is 2.5 ml/l (Galtsoff and Whipple, 1931), and in the *Magaki*, 1.5 ml/l (Ishida, 1935). Regarding the influence of water temperature, Mitchell (1912) carried out a study on the Virginia oyster, and Nozawa (1929) and Takatsuki (1949) did the same on *Ostrea circumpicta*. When the temperature was increased, a linear increase in the oxygen uptake occurred; the range of temperature for such increase was between 19 to 28°C in the Virginia oyster, and 10 to 25°C in the *Ostrea circumpicta*. Above 25°C, there was a gradual decrease in the oxygen uptake. According to the experiments carried out by Mori *et al.* (unpublished) on *Magaki* of Onagawa Bay a number of oysters showed a linear increase in oxygen uptake even up to 32°C, just before the spawning period when physiological activities are sluggish. Very little work has been done on the influence of high salinity, but considerable work has been carried out on the effect of low salinity. Various reports such as Ishida's (1935) and Galtsoff's (1964) reveal an insignificant difference between the oxygen uptake in oysters reared for three days in sea water with low salinity (24.1‰) and those grown in normal sea water

(31.58 ‰). As shown in Figure 2.34, pH exerts a strong influence on respiration (Galtsoff, 1964). The influence of various factors on respiration was also discussed and one should keep in mind that these might change according to seasonal variations (Mori *et al.*, 1965a).

Compared to the reports on the respiration of the body, there are very few on tissue respiration (Jodrey and Wilbur, 1955; Hoshi, 1958; Kawai, 1959; Hoshi and Taguchi, 1960; Usuki, 1962; Mori, 1968a). According to the results of experiments on *Magaki* (2 to 3-year-old), the oxygen uptake (measured at 22 to 32°C) of the tissues is in the order of digestive diverticulum > gills > pallial margin > smooth part of the adductor muscles (Mori, unpublished; Mori, 1968a). The order is independent of seasonal variations. On the other hand, according to Kawai (1957), the oxygen uptake by the tissues in *Akoyagai (Pteria martensii)* is in the order of gills > pallial margin > digestive diverticulum. The consumption is

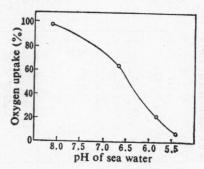

Figure 2.34. Lowering of oxygen uptake in Virginia oyster by pH. Data collected in summer. Graph drawn on the basis of results of experiments carried out at 25°C at Woods Hole (Galtsoff, 1964).

exceptionally high in the gills. The difference in the results of the two investigators shows that the degree of oxidization in the tissues differs from species to species among the bivalve group. The value of RQ not only depends upon the chemical properties of the substances oxidized but also on the mutual exchange of substances. Hence, even in the case of oyster tissues, by measuring the value of RQ it would be possible to find the types of respiratory bases and their exchange. Experiments along this line are already under progress (Mori, 1968a). Table 2.16 shows the measured values obtained for the digestive diverticulum. The measurements were taken at 25°C by using Warburg's manometer. In order to convert CO_2 in a liquid phase into a gaseous form, H_2SO_4 was injected from the side chamber into the main chamber and KOH from the central well was used for the absorption of the gaseous CO_2. The amount of O_2 absorbed in 20 minutes was 19.8 μl, and the amount of CO_2 given out during this time was 19.2 μl. Hence, the value of RQ is 0.97. The results of investigations on the seasonal variations of RQ will be detailed under section 5.8.2 "Physiology of Reproduction."   (Mori)

5.5 Energy dispersion

With regard to the carbohydrate breakdown occurring in the adductor muscles, Humphrey (1944 to 1950) carried out a series of investigations. He found that glycogen present in the adductor muscles is broken by potassium, magnesium, DPN, etc., and that piruvic acid and lactic acid form. He also found that the formation of these acids is checked by fluorides and iodine acetate, and that there is a possibility of glycogen forming from glucose-1-phosphoric acid. However, the capacity of the adductor muscles of oysters for breaking carbohydrates is far less than the capacity of rabbit muscles. Usuki (1956) studied the influence of iodine acetate and

Table 2.16. Measurement of RQ in digestive diverticulum of *Magaki*[1] (Warburg's manometer, 25°C)

Manometer	I	II	III	Temperature
	k_{O_2}[2] = 1.373, k_{CO_2} = 1.520	k_{CO_2} = 1.578	k_{O_2} = 1.421	
Tissues	100.2 mg	100.5 mg	100.0 mg	
Main chamber: Artificial sea water	1.8 ml	1.8 ml	1.8 ml	2.0 ml
Side chamber: H_2SO_4	0.2 ml	0.2 ml	—	
Central well: 17% KOH	—	—	0.3 ml along with filter paper	

Time (minutes)	I			II			III			Temperature	
	R[3]	D[4]	C[5]	R	D	C	R	D	C	R	D
0 (injection of H_2SO_4 into the main chamber of manometer II)	116.0	—	—	123.0	—	—	176.7	—	—	150.8	—
10	110.0	−6.0	−5.2	182.5	+59.5	+60.3	168.7	−8.0	−7.2	150.0	−0.8
20 (injection of H_2SO_4 into the main chamber of manometer I)	105.0	−11.0	−10.2	—	—	—	−62.0	−14.7	−13.9	150.0	−0.8
30	177.0	+61.0	+60.8	—	—	—	—	—	—	151.0	+0.2

[1] Oyster grown in Onagawa Bay collected in early September. [2] k:volume constant (mm²). [3]:Reading (mm). [4]:Difference in readings. [5]Adjusted values.

fluorides on the ciliary activity of the gills. The results showed that ciliary activity is affected by both these substances, and that the effect becomes more and more prominent with an increase in the acidity of sea water (Figures 2.35, 2.36, and 2.37).

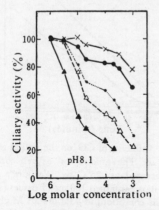

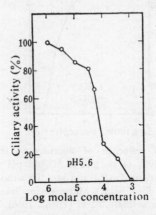

Figure 2.35. Influence of monoiodine acetate on the ciliary activity of the gills in *Magaki*. pH of water was adjusted to 8.1. Measurements taken in winter at $25.0 \pm 0.1°C$ (Usuki, 1956).

×—after 10 minutes; △—after 3 hours; ●—after 1 hour; ▼—after 6 hours; . —after 2 hours.

Figure 2.36. Influence of monoiodine acetate on the ciliary activity of the gills in *Magaki*. pH of sea water was adjusted to 5.6. Treatment was carried out for 1 hour. Measurements taken in winter at $25.0 \pm 0.1°C$ (Usuki, 1956).

The effect of iodine acetate is controlled by cysteine. The effect of fluoride is partly controlled by piruvic acid and partly by succinic acid but not by glucose. Usuki and Okamura (1956) found almost all of the neutral substances of the Embden-Meyerhof system present—lactic acid, piruvic acid, citric acid, etc.—in the gills, and that their quantities were greatly affected by monoiodine acetate, fluorides, 2,4-dinitro phenol, azide, malonic acid, etc. These show the presence of the Embden-Meyerhof system as well as a phosphorization reaction in the respiratory system.

There is no doubt that there is a TCA cycle and a cytochrome system in oysters. Cleland (1951) assumes that most of the enzymes of the TCA cycle are present in the eggs of oysters. This is because respiration in the oyster is accelerated by adding the neutral substances of the TCA cycle to the homogenate of unfertilized eggs. Jodrey and Wilbur (1955) studied the respiratory metabolism of the pallium and found various enzymes of the TCA cycle present, such as anhydrous isocitraze, anhydrous succinaze, fumaraze, anhydrous malaze, etc., and the cytochrome oxidase in the tissues. But it was found that the cytochrome system does not play an important role in the respiratory system of the pallium. Later it was found that the cytochrome system was present in various tissues of oysters, and that the oxidizing metabolism greatly depended upon this system (Kawai, 1959; Hoshi, 1958; Hoshi and Taguchi, 1960). In Figures 2.38 and 2.39 the influence of CO_2, cyanide, or methylene blue on the respiration of gills and pallium is shown. Black (1962a, b) carried out detailed physiological studies on the changes in the rate of respiration during early development, the electron-transporting enzyme activity, the activity of the enzymes in the TCA cycle, etc. The results were highly interesting. A study of

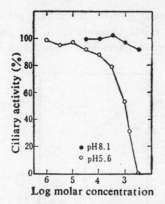

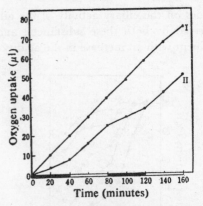

Figure 2.37. Influence of fluorides on the ciliary activity of the gills in *Magaki*. pH of sea water adjusted to 8.1 and 5.6. Treatment carried out for 1 hour. Measurements taken in winter at $25.0 \pm 0.1°C$ (Usuki, 1956).

Figure 2.38. Influence of CO_2 on the pallial respiration of *Magaki*. The amount of oxygen consumption was measured at 25°C by Warburg's manometer. Each flask contained 100 mg of tissues (weight taken in fresh condition) in 5 ml of 0.03 M glyceroglycine buffered sea water. Blank lines at bottom of graph indicate the duration of brightness (Kawai, 1959).

the distribution of anhydrous succinaze activity and anhydrous malaze activity in the tissues of oysters during sexual maturation, showed that the activity of anhydrous succinaze was strongest in ova and sperms. The activity was less strong in renal ducts and digestive tissues. But the activity of anhydrous malaze is very strong in the connective tissues (Mori, 1967). (Mori)

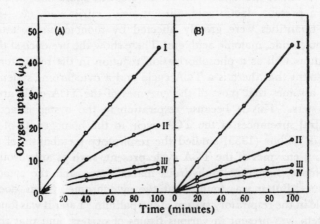

Figure 2.39. Influence of cyanide and methylene blue on respiration in the pallium and the gills of *Magaki*.

(A)—gills; (B)—pallium; I—control; II—10^{-3}M NaCN+6×10^{-5}M methylene blue; III—10^{-3}M NaCN+10^{-5}M methylene blue; IV—10^{-3}M NaCN.

Oxygen uptake was measured at 25°C by Warburg's manometer. When gills were studied, the flask contained 50 mg of gill tissues (fresh condition) in 1.5 ml of 0.03 M glycine buffered sea water. When the pallium was studied, 100 mg of pallial tissue was placed in the flask (Kawai, 1959).

5.6 Excretion

It is already known that the excretory system of bivalves involves nephridia, the pericardial gland, amoebocytes, the mucus cells of the pallial epithelium, etc. At present, it is impossible to determine the extent of excretory action in these tissues and cells. Since the structure of nephridia in oysters is very complicated, the excretory function is not yet clear. Excretion due to diapedesis is found to be very important, but the details are not completely understood. It is assumed that tissues and cells excrete waste materials through a combined action (Galtsoff, 1964).

There are a number of reports on the structure of nephridia, but all of them differ in details (Takatsuki, 1949). For example, while Galtsoff (1964) reports the presence of an internephridial passage connecting the nephridia in the Virginia oyster, Fingerman and Fairbanks (1958) did not find one in this species. Hence data on the same species conflict. However, there are several common points, *viz*. nephridia consist of fine tubules symmetrically arranged on the right and left sides, with blind vesicles at the end of the tubules and, in many oysters, the right nephridia develop earlier than the left.

Few investigations have been carried out on the excretory products of oysters. Uric acid which is usually absent in bivalves (Brunel, 1938), is excreted in small quantity by the Portuguese oyster, *Crassostrea angulata* (Delaunay, 1931). Among the nitrogenous products of excretion, amino-nitrogen is maximal (13.2%), followed by ammonia nitrogen (7.2%), urea nitrogen (3.2%), and uric acid nitrogen (1.6%) (Spector, 1956).

Although Kumano (1929) as well as Fingerman and Fairbanks (1957, 1958) have found that the excretory system adjusts the osmotic pressure of the oyster, several details are still lacking, and further studies are awaited.   (Mori)

5.7 Sensory perception

The nervous system of the oyster (Figure 2.40) is relatively simple. Cerebral and visceral ganglia are present in the adult, but not pedal ganglia. As in other mollusks however, pedal ganglia are present in the larval stage. Hence, degeneration of pedal ganglia is said to be associated with the sessile mode of life in the adult stage. The cerebral ganglia and the visceral, are connected by nerves which constitute the central nervous system. Several nerves branch from these ganglia to various parts of the body to form the peripheral nervous system. While the anatomical aspect of the nervous system is well-studied, there has been almost no experiment related to the physiological aspect of the nervous system.

Tentacles along the margin of the pallium and the presence of sense organs on both the right and left sides near the adductor muscles, around the posterior pallial nerve, should be mentioned. The tentacles function as photoreceptors, chemoreceptors, and also as thermoreceptors (Hopkins, 1932a, b; Galtsoff, 1964). The function of the sense organ on the ventral region is not yet clear. However, it is assumed to be sensitive to the movement of environmental water (Galtsoff, 1964). During the larval stage, there are eye spots and statocysts, but these degenerate during larval setting (Takatsuki, 1949).   (Mori)

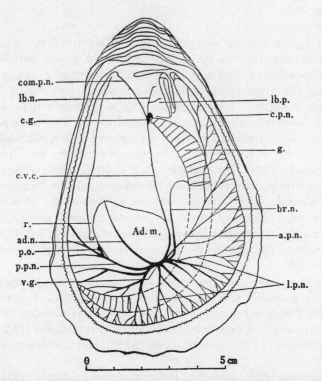

Figure 2.40. Nervous system of the Virginia oyster (viewed from the right).

Ad. m.—adductor muscle; ad. n.—adductor nerve; a. p. n.—anterior pallial nerve; br. n.—branchial nerve; c. g.—cerebral ganglion; c. p. n.—circum-pallial nerve; com.—commissure; c. v. c.—cerebro-visceral connective; g.—gills; lb. n.—labial nerve; lb. p.—labial palp; l. p. n.—lateral pallial nerve; p. o.—ventral sensory organ; p. p. n.—posterior pallial nerve; r.—rectum; v. g.—visceral ganglion (Galtsoff, 1964).

5.8 Reproduction

5.8.1 MORPHOLOGY

The reproductive system of *Magaki* consists of follicles, gonads, gonoducts, and urinogenital apertures. The organs are found in the connective tissues between the digestive diverticulum and the body wall (Figure 2.41). The inner structure of the reproductive organs is highly complicated and consists of a number of branching ducts connected with follicles. They occupy a wide area and cover most of the viscera. They are found from the upper part of the esophagus in the head, down to the anterior edge of the pyloric protrusion of the mid gut.

The gonads are not enclosed in sacs and the boundary is not clear. During the maturation period, the body wall becomes very thin and the network structure of the gonads can be seen. The gonads gradually thicken and connect with the gonoducts formed by the ciliary epithelium. The gonoducts open at the anterior part of the opening of the urinary ducts, on the ventral side of the adductor muscles.

The development of the gonads can be ascertained from the gradual thickening.

Gonadal development and distribution are not the same in all parts of the body, and differ according to age, type, and environment (Loosanoff and Engle, 1942; Korringa, 1952; Galtsoff, 1964). There is no morphological difference in the structure of either the testis or the ovary.

1. *Spermatogenesis:* The testis is not an organ which develops independently. It has seminiferous tubules surrounded by a thin basal membrane and collagenous fibers. Hence, it seems to be directly connected to the neighboring connective tissues (Figures 2.42 and 2.43). The reproductive cells inside the seminiferous tubules can be divided into supporting cells, spermatogonia, spermatocytes, spermatids, and sperms, on the basis of the shape and size of the cells, color, affinity to stain, size of the nucleus, con-

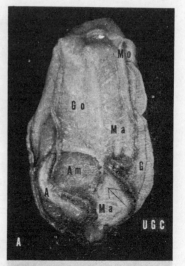

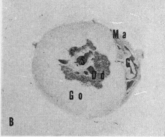

Figure 2.41. A—soft parts of an oyster; B—transverse section of the same.

A—anus; Am—adductor muscle; Dd—digestive diverticulum; G—gills; Go—gonads; Ma—mantle; Mo—mouth; S—stomach; UGC—urinogenital aperture (present on the inner side of the mantle) (Sugiwara).

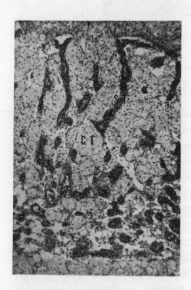

Figure 2.42. Seminiferous tubules of *Magaki* (Sugiwara).
CT—connective tissue;
ST—seminiferous tubules.

dition of the nucleolus, and distribution in the tubules (Figure 2.44).

The supporting cells show a positive PAS reaction. Since the supporting cells contain granules which are osmiophilic, they resemble the interstitial cells in the tubules. As shown in Figure 2.44, the boundary of the supporting cells is not clear.

The spermatogonia are in close contact with the basal membrane of the tubules. Despite its poor stainability, the nucleus, when stained, is clearly distinguishable from the other parts of the cell. Moreover, the cytoplasm shows a characteristic affinity for pyronin.

Spermatocytes can be classified according to the stainability of their nuclei. In a stained preparation, one often observes the leptotene stage, zygotene stage, pachytene stage, and the diplotene stage of prophase, but never a metaphase or anaphase. The size of the spermatocyte is maximal from the pachytene to the diplotene

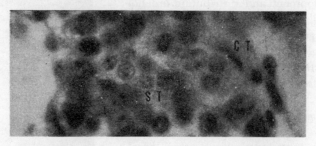

Figure 2.43. Magnified view of Figure 2.42 (Sugiwara).
CT—connective tissue; ST—seminiferous tubules.

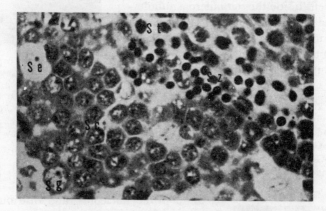

Figure 2.44. Section of seminiferous tubules (Sugiwara).
The cells from the wall of the seminiferous tubules are arranged in the order of:
Sg—spermatogonia; Sc—spermatocytes; St—spermatids; Sz—spermatozoa.
Cells of Sertoli (Se) are present in between these cells.

stage of prophase. A study of the structure of these cells reveals mitochondria, endoplasmic reticulum, and a few osmiophilic granules (Figure 2.45).

Spermatozoa can be classified into three types: those in which the acrosome granule appears, those in which the headcap develops, and those in which the acrosome forms. In all three types, the nucleus shows a strong Feulgen reaction and can be stained with methyl-green. It is during the course of spermatogenesis that a maximum number of changes occur. In other words, substances with heavy electron density are stored in acrosomal vesicles and give rise to acrosome. Mitochondria increase in size and become *neben kern* with well-developed cristae.

The head of the spermatozoan is elliptical in shape and measures $1.5 \times 2.0\,\mu$. The tail is long and measures 25 to 35 μ. An osmiophilic acrosome with a high electron density is present at the anterior end of the nucleus in the head region. A centromere is found between the head and the middle piece. From here the axial fiber of the tail develops. Four mitochondria from the middle part, arranged around the basal part of the axial fiber. The tail consists of a central strand without a peri-

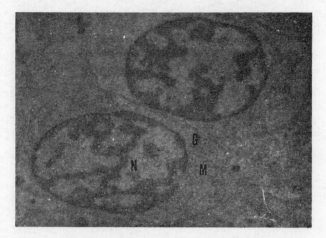

Figure 2.45. Electron microscope view of spermatocyte (Sugiwara).
G—osmiophilic granules; M—mitochondria; N—nucleus.

pheral strand. The central strand is made up of a single set of axial filaments covered by nine sets of peripheral filaments (Figures 2.46 and 2.47).

The structure of the axial fiber in the tail part of the spermatozoa is not the same in all animals. In fish, frogs (Higashi, 1964), *Ikechogai* (*Hyriopsis schlegelii*) (Higashi, 1964), the Virginia oyster (Galtsoff and Philpott, 1960), etc., in which fertilization is external, the axial fiber resembles that of *Magaki* in comprising only the central strand and lacking the peripheral. It seems that the structure is simple because the union of sperm and ovum does not pass through a complicated course. In those animals in which fertilization is internal, the axial fiber of the tail differs in having a central strand covered by a peripheral strand of high electron density (Yasudo, 1957).

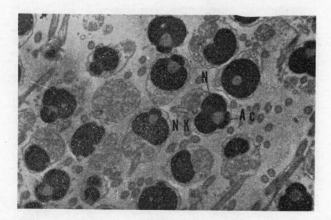

Figure 2.46. Minute structure of *Magaki* spermatozoa (Sugiwara).
AC—acrosome; N—nucleus; NK—*neben kern*.

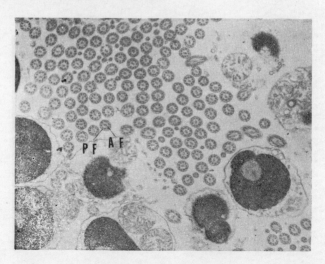

Figure 2.47. Minute structure of the tail part of spermatozoa of *Magaki* (Sugiwara).

Transverse section of the tail shows one set of (AF) axial filaments at the
center, and nine sets of (PF) peripheral filaments at the periphery.

2. *Oogenesis:* Like the testis, the ovary is also not simple in structure. It has a
number of follicles with interstitial tissues between them. The follicles contain
nuclei and follicular cells with characteristic cytoplasm and female reproductive cells.

The follicular cells seem to support the female reproductive cells and supply nutrition. Unlike those of higher animals, these do not form a granular layer; a part of
them forms the follicular epithelium, while another part directly adheres to the
oocytes. The nucleus is narrow and long, and the nucleolus is indistinct. The cytoplasm is bright and the cell boundary is not clear (Figure 2.48). The nucleus reacts
strongly to Feulgen and the cytoplasm shows an affinity to pyronin and gives a positive
PAS reaction.

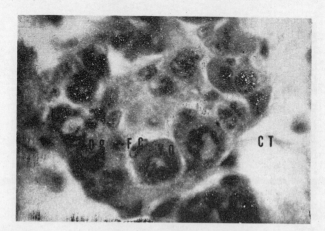

Figure 2.48. Cells of an ovary in the early stage of development (Sugiwara).
Og—Oogonia; YO—primary oocyte; FC—follicular cells.

In a resting stage, the oogonia remain in the follicular epithelium and have distinct nucleoli. The nucleus is relatively large in the cytoplasm, which shows an affinity to basic pigments and a mild PAS reaction. The nucleus is circular in shape and measures 6 to 7 μ. The shape of the cell is not very clear because its boundary is indistinct. The oogonia multiply by repeated mitotic divisions. Within a fixed period, they enter the growth phase (Figure 2.48).

The primary oocytes formed by the division of oogonia do not undergo maturation division until they are liberated, and hence during this period they only grow in size. As growth occurs, the nucleus or germ vesicle looks like a large sac, and the nucleolus or germ spot becomes distinct (Cleland, 1947). In its early stage, the cytoplasm of the primary oocytes is basophilic, but following growth in the oocytes, yolk granules of protein and fat collect in the cytoplasm, microvilli form from it and so does the chorion in which the microvilli become embedded (Figure 2.49). Pinocytic vesicles

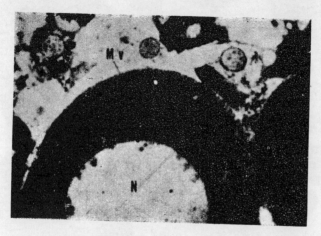

Figure 2.49. Microvilli (Mv) and yolk granules in cytoplasm of primary oocytes (Sugiwara).

are seen at the base of the microvilli, which indicate that nutrition is absorbed from outside the cells. When the primary oocytes are on the verge of liberation, the following changes occur: the nucleolus divides into a basophilic part (*Karyosome*) containing RNA and an acidophilic (*Plasmosome*) containing a PAS positive substance (Figure 2.50); inside the nucleus, the PAS substance collects and gradually spreads in and around the nucleolus, the entire nucleus, and the nuclear membrane, while the cytoplasm fills with yolk granules, and the microvilli become more distinct.

3. *Periodic changes in the gonads:* (a) *Multiplication of germ cells:* In Onagawa Bay, spermatogonia (oogonia) in the gonads, which had constricted during the previous winter, arrange themselves in the germinal epithelium and multiply from March to April when the temperature of the sea water is 8 to 10°C (Figure 2.51). A part of the germinal epithelium becomes extended in a direction perpendicular to the body surface. Follicles or secondary gonads develop in the connective tissues (a large amount of glycogen is stored in this connective tissue and Bargeton calls it the glycogen-bearing tissue) between the digestive organ and the gonads. The follicles

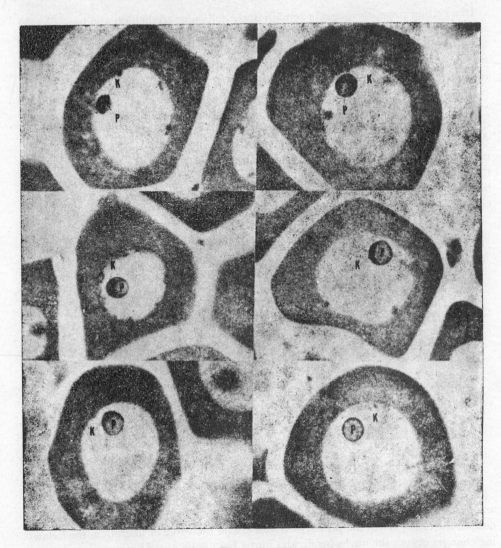

Figure 2.50. Primary oocytes. The nucleolus is divided into two parts, *viz.*
(K) *Karyosome* and (P) *Plasmosome* (Sugiwara).

branch and spread inside the connective tissue when the spermatogonia (oogonia) multiply. The spermatogonia grow to form sperms (ova) as explained earlier. The follicles increase and simultaneously the interstitial tissue in the gonads rapidly decreases (Figures 2.52 and 2.53). In July, when the temperature of the sea water reaches 20°C, the testis as well as the ovary is so well developed that all the soft parts of the animal, except the digestive system, are covered with gonads. In the males, the follicles are filled with sperms in their final stage of development, and in the females, the primary oocytes are fully grown (Figures 2.54 and 2.55).

(b) *Liberation of sperms and ova:* In Onagawa Bay, sperms and ova are released in August to September when the temperature of the sea water is 23°C. Since the primary

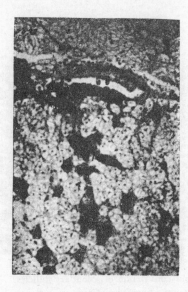

Figure 2.51. Early stage in the formation of germ cells.

Spermatogonia and oogonia of primary follicles (gonads) (GD) multiply
and permeate the connective tissue
(CT) on the sides of the viscera,
and give rise to secondary follicles
(F) (Sugiwara).

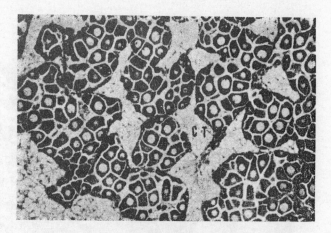

Figure 2.53. Secondary follicles showing rapid
development (Sugiwara).

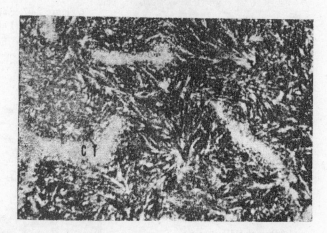

Figure 2.54. Well-developed seminiferous tubules; connective tissues (CT) greatly decreased (Sugiwara).

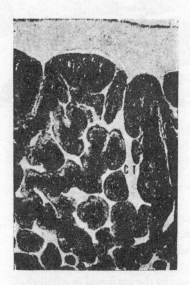

Figure 2.52. Seminiferous tubules
showing rapid development
(Sugiwara).

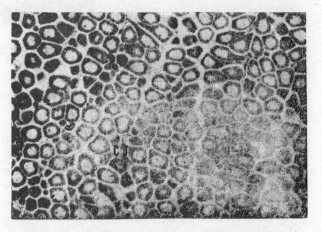

Figure 2.55. Well-developed secondary follicles (Sugiwara).

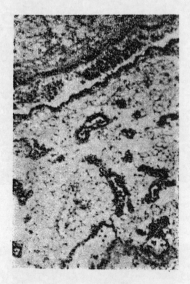

Figure 2.56. Rapid constriction of seminiferous tubules after the liberation of sperms.

ST—Seminiferous tubules (Sugiwara)

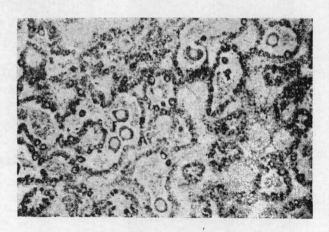

Figure 2.57. Rapid constriction of secondary follicles (F) after the liberation of ova.

Arrows show the penetration by amoebocytes inside and outside the follicles (Sugiwara).

oocytes take much time to grow, ova are released after a considerable period. Sperms however grow rapidly and are liberated within a short time.

After the liberation of sperms and ova, the gonads rapidly reduce in size. The unreleased reproductive cells degenerate or are absorbed by amoebocytes. The cells of the connective tissue systems are activated and the animal looks like what is popularly known as a "water oyster" (*mizugaki*) in Japan (Figures 2.56, 2.57, and 2.58).

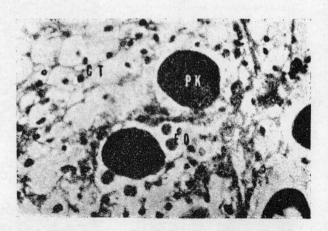

Figure 2.58. Degeneration of the unreleased ova (Sugiwara).

PK—nucleus greatly condensed; FO—cytoplasm reduced to small bits; CT—rapid regeneration of interstitial tissue.

(c) *Resting period following fertilization:* After fertilization, the cells of the interstitial connective tissue in the gonads multiply and rapidly increase in number. They give rise to new cells which fill the spaces left by the gonadal tissues. They collect and store glycogen and fat, and become the so-called "glycogen-bearing tissue." Meanwhile, the germ cells remain as spermatogonia and oogonia in the germinal epithelium until the following year when they repeat the foregoing activities. As explained by Loosanoff (1942) and Bargeton (1942), glycogen and other materials stored in the interstitial connective tissues of the gonads are utilized during the multiplication and growth of germ cells. Again, amoebocytes seem to play an important role in the metabolism essential for the multiplication and growth of reproductive cells, and also in the absorption of unreleased reproductive cells after fertilization.

(Sugiwara)

5.8.2 PHYSIOLOGY OF REPRODUCTION

To study the physiological aspects of reproduction, *Magaki* were collected from two fishing centers, Onagawa Bay and Matsushima Bay below the Miyagi Prefecture, where ecological conditions differ during reproduction. With specimens from two different places, a comparative analysis of the physiological aspects of reproduction during the breeding period could be made, and the seasonal variations in water temperature in these bays ascertained (Figure 2.59).

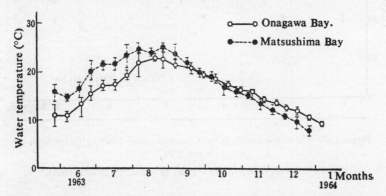

Figure 2.59. Water temperature (surface water) at Onagawa Bay and Matsushima Bay. The graph shows mean, maximum, and minimum values. (Data collected by Onagawa Fisheries Experiment Station attached to the Faculty of Agriculture, Tohoku University, and by the Sendai Power Station of Tohoku.)

1. *Analysis of reproduction period on the basis of seasonal variations in the soft parts:* Severe morphological changes in the soft parts of *Magaki* from both bays during the period of reproduction, can be seen with the naked eye and also determined through weight. (Mori *et al.*, 1965a; 1965b). It is clear that these changes are directly related to the formation and release of sperms and ova. Chemical analysis (Mori *et al.*, 1965b) also shows that the formation and release of sperms and ova greatly influence the other parts of the body. Such changes are not peculiar to oysters, being found in other animals also and referred to as the "seasonal cycle." But these changes are rather striking in *Magaki*.

From the data collected by Mori *et al.*, it is clear that the sexual maturity of *Magaki* differs according to the environmental conditions and age of the animal. However, sexual maturity begins mostly in March to April and fertilization occurs in August to September. Hence the period of reproduction can be divided into the following four stages:

Stage I: After the release of ova and sperms and before the beginning of sexual maturation (October to March);

Stage II: Course of sexual maturation (April to June);

Stage III: Just before the release of ova and sperms (July to August);
Stage IV: Period of fertilization (August to September).

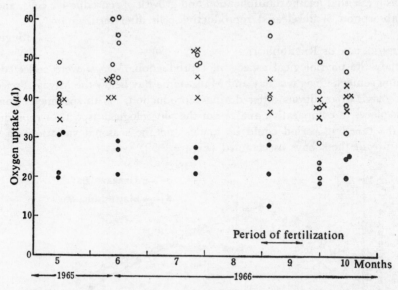

Figure 2.60. Tissue respiration in the Onagawa oyster (μl O_2/100 mg fresh
weight/hour, 25°C)

O = digestive diverticulum; × = gills; ● = pallium (Mori, 1968).

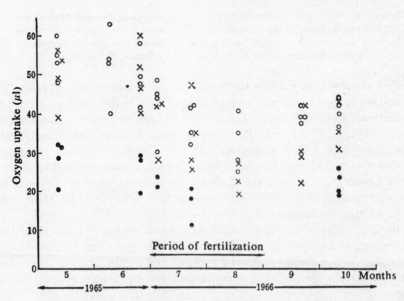

Figure 2.61. Tissue respiration of the Matsushima oyster (μl O_2/100 mg
fresh weight/hour, 25°C)

O = digestive diverticulum; × = gills; ● = pallium (Mori, 1968).

Table 2.17. Quantitative morphological analysis of sexual maturation (Mori et al., 1965b)

ONAGAWA OYSTER

Date of collection (Number of specimens)	1964			
	May 13 (8)	June 24 (10)	July 30 (6)	August 31 (8)
A, % (maximum-minimum)	45.1 (57.6-36.2)	42.3 (68.6-16.2)	75.5 (89.5-64.3)	71.4 (79.4-53.7)
B, % (maximum-minimum)	41.3* (65.0-23.7)	61.6 (92.5-40.0)	98.2 (99.5-97.0)	97.4 (99.5-94.0)
C, %	18.6	26.1	74.1	70.0

MATSUSHIMA OYSTER

Date of collection (Number of specimens)	1964							
	May 8 (9)	June 2 (7)	July 2 (6)	July 15 (8)	July 25 (5)	August 5 (7)	August 12 (10)	August 19 (6)
A, % (maximum-minimum)	45.6 (54.3-35.8)	61.9 (76.6-46.7)	57.9 (69.1-37.3)	57.7 (68.0-47.9)	60.2 (68.0-54.4)	72.9 (80.7-65.2)	/	55.3 (68.6-43.6)
B, % (maximum-minimum)	57.2 (82.3-41.7)	91.7 (97.0-81.5)	96.2 (98.5-95.0)	84.8 (99.0-60.0)	95.7 (96.5-90.0)	96.2 (99.0-95.0)	91.5 (97.0-87.0)	76.9 (93.0-56.0)
C, %	26.1	56.8	55.7	48.9	57.6	70.1	/	42.5

A—Percentage of gonads in transverse section of visceral part

B—Percentage of the substance and germ cells

C—Percentage of gonads in transverse section of central viscera expressed as $A \times \dfrac{B}{100}$

* Although it is not possible to differentiate oocytes and spermatocytes by the naked eye, they can be differentiated under the microscope.

In Stage I, although some gonads still remain, most of them cannot be differentiated from the surrounding connective tissues. In Stage II, the germinal epithelium and the follicles begin to form in the connective tissue referred to in Stage I. Germ cells begin to form in the wall of the follicles and gametogonial cells are produced. In Stage III, the gonads increase in size and thickness, lodging a large number of spermatocytes or oocytes. Consequently, the connective tissues around the gonads decrease. In considering the histology of the tissues during these stages, the morphology of the reproductive organs (5.8.1) should be taken into account.

According to the results (Table 2.17) of the study of the quantitative morphological changes during sexual maturation by Chalkley's method (1943), an analysis of Stage I is highly complicated. In Stage II, 20 to 50% of the viscera consists of gonads when a transverse section is studied. In Stage III, the percentage of gonadal tissues constitutes more than 70%. In Stage IV, which is soon after the fertilization, this decreases to 40%. Hence in Stage III and Stage IV, considerable physiological activity occurs in oysters. It would be interesting to probe into such problems as the effect of this physiological activity on metabolism and on the controlling factors of the latter. In the case of oysters collected from Onagawa fishing grounds, spawning occurred only once in a year, but the oysters collected from Matsushima Bay showed rapid sexual maturation, and spawning occurred twice in a year. Hence it must be assumed that physiological activity is greater in the Matsushima oysters.

2. *Tissue respiration and changes in tissue RQ during reproduction:* In order to study the physiological activity of different tissues during reproduction, seasonal changes in the respiratory metabolism of the main organs of the body (digestive diverticulum, gills, pallium) were observed with the help of Warburg's manometer. In both Onagawa and Matsushima Bay the O_2 uptake of the tissues of the oyster during the entire period of the experiment decreased in the order of digestive diverticulum > gills > pallium. In view of the fact that the physiological activity in the Matsushima oyster is greater, various tissues of this oyster showed a greater decrease in the respiratory coefficient during and soon after fertilization (Figures 2.60 and 2.61) than did the oysters of Onagawa Bay.

As far as the gills are concerned, the RQ value of the oysters of both bays was in the vicinity of 1.0 regardless of the course of reproduction. However, in the digestive diverticulum and pallium, the value of RQ during Stages I and II was 1.0 or more, and in III and IV, 0.7 or less (Figures 2.62 and 2.63). Hence chemical changes in the respiratory base of the two tissues (digestive diverticulum and pallium) seem to occur during reproduction. In higher animals, the RQ is about 1.0 when the source of energy is carbohydrate, and about 0.7 when the source of energy is fat. Hence the respiratory base of the gills is carbohydrate, but apparently in the digestive diverticulum and pallium, a change from carbohydrate to fats occurs, during the course of sexual maturation.

The seasonal relationship between O_2 consumption and RQ shows, as explained earlier, that the O_2 consumption in the gills of Matsushima oysters varies greatly, but the RQ is usually 1.0. This seems to indicate that fat cannot be utilized as the basic substance for cell respiration in the gills, or that the energy production required for the physiological functions of the gills is not sufficient because of the low rate of fat metabolism. The minimum value of oxygen consumption by the digestive diverticulum and pallium of Matsushima oysters is seen when the RQ is below 0.7. At the same time, the rate of respiration is high in the digestive diverticulum when the RQ is 1.0 or more. Hence, as seen in the gills, carbohydrates are also more effective than fats as a source of energy in the digestive diverticulum and the pallium.

3. *Changes in the distribution of fat and glycogen during fertilization:*   In order to prove that during the course of sexual maturation there is a change from carbohydrates to fat, it is essential to carry out a histochemical study related to seasonal changes in the distribution of chemical components in the body. As explained earlier under section 5.2, "Digestion and Nutrition," there is a reduction in the glycogen and fat of the connective tissues during sexual maturation (Figure 2.31). On the other hand, there is an accumulation of fat in the visceral epithelium, for example, the epithelium of the digestive diverticulum. Hence it seems that during the course of sexual maturation, there is a conversion from glycogen to fat as well as the movement of fat itself. The seasonal changes in the distribution of such components are much more pronounced in the Matsushima oyster than in the Onagawa oyster. Hence it is clear that seasonal changes are related to reproduction (Mori *et al.*, 1965b).

Histological studies carried out simultaneously with the aforementioned observations show that the large amoebocytes of Sudan III positive, penetrate the connective tissues around the digestive diverticulum and intestine, during the period from maturation to spawning. This has already been shown in the photograph (Figure 2.32) in section 5.3, "Circulation," and would seem to indicate that the movement of fat by the phagocytes is very active during the reproductive period.

4. *Reproduction and steroid metabolism:*   In vertebrates, steroids play an important part as the so-called sex hormone in reproduction, and carbohydrate and fat metabolisms are thought to participate in its functional mechanism. But this aspect is not clear in invertebrates. As explained earlier, there are severe seasonal variations in carbohydrate and fat metabolism in oysters. Steroids are assumed to control these metabolism. It has been reported that glycogen is utilized during the synthesis of sex hormones in higher animals (Figure 2.64). From the results explained earlier, it is clear that there is characteristic decrease in glycogen in the connective tissue system of the oyster, accompanied by fat accumulation in the visceral epithelium during the course of reproduction. Hence the synthesis of steroids at the cost of glycogen can be expected even in the oyster.

The presence of a gonadal steroid has been confirmed even among marine invertebrates including bivalves. Since estrogen is also found in plants, there is a possibility of marine invertebrates deriving estrogen through food materials from these plants. Hence to say that a gonadal steroid is present in the body does not necessarily mean that there is a mechanism for synthesizing it (Hoar, 1965). Therefore, it is necessary to trace the course of steroid metabolism in relation to reproduction, either by enzymological study or by physiological study by considering the special features of the organs and the tissues. This type of work has not been done in any invertebrates, let alone the oyster.

A histological study was done on the distribution of Δ^5-3β-hydroxysteroid dehydrogenase (3β-DH) and 17β-hydroxysteroid dehydrogenase (17β-DH), well-known to play an important role in the synthesis of the sex steroid in various tissues. The results showed that 3β-DH activity was found in the visceral ganglia, the epithelial

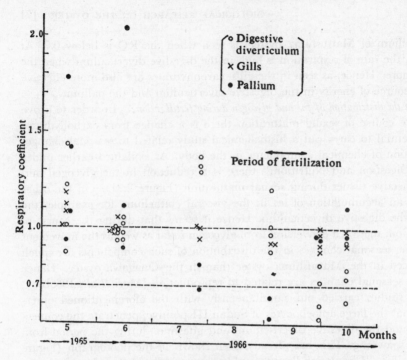

Figure 2.62. Respiratory coefficient (RQ) in the tissues of Onagawa oysters (measurement taken at 25°C) (Mori, 1968).

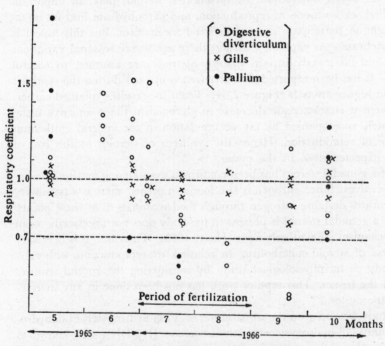

Figure 2.63. Respiratory coefficient (RQ) in the tissues of Matsushima oysters (measurement taken at 25°C) (Mori, 1968).

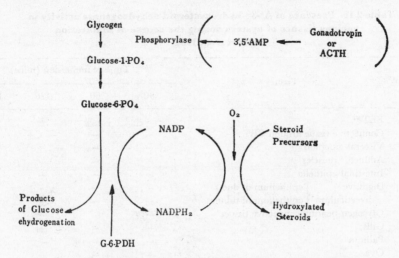

Figure 2.64. Correlation of glycogen breakdown, G-6-PDH system, and steroid metabolism (Haynes *et al.*, 1960; Kobayashi, 1966). The steps shown in parentheses have not yet been confirmed in oysters.

tissues (EET) near the adductor muscles, the interstitial cells of the glycogen-bearing connective tissues between the gonads and the digestive diverticulum, and the interstitial cells in the gonads, etc., of mature oysters (Table 2.18). On the other hand, 17β-DH activity was seen in the nephridia, the digestive diverticulum, and the epithelium of the intestine of the animals undergoing maturation (Table 2.19). A seasonal study of 17β-DH activity (Table 2.20) showed that it was absent in Stage I of reproduction, but became strong in Stage II. The activity was at its maximum in Stage III, but gradually disappeared after fertilization. This type of seasonal change was seen in both the Onagawa and Matsushima oysters. As mentioned earlier, in Matsushima Bay oysters in which sexual maturation as well as spawning is repeated, the activity in the nephridia occurs much earlier than in the oysters of Onagawa Bay. Hence it is clear that 17β-DH activity varies in relation to sexual maturation. It can be said that steroid synthesis begins in EET and in the interstitial cells of gonads in Stage II, and continues up to the beginning of Stage IV.

Glucose-6-phosphate dehydrogenase (G-6-PDH) activity, the endocrinological importance of which is known in the production of nicotinamide-adenine dinucleotide phosphate (NAPDH$_2$), is seen even in the epithelia of nephridia, the digestive diverticulum, and the intestine of oysters (Table 2.21). This substance (NADPH$_2$) is essential in several reactions including steroid synthesis. On the other hand, as mentioned in Table 2.19, the activity of 17β-DH involved in steroid metabolism was found in these tissues. Hence, these tissues are considered to be very important for steroid metabolism. Histological investigations were carried out in order to study in detail the localized G-6-PDH activity and its significance, and the activities of succinate dehydrogenase (S-DH) and malate dehydrogenase (M-DH) in the TCA cycle. The distribution of these two activities differs considerably from that of G-6-PDH (Table 2.21). Thus the G-6-PDH reaction is negative in the reproductive cells, the connective tissues around the nephridia, the adductor muscle, and the

Table 2.18. Presence of Δ^5-3β-hydroxysteroid dehydrogenase activity in various tissues of oysters during the course of maturation
(Mori et al., 1964)

Animals studied	Tissues		Time of immersion (mins)			
			30	60	120	120*²
Oyster	EET*1		—·±	+·#	#	—
	Connective tissue around EET		—	—	—·+	—
	Visceral ganglion		—	—	—	—
	Adductor muscles		—	—	—	—
	Intestinal epithelia		—	—	±	±
	Digestive diverticulum	epithelium of ducts	—	—	±·+	±
		epithelium of tubules	—	—	±	±
	Glycogen-bearing connective tissues		—	—	—	—
	Gills		—	—	—	—
	Pallium		—	—	—	—
	Ova		—	—	—	—
	Sperms		—	—	—	—
	Interstitial cells		—	+	#	—
	Large amoebocytes		—	—	—	—
White rats*³	Adrenation body	Corpuscular layer	+	+·#	#	—
		Columnar and network	#	#·#	#	—
		Marrow	—	—	—	—
	Liver		—	—	—	—
	Kidney		—	—	—	—
	Testis	Sperms	—	—	—	—
		Interstitial cells (Leidig Cell)	—	±·+	+	—

*¹Narrow epithelial tissues near the adductor muscles and visceral ganglia.
*²Control experiment.
*³This experiment was carried out in order to study the reliability of this method.

glycogen-bearing connective tissues. But M-DH and S-DH reactions are positive in these tissues. This type of difference in reactions could also be seen between 3β-DH and 17β-DH, and in the enzymes related to the TCA cycle. Hence, as in the case of mammals, the activity of enzymes related to steroid metabolism and G-6-PDH is localized even in oysters (Cohen, 1959; Wattenberg, 1958). S-DH and M-DH activities do not change much either before or after spawning, but the G-6-PDH activity, like the 17β-DH, decreases in nephridia, the digestive diverticulum, and the intestine (Table 2.22). Hence, it is clear that the G-6-PDH system in the tissues is as closely related to reproduction as it is to steroid metabolism.

The seasonal variations in the distribution of neutral fat, glycogen, 17β-DH, and G-6-PDH in the body of the oyster are shown in Figure 2.65. It is clear from this figure that the changes in the activity of 17β-DH snd G-6-PDH related to steroid synthesis, show a negative correlation to the changes in the amount of glycogen. Hence it is possible to assume that at least a part of the glycogen consumption is utilized in the steroid synthesis.

A study of the physiological influence of the so-called sexual steroid shows that

Table 2.19. Histochemical evidence of the presence of 17β-hydrosteroid dehydrogenase activity in the tissues of oysters during the course of maturation (Mori *et al.*, 1965)

Time of immersion: 120 minutes

Animals studied	Tissues/substrate and enzymes		Estradiol-17 +NAD[2]	without NAD	Testosterone[1] +NAD	without NAD	Estrone +NAD	With NAD only
Oyster	Nephridial epithelium		+++	−	+	−	−	−
	Connective tissue around nephridia		−	−	−	−	−	−
	Adductor muscles		−	−	−	−	−	−
	Visceral ganglia		−	−	−	−	−	−
	Intestinal epithelium		± - +	−	— - ±	−	−	−
	Digestive diverticulum	Epithelium of duct	± - +	−	— - ±	−	−	−
		Epithelium of tubules	++ *3	−	± - + *3	−	−	−
	Glycogen-bearing connective tissues		−	−	−	−	−	−
	Gills		−	−	−	−	−	−
	Pallium		−	−	−	−	−	−
	Gonads	Ova						
		Sperms						
		Interstitial cells						
White rats[7]	Kidney							
	Liver		+ - ++ *5	−	+ - ++ *5	−	−	−
	+ = Duodenum	Ciliated epithelium	+ - ++ *6	−	++ *6	−	−	−
		Crypts of Lieberkühn	−	−	−	−	−	−
		Glands of Brunner	−	−	−	−	−	−
	Adrenalin body		−	−	−	−	−	−
	Ovary	Yolk	−	−	−	−	−	−
		Follicular epithelium	−	−	−	−	−	−
		Follicular membrane	−	−	−	−	−	−
		Interstitial cells	+++	−	++ - +++	−	−	−

[1]Propionate.

[2]NAD—nicotinamide adenine dinucleotide.

[3]The reaction is either weak or nil in dark cell (or crypt cell).

[4]Immersion time=60 minutes.

[5]Cytoplasm of the hepatic cells show reaction.

[6]Reaction absent in striated border.

[7]This experiment was carried out in order to study the reliability of this method.

estradiol-17β accelerates the maturation of female oysters (Figures 2.66 and 2.67). At the same time, the tissue respiration of the ovaries in the course of maturation and the digestive diverticulum in which 17β-DH activity is present (female oysters), is also accelerated (Table 2.23). However, estradiol 17β does not seem to affect any other tissues, but seems to play a selective role in steroid metabolism. This estradiol 17β influences fertilization and development and, according to the results of

Table 2.20. Seasonal variations in 17 β-hydroxysteroid dehydrogenase activity (Mori et al., 1966)

ONAGAWA OYSTERS   1965

Tissue	May 13	June 11		July 9			July 24	August 19	September 27			November 24	
Sex	?	♀	♂	♀	♂	?	♀	♀	♀	♂	?	♂	?
Nephridial epithelium	±*&−	±*&+	+	‡*&±	+−‡	±*&−	+−‡	‡	±*&−	±*&−	±*&−	±*&−	±*&−
Intestinal epithelium	−	−	−	‡	−	−	+−‡	±−‡	−	−	−	−	−
Digestive diverticulum — Epithelium of ducts	−	−±	−±	‡	−±	−	−±	±−‡	−±±	−−+	−	−	−
Digestive diverticulum — Epithelium of tubules	−	+−‡	±−+	±	+−‡	−	±−‡	+−‡	+−+	‡	−	−	−

MATSUSHIMA OYSTERS   1965

Tissue	May 13	May 31	July 7	July 24	August 18			Septem- ber 8			September 27	November 17	December 16
Sex	?	♀	♂	♀	♀	♀	♂	♂	♀	♂	♂	♀	?	?
Nephridial epithelium	+*&−	‡	‡	‡*&+	‡	−	+−‡	+	‡	−+‡	+*−‡	±−‡	±*&−	±*&−
Intestinal epithelium	−	−	−	−	−	−	±−‡	−	−	−	−	−	−	−
Digestive diverticulum — Epithelium of ducts	−	±−+	−	±−+	±−+	−	−	−	−	−	−	−	−	−
Digestive diverticulum — Epithelium of tubules	−	+−+	+−+	±−+	−	−	±−+	−	−	−	−	±	−	−

* Parts showing reaction marked with * are localized.
** Rarely found.

Table 2.21. Glucose-6-phosphate dehydrogenase activity and anhydrous enzyme activity related to TCA cycle in various tissues of oysters during the course of maturation (conditions of immersion: 60 minutes at 37°C) (Mori, 1967)

Tissue	Substrate and enzyme				
	Glucose-6-phosphate[1]		NADP without substrate	Succinate[3] without enzyme	L-Malate[3] +NAD[4]
	+NADP[2]	without NADP			
Nephridial epithelium	++	—	—	++ - +++	+ - ++
Connective tissues around nephridia	—	—	—	+	++ - +++
Adductor muscles	—	—	—	+	+ - ++
Visceral ganglia	— - ±	—	—	±	+
Cerebro-visceral connective	± & ++	—	—	±	+
Intestinal epithelium	± - +	—	—	— - ++	++
Digestive diverticulum { Epithelium of ducts	++ - +++ *5	—	—	++ - +++	+ - ++
Epithelium of tubules	— - ±	—	—	+	+ - ++
Glycogen-bearing connective tissue	—	—	—	— - ±	+++
Gonads { Ova	—	—	—	+++	++
Sperms	—	—	—	+++	
Gonoducts	— - +	—	—	±	++
Transporting ducts	— - ++	—	—	±	+ - +++

*1Barium salt.   *2NADP=nicotinamide-adenine dinucleotide phosphate.   *3Sodium salt.
*4NAD=Nicotinamide-adenine dinucleotide.   *5Reaction is either weak or nil in the basal part.

Table 2.22. Histochemical evidence of the presence of glucose-6-phosphate dehydrogenase activity and anhydrous enzyme activity in the TCA cycle in the tissues of oysters after spawning (conditions of immersion: 60 minutes at 37°C) (Mori, 1967)

Tissue	Substrate and enzyme				
	Glucose-6-phosphate		NADP without substrate	Succinate without enzyme	L-Malate +NAD
	+NADP	without NADP			
Nephridial epithelium	—	—	—	++	++
Connective tissue around nephridia	—	—	—	— - +	+ - ++
Adductor muscle	—	—	—	+	+ - ++
Visceral ganglia	— - ±	—	—	±	+
Cerebro-visceral connective	— - ±	—	—	±	±
Intestinal epithelium	++ *1 & —	—	—	— - ++	++
Digestive diverticulum { Epithelium of ducts	— - ++	—	—	++ - +++	+
Epithelium of tubules	—	—	—	+	+
Glycogen-bearing connective tissue	—	—	—	—	+ - ++
Gonads { Ova	—	—	—	+	— - ±
Sperms	—	—	—	±	
Gonoducts	± - ++ & — *2	—	—	±	++
Transporting ducts	+ - ++ & — *2	—	—	±	++

*1In rare cases.

*2 3 out of 7 showed negative reaction.

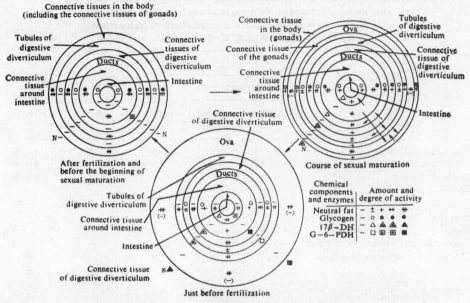

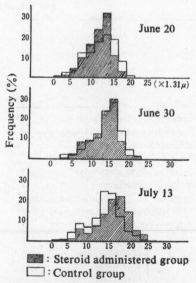

Figure 2.65 Seasonal variations in the distribution of neutral fat, glycogen, 17β-DH, and G-6-PDH in the tissues of oysters.

N=nephridium. Arrows between the tissues indicate the assumed direction of migration of stored fat.

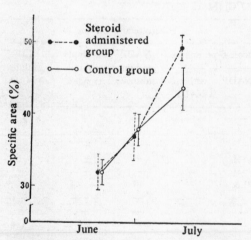

Figure 2.66 Transition in the percentage (specific area) of gonads in a transverse section of the central soft part of a female oyster after the administration of estradiol benzoate, according to the Chalkley (1943) method. Reliability 95% (Mori, 1969).

Figure 2.67. Transition in the frequency of egg nuclei after the administration of estradiol benzoate (Mori, 1969).

Table 2.23. Influence of estradiol-17β[1] on the tissue respiration of oysters during the course of maturation (Mori, 1968b)

Tissues[2]	Experiment number	Respiration rate before adding[3]	Respiration rate 30 minutes after adding[3]	Increase in respiration rate (%)
Ovary	1	22.8	30.6	34.2
	2	18.0	27.6	53.3
	3	25.6	32.0	25.0
	4	20.3	29.2	43.8
	5	20.0	29.0	45.0
	6	26.8	29.5	10.1
	7	26.2	30.1	14.9
	mean			32.3
Testis	1	110.4	235.2	113.0
	2	122.4	187.2	52.9
	3	120.0	233.2	94.3
	4	114.4	174.0	52.1
	5	140.4	220.0	56.7
	6	104.8	200.0	90.8
	mean			76.6
Digestive diverticulum (female)	1	51.6	57.6	11.6
	2	34.3	42.9	25.1
	3	57.1	57.1	0
	4	40.3	52.3	29.8
	5	50.4	56.0	11.1
	6	48.1	52.2	8.5
	mean			14.4
Digestive diverticulum (male)	1	55.9	46.0	−17.7
	2	52.4	57.5	9.7
	3	39.0	37.7	−3.3
	4	43.8	45.0	2.7
	mean			−2.2

[1]Estradiol-3-benzoate, 0.1 mg (suspended in water).

[2]It was difficult to observe the effect in the gills and the pallium.

[3]μlO_2 100 mg fresh weight/hour (measurement taken at 27°C).

experiments (Table 2.24), the rate of fertilization increases from 35% to 94.4% when 3 µg/ml of the substance is administered. In some cases the rate of development increases by even five times. A study of the influence of estradiol-17β on the sex change of two-year-old *Magaki* in Onagawa Bay according to the method described in Table 2.25, showed that the effect was not seen when the administration was started in late May, a time when the gonads had enlarged, but when the administration was started in March to April, the time of early maturation, the number of females in the steroid administered group was more than the number found in any control group (Table 2.26). Hence, some of the male specimens at the early stage of maturation had changed to females by the administration of estradiol-17β. This then can be viewed as the function of the female gonad steroid in oysters.

Table 2.24. Influence of estradiol-17β on fertilization and development of oysters (Mori, 1968c)

Experiment number	Amount of estradiol administered μg/ml	Duration between insemination and beginning of observation	Temperature of water (°C)	Rate of fertilization (%)	Rate of development (%)			Remarks
					B	G	B+G	
1	0	4 hrs 15 mins	25.0	82.5	10.2	0	10.2	Egg concentration: 3,000/ml
	10	,,	,,	93.0	18.0	0	18.0	Material: 2-year-old
	20	,,	,,	94.5	50.0	0	50.0	Magaki collected from Onagawa Bay June 13
2	0	5 hrs 0 mins	22.0	80.6	7.1	2.4	9.5	Egg concentration: 2,600/ml
	10	,,	,,	98.0	35.2	7.1	42.3	Material: same as above
	20	,,	,,	100.0	25.5	11.9	37.4	
	30	,,	,,	94.6	22.9	25.7	48.6	
3	0	6 hrs 15 mins	27.0	35.0	57.1	7.1	64.2	Egg concentration: 130/ml
	3	,,	,,	94.4	17.6	52.9	70.5	Material: 2-year-old
	7	,,	,,	68.8	30.4	42.2	72.6	Magaki collected June 6 at Onagawa Bay and reared for nearly 2 weeks in tanks at 17°C
	10	,,	,,	64.3	44.4	22.3	66.7	
4	0	6 hrs 0 mins	27.5	62.6	13.9	8.0	21.9	Egg concentration: 1,500/ml
	3	,,	,,	93.7	29.0	27.7	56.7	Material: same as above
	10	,,	,,	97.1	28.2	26.3	54.5	

Note: $B = \dfrac{\text{number of blastula}}{\text{number of fertilized eggs}} \times 100$, $G = \dfrac{\text{number of gastrula}}{\text{number of fertilized eggs}} \times 100$

Salinity of sea water in all the experiments was 27.3‰.

Table 2.25. Influence of estradiol-17β on the sex change of 2-year-old
Magaki **of Onagawa Bay (Mori et al., 1969)**

Experiment number	Period of experiment	Interval of injections (days)	Amount of estradiol-3-benzoate injection		Date of collection
			mg/time/ specimen	Total weight mg/specimen	
1	March 12 to May 21, 1968	9 - 16	0.10	0.20 - 0.60	April 10
2	April 10 to June 21, 1968	5 - 12	0.02	0.04 - 0.16	April 20
3	May 26 to July 13, 1967	10 - 14	0.10	0.20 - 0.40	June 20

Table 2.26. Influence of Estradiol-17β on the sex change of 2-year-old
Magaki of Onagawa Bay (**Mori** *et al.*, **1969**)

Experiment number	1			2			3		
	S	C	N	S	C	N	S	C	N
Treatment									
Total number	36	13	10	44	20	20	44	—	83
♀	28	9	7	29	12	13	33	—	60
♂	6	4	3	8	5	6	11	—	22
♀♂	2	0	0	7	3	1	0	—	1
♀/♂	4.67	2.25	2.33	3.63	2.40	2.17	3.00	—	2.73

S=steroid injection, C=injection of aseptic sea water (control), N=untreated.

It is clear that steroid metabolism is closely related to reproduction. As in mammals, the steroid plays an important role in the sexual maturation of oysters. As already explained, the steroid seems to control the seasonal variations in carbohydrate and fat metabolism during reproduction.

Figure 2.68 shows the relationship between the changes in carbohydrate and fat during reproduction and physiological activity. By vigorous feeding just before sexual maturation (I), oysters collect and store glycogen and fat in the connective tissues. Glycogen is consumed when sexual maturation begins and the steroid metabolism as well as the G-6-PDH system progress. Just before fertilization (III), there is a decrease in the physiological activity along with the changes in the respiratory base (the dependence on fat increases) (Mori *et al.*, 1965a). On the other hand, the fat shifts toward the ova and the epithelial system of the viscera when the gonads are formed. Just before fertilization (III), fat is utilized as the main respiratory base.

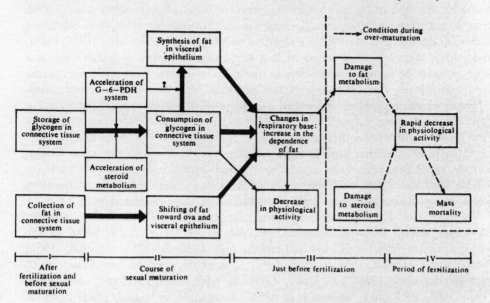

Figure 2.68. Shifting of carbohydrate and fat, and changes in
the physiological activities during reproduction in oysters.

During this time, i.e. until fertilization begins (III), considerable energy is utilized which is expressed in the form of a physiological burden. When there is over-maturation as in Matsushima oysters (Tamate *et al.*, 1965), the fat metabolism is adversely affected (Mori *et al.*, 1965b), as is the steroid metabolism (Mori *et al.*, 1966). This brings about a further decrease in the physiological activity (Mori *et al.*, 1965a). At times this results in a mass mortality of oysters.

In the present investigation, the relationship between reproduction and steroid metabolism in the oyster becomes clear. At the same time, on the basis of the action of the sexual steroid in maturation, the condition of maturity, the period of spawning, and the number of spawnings, over-maturation, or defective spawning, could be determined. Furthermore, the accelerating effect of this action on fertilization and development was also noted, and the authors would like to point out that these data may be useful in further improving the collection of oyster seedlings.

(Mori)

6. Phenomenon of mass mortality

6.1 General outline and pathological and histological studies

Mass mortality of oysters has been reported in Japan from the end of the 19th century to the present, and in other countries from time to time. The common features of mass mortality in various countries are that the oysters which die are parent of 2-year-old or more, and the mass mortality occurs during the period from the liberation of eggs and sperms to the post-fertilization in summer. The mortality rate is higher among fully grown oysters. The primary factors contributing to mass mortality are said to be high water temperature, high salinity, water pollution by organic materials, adverse conditions of the sea bottom, parasites, and predatory animals (Ogasawara *et al.*, 1962). According to Kusakabe (1931), Seno (1935, 1936), and others, mass mortality occurs in that age group of oysters in which there is a large amount of follicles. According to these investigators, mass mortality occurs due to faulty spawning. In view of the fact that mass mortality occurred simultaneously in all the regions of western Europe, and that neither an infection nor a parasite was detected, Spark (1960) assumed that death was due to some pathological defect in the oysters themselves. Imai *et al.* (1965, 1968), who studied the mass mortality at Matsushima Bay, thought that the rapid increase in the meat weight, and the rapid development of gonads due to rich nutrition from organic waste, as well as high water temperatures, contributed to an unusual metabolic acceleration which brought about some pathological changes in these oysters (Figure 2.69).

Various types of pathological organisms, for example, *Dermocystidium marinum*, *Bucephalus cuculus*, *Nematopsis ostrearum*, *Hexamita* sp., MSX, etc., are found in the Virginia oyster (Prytherch, 1940; Landau and Galtsoff, 1951; Mackin, 1951; Andrews and Hewatt, 1957; Burton, 1961). Amoeboid parasites are found in *C. gigas* (Sindermann, 1966). In Japan, Takeuchi, Matsubara, Hirokawa, and others (1955, 1956, 1957) and Matsuo *et al.* (1957) were able to identify three types of bacteria belonging to the gram-positive group of *Achromobacter* in the digestive organs and gonads of oysters which died on a mass scale at Matsushima Bay.

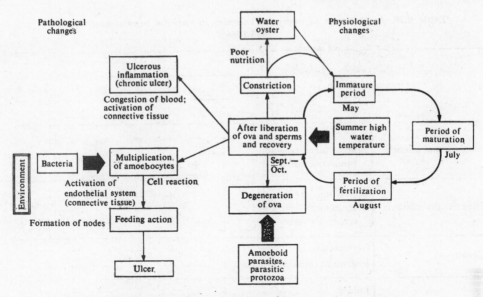

Figure 2.69. Pathological aspects of the mass mortailty of oysters at
Matsushima Bay (1968).

Hence, they inferred that the mass mortality occurred as a result of infection by
these pathological organisms. On the other hand, as shown in Tables 2.27 and 2.28,

Table 2.27. Frequency of bacterial colonies appearing in *Magaki*
of Matsushima Bay (1965-1966)

Items	Months											
	4	5	6	7	8	9	10	11	12	1	2	3
Number specimens examined	20	10	20	40	40	30	20	10	10	10	10	10
Number specimens in which bacteria detected	1	0	0	0	1	5	4	1	1	3	0	2
Infection rate, %	5	0	0	0	2.5	16.7	20	10	10	30	0	20
Mortality, %	0	0	0	0	23	16	8	0	0	0	0	0

Note: Mortality herein refers to that at the time of the investigation.

and Figures 2.70, 2.71, 2.72, and 2.73, oysters infected by bacteria as well as amoe-
biasis were found at Matsushima Bay. However, no clear relationship was found
between the period of mass mortality and the period of infection. The frequency
of such infections was also low and, at present, it is not possible to conclude that
they were the primary causal factors of the mass mortality (Numachi *et al.*, 1965;
Sugiwara *et al.*, 1966; Imai *et al.*, 1968).

Several investigations are being carried out on the large-scale mortality occurring
in various countries. In order to know the actual conditions of mass mortality, it
is necessary to study the pathological and histological changes, whether the mor-

Table 2.28. Appearance of bacterial colonies in various organs and tissues (1965-1966)

Organs and tissues		4	8	9	10	11	12	1	3
						Months			
Stomach and intestine	Epithelium	−	−	+	−	−	−	−	−
	Mucous membrane and connective tissue	+	+	⧺	⧺	+	+	+	⧺
Digestive diverticulum	Epithelium	−	−	+	−	−	−	−	−
	Mucous membrane and connective tissue	+	+	⧺	+	−	−	+	+
Gonads	Germplasm	−	−	+	+	−	−	−	−
	Interstitial connective tissue	+	+	+	+	−	+	+	+
Blood vessels	Lumen	+	−	+	+	−	−	+	+
	Connective tissue around blood vessel	+	−	+	+	−	−	+	+
Skin and pallium		−	−	+	+	−	−	−	+

− = Absent, + = Present, ⧺ = Present in large amount.

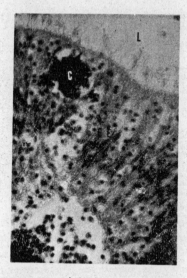

Figure 2.70. Bacterial colonies (C) found in the gastric epithelium (GE).

L—lumen of the stomach (Imai et al., 1968).

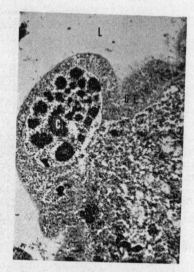

Figure 2.71. Bacterial colonies (C) in the mucous membrane of the stomach. Many bacterial colonies are engulfed by phagocytes and give rise to ulcers.

GE—gastric epithelium (Imai et al., 1968).

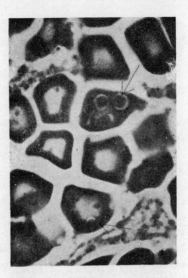

Figure 2.72. Bacterial colonies (C) in the connective tissues around the blood vessels (BV). Bacterial colonies seen where the phagocytes form groups (Imai *et al.*, 1968).

Figure 2.73. Amoeboid parasites (shown by arrows) entering ova (Imai *et al.*, 1968).

tality is due to physiological abnormalities or infection by some pathological organism.

A study of the pathological and histological changes during mass mortality showed that the pathological changes resulted from the infiltration of amoebocytes, reproduction, and changes in the tissues. Localized inflammations were seen. Pathological changes occurred mainly in the digestive diverticulum, stomach, intestine, etc. Among these, the digestive diverticulum was the most affected part and in an infected specimen, a severe ulcerous condition was seen. However, symptoms of poisoning resulting from poisonous substances such as hydrogen sulfide or low oxygen in the over-used medium were not seen (Fujita *et al.*, 1953, 1955; Tamate *et al.*, 1965; Araki *et al.*, 1965, 1966; Numachi *et al.*, 1965; Sugiwara *et al.*, 1966; Imai *et al.*, 1968). Whether a relationship between these pathological changes and mortality exists is presently under discussion; to date no common view has been reached.

As a countermeasure to mass mortality, the rearing of 2 to 3-year-old oysters, the usual victims of mass mortality, has been stopped and the rearing of one-year-olds, which are less susceptible to mass mortality, taken up (Ogasawara *et al.*, 1962; Araki *et al.*, 1965). Simultaneously, pathological studies to determine whether mass mortality is in any way related to neural poison, a faulty circulatory system, or diseases of the digestive diverticulum are in progress.

(Sugiwara)

6.2 Pathological and physiological studies

The relationship between the shifting of carbohydrate and fat and the changes in the physiological activity during reproduction, was already discussed under section 5.8.2 "Reproduction." The discussion was supplemented by Figure 2.68. From this figure it is clear that there are parts in the oyster which do not show any change during maturation (parts shown by a dotted line). In Matsushima oysters in which mass mortality occurs, the changes take place rapidly and, moreover, the conditions during over-maturation are distinctly expressed. Hence, a discussion on over-maturation, faulty fat metabolism, faulty steroid metabolism, and a sudden decrease in the physiological activity, seems relevant.

6.2.1 OVER-MATURATION

This is a pathological condition which occurs when the ripened reproductive units remain inside the gonads for a considerable period. An abnormal increase of glycogen and free fatty acids in the ova results, together with a decrease in the activity of anhydrous succinaze in the sperms, and a low polymerization of nuclear DNA (Tamate et al., 1965). The size of the nucleus in oocytes measured in the oysters grown in Matsushima Bay (Annual Report of 1964) is shown in Figures 2.74 and 2.75. In early June, the nucleus has already reached an average of 30 μ. A duplicate nucleolus characterizes the nucleus of the ovum when it has completed the maturation phase. Since spawning does not occur until early July, the ova which have matured by that time are forced to remain in the ovary for nearly a month. This leads to

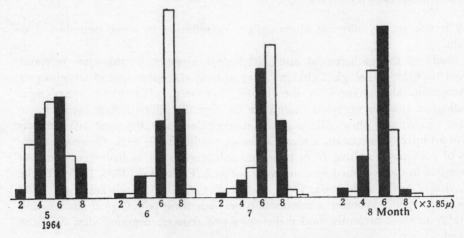

Figure 2.74. Changes in the size of the egg nucleus of 2-year-old
Matsushima Bay oysters (Tamate et al., 1965).

the condition known as over-maturation. During this period, there is a concentration of ova coupled with a heavy physiological burden on the oysters during the time just before spawning. The measurement of the surface area by Chalkley's method shows that in early June the transverse section of the middle part of the body contains nearly 60% of the gonadal tissues. This condition continues up to mid-July when spawning begins, at which time there is a slight reduction in the percentage of gonadal tissues, but this again recovers and reaches 70%. In early August (mid-August), along with large-scale spawning, a decrease in the amount of gonadal tissues recurs. Thus, it is clear that a gonadal overgrowth occurs during the spawning period, since the gonads cover 60% one month before spawning and reach 70% by the end of July.

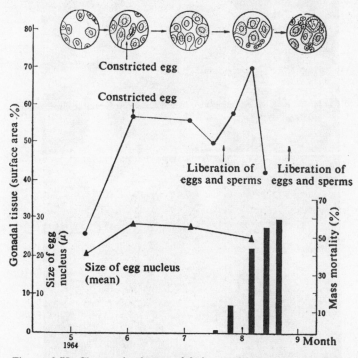

Figures 2.75. Changes in the gonadal tissues of 2-year-old Matsushima Bay oysters (Tamate *et al.*, 1965).

6.2.2 FAULTY FAT METABOLISM

The histology of the epithelium of the digestive diverticular duct in which a large amount of free fatty acids was detected in some oysters of Matsushima Bay after sexual maturation, showed that there was localized degeneration (Figures 2.76 and 2.77) (Mori *et al.*, 1965b). When a large amount of fat accumulates in the epithelial system of higher animals, it is considered a pathological condition. If it is accompanied by tissue degeneration, it results in mortality, and mass mortality is often inevitable.

6.2.3 FAULTY STEROID METABOLISM

As discussed earlier and shown in Table 2.20, the 17β-hydroxysteroid dehydrogenase activity found in the epithelium of the digestive diverticulum of the oyster (Matsushima Bay) almost ceases in late July after sexual maturation. The absence of this activity seems to indicate a pathological and/or physiological condition (Mori *et al.*, 1966) a conclusion supported by the fact that the inflammation as well as the constriction of the digestive diverticulum is found in summer in the oysters collected from the same bay (Tamate *et al.*, 1965). Figure 2.78 shows the changes in the surface area of the epithelium of the digestive diverticular duct of Matsushima

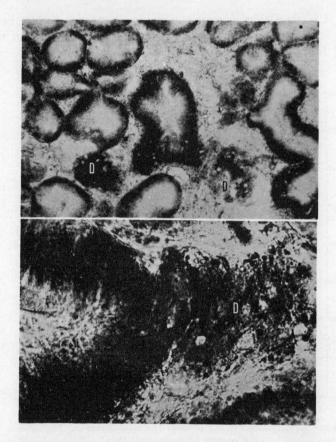

Figures 2.76 (top) and 2.77 (bottom). Epithelium of digestive diverticular duct (D) in which a large amount of free fatty acids was detected, which resulted in the localized degeneration of the tissue. Figures show sections of gelatin-embedded tissues of a 2-year-old Matsushima Bay oyster during sexual maturation (Mori *et al.*, 1965b).

oysters, measured by Chalkley's method. When spawning begins, the epithelium shows constriction and degeneration.

6.2.4 SEVERE DECREASE OF PHYSIOLOGICAL ACTIVITY

Mass mortality does not occur at Onagawa Bay and the growth and meat of these oysters are very good. Hence this bay is good for oyster culture. The physiological activity of 2-year-old oysters (seedlings collected from Mangoku-Ura in the summer of 1961) grown in this environment, shows seasonal variations (Figure 2.79). In other words, as sexual maturation progresses, there is a mild decrease in the beating rate of ctenidia. On the other hand, when the oysters are transplanted to Matsushima Bay, there is a rapid decrease in the physiological activity during the fertilization period (Figure 2.80). This decrease is not noticed in Onagawa Bay. The same trend is also observed on the basis of the environmental resistance of ctenidia to sea water with high salinity (Figures 2.81 and 2.82), an insufficient oxygen supply, acidity of sea water (Mori *et al.*, 1965a), etc. Seasonal variations in the physiological

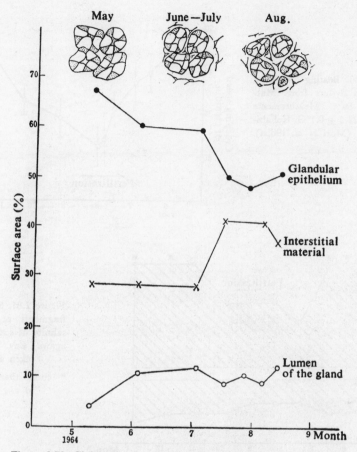

Figure 2.78. Changes in the digestive diverticular duct of a 2-year-old Matsushima Bay oyster (Tamate *et al.*, 1965).

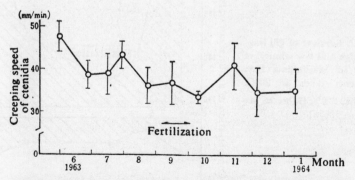

Figure 2.79. Beating rate of ctenidia in oysters from Onagawa Bay. Measurements taken at $22.5 \pm 0.1°C$. Reliability 95% (Mori, 1965a).

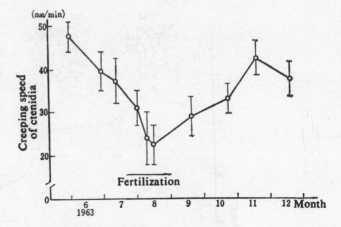

Figure 2.80. Beating rate of ctenidia in oysters from Matsushima Bay. Measurements taken at $22.5 \pm 0.1°C$. Reliability 95% (Mori *et al.* 1965a).

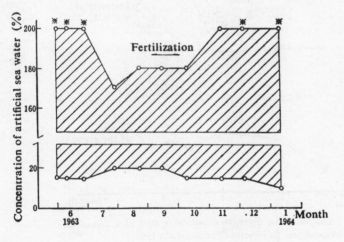

Figure 2.81. Survival of gill fragments at high and low salinity of sea water in Onagawa Bay. Measurements taken at 20°C.

*—more than 200% (Mori *et al.*, 1965a).

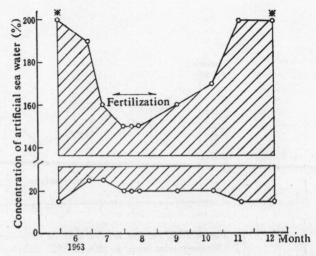

Figure 2.82. Survival of gill fragments at high and low salinity of sea water in Matsushima Bay. Measurements taken at 20°C.

*—more than 200% (Mori *et al.*, 1965a).

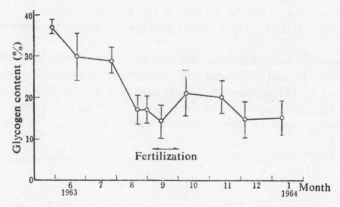

Figure 2.83. Glycogen content in all the soft parts except the adductor muscles of oysters from Onagawa Bay. Dry weight %. Reliability 95% (Mori *et al.*, 1965a).

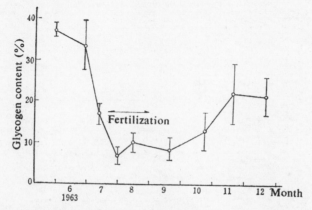

Figure 2.84. Glycogen content in all the soft parts except the adductor muscles of oysters from Matsushima Bay. Dry weight %. Reliability 95% (Mori *et al.*, 1965a).

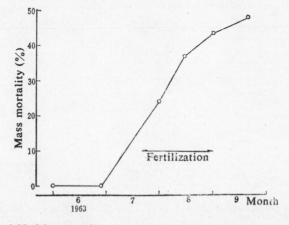

Figure 2.85. Mass mortality of oysters from Matsushima Bay. Observations made for groups of 80 oysters (Mori *et al.*, 1965a).

activity show a trend similar to that of the glycogen content in all the soft parts (Figures 2.83 and 2.84), which proves that there is a close relationship between the two.

Mass mortality studied in these two bays showed just a 6% mortality in Onagawa Bay between the end of August and mid-September, but when the oysters were transplanted to Matsushima Bay, mass mortality increased (Figure 2.85). Hence, it is clear that both the time and the degree of mass mortality are closely related to the decreasing trend of physiological activities.

(Mori)

Bibliography

1. AMEMIYA, I., 1926a. Hermaphroditism in the Portuguese Oyster, *Proc. Roy. Phys. Soc.*, **21,** 97.

2. ——, 1926b. Notes on Experiments on the Early Developmental Stage of the Protuguese, American, and English Native Oyster, with Reference to the Effect of Varying Salinity, *J. Mar. Biol. Ass.*, **14,** 61.

3. ——, 1928. Ecological Studies of the Japanese Oyster, with Special Reference to the Salinity of Their Habitats, *J. Coll. Agr. Univ. Tokyo.* **9,** 333.

4. ANDREWS, J. D., and W. G. HEWATT, 1957. Oyster Mortality Studies in Virginia: II. The Fungus Diseases Caused by *Dermocystidium marinum* in the Oysters of Chesapeake Bay, *Ecol. Monogr.*, **27,** 1-26.

5. ARAKI, F., T. TAKEUCHI, T. KIMURA, and Y. MIYOSHI, 1965. Hiroshima-kenka no kaki heishi no byorisoshikigakuteki kenkyu ni tsuite (On Pathological Studies of Mass Mortality of Oysters in Hiroshima Prefecture), Report on the Pathological Studies Related to Mass Mortality of Oysters, 2-5.

6. ——, ——, ——, ——, 1966. Umida-Wan ni okeru suishitsu no ijo ni yoru kaki heishi no soshikigakuteki kenkyu (Histological Studies Related to Mass Mortality of Oysters Due to Abnormal Characteristics of the Water in Kaida Bay), *Ibid.*, 5-6.

7. AWATI, P. N., and H. S. RAI, 1931. *Ostrea cucullata* (The Bombay Oyster), *Indian Zool. Mem.*, **3,** 1-107.

8. BANGHMAN, 1948. An Annotated Bibliography of Oysters, with Pertinent Material on Mussels and Other Shellfish, and an Appendix on Pollution, Tex. A. & M. Res. Found., College Station.

9. BARGETON, M., 1942. Les variations saisonières du tissu conjonctif vésiculeux de l'huifre, *Bull. biol.*, **76,** 175-191.

10. BLACK, R. E., 1962a. Respiration, Electron-transport, and Krebs-cycle Enzymes in Early Developmental Stages of the Oyster, *Crassostrea virginica, Biol. Bull.*, **123,** 58-70.

11. ——, 1962b. The Concentrations of Some Enzymes of the Citric Acid Cycle and Electron-transport System in the Large Granule Fraction of Eggs and Trochophores of the Oyster, *Crassostrea virginica, Ibid.*, **123,** 71-79.

12. BOUCHON-BRANDELY, 1882. Sur la sexualite de l'*Ostrea edulis* et *l'Ostrea angulata*, et ficondation artificielle l'huitre de Portugal, *C. R. Acad. Sci. Paris.* 95-256.

13. BROOKS, W. K., 1879. Abstract on Observations upon the Artificial Fertilization of Oyster Eggs and on the Embryology of the American Oyster, *Amer. J. Sci.*, **43.**

14. ——, 1880. The Development of the American Oyster, *Studies Biol. Lab. John Hopkins Univ.*, **1,** 1.

15. BRUNEL, A., 1938. Su la dégradation des substances dórigine purique chez les Mollusques Lamellibranches, *Comptes Rendus Hebdomadaires des Séances de*

l'Académie des Sciences, **206**, 858-860.

16. BURTON, R. W., 1961. Distribution of Oyster Microparasites in Chesapeake Bay, *Maryland Nat. Shell. Ass.*, **52**, 65-74.

17. CASPERS, H., 1950. Die Lebensgemeinschaft der Helgolander Austerbank, *Helgoland wiss. Meeresunters*, **3**, 119-169.

18. CHAET, A. B., and R. A. McCONNAUGHY, 1959. Physiologic Activity of Nerve Extracts, *Biol. Bull.*, **117**, 407-408.

19. CHALKLEY, H. W., 1943. Method for the Quantitative Morphologic Analysis of Tissues, *J. Nat. Cancer Inst.*, **4**, 47-53.

20. CHESTNUT, A. F., 1946. Some Observations on the Feeding of Oysters with Special Reference to the Tide, *Conv. Nat. Shellfish. Ass.*, 22-27.

21. CLELAND, K. W., 1947. Some Observations on the Cytology of Oogenesis in the Sydney Rock Oyster *(Ostrea commercialis* I and R*)*, *Proc. Linn. Soc., N.S.W.*, **72**, 159-182.

22. ——, 1951. The Enzymatic Architecture of the Unfertilized Oyster Egg. *Australian J. Exp. Biol. Med. Sci.*, **29**, 35-45.

23. COE, W. R., 1931a. Sexual Rhythm in the Californian Oyster *(Ostrea lurida)*, *Science*, **74.**

24. ——, 1931b. Spermatogenesis in the Californian Oyster *(Ostrea lurida)*, *Biol. Bull.*, **61**, 309-315.

25. ——, 1932a. Sexual Phases in the American Oyster *(Ostrea virginica)*, *Ibid.*, **63,** 419-441.

26. ——, 1932b. Development of the Gonads and the Sequence of the Sexual Phases in the Californian Oyster *(Ostrea lurida)*, *Bull. Scripps Inst. Oceangr. Tech. Ser.*, **3**, 119-441.

27. ——, 1932c. Histological Basis of Sex Changes in the American Oyster *(Ostrea virginica)*, *Science.* **76**, 125-127.

28. ——, 1934. Alternation of Sexuality in Oysters, *Am. Nat.*, **68**, 236-251.

29. ——, 1936. Environment and Sex in the Oviparous Oyster, *Ostrea virginica*, *Biol. Bull.*, **71**, 353-359.

30. COHEN, R. B., 1959. Histochemical Localization and Metabolic Significance of the Glucose-6-phosphate Dehydrogenase System in the Adrenal Cortex, *Proc. Soc. Exp. Biol. Med.*, **101**, 405-407.

31. COLE, H. A., 1936. Experiments in the Breeding of Oysters *(Ostrea edulis)* in Tanks, with Special Reference to the Food of the Larva and Spat, *Fish. Invest.*, Series II, 15, London.

32. ——, 1938a. New Pallial Sense Organs in the Early Fixed Stage of *Ostrea edulis, Nature*, **141**, 161.

33. ——, 1938b. The Fate of the Larval Organs in the Metamorphosis of *Ostrea edulis, J. Mar. Biol. Ass.*, **22**, 469-485.

34. ——, 1939. Further Experiments in the Breeding of Oysters in Tanks, *Fish. Invest.*, Series II, 16, London.

35. ——, and E. W. KNIGHT JONES, 1939. Some Observations and Experiments on the Setting Behavior of Larvae of *Ostrea edulis, J. du Conseil*, **14**, 86-105.

36. COLLIER, A., S. RAY, and W. MAGNITZKY, 1950. Preliminary Note on the Naturally Occurring Organic Substances in Sea Water Affecting the Feeding

of Oysters, *Science*, **111**, 151-152.

37. DALIDO, P., 1948. *L'huitre du Morbihan: Etude Economique et Sociale*. Marcel Rivière & Cie., Paris, 149 p.

38. DANNEVIG, A., 1951. Lobster and Oyster in Norway, *Rapp. Cons. int. Explor. Mer.*, **128**, 92-96.

39. DAVIS, H. C., 1949. On the Cultivation of Larvae of *Ostrea lurida*, *Anat. Rec.*, **105**, 591.

40. ——, and A. D. ANSELL, 1962. Survival and Growth of Larvae of the European Oyster, *O. edulis*, at Lowered Salinities, *Biol. Bull.*, **122**, 33-39.

41. DELAUNAY, H., 1931. L'excrètion azotèe des invertèbrès, *Biological Reviews of the Cambridge Philosophical Society*, **6** (3), 265-301.

42. ERDMANN, W., 1934. Aufzucht von Austern Larven, *Der Fischmarkt.*, *N. F.*, **1**, 340-341.

43. FINGERMAN, M., and L. D. FAIRBANKS, 1957. Investigations of the Body Fluid and "Brown-spotting" of the Oyster, *Proc. Nat. Shellfish. Ass.*, **47**, 146-147.

44. ——, ——, 1958. Histo-physiology of the Oyster Kidney, *Ibid.*, **48**, 125-133.

45. Fujimori, S., 1949. Kaki hassei no seiikujoken tsuki suminoegaki-gata oyobi magaki-gata no ido ni tsuite (Variations in *C. rivularis* and *C. gigas* in Relation to the Conditions of Oyster Development), *Ariake-Kai Higata Riyo Hokoku*, 351-444.

46. FUJITA, T., T. MATSUBARA, Y. HIROKAWA, and F. ARAKI, 1953. Hiroshima-Wan San magaki ni mirareru ensho-sei byozo no byori soshikigakuteki kenkyu I (Pathological Aspects of Inflammation Found on *Ostrea gigas* of Hiroshima Bay I), *Nihon Suisan Gakkaishi (Bulletin of the Japanese Society of Scientific Fisheries)*, **19** (6), 766-770.

47. ——, ——, ——, ——, 1955. *Ibid.*, II, *Ibid.*, **20** (12), 1063-1065.

48. GALTSOFF, P. S., 1930. The Role of Chemical Stimulation in the Spawning Reaction of *Ostrea virginica* and *Ostrea gigas*, *Proc. Nat. Acad. Sci. Washington*, **16**, 555-559.

49. ——, 1931. Specificity of Sexual Reactions in the Genus *Ostrea*, *The Collect. Net.*, **6**, 277.

50. ——, 1932. Spawning Reactions of Three Species of Oysters, *J. Wash. Acad. Science*, **22**, 65-69.

51. ——, 1935. Physiology of Ovulation and Ejaculation in the Oyster, *The Collect. Net.*, **10**, 261.

52. ——, 1938. Physiology of Reproduction of *Ostrea virginica*: I. Spawning Reactions of the Female and Male, *Biol. Bull.*, **74**, 461-486.

53. ——, 1938. Physiology of Reproduction of *Ostrea virginica*: II. Stimulation of Spawning in the Female Oyster, *Ibid.*, **75**, 286-307.

54. ——, 1940. Physiology of Reproduction of *Ostrea virginica*: III. Stimulation of Spawning in the Male Oyster, *Ibid.*, **78**, 117-135.

55. ——, 1964. The American Oyster *Crassostrea virginica* Gmelin, *Fish. Bull.*, **64**, 1-480.

56. ——, and D. V. WHIPPLE, 1931. Oxygen Consumption of Normal and Green Oysters, *Bull. U. S. Bureau Fisheries.* **46**, 489-508.

57. ——, and R. O. SMITH, 1932. Stimulation of Spawning and Cross-fertilization

between American and Japanese Oysters, *Science,* **76.**

58. GALTSOFF, P. S., and D. E. PHILPOTT, 1960. Ultrastructure of the Spermatozoon of the Oyster, *Crassostrea virginica, J. Ultrastructure Res.,* **3,** 241-253.

59. GEORGE, W. C., 1952. The Digestion and Absorption of Fat in Lamellibranches, *Biol. Bull.,* **102,** (2), 118-127.

60. GOULD, A. A., 1850. Mollusks from the South Pacific, *Proc. Boston Soc. Nat. Hist.,* **3,** 340-346.

61. GROSS, F., and J. C. SMYTH, 1946. The Decline of Oyster Populations, *Nature,* **157,** 540-542.

62. GUNTER, G., 1950. The Generic Status of Living Oysters and the Scientific Name of the Common American Species, *Amer. Midl. Nat.,* **43,** 438-449.

63. HAGMEIER, A., 1941. Die intensive Nutzung des nordfriesischen Wattenmeeres durch Austern und Muschelkultur, *Z. Fisch.,* **39,** 105-165.

64. HASKIN, H. H., L. A. STAUBER, and J. A. MACKIN, 1966. *Minchinia nelsoni* n. sp. *(Haplosporida, Haplosporidiidae)*: Causative Agent of the Delaware Bay Oyster Epizootic, *Science* **153,** 1413-1416.

65. HAYNES, R. C. JR., E. W. SUTHERLAND, and T. W. RALL, 1960. The Role of Cyclic Adenylic Acid in Hormone Action, *Recent Progr. Hormone Research,* **16,** 121-138.

66. HEWATT, W. G., and J. D. ANDREWS, 1956. Temperature Control Experiments on the Fungus Disease, *Dermocystidium marinum* of Oysters, *Proc. Nat. Shellfish. Ass.,* **46,** 129-132.

67. HIGASHI, N., ed., 1964. *Igaku Seibitsugaku-Yo Denshi Kenbikyo-gaku* (Electron Microscope Study), Tokyo: Bunkodo.

68. HIGASHI, S., 1964. Tansuisan nimaigai ikechogai no seishi keisei katei no denshikenbikyoteki kenkyu [Electron Microscope Study of Spermatogenesis of Fresh Water *Hyriopsis schlegelii* (Ikechogai)], *Nihon Suisan Gakkaishi (Bulletin of the Japanese Society of Scientific Fisheries),* **30** (7), 564-569.

69. HILLMAN, R. E., 1964. Chromatographic Evidence of Intraspecific Genetic Differences in the Eastern Oyster, *Crassostrea virginica, Systematic Zool.,* **13,** 12-18.

70. HIRASE, S., 1930. On the Classification of the Japanese Oyster, *Jap. J. Zool.,* **3,** 1-65.

71. HOAR, W. S., 1965. Comparative Physiology: Hormones and Reproduction in Fishes, *Annual Review of Physiology,* **27,** 51-70.

72. HOPKINS, A. E., 1932a. Chemical Stimulation by Salts in the Oyster, *Ostrea virginica, J. Exptl. Zool.,* **61**(1), 13-28.

73. ——, 1932b. Sensory Stimulation of the Oyster, *Ostrea virginica,* by Chemicals, *Bull. U. S. Bureau Fisheries.* **47,** 249-261.

74. ——, 1934. Accessory Hearts in the Oyster, *Science,* **80,** 411-412.

75. ——, 1937. Experimental Observations on Spawning, Larval Development, and Setting in the Olympia Oyster, *Ostrea lurida, Bull. U. S. Bureau Fisheries,* **48,** 439-503.

76. HOSHI, T., 1958. Physiology of the Oyster: II. Mode of Occurrence of Cytochrome in the Oyster, *Sci. Rep. Tohoku Univ.,* Ser. IV, **24,** 131-136.

77. ——, and K. TAGUCHI, 1960. Contributions to the Marine Biology from the

Japan Sea: Physiology of the Oyster: III. A Relation between O_2 Consumption and Ciliary Activity in the Gill of the Oyster through the Action of Azide and Cyanide, *J. Fac. Sci. Niigata Univ.*, Ser. II, **4** (1), 33-46.

78. HUMPHREY, G. F., 1944. Glycolysis in Extracts of Oyster Muscle, *Australian J. Exptl. Biol. Med. Sci.*, **22**, 135-138.

79. ——, 1946. The Endogenous Respiration of Homogenates of Oyster Muscle, *Ibid.*, **24**, 261-267.

80. ——, 1949. Adenosinetriphosphatases in the Adductory Muscle of *Saxostrea commercialis, Physiologia Comparata et Oecologia*, **1**, 366-375.

81. ——, 1950. Glycolysis in Oyster Muscle, *Australian J. Exptl. Biol. Med. Sci.*, **28**, 151-160.

82. IMAI, T., 1960. Kaki yoshokujo ni okeru seisan teika gensho to sono taisaku (Decline in the Production of Oysters and Some Counter Measures), Investigations Financed by the Ministry of Education, Government of Japan, 3-7.

83. ——, and M. HATANAKA, 1949. On the Artificial Propagation of the Japanese Oyster, *Ostrea gigas* Thunberg, by Non-colored, Naked Flagellates, *Bull. Inst. Agr. Res. Tohoku Univ.*, **1**, 33-46.

84. ——, and S. SAKAI, 1961. Study of Breeding the Japanese Oyster, *Crassostrea gigas, Tohoku J. Agr. Res.*, **12**, 125-171.

85. ——. ——, and H. OKADA, 1953. Transplantation of the European Flat Oyster, *O. edulis*, into Japanese Water and Its Breeding in a Tank, *Tohoku J. Agr. Res.*, **3**, 311-320.

86. ——, M. HATANAKA, R. SATO, and S. SAKI, 1951. Ecology of Mangoku-Ura Inlet with Special Reference to Seed Oyster Production, *Sci. Rep. Res. Inst. Tohoku Univ.*, **1-2**, 127-156.

87. ——, K. NUMACHI, J. OIZUMI, and S. SATO, 1965. Matsushima Wan ni okeru kaki no tairyoheishi ni kansuru kenkyu II. Ishoku shiken ni yoru heishi yoin no tankyu to bogyosaku no kento (Study on the Mass Mortality of Oysters Occurring at Matsushima Bay: II. Factors Causing Mass Mortality Traced through Transplanting Experiments and Measures to Prevent Mass Mortality), *Tohoku Suisan Kenkyujo Kenkyu Hokoku (Bulletin of Tohoku Regional Fisheries Research Laboratory)* (25), 27-38.

88. ——, M. HATANAKA, R. SATO, S. SAKAI, and R. YUKI, 1950. Artificial Breeding of Oysters in Tanks, *Tohoku J. Agr. Res.*, **1**, 69-86.

89. ——, K. MORI, Y. SUGAWARA, H. TAMATE, J. OIZUMI, and O. ITIKAWA, 1968. Studies on the Mass Mortality of Oysters in Matsushima Bay: VII. Pathogenetic Investigation, *Tohoku J. Agr. Res.*, **19** (4), 250-265.

90. ISHIDA, S., 1935. On the Oxygen Consumption in the Oyster, *Ostrea gigas* Thunberg, under Various Conditions, *Sci. Rep. Tohoku Imp. Univ.*, Ser. IV, **10**, 619-638.

91. ITO, S., and T. IMAI, 1955. Ecology of the Oyster Bed: I. On the Decline of Productivity Due to Repeated Cultures, *Tohoku J. Agr. Res.*, **5**, 9-26.

92. JODREY, L. H., and K. M. WILBUR, 1955. Studies on Shell Formation: IV. The Respiratory Metabolism of the Oyster Mantle, *Biol. Bull.*, **108**, 346-358.

93. JULLIEN, A., 1936. Effects chronotropes, inotropes et tonotropes de la cicutine, du curare, de la vèratrine et de la digitaline sur le coeur de l'huitre, *Comptes*

Rendus Hebdomadaires des Séances et Mèmoires de la Sociètè de Biologie et de ses Filiales et Associèes, **121**, 1002-1004.

94. KANATANI, H., 1964. Spawning of Starfish: Action of Gamete-shedding Substance Obtained from Radial Nerves, *Science,* **146**, 1177-1179.

95. KANNO, H., M. SASAKI, Y. SAKURAI, T. WATANABE, and K. SUZUKI, 1965. Matsushima-Wan ni okeru kaki no tairyoheishi ni kansuru kenkyu: I. Tairyoheishi no jokyo to kankyo ni tsuite (Study on the Mass Mortality of the Oyster Occurring at Matsushima Bay: I. On the Conditions of Environment during Mass Mortality), *Tohoku Suisan Kenkyujo Kenkyu Hokoku (Bulletin of Tohoku Regional Fisheries Laboratory)* (25), 1-26.

96. KAWAI, K., 1957. Akoyagai no busshitsu taisha ni kansuru kenkyu: III. Soshikikokyu ni tsuite [Study on the Metabolism of *Pteria martensii* (Akoyagai): III. On Tissue Respiration], *Nihon Suisan Gakkaishi (Bull. of the Jap. Soc. of Sci. Fish.),* **22** (10), 626-630.

97. ———, 1959. The Cytochrome System in Marine Lamellibranch Tissues, *Biol. Bull.,* **117**, 125-132.

98. KINCAID, T., 1951. The Oyster Industry of Willapa Bay, Washington, Calliostoma Co., Seattle, 45 p.

99. KNIGHT JONES, E. W., 1951. Aspects of the Setting Behavior of Larvae of *Ostrea edulis* on Essex Oyster Beds, *Rapp. Cons. Explor. Mer.,* **128** (II), 30-34.

100. KOBAYASHI, T., 1966. Gonadotorobin to seihorumon (Gonadotropin and Hormones), *Taisha (Metabolism and Disease),* **3** (1), 32-38.

101. KOGANEZAWA, A., N. ISHIDA, and T. IMAI, 1964. Miyagi-ken Ojika Hanto ni okeru tanegaki seisan no seitaigakuteki chosa (Ecological Studies of Seed Oysters in Ojika Hanto, Miyagi Prefecture), Proceedings of the 1964 Congress of the Japanese Society of Scientific Fisheries, p. 49.

102. KORRINGA, P., 1941. Experiments and Observations on Swarming, Pelagic Life, and Setting in the European Flat Oyster, *Ostrea edulis* L., *Arch. neerl. Zool.,* **5**, 1-249.

103. ———, 1947. Relations between the Moon and Periodicity in the Breeding of Marine Animals, *Ecol. Monogr.,* **17**, 347-381.

104. ———, 1952. Recent Advances in Oyster Biology, *Quart. Rev. Biol.,* **27**, 266-308, 339-365.

105. KRIJGSMAN, B. J., and G. A. DIVARIS, 1955. Contractile and Pacemaker Mechanisms of the Heart of Mollusks, *Biological Review of the Cambridge Philosophical Society,* **30** (1), 1-39.

106. KUMANO, M., 1929. Chemical Analysis on the Pericardial Fluid and the Blood of *Ostrea circumpicta* Pils, *Sci. Rep. Tohoku Imp. Univ.,* Ser. IV, **4** (1), 281-284.

107. KURODA, T., 1931. Nihonsan Kairui mokuroku kaki-ka (List of Japanese Shell-bearing Mollusks), *Venus,* 2.

108. KUSAKABE, D., 1931. Suikashiki yoshoku kaki no heishi ni tsuite (On the Mortality of Oysters during Hanging Culture), *Suisan Butsuri Danwa Kaiho,* **22**, 305-317.

109. LAMBERT, L., 1946. Les huîtres desc otes francaises, *Pêche Marit.,* **29**, 31-33.

110. ———, 1946b. Lòstréiculture: le captage, *Ibid.,* **29**, 81-83, 99-100.

111. LANDAU, H., and P. S. GALTSOFF, 1951. Distribution of Nematopsis Infection

on the Oyster Grounds of the Chesapeake Bay and in Other Waters of the Atlantic and Gulf States, *Texas J. Sci.,* **3**, 115-130.

112. LI, M. F., C. FLEMMING, and J. E. STEWART, 1967. Serological Differences between Two Populations of Oysters *(Crassostrea virginica)* from the Atlantic Coast of Canada, *J. Fish. Res. Bd. Canada,* **24**, 443-446.

113. LISCHKE, C. E., 1869. Japanische Meeresconchylien, I.

114. ——, 1871. Japanische Meeresconchylien, II.

115. LOOSANOFF, V. L., 1942. Seasonal Gonadal Changes in the Adult Oyster, *Ostrea virginica,* of Long Island Sound, *Biol. Bull.,* **82**, 195-206.

116. ——, 1945. Precocious Gonad Development in Oysters Induced Midwinter by High Temperature, *Science,* **102**, 124-125.

117. ——, 1949a. On the Food Selectivity of Oysters, *Ibid.,* **110**, 122.

118. ——, 1949b. Vertical Distribution of Oyster Larvae of Different Ages during the Tidal Cycle, *Anat. Rec.,* **105**, 591-592.

119. ——, 1950a. Rate of Water Pumping and Shell Movements of Oysters in Relation to Temperature, *Ibid.,* **108**, 229.

120. ——, 1958. Challenging Problems in Shellfish Biology in *Perspectives in Marine Biology* edited by A. A. Buzzati-Traverso. Univ. of California Press, Berkeley and Los Angeles, 383-395.

121. ——, and J. B. ENGLE, 1940. Spawning and Setting of Oysters in Long Island Sound in 1937, and a Discussion of the Method for Predicting the Intensity and Time of Oyster Setting, *Bull. U. S. Bur. Fish.,* **33**, 217-255.

122. ——, ——, 1942a. Effect of Different Concentrations of Plankton Forms upon Shell Movements, Rate of Water Pumping, and Feeding and Fattening of Oysters, *Anat. Rec.,* **84**, 86.

123. ——, ——, 1942b. Accumulation and Discharge of Spawn by Oysters Living at Different Depths, *Biol. Bull.,* **82**, 413-422.

124. ——, ——, 1947a. Feeding of Oyster in Relation to Density of Microorganisms, *Science,* **105**, 260-261.

125. ——, ——, 1947b. Effect of Different Concentrations of Microorganisms on the Feeding of Oysters *(O. virginica),* *Bull. U. S. Bur. Fish.,* **42**, 31-57.

126. ——, and F. P. TOMMERS, 1948. Effect of Suspended Silt and Other Substances on Rate of Feeding of Oysters, *Science,* **107,** 69-70.

127. ——, ——, 1948. Effect of Low pH upon the Rate of Water Pumping of Oysters, *Ostrea virginica,* *Anat. Rec.,* **99**, 668-669.

128. ——, and H. C. DAVIS, 1949. Gonad Development and Spawning of Oysters at Several Constant Temperatures, *Anat. Rec.,* **105**, 592.

129. ——, and C. A. NOMEJKO, 1949. Growth of Oysters, *O. virginica,* during Different Months, *Biol. Bull.,* **47**, 82-94.

130. ——, and H. C. DAVIS, 1950. Spawning of Oyster at Low Temperatures, *Science,* **111**, 521-522.

131. ——, and C. A. NOMEJKO, 1951a. Spawning and Setting of the American Oyster, *O. virginica,* in Relation to Lunar Phases, *Ecology,* **32**, 113-134.

132. ——, ——, 1951b. Existence of Physiologically Different Races of Oysters, *Crassostrea virginica, Biol. Bull.,* **101**, 151-156.

133. ——, and H. C. DAVIS, 1963. Shellfish Hatcheries and Their Future, *Com-*

mercial Fish. Rev., **25**, 1-11.

134. MACKIN, J. G., 1946. A Study of Oyster Strike on the Seaside of Virginia, *Contr. Va. Fish. Lab.* (25), 18.

135. ——, 1951. Histopathology of Infection of *Crassostrea virginica* (Gmelin) by *Dermocystidium marinum* Mackin, Owen and Collier, *Bull. Mar. Sci. Gulf and Caribbean*, **1**, 72-87.

136. ——, 1956. *Dermocystidium marinum* and Salinity, *Proc. Nat. Shellfish. Ass.*, **46**, 116-128.

137. ——, and J. L. BOSWELL, 1956. The Life Cycle and Relationships, *Dermocystidium marinum*, *Ibid.*, **46**, 112-115.

138. MacGINITIE, G. E., 1941. On the Method of Feeding of Four Pelecypods, *Biol. Bull.*, **80**, 18-25.

139. ——, 1945. The Size of the Mesh Openings in Mucous Feeding Nets of Marine Animals, *Ibid.*, **88**, 107-111.

140. MATSUO, Y., 1957. Hiroshima-Wan magaki no i joheishi ni tsuite—kotoni saikingaketeki tachibakara [Abnormal Mortality of *Magaki (Gryphaea gigas)* at Hiroshima Bay with Special Reference to the Bacteriological Aspects of the Mortality], *Hiroshima Igaku (Journal of the Hiroshima Medical Association)*, **5**, 726-736.

141. MENZEL, R. W., and S. H. HOPKINS, 1955. The Growth of Oysters Parasitized by the Fungus *Dermocystidium marinum* and the Trematode *Bucephalus cuculus*, *J. Parasitol.*, **41**, 333-342.

142. MILLAR, R. H., 1951. Scottish Research on Oyster Fisheries, *Rapp. Cons. Epplor. Mer.*, **128**, 18.

143. MORI, K., 1967. Histochemical Study on the Localization and the Physiological Significance of Glucose-6-phosphate Dehydrogenase System in the Oyster during the Stages of Sexual Maturation and Spawning, *Tohoku J. Agr. Res.*, **17** (4), 287-301.

144. ——, 1968a. Changes of Oxygen Consumption and Respiratory Quotient in the Tissues of Oysters during the Stages of Sexual Maturation and Spawning, *Ibid.*, **19** (2), 136-143.

145. ——, 1968b. Effect of Steroid on Oysters: I. Activation of Respiration in Gonad by Estradiol-17β. *Bull. Jap. Soc. Sci. Fish.*, **34** (10), 915-919.

146. ——, 1968c. Effect of Steroid on Oysters: II. Increase in Rates of Fertilization and Development by Estradiol-17β, *Ibid.*, **34** (11), 997-999.

147. ——, 1969. Effect of Steroid on Oysters: IV. Acceleration of Sexual Maturation in Female *Crassostrea gigas* by Estradiol-17β, *Ibid.*, **35** (11), 1077-1079.

148. ——, H. TAMATE, and T. IMAI, 1964. Presence of Δ^5-3β-hydroxysteroid Dehydrogenase Activity in the Tissues of Maturing Oysters, *Tohoku J. Agr. Res.*, **15** (3), 269-277.

149. ——, ——, ——, 1965. Presence of 17β-hydroxysteroid Dehydrogenase Activity in the Tissues of Maturing Oysters, *Ibid.*, **16** (2), 147-157.

150. ——, ——, ——, 1966. Histochemical Study on the Change of 17β-hydroxysteroid Dehydrogenase Activity in the Oyster during Stages of Sexual Maturation and Spawning, *Ibid.*, **17** (2), 179-191.

151. ——, T. MURAMATSU, and Y. NAKAMURA, 1969. Effect of Steroid on Oysters:

III. Sex Reversal from Male to Female in *Crassostrea gigas* by Estradiol-17β, *Bull. Jap. Soc. Sci. Fish.*, **35** (11), 1072-1076.

152. MORI, K., T. IMAI, K. TOYOJIMA, and I. USUKI, 1965a. Matsushima Wan ni okeru kaki no tairyoheishi ni kansuru kenkyu: IV. Seiseijuku oyobi sanran ni tomonau kaki no seiriteki kassei to togenryo no henka (Study on the Mass Mortality of Oysters Occurring at Matsushima Bay: IV. Changes in the Physiological Activities and Carbohydrate Content during Sexual Maturation and Spawning), *Tohoku Suisan Kenkyujo Kenkyu Hokoku (Bulletin of Tohoku Regional Fisheries Research Laboratory)*, (25), 49-63.

153. ——, H. TAMATE, T. IMAI, and O. ICHIKAWA, 1965b. Matsushima Wan ni okeru kaki no tairyoheishi ni kansuru kenkyu: V. Seiseijuku oyobi sanran ni tomonau kaki no shishitsu oyobi totaisha no henka (Study on the Mass Mortality of Oysters Occurring at Matsushima Bay: V. Changes in the Fat and Carbohydrate Metabolism of Oysters during Sexual Maturation and Spawning), *Ibid.*, (25), 65-68.

154. NEEDLER, A. W. H., 1940. Helping Oyster Growers to Collect Spat by Predicting Sets, *Progr. Rep. Atl. Biol. Sta.*, **27**, 8-10.

155. ——, 1941. Oyster Farming in Eastern Canada, *Bull. Fish. Res. Bd. Can.*, **60**, 1-83.

156. NELSON, T. C., 1918. On the Origin, Nature and Function of the Crystalline Style of Lamellibranches, *J. Morph.*, **31**, 53.

157. ——, 1921. The Mechanism of Feeding in the Oyster, *Proc. Soc. Exp. Biol. Med.*, **21**, 90.

158. ——, 1926. Ciliary Activity of the Oyster, *Science*, **64**, 72.

159. ——, 1927. On the Relation of Spawning of the American Oyster to Temperature, *Anat. Rec.*, **29**, 97.

160. ——, 1928. On the Distribution of Critical Temperatures for Spawning and for Ciliary Activity in Bivalve Mollusks, *Science*, **67**, 220.

161. ——, 1936a. Water Filtration by the Oyster and New Hormone Effect upon the Rate of Flow, *Proc. Soc. Exp. Biol. Med.*, **34**, 189.

162. ——, 1936b. On the Effects of Hormones of Spermatic Fluid upon the Oyster, *Anat. Rec.*, **67**, 71.

163. ——, 1938. The Feeding Mechanism of the Oyster: I. On the Pallium and Branchial Chambers of *Ostrea virginica*, *O. edulis*, and *O. angulata*, with Comparisons with other Species of the Genus, *J. Morph.*, **63**, 1.

164. ——, 1942. The Oyster: The Boylston Street Fishweir, *Pap. Peabody Fdn. Archeol.*, **2**, 49-64.

165. ——, and J. B. ALLISON, 1937. On the Nature and the Effects of a Hormonelike Substance Produced on the Spermatozoa of the Oyster, *Anat. Rec.*, **70**, 124.

166. NUMACHI, K., 1959. Serological Studies on Relationships among Oysters of Different Genus, Species and Races, *Tohoku J. Agr. Rec.*, **10**, 313-319.

167. ——, 1962. Serological Studies of Species and Races in Oysters, *Am. Naturalist*, **96**, 211-217.

168. ——, 1965. Kozatsushu no kakuninho (Hybridization of Oysters), Proceedings of the Autumn Meeting of the Japanese Society of Scientific Fisheries Held in 1965. Abstract of the Symposium on the Breeding of Marine Orga-

nisms, 1966, 15-21. Also, *Suisan Zoshoku (Aquaculture)*, **13**, 126-129.

169. NUMACHI, K., J. OIZUMI, S. SATO, and T. IMAI, 1965. Matsushima-Wan ni okeru kaki no tairyo heishi ni kansuru kenkyu: III. Guramuyoseikin ni yoru kaki no byohen to shutsugenhindo (On the Mass Mortality of Oysters Occurring at Matsushima Bay: III. Diseases Caused by Gram-positive Bacteria), *Tohoku Suisan Kenkyujo Kenkyu Hokoku (Bull. of Tohoku Reg. Fish. Res. Lab.)* (25), 39-47.

170. OGASAWARA, Y., U. KOBAYASHI, R. OKAMOTO, A. FURUKAWA, M. HISAOKA, and K. NOGAMI, 1962. Kaki yoshoku ni okeru yokusei shubyo no shiyo to sono seisanteki igi (Control of Seedlings during Oyster Culture), *Naikaiku Suisan Kenkyujo Kenkyu Hokoku (Bull. of Naikai Reg. Fish. Res. Lab.)* (19), 1-153.

171. ORTON, J. H., 1921. Sex Change in the Native Oyster *(Ostrea edulis)*, *Nature*, 107.

172. ——, 1922a. The Blood Cells of the Oyster, *Ibid.*, **109**.

173. ——, 1922b. The Phenomena and Conditions of Sex Change in Oysters *(O. edulis)* and *Crepidula*, *Ibid.*, **110**.

174. ——, 1923. Summary of an Account of an Investigation into the Cause or Causes of Unusual Mortality among Oysters in English Oyster Beds during 1920 and 1921, *J. Mar. Biol. Ass.*, **13**.

175. ——, 1926. On Lunar Periodicity in Spawning of Normally Grown Falmouth Oyster *(O. edulis)* in 1925, with a Comparison of the Spawning Capacity of Normally Grown and Dumpy Oysters, *Ibid.*, **14**, 199.

176. ——, 1927. Researches on the Sex Change of the European Oyster, *Nature*, **119**, 117.

177. ——, 1927. The So-called Viscoid Secretion in Spawning Oysters, *Ibid.*, **120**, 843.

178. ——, 1937. *Oyster Biology and Oyster Culture.* London: E. Arnold & Co.

179. ——, and P. R. AWATI, 1926. Modification by Habitat in the Portuguese Oyster, *Ostrea (Gryphaea) angulata*, *J. Mar. Biol. Ass.*, **14**, 227.

180. ——, and C. AMIRTHALINGAM, 1927. Notes on Shell Depositions in Oysters, with a Note on the Chemical Decomposition of "Chalky" Deposits in Shells of *O. edulis*, *J. Mar. Biol. Ass.*, **14**, 935.

181. PRYTHERCH, H. F., 1934. The Role of Copper in the Setting, Metamorphosis, and Distribution of the American Oyster, *Ostrea virginica*, *Ecol. Monogr.*, **4**, 49-107.

182. ——, 1940. The Life Cycle and Morphology of *Nematopsis ostrearum* sp. nov., a Gregarine Parasite of the Mud Crab and Oyster, *J. Morph.*, **56**, 39-65.

183. QUOY, J. R. T., and P. GAIMARD, 1835. Voyage de l'Astrolabe, *Zoologie*, 3 p.

184. RANSON, G., 1948a. Prodissoconques at classification des Ostréidés vivants, *Bull. Mus. Hist. nat. Belg.*, **24**, 1-12.

185. ——, 1948b. Ecologie et vépartition géographique des Ostréidés vivants, *Rev. Sci.*, Paris, **86**, 469-473.

186. ——, 1948c. *Gryphaea angulata* Lmk. est l'espéce "type" du genre *Gryphaea* Lmk., *Bull. Mus. Hist. nat. Paris*, **20**, 514-516.

187. ——, 1950. La chambre promyaire et la classification zoologique des Ostréidés,

J. Conchyliol., **90**, 195-200.

188. RANSON, G., 1951. Observation, morphologiques, biologiques, biogéographiques, géologiques et systématiques sur une espéce d'huître de Madagascar et d'Afrique du Sud: *G. margaritacea, Bull. Inst. Oceanogr. Monaco*, **48**, 1-20.

189. ——, 1952. Les huitres: biologie, culture, bibliographie, *Ibid.*, **1001**, 1-134.

190. ——, 1960. Les prodissoehonques (Coquilles larvaires) des Ostréidès vivants, *Ibid.*, **1183**, 1-41.

191. REGNAULT, and REISET, 1849. Recherches chimiques sur la respiration des animaux, *Ann. de chim. et de physiol.*, **26**.

192. ROUGHLEY, T. C., 1933. The Life History of the Australian Oyster (*Ostrea commercialis*), *Proc. Linn. Soc. N. S. W.*, **58**, 279-333.

193. SCHAFFER, M. B., 1937. Attachment of the Larvae of *Ostrea gigas*, the Japanese Oyster, to Plane Surfaces, *Ecol.*, **18**, 523.

194. SENO, H., 1935. Showa 9-nendo Misaki sho-Iso-wan ni okeru kaki heishi ni tsuite [Mortality of Oysters Occurring at Misaki Bay (Annual Report of 1934)], *Dozatsu (Zoological Magazine)*, **47**, 556.

195. ——, 1936. Kako 5-nen ni Wataru Misaki sho-Iso-wan ni okeru kaki no heishi ni tsuite [On the study of Oyster Mortality Occurring at Misaki Bay (Covering a Period of 5 Years)], *Ibid.*, **48**, 4.

196. ——, and J. HORI, 1929. Magaki kegaki, iwagaki no sogojusei ni kansuru kenkyu [Study on the Cross-fertilization among *C. gigas* (*Magaki*), *O. echinata* (*Kegaki*), and *C. nippona* (*Iwagaki*)], *Ibid.*, **41**, 490-491.

197. ——, ——, and D. KUSAKABE, 1926. Effects of Temperature and Salinity on the Development of the Eggs of the Common Japanese Oyster, *Ostrea gigas* Thunberg, *J. Fish. Inst. Tokyo*, **22**, 41.

198. SINDERMANN, C. J., 1966. Manuscript Report No. 66-11, *U. S. Bureau of Commer. Fisheries Biol. Lab.*

199. ——, 1966. Manuscript Report No. 66-13, *Ibid.*

200. SPARK, R., 1951. Fluctuations in the Stock of Oysters (*Ostrea edulis*) in Limfjord in Recent Times, *Rapp. Cons. Explor. Mer.*, **128**, 27-29.

201. ——, 1960. Investigations on the Biology of the Oyster: XII. On the Fluctuations in the Oyster Stock of Northwestern Europe, *Rep. Danish Biol. Sta.*, **52**, 41-45.

202. SPECTOR, W. S. (ed.), 1956. *Handbook of Biological Data*. Philadelphia: W. B. Saunders Company.

203. STAUBER, L. A., 1950. The Problem of Physiological Species with Special Reference to Oysters and Oyster Drills, *Ecology*, **31**, 109-118.

204. SUGIWARA, Y., J. OIZUMI, O. ICHIKAWA, and T. IMAI, 1966. Saikin kansen gaki no byorisoshikigakuteki kenkyu (On the Pathological Aspects of Bacterial Infection in Oysters), *Pathological Studies Related to Oyster Mortality*, 8-14.

205. TAKATSUKI, S., 1934. On the Nature and Functions of the Amoebocytes of *Ostrea edulis, Quart. J. Micro. Sci.*, **76**, 379-413.

206. ——, 1949. *The Oyster (Kaki)*. Tokyo: Gihodo.

207. TAKEUCHI, T., T. MATSUBARA, K. HIROKAWA, and A. TSUKIYAMA, 1955. Hiroshima-Wan magaki no ijoheishi ni kansuru Saikingakuteki kenkyu: I

[Bacteriological Study of Abnormal Mortality of *Ostrea gigas* (*Magaki*) in Hiroshima Bay: I], *Nihon Suisan Gakkaishi* (*Bulletin of the Japanese Society of Scientific Fisheries*), **20**, 1066-1070.

208. TAKEUCHI, T., T. MATSUBARA, K. HIROKAWA, and A. TSUKIYAMA, 1956. *Ibid.*, II, *Ibid.*, **21**, 1199-1203.

209. ——, ——, ——, and Y. MATSUO, 1957. *Ibid.*, III, *Ibid.*, **23**, 19-23.

210. TAMATE, H., K. NUMACHI, K. MORI, O. ICHIKAWA, and T. IMAI, 1965. Matsushima-Wan ni okeru kaki no tairyoheishi ni kansuru kenkyu: VI. Byorisoshikigakuteki kenkyu (Study on the Mass Mortality of Oysters Occurring at Matsushima Bay: VI. Pathological Investigations), *Tohoku Suisan Kenkyujo Kenkyu Hokoku (Bulletin of Tohoku Regional Fisheries Research Laboratory)* (25), 89-104.

211. TAMURA, T., 1957. *Suisan Zoshoku (Aquaculture)*. Tokyo: Kigensha.

212. TANAKA, K., 1964. *Jitsuyo kairui ketsuekigaku (Serological Studies on Mollusks)*. Takamatsu City: Ueda Publishers.

213. TANAKA, Y., 1948. *Dobutsu no ikushuidengaku (Genetics and Breeding of Animals)*. Tokyo: Yokendo.

214. TATSU, I., 1933. Ariake-Kai-San suminoegaki oyobi magaki no shu no ido ni kansuru Fujimori, Saburoshi no hokokusho o yomite (A Review of the Report Concerning the Specific Difference between *C. gigas* and *C. rivularis* in Ariake-Kai by Mr. S. Fujimori), *Venus*, **3**, 365-377.

215. THOMSON, J. M., 1950. The Effect of the Orientation of Cultch Material on the Setting of the Sydney Rock Oyster, *Aust. J. Mar. Freshw. Res.*, **1**, 139-154.

216. ——, 1954. The Genera of Oysters and the Australian Species, *Ibid.*, **5**, 132-167.

217. THUNBERG, C. P., 1793. Tekning Och beskrifning pa en stor Ostronsort ifran Japan, *K. Vetensk Akad-Handl.*, **14**, 140-142.

218. USUKI, I., 1956. Effects of some Inhibitors of Anaerobic Glycolysis on the Ciliary Activity of the Oyster Gills, *Sci. Rep. Tohoku Univ.*, Ser. IV, **22**, 49-56.

219. ——, 1962. Energy Source for the Ciliary Movement and the Respiration and Metabolism in Oyster Gills, *Ibid.*, **28**, 59-83.

220. ——, and N. OKAMURA, 1956. Glycolytic Intermediates in the Oyster Gills, *Sci. Rep. Tohoku Univ.*, Ser. IV, **22**, 225-232.

221. WAKIYA, Y., 1915. Oysters of Korea, *Rep. Fish. Invest., Government General of Korea*.

222. ——, 1929. Japanese Food Oysters, *Jap. J. Zool.*, **2**, 359-367.

223. WATTENBERG, L. W., 1958. Microscopic Histochemical Demonstration of Steroid-3β-ol Dehydrogenase in Tissue Sections, *J. Histochem. and Cytochem.*, **6**, 225-231.

224. WILSON, O. P., 1941. Oyste Rearing on the Yealm River, *J. Mar. Biol. Ass.*, **25**, 125-127.

225. YASUDO, G., 1957. Seishi no denshikenbikyoteki kenkyu (Electron Microscope Study of Sperms), *Denshi Kenbikyo (Electron Microspcope)*, **5**, 14-30.

226. YONGE, C. M., 1926. Structure and Physiology of the Organs of Feeding and Digestion in *Ostrea edulis*, *J. Mar. Biol. Ass.*, **14**, 295-386.

227. ——, 1946. Digestion of Animals by Lamellibranches, *Nature*, **157**, 729.

228. ——, 1960. *Oysters*. London: Collins.

CHAPTER II

Technique of Oyster Culture

1. Introduction

The "hibitate" culture technique (culture in nets fixed to bamboo poles) was developed in Hiroshima in the 17th century and since then, over a period of about 300 years, there has been steady progress in the cultivation of oysters. However, really significant developments in oyster culture have only occurred in the last 50 years. Seno and Hori (1927) developed "hanging culture" at the end of the Taisho era (1912-1925), and about this time, experiments on seed production by the hanging culture method were carried out and developed at the Miyagi and Kanagawa Prefectures. This development was a landmark in the history of oyster culture in Japan. Compared to the "hibitate" method and the sowing method, the hanging culture method is far more efficient in several respects, for example, manageability, better utilization of fishing grounds, rate of production, etc. The method is adapted in different regions to the rack type, the raft type, and the long-line type of hanging culture. Its application to seed collection furthered its efficiency.

The hanging culture method has gradually spread to various parts of Japan since 1930 and just before the Second World War, i.e. in 1941, oyster production in Japan reached a maximum 60,377 tons (including shell weight). At present, the raft method is followed in Kesen-numa and Ogihama Bay of the Miyagi Prefecture, and the rack method in Hamana-Ko of the Shizuoka Prefecture. All these places are very popular as oyster-growing regions. There was no oyster culture until 1929 in several regions such as Obunato Bay of the Iwate Prefecture, Kamo-ko of the Niigata Prefecture, Nanao Bay of the Ishikawa Prefecture, Matoya Bay of the Mie Prefecture, and Naruto of the Tokushima Prefecture, etc. (Ogasawara et al., 1962).

Oyster culture in Japan advanced rapidly in the 1950's, stimulated by the food shortage in the country, and by an increased demand for canned food for export, particularly from Hiroshima. The economic recovery of the country in the following years also influenced the development of oyster culture. Production in 1967 reached 232,200 tons (including shell weight) or 38,037 tons (meat weight only), and the total capital invested in the business was 6,100,000,000 yen ($1=250 yen in 1972). The production in 1967 was about fifty times more than the production in 1920, and about four times greater than that of 1941 just before World War II (Figure 2.86).

The main factors responsible for the remarkable growth in oyster culture were the changeover from the simple hanging culture method developed in Hiroshima in 1950 to the raft type, and spreading the culture areas to the seas. The rearing of one-year-old oysters was also an important factor (Ogasawara, 1962). The long-line culture method, which has become common in the Miyagi Prefecture since 1952, greatly influenced culture techniques along the Pacific coasts and should be deemed

205

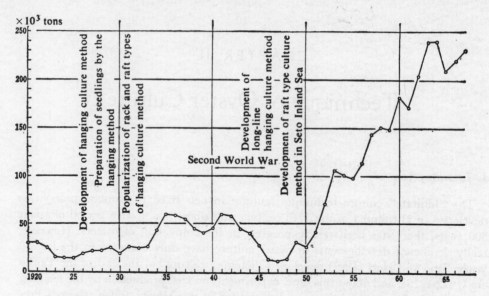

Figure 2.86. Transition in the production of cultured oysters (including shells).

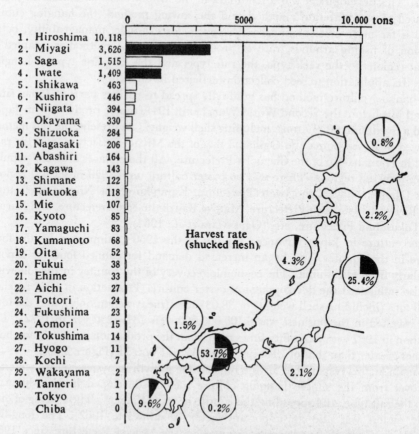

Figure 2.87. Harvest rate of oysters (shucked flesh) in various Prefectures and sea zones (mean values taken for 10 years between 1953 and 1962).

an important contributory factor in the progress of the oyster industry.

Oyster culture in Japan is at present carried out mainly in two regions, namely, the Pacific coast in the Tohoku region with the Miyagi Prefecture as its center, and the Seto Inland Sea with the Hiroshima Prefecture as its center. Figure 2.87 shows the rate of harvest of oysters cultured in various regions during the decade 1953-1962. A study of the number of enterprises involved in oyster culture (Figure 2.88) shows that 54% of the oyster production of the entire country is covered by the Seto Inland Sea (Tōhoku Fisheries Experiment Station, *Tōhoku Suiken,* 1965), and this percentage has increased since 1965. The Pacific coastal zones with Miyagi and Iwate as the

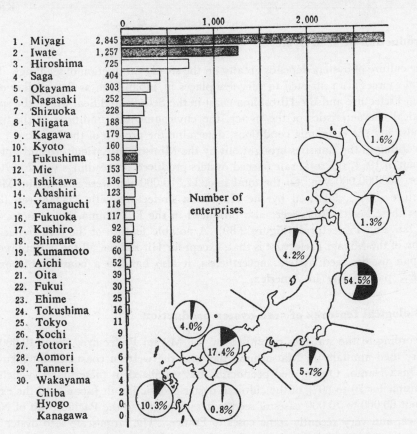

Figure 2.88. Number of enterprises in various Prefectures and sea zones (mean values taken for 9 years between 1954 and 1962).

main centers are next, accounting for 25% of the total production. On the other hand, 56% of the total number of concerns involved in oyster production are located in the northern zones of the Pacific coast, followed by the Seto Inland Sea zones with Hiroshima and Okayama as the main centers (17%). Production per unit is 8 tons in the Seto Inland Sea zones, and 1 ton in the northern zones of the Pacific coast. Thus, there is a wide difference between the two growing centers. It seems that the zones in the Seto Inland Sea were developed as large culture regions for edible oysters, while the centers along the Pacific coasts of the Tohoku regions were developed as small culture beds for the production of seed oysters.

Oyster culture, which up to now has made great progress, is currently facing a number of problems, *viz.* competition for fishing grounds, loss or degeneration of oyster beds due to industrial pollution, aging of the fishing grounds, loss of seed-producing grounds, a shift to other cultures, a decrease in the number of workers in the industry, etc. Hence, large-scale cultivation, full utilization, revival of aging fishing grounds, enlargement of the seed-producing grounds, etc. have to be carefully studied (Office of Science Technology, Government of Japan, 1964, 1968).

(Kanno)

2. Production of seed oysters

The culture of oysters depends greatly on the steady production of seedlings. Their natural sources are restricted to very few places in Japan such as Sendai Bay of the Miyagi Prefecture and the Hiroshima coast in the Seto Inland Sea, where the special biological characteristics of the oyster, the environmental conditions of the fishing grounds, and the economic conditions, determine the growth of the industry.

According to the statistics brought out by the Ministry of Agriculture and Forestry of Japan in 1964, the total sale of seed oysters produced for seedlings was 20,070,000 ren for 300,000,000 yen. Of the total sale, 17,510,000 ren were sold at the Miyagi Prefecture (88%), followed by the Hiroshima Prefecture with 153,000 ren (8%). Besides these, seed oysters were also produced in the Fukushima, Kumamoto, Mie, and Okayama Prefectures (Figure 2.89). A notable feature of the seed oyster production of the Miyagi Prefecture is that, except for Hiroshima, all the culture regions in Japan use its seed oysters; nevertheless, it also handles a considerable export business, particularly in America.

2.1 Biological features of seed-oyster production

According to the statistics published by the Miyagi Prefecture, about 40,000,000 ren are used annually in Matsushima Bay and the western coast of Mangoku-Ura and Ojika Hanto. Oyster seed production has stabilized in these places, which are responsible for 70 to 80% of the cultured oyster production in Japan, and the export of about 60,000 to 70,000 cases of seed oysters annually to the Pacific coast of North America, and very recently some cases to France. The large-scale seed oyster production around Sendai Bay is greatly influenced by biological and environmental conditions. Some of the favorable conditions are the amount of adult stock, a suitable rise in water temperature in summer, a healthy growth of larvae, and physical factors which contribute to the dispersal and accumulation of these larvae, in addition to favorable larval stock basins in the open seas, a water surface suited for seedlings on the coastal parts, a water surface suited for floor settlements, etc.

2.1.1 GEOGRAPHICAL CONDITIONS

Geographical conditions are shown for Sendai Bay and Matsushima Bay in Figures 2.90 and 2.91. Matsushima Bay is on the northwest side of Sendai Bay,

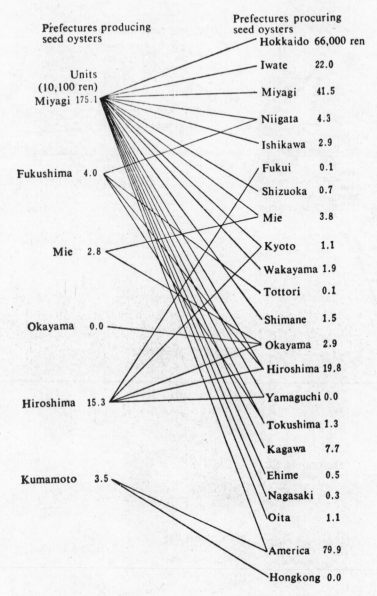

Figure 2.89. Production and sale of seed oysters (1964).

and has an area of about 43 km² with an average depth of 2 m. It is connected with Sendai Bay by five straits—Shirogasaki, Kaneshima, Ishihama, Sabusawa, and Wanigafucho—and by a waterway formed by the diggings in Sengaura to connect it with the mainland.

Mangoku Ura is a creek on the northeast edge of Sendai Bay, at the base of Ojika Hanto. It has an area of about 7.1 km² with an average depth of 2 m. It is connected with Sendai Bay by a single, narrow, and short strait. The tides which flow through

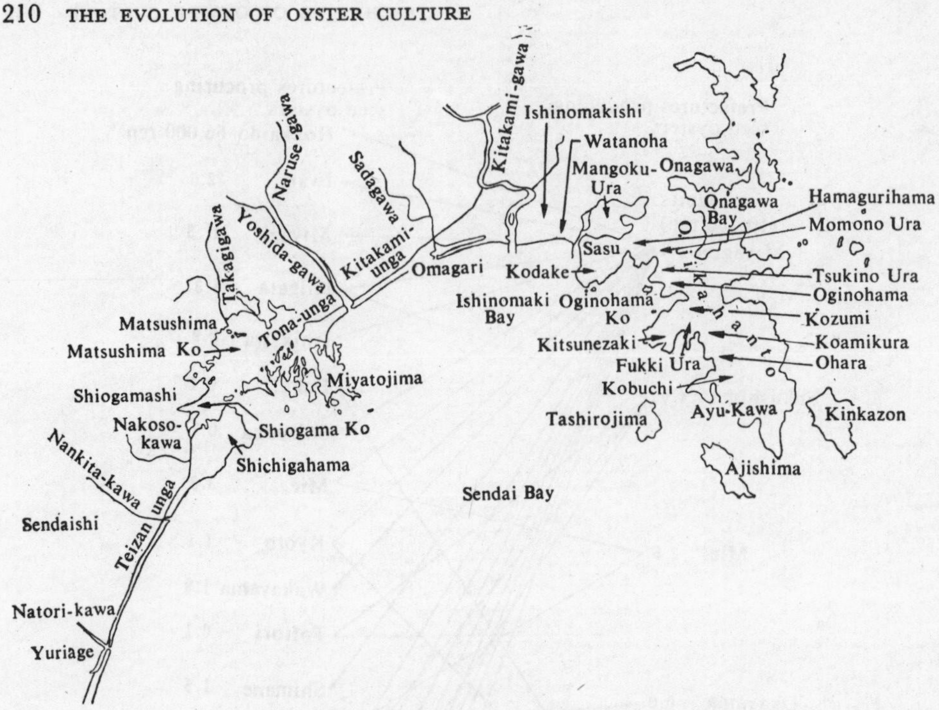

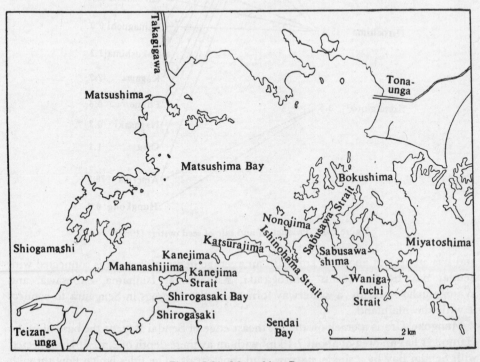

Figure 2.90. Sendai Bay.

Figure 2.91. Matsushima Bay.

this strait control the seed oyster production in the two bays. The tidal current also plays an important role in the dispersal and accumulation of oyster larvae in the bays. This has resulted in the formation of seedling grounds around the mouth of the bays where the population of oyster larvae is very dense.

2.1.2 CHANGES IN WATER TEMPERATURE AND DENSITY

In the year taken for study, the surface water temperature in both bays exceeded 10°C in early April, but gradually increased to 20°C by mid-June. With the advent of the rainy season in mid- and late July, the temperature suddenly rose to about 25°C. A maximum temperature of 27 to 28°C was attained in mid-July and the temperature hovered around 25°C until early September when it started to drop. The rise in water temperature and the environmental conditions during this period, accelerated the development of the reproductive organs in the oysters and influenced spawning and larval growth.

Since neither bay receives any major rivulets (estuaries), there is not much change in water density except for a slight decrease during the rainy season, i.e. between mid-June and mid-July, and the decrease is not significant unless the rainfall is very heavy.

2.1.3 LIFE CYCLE AND ECOLOGICAL CHARACTERISTICS OF ZOSTERA MARINA (AMAMO)

Zostera marina grows abundantly in about 80% of the entire area of both bays. The growth of plumules and their blooming in spring (luxuriant period), their leaf fall and degeneration in summer (period of degeneration), and the appearance of new plumules and their growth in autumn, are closely related to the changes in the water pH and the amount of dissolved oxygen and salt-nutrients. Hence, *Zostera marina* plays a large role in seed oyster production. The relationship between the changes in the environmental conditions and the life cycle of *Zostera marina* is shown in Figure 2.92.

It is estimated that about 24,000 tons (dry weight) of *Zostera marina* is present during the luxuriant period in Matsushima Bay, which reduces to about 4,500 tons during the period of degeneration; of the remaining 20,000 tons *Zostera marina*, some degenerate in the summer and the rest is used as food by the flagellates. In Mangoku Ura, about 1,400 to 2,200 tons of *Zostera marina* degenerate. Colorless flagellates (*Monas* sp.) are found in both bays at the rate of about 300 to 800 per 1 m*l* of sea water during the winter. In July, or the drifting period of *Zostera marina*, flagellates increase to about 3,000 and in August to 6,600 per 1 m*l* of sea water. In September, the figure drops to 3,300. It has been found that oyster larvae feed on colorless flagellates and when a coincidence in larvae and flagellate production occurs, the seed oyster crop is large (Imai *et al.*, 1951).

2.1.4 CURRENT AND TIDES

When the oyster larvae enter the planktonic stage, they are carried away by tidal currents from the inner bays, which are the primary spawning areas. The larvae collect in water basins in Sendai Bay and then are carried by the coastal current or wind to the coastal regions of Ojika Hanto, or the mouths of both bays, where they collect in the straits. Hence water current plays an important role in the dis-

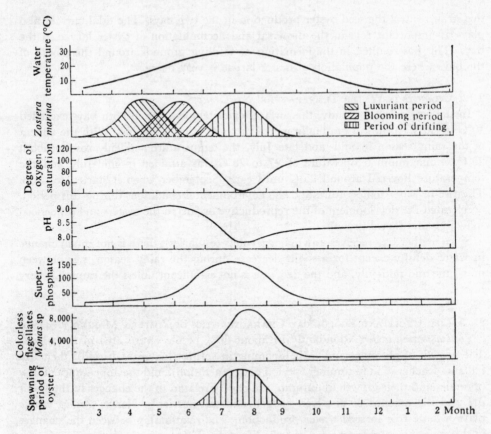

Figure 2.92. Life cycle of *Zostera marina* (Imai *et al.*, 1951).

tribution of larvae at the time of setting. Kikuchi *et al.* (1960, 1961) found an extremely dense population of oyster larvae, a 2 to 6 m layer, in the water basins of Sendai Bay. Distribution depends upon the number of eggs produced, the number of days in drifting, and the environmental conditions after the larvae have been carried to Okiai (offshore). Kikuchi also found that the centers of distribution shifted toward Okiai as the larvae grew from very small to medium size and larger. On the other hand, Koganesawa (1964) found that the water basins in which the oyster larvae were found in high density in Sendai Bay, moved leftwards within a relatively short period. He observed that the seedlings were not only confined to the coastal regions of Ojika Hanto (western coast) but also to the offshore areas, which meant that larvae could be collected thereafter from both places.

As the water containing a high density of oyster larvae shifts, flagellates simultaneously appear in large numbers toward Sendai Bay. During this time, 1,000 to 1,600 flagellates per 1 m*l* of sea water were detected, an amount which seems to have been sufficient food for 1.3 kg/*l* of larvae for 2 weeks.

According to Sato, water favorable for the growth of oyster larvae is found around the outer side of the coastal line of Matoya Bay. Oyster larvae grow in water areas where coastal water and water from the open seas mix; they are then carried by high tides to the coastal regions where they settle. The Miyato and Urato zones, which form the outer side of Matsushima Bay, are the stocking places for larvae just before setting, and hence ideal seedling places unless the weather and sea conditions are extremely adverse to a large larvae stock.

2.1.5 SPAWNING GROUP AND SPAWNING RATE

A spawning group consists of one-year-old and two-year-old cultured and naturally grown oysters. In the case of cultured oysters, the rate of spawning and the number of viable eggs differs according to the number of culture installations, density per installation, and the age of the oysters. However, since the number of installations depends upon the supply and demand position of oysters and other factors, the number in the parent stock fluctuates. Neither the spawning rate nor the amount of eggs produced is related to the amount of viable eggs capable of development. According to Koganezawa (1958), two-year-old oysters produce more eggs than one-year-olds and, furthermore, the number of their viable eggs is also greater. The peak period of spawning among two-year-old oysters is seen at the initial stage of a temperature increase, and their larval growth is consequently steady. The spawning peak among one-year-old oysters is at the end of their seedling period, and is immediately followed by a decrease in temperature, resulting in the loss of their larvae.

Table 2.29 shows the number of eggs produced by mother oysters in Matsushima Bay as estimated by Imai *et al.* (1955). The authors pointed out that variations occurred in the number of mother oysters from year to year, and that their age composition was responsible for the fluctuations in seedling production.

Table 2.29. Composition of spawning mother oyster groups

		1953		1954	
		Number of oysters	Number of eggs	Number of oysters	Number of eggs
Cultured oyster	One-year-old	$16,827 \times 10^4$	$1,159 \times 10^{12}$	$14,821 \times 10^4$	$1,354 \times 10^{12}$
	Two-year-old	$2,962 \times 10^4$	643×10^{12}	374×10^4	130×10^{12}
Naturally grown oyster		$1,297 \times 10^4$	26×10^{12}	$1,297 \times 10^4$	43×10^{12}
Total		$21,086 \times 10^4$	$1,828 \times 10^{12}$	$16,492 \times 10^4$	$1,527 \times 10^{12}$

The number of oyster culture installations in Matsushima Bay has decreased because of recent mass mortality, water pollution, and competition with laver cultivation (Table 2.30). At present, the number of installations is even less than given in the Table. Since the number of cultivating rafts has increased in the coastal region of Ojika Hanto, the number of mother oysters capable of producing about 4,000,000 ren of seedlings per annum, is maintained in the whole of Sendai Bay.

The natural conditions and environmental factors related to seed oyster production at Sendai Bay have been discussed. These conditions and factors do not remain the same every year, and so the biological cycle of the seed oysters continually changes; in other words, these conditions and factors collectively and individually exert an influence on seed production.

Table 2.30. Oyster production in Sendai Bay
(Data collected at the Statistical Division of Miyagi Prefectural Administration
under the Ministry of Agriculture and Forestry)

Year	Number of enterprises and types of installations	Regions			
		Ojika	Ishinomaki	Matsushima Bay	Total
Showa 33 (1958)	Enterprises	102	507	1,052	1,661
	Rafts	49	60	—	109
	Long line	537	849	—	1,386
	Simple hanging method (m²)	—	10,550	186,632	197,182
	Sowing (m²)	—	41,877	511,500	553,377
	Production (k)	65,913	400,763	1,392,757	1,859,433
Showa 34 (1959)	Enterprises	96	548	1,006	1,650
	Rafts	51	80	—	131
	Long line	548	911	—	1,459
	Simple hanging method (m³)	—	33,594	259,446	293,040
	Sowing (m²)	—	42,021	379,500	421,521
	Production (k)	261,522	508,136	1,456,783	2,226,441
Showa 35 (1960)	Enterprises	118	589	1,004	1,711
	Rafts	22	146	—	168
	Long line	806	1,000	—	1,806
	Simple hanging method (m²)	—	39,139	252,041	291,180
	Sowing (m²)	—	67,221	363,500	430,721
	Production (k)	115,803	654,091	1,099,896	1,869,780
Showa 36 (1961)	Enterprises	119	546	1,036	1,701
	Rafts	82	67	—	149
	Long line	744	1,270	—	2,014
	Simple hanging method (m²)	—	41,580	213,809	255,389
	Sowing (m²)	—	60,390	363,000	423,390
	Production (k)	139,660	783,556	1,083,436	2,006,657
Showa 37 (1962)	Enterprises	102	651	1,043	1,796
	Rafts	42	48	—	90
	Long line	807	1,808	—	2,615
	Simple hanging method (m²)	—	47,025	229,316	276,341
	Sowing (m²)	—	66,990	297,000	363,399
	Production (k)	133,328	1,127,130	657,208	1,917,666
Showa 38 (1963)	Enterprises	127	677	748	1,552
	Rafts	15	48	—	63
	Long line	740	1,908	—	2,648
	Simple hanging method (m²)	—	55,473	214,670	270,143
	Sowing (m²)	—	54,780	132,000	186,780
	Production (k)	287,507	1,320,772	966,422	2,574,701

(Contd.)

Table 2.30—*Contd.*

	Enterprises	122	429	657	1,208
	Rafts	8	12	—	20
Showa 39	Long line	945	1,857	—	2,802
(1964)	Simple hanging method (m²)	—	53,559	175,329	228,888
	Sowing (m²)	—	54,780	—	54,780
	Production (k)	491,520	1,645,883	525,002	2,662,405
	Enterprises	116	616	726	1,458
	Rafts	4	10	—	14
Showa 40	Long line	644	1,917	—	2,561
(1965)	Simple hanging method (m²)	—	198,825	994,915	1,193,740
	Sowing (m²)	—	51,480	—	51,480
	Production (k)	301,940	909,750	341,414	1,553,104

2.2 Evolution in seed-oyster production

2.2.1 History of Seedlings

In 1670, Goroemon Kobayashi of Hiroshima reportedly placed bamboo poles with twigs and nets in sea water, collected the young oysters that settled there, and tried to culture them. Kobayashi's efforts mark the beginning of oyster seedling culture. Around the same time, Utsumi Shozaemon of Nonoshima decided that if the oysters at Matsushima Bay were left to nature, their number might decrease. He determined the period of oyster culture and tried to grow them by collecting the spat around the Island and releasing these in sea water he considered suitable. After allowing them sufficient time to attain maturity, he collected them, retained the grown oysters, and returned the shells and spat to the sea.

During the early Meiji era (1868-1911) efforts were made to encourage seedlings to adhere to racks of about 2.5 m made of pine poles fixed in sea water. The young oysters were then transplanted the following year to freshly prepared beds scattered with empty shells. In the later Meiji era, seedlings were collected on twigs and frames affixed to bamboo poles, and either reared directly or after sowing.

In Hiroshima, a method with a history of over 300 years, now popularly known as the "Hiroshima method" is currently used. Bamboo rafts, commonly called *hibi*, are installed in the open sea or seed-oyster collecting places, and the seedlings allowed to settle. When grown into young oysters, these are transplanted to summer grounds (*natsu-okiba*) or cultivation grounds (*toyaba*) for further rearing. This method was gradually modified from year to year and is said to yield good results.

From 1887 to 1891 (Meiji 20), Matsushima Bay procured bamboo blinds from Hiroshima and prepared a model cultivation ground on the lines of the Hiroshima method. As a result, the Hiroshima technique and that already present in Matsushima Bay merged.

In order to collect seed oysters, materials easily available in each region were used, e.g., bamboo, pine branches, twigs, tiles, shells of oysters and other mollusks, slate, stones, pebbles, earthen pipe, rope, etc. The conditions were more or less the same even in foreign countries; while ropes intertwined with twigs were used in Italy, metal net baskets or triangular wooden frames with empty shells inside, were used in North America.

In a follow-up study on oysters exported from Mangoku Ura to the west coast of North America in 1919, Miyagi and Abe found that the consignment contained a large number of young oysters, which survived the journey and grew well, while all the parent oysters were dead on arrival. Encouraged by this discovery, the two authors, who were supervising the cultivation of oysters in America at that time, successfully adapted the hanging method of seedling production by growing seedlings using oyster shells (1924). Understandably, only young oysters have been exported since 1924. Around that time, too, oyster cultivation in Japan shifted to the hanging culture method suggested by Seno and Hori which, since it can be either a raft or a simple wood frame, spread from the surface of shallow waters to deep waters. The results were very encouraging as the technique stabilized, and this method soon became popular in various oyster-growing regions.

The hanging method of producing seedlings consists of the following steps: holes are made in the center of empty shells of oysters or abalones, the shells strung to a metal wire of about 2 m length, and the string hung vertically in sea water. Depending upon the size of the shell, a string may have from 60 to 100 pieces. Each string is called a unit, "ren" in Japanese. The oyster larvae are allowed to settle on these units. Oyster shells are used in making "ren" in Japan. Since they break easily, in America they are used for packing and are scattered for sowing. Scallop shells attached to coal-tar painted ropes are stronger, i.e. do not break so easily, and therefore do not detach from the ropes as quickly. In Hiroshima, shells of large pectens, abalones, and clams are also widely used.

2.2.2 EQUIPMENT FOR SEEDLING COLLECTION (Figure 2.93)

In Matsushima Bay and Mangoku Ura, racks of about 15 m length and about 1.5 m breadth are kept in sea water until mid-July to collect seedlings. This method resembles the simple wood frame method of cultivation. After folding once, units (strings of shells) are hung vertically when the time has come for seedling collection. When the water depth is shallow and the width of the rack narrow, the units are laid across the rack. When rafts are used in open seas, or seedlings collected while rearing is still in process on the rafts or racks, 5 or 6 units are grouped together and the bundles hung vertically outside the rafts or racks.

Seedling collecting equipment is installed in the mouth of bays, water straits, or water paths. Depending upon the topography of the ground, a natural formation of eddies may be found in some areas. If the equipment is installed in these areas, a dense group of oyster larvae is possible. Eddies can be caused by stopping the tidal current for a temporary period in places where the units are hung from the racks or rafts. Since these places have a relatively higher salinity, the oyster larvae collect near the surface of the water. Hence the collecting equipment is usually placed about 1 m deep in the water. Despite reports that larvae are unable to settle at the time of high tidal currents, in Shirogesaki where the usual flow of sea water is 50 cm per second at the cold water intake end of the Sendai Power Generating Station, a large number of oysters were found attached to the concrete wall. One of the conditions for oyster setting suggested by Seno, that the current speed should be high

(60 cm per second), has therefore been substantiated. Since oysters were found only on the rough areas of the concrete wall, it was concluded that eddies formed in these

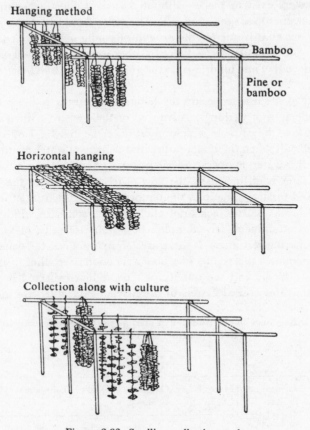

Figure 2.93. Seedling collecting racks.

places, which encouraged the attachment of oyster larvae which, in turn, caused further eddies that resulted in the attachment of still further larvae.

When the hanging method was first adopted in Matsushima Bay and Mangoku Ura, it was possible to collect the seedlings even from the corner parts of the bays, but when the number of installations was increased, the collection declined every year, forcing the migration of the installations toward the far seas. Since research on the ecology of oyster larvae has rapidly progressed, the seed tracks (*taneba*) have now been shifted to water basins in the mouth of the gulf.

To collect seedlings by the hanging method, the conditions of the seas during the peak period of spawning and the distribution conditions of water temperature and density, must be kept under constant surveillance. Since the larvae of *Balanus* appear simultaneously with the larvae of large oysters and have an adverse effect on the latter, this condition must also be observed. Fortunately in Matsushima Bay, the larvae of *Balanus* appear about two weeks earlier than those of the oyster, and thus the seedling collecting strings can be safely hung two weeks later.

Oyster larvae prefer surfaces which are somewhat dirty to those which are clean when setting, and also have a tendency to group. Larvae likewise prefer places where young oysters exist to places without oysters. Collecting equipment should be free of *Balanus*, diatoms, sand, and silt. The units should be hung only during a suitable period, one that is safe for the growth of young oysters which have already overcome many difficulties before setting. In other words, the period of unit installations should be devoid of any hindrance to the growing larvae and free from dangerous pollutants.

This selection of a suitable period for hanging the units is the most important factor in the collection of seedlings. Advanced or delayed by a single day, it might result in a great loss, the cause of which might never be known. Collecting seedlings becomes very difficult when the temperature and density of water continue to be low; in some years, the water temperature even during the summer is below normal; sometimes the water is disturbed by the flow of fresh water from the land, or from the open seas; and finally, sometimes the temperature and density of water continue to be at a high level due to black tides and a strong southern wind.

The amount of equipment, for example, the number of rafts, racks, and units, is important in preparing seedlings. If there are too many units, the water cannot pass freely through them and this results in a poor collection of seedlings; simultaneously, control becomes difficult and the quality of the product collected is often inferior.

The production rates of seed oysters in the Miyagi Prefecture are shown in Table

Table 2.31. Differences in the number of strings used for collecting seed oysters (Unit=1,000 ren)

Year	Matsushima Bay	Mangoku Ura	Ogihama Bay	Obara	Ishinomaki	Others	Total
Showa 21 (1946)	1,195	784	60	15	30		2,084
22 (1947)	778	560	30	15	20		1,403
23 (1948)	874	560	30	13	20		1,497
24 (1949)	1,108	352	60	16	38		1,574
25 (1950)	1,290	256	35	16	23		1,620
26 (1951)	1,672	622	131	65	50		2,540
27 (1952)	2,087	633	97	71	45		2,933
28 (1953)	2,833	623	104	60	44		3,664
29 (1954)	2,362	613	209	78	25		3,287
30 (1955)	2,425	976	864	134	42		4,441
31 (1956)	1,852	520	856	268	30		3,526
32 (1957)	2,209	1,280	1,159	80	68		4,796
33 (1958)	2,230	589	1,267	329	—		4.415
34 (1959)	1,975	650	866	444	69		4,004
35 (1960)	2,254	860	822	493	425	15	4,815
36 (1961)	1,959	1,265	851	315	79	24	4,493
37 (1962)	1,604	1,035	1,223	263	274	28	4,427
38 (1963)	1,103	1,094	1,476	261	476	47	4,457
39 (1964)	772	974	1,399	250	189	60	3,644
40 (1965)	896	1,026	1,316	227	211	40	3,716

2.31. The difference in the amount of seed oysters exported and that kept for local consumption, is shown in Table 2.32.

Table 2.32. Differences in the amount of seed oysters retained for local consumption and exported to North America

Year	Local consumption	To America	Year	Local consumption	To America	Year	Local consumption	To America
	1,000 ren	Number of boxes		1,000 ren	Number of boxes		1,000 ren	Number of boxes
Taisho 12 (1923)		492	13		18,529	30	566	99,792
13		840	14		14,839	31	629	59,824
14		1,403	15		15,629	32	634	59,938
15 (1926)		4,050	16 (1941)	523		33	581	60,465
Showa 2 (1927)		2,800	17	923		34	580	47,802
3		8,000	18	933		35	763	35,630
4		14,000	19	828		36	500	40,308
5		16,000	20	199		37	1,687	56,735
6		32,453	21	788	56,704	38	1,463	39,270
7		47,392	22 (1946)	704	33,359	39	1,027	36,929
8		68,103	23	485	45,873	40 (1965)	821	14,126
9		71,787	24	1,190	45,993			
10		42,953	25	1,424	52,534			
11		32,956	26	397	83,825			
12		17,788	27	479	72,011			
			28	514	66,958			
			29	528	53,788			

2.3 Ecology of seedling culture

2.3.1 SPAWNING

C. gigas is usually unisexual and hermaphroditic types are rarely found. It is oviparous and the eggs are laid in the primary oocyte stage when the water temperature is around 25°C. A sudden rise in the water temperature, a sudden decrease in the density of water due to rain, spermatic saline, etc., are said to stimulate spawning. However, any form of mild stimulus will induce spawning if the eggs are in a mature condition in the gonads and the animal is capable of spawning.

Seasonal changes in the gonads of oysters, studied through histological sections of the same, show that in winter when the water temperature is below 10°C the cells are in the form of primordial germ cells, which cluster together here and there in the connective tissues of the gonads.   It is very difficult to differentiate the male from the female.   In April, when the water temperature exceeds 10°C, the germ cells divide and in the females become oocytes; these grow and occupy a considerable space in the connective tissues.   In the males, these germ cells become spermatozoa. In May to June, when the water temperature gradually increases to about 20°C, the gonads are well developed with the males rich in spermatozoa, and the females equally rich in oocytes.   When the viscera become covered with gonads, the animals are ready to spawn.   While the development of gonads can be accelerated by a steady increase in water temperature, this would not lead to healthy spawning, unless it coincides with the release of spermatozoa.

Sato (1955) estimated the maturation temperature exponent of the gonads which is required for spawning in *C. gigas* $[T; \sum (T_i - \theta)]$, and studied the relationship between the changes in the period of spawning and the water temperature in Matsushima Bay.   By assuming θ of *C. gigas* to be 17°C, he found that in 1953, 1954, and 1955, spawning occurred when the maturation temperature exponent reached about 3,800. In the experiments carried out by Loosanoff on the Virginia oyster (*C. virginica*), θ was 10°C, which was also the critical temperature for maturation, and the maturation temperature exponent required for spawning in *C. virginica* was reported to be about 2,600.

In the investigations carried out on the *C. gigas* of Kesen-numa Bay, Sasaki (1966) found that the integrated temperature from the time when it exceeded 10°C to the time of natural spawning, was the same as the integrated temperature when the latter was suddenly increased to 23°C and spawning thus induced.   Hence, according to Sasaki, spawning may be induced if the integrated temperature is raised to a particular level.

In this regard, it is very important to determine the standard line of temperature. By considering the water temperature during the initial period of the activities, judged from the formation of reproductive cells, it seems that 10°C is suitable for *C. gigas*. Results of the investigations carried out on *C. gigas* reared by the hanging culture method near the cold water intake end of the Sendai Power Station in Shirogesaki of Matsushima Bay (Table 2.33), showed that spawning occurred when the integrated water temperature was about 600°C with the standard temperature of water at 10°C. Since the hanging culture was carried out on floating bridges, it did not seem to

Table 2.33. Integrated water temperature and spawning days of oysters in Shirogesaki in Matsushima Bay

Year of observation	Dates when water temperature exceeded 10°C	Dates when integrated water temperature reached 600°C	Number of days	Date of actual spawning	Number of days of actual spawning
Showa 35 (1960)	April 9	July 10 High (Tide)	92	July 3 Low (Tide)	85
36 (1961)	April 4	July 8 Low	95	July 6 Low	93
37 (1962)	April 7	July 12 Low	96	July 12 Low	96
38 (1963)	{ April 5, April 11	July 14 Low	100 (94)	July 13 Low	99 (93)
39 (1964)	April 1	July 9 High	99	July 13 Medium	103
40 (1965)	April 15	July 22 Low	98	July 22 Low	98
41 (1966)	April 7	July 7 Medium	91	July 11 Low	95
42 (1967)	April 16				

be affected by the tides; nevertheless spawning usually occurred on days of low tides.

Judging from the relationship between the spawning of oysters and the water temperature, the spawning and setting of *C. gigas* can be seen throughout the year in subtropical zones like Taiwan where the water temperature is 15.8 to 32.0°C (Matsui, 1934). However, a collection made by Imai in December, 1966 from the same fishing ground, consisted of oysters in various stages of pre- and post-maturation conditions. For example, in some females there were gonads with oogonial cells, in others the cell divisions of the gonads had well advanced, and still others were in a completely matured condition; "spent" oysters, indicating a post-spawning stage, were likewise present, and most of the specimens were small in size. All of the male specimens, on the other hand, were found to have formed spermatozoa and gonadal variations were comparatively very few. However in both the male and female specimens, the gonads were not thick and the number of eggs or spermatozoa in them was not large. The foregoing would seem to indicate that the eggs in these oysters do not mature throughout the year, and that the reproductive cycle differs in each individual.

In Ariake-Kai spawning begins in early May when the water temperature is 20°C. Spawning is maximal in June to August when the temperature is 23 to 24°C, but continues until early November with four peak periods. In Hiroshima Bay the first round of spawning occurs in mid-June when the water temperature is 23°C, and recurs three times until September. Spawning in the *C. gigas* of Matoya Bay— a strain between the Hiroshima strain and the Sendai—begins in early May when the temperature is 20°C, attains its maximum in late June when the temperature increases to 22 to 23°C, and declines from July onwards. Around Sendai Bay, spawning begins in mid-July when the water temperature is 23 to 25°C, and continues up to late August with two or three peak periods. The fact that planktonic larvae of *C. gigas* are caught in the plankton nets in mid-June, and that young oysters are found setting on the hanging plates in late June, shows that spawning is induced by some unknown stimulus or stimuli even before the critical temperature of spawning, i.e. 23 to 25°C, is reached.

In Onagawa Bay and Kesen-numa Bay on the northern side of Ojika Hanto, the rise in water temperature is delayed, and spawning in some areas occurs from late August to early September. Once the water temperature begins to fall, the growth of larvae is also delayed and the number of young oysters that settle is consequently extremely low.

Moving from the southern regions with high water temperatures to the northern regions with low water temperatures, reveals that the spawning period becomes more and more restricted, and the number of spawnings fewer, the further north one goes. On the other hand, the size of the gonads gradually increases as does the number of eggs laid per spawning. Furthermore, within the same fishing ground, spawning in shallow places is somewhat earlier than that in the deeper places. The larger the oysters, the greater the number of eggs produced, and the more rapid their maturation. Imai and Sakai (1961) have reported that the growth of the Sendai oyster is more rapid than the Hiroshima oyster, that it produces a larger number of mature ova in one lot, and that its gonads are thicker. Because the Hiroshima oysters spawn many times, their gonads are thin.

Figure 2.94 shows the seasonal variations in the size of eggs of cultured oysters from Matsushima Bay and Onagawa Bay. The water temperature of Matsushima Bay in January to February is lower than that of Onagawa Bay and hence the gonadal development is delayed. With an increase in water temperature in March,

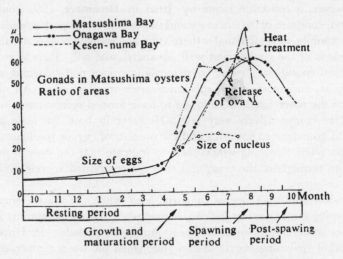

Figure 2.94. Seasonal variations in the size of eggs of oysters from
Matsushima Bay and Onagawa Bay.

cell division begins, and by April it crosses the level of growth in Onagawa Bay. The growth after this period is rapid. The size of the eggs in mid-July is 60 μ. The development curves of the eggs in these two bays are similar to the curves for water temperature between April and August.

For purposes of comparison, the ratio of the size of the egg nucleus and the gonadal area in oysters grown in Matsushima Bay as studied by Tamate et al. (1965), has been included in Figure 2.94, as have the measurements taken by Sasaki at Kesen-numa Bay, which were found to follow the same course as those of Onagawa Bay. However, in the experiments carried out in mid-July at higher than the usual temperatures, the egg size was 68 μ due, it was thought, to over-maturation.

The first round of spawning is observed in late July when one-third of the eggs are released. The second round occurs in mid-August when half the eggs are released, and in late August the balance are released. But in September a considerable number of eggs are still found in the gonads, and in Onagawa Bay a number of oysters had eggs even in October. According to histological observations, the residual eggs in Matsushima oysters degenerate or are absorbed in mid-September, but in Onagawa oysters the degeneration and absorption of the residual eggs occur in late September to mid-October.

It would seem therefore that the spawning of oysters continues for a fairly long period. The condition of the gonads, as observed by histological preparations, clearly shows that they vary from type to type, depending upon the environmental and nutritional conditions, and while there are specimens which show early gonadal maturation, there are also some which show late maturation. Although the difference is not as much as that of the oysters of Taiwan, the dates of spawning do vary to some extent, and because of this variation the free-living larvae are found for a fairly long period.

2.3.2 MICROSCOPIC EXAMINATION OF LARVAE, PREDICTION OF SEEDI INGS, AND CHECKING OF SEED OYSTERS

The eggs and sperms released in sea water by *C. gigas* unite, and after fertilization give rise to a totally independent organism. A detailed account of the development of eggs has been published by Fujita, Amemiya, Hori, and Seno. Galtsoff (1964) has recently published a detailed report on the Virginia oyster which closely resembles *C. gigas*. When the eggs are released, they are still in the primary oocyte stage, and the germ cells have not yet disappeared. When the spermatozoa approach the eggs, a fertilization membrane forms on the egg surface, and it undergoes a maturation division to produce two polar bodies. It is at this time that the egg nucleus is fertilized and divides, and active life begins. The division continues and the morula forms. Following this, cilia develop on the surface of the morula and the embryo becomes a blastula showing rotating movements. The embryo soon develops into a gastrula with a shell-line on the dorsal surface and a blastopore on the ventral surface. It then emerges as a trochophore larva, which ultimately develops a plate-like shell, after which it is called the veliger larva. These veliger larvae float near the surface of the water and are often caught in plankton nets. When other soft parts in the shell develop, the larva is referred to as type D. The various organs then develop one by one. According to Amemiya (1928) and Numachi (1960), there is a time difference in the early development of oysters owing to the differences in water temperature and chlorine concentration in places where the parent oysters live. One of the controlling factors in the development of *C. gigas* is the water temperature. Seno *et al.* (1926) observed that the optimum water temperature for the development of *C. gigas* is 23 to 26°C, and when the temperature is below 15°C or above 30°C, the number of parasites increases rapidly. Regarding the chlorine content, Seno has reported that this should be 1.017 to 1.021, and according to Amemiya (1921), 1.014 to 1.021. When the chlorine concentration deviates too much from this range, the rate of development is retarded.

After one or two days, the type D larvae attain a size of 70 to 80 μ and put forth velum between the shells. By collecting food with the help of cilia, they grow further. As the shells gradually grow, their vertexes protrude, that of the left shell being more than that of the right. Hence, the right and left shells become unequal in size. The larvae in this stage are known as umbo larvae, and the fact that their two shells are unequal in shape helps in the separation of this variety from other bivalves.

After leading a pelagic life for two to three weeks, the larvae adhere to some substratum and enter a sessile life. At this stage, the length and height of the shells is about 300 μ. In order to collect seedlings, the setting period must be estimated and the collecting implements placed at the right time.

Information on the conditions of sea water such as the water temperature, density, etc., as well as the spawning condition and the emergence of the larvae is made available in Hiroshima from time to time by the Hiroshima Fisheries Experiment Station. The period best suited for the installation of seedling collecting equipment can then be forecast. At Sendai Bay, information of a similar nature is given by the Oyster Fisheries Association of the Miyagi Prefecture, after conditions in the vicinity of the various seedling grounds in Matsushima Bay and the coastal regions of Ojika Hanto have been observed every third day in July and August. The Fisheries Experiment Station in the Miyagi Prefecture also cooperates with this Association, and with the further cooperation of the Tohoku Fisheries Experiment Station, the Association carries out a survey in Sendai Bay, which is the growing bed for the larvae of *C. gigas*.

In Hiroshima, the emerging conditions of the larvae of *C. gigas* are surveyed at selected places during full tides every two to three days during the period of June to August. Plankton nets of XX16 with a diameter of 30 cm at the opening, are used in this survey. The nets are hung 3 m under water and oyster larvae at this depth are collected and the specimens examined under a microscope. They are then fixed in formalin, measured with the help of a micrometer, and divided into groups according to size as shown in Table 2.34. By carefully studying the Table, the setting period can be estimated. At Sendai Bay also, nets are placed in water 2 m deep or, if necessary, at a depth of 5 m. The collected materials are separated with the help of sieves and the number of larvae counted according to size. Under the guidance of experts, a number of high school students from the Fisheries School at Miyagi Prefecture take part in the survey carried out by the Seed Oyster Association; these youngsters display a great deal of interest in the collection method, observations, microscopic examination, and groupings by sieves. The number of young people joining the Youth Association of Fisheries is increasing as the activities in this field expand. Students in those Middle Schools equipped with seed tracts, form clubs to study the behaviour of larvae and the conditions of sea water, and under expert guidance frequently contribute useful data.

When the setting period nears, the seedling units (prepared with scallop shells)

Table 2.34. Larval groups and their sizes in various regions

Larval groups	Hiroshima Fisheries Experiment Station (Hiroshima Suiken)		Miyagi Seed Collecting Association		Miyagi Fisheries Experiment Station (Miyagi Suiken)
	Size	Number of days	Size	Number of days	Number of days up to setting
1. D-larvae	Below 90 μ	1			12-14
2. Small pre-umbo larvae	Below 110 μ	2-3	1. Below 100μ	Less than 4 days	10-12
3. Small post-umbo larvae	Below 150 μ	4	2. 100-150μ	4-6	9-10
4. Medium size umbo larvae	Below 220 μ	7	3. 150-200μ	7-9	7-8
5. Large umbo larvae	Below 270 μ	10	4. 200-250μ	10-11	3-4
6. Full grown larvae	Above 270 μ	12	5. Above 250μ	More than 12 days	1-2

are hung vertically in the water at selected points, and the number of spat at a particular water level calculated. On the basis of this calculation, the setting day is forecast, and the day when the seedling collecting implements should be installed is announced. At some places, the workers themselves are able to calculate the period of seedling collection by making such observations in their fields. In order to observe the seedlings, experimental units are hung every day on the racks, removed the following day, and the number of seedlings attached counted. For this purpose, it is convenient to use a magnifying glass capable of magnifying 10 times, such as a watch maker's lens, since one must be able to differentiate the oysters from barnacles and small clams. Since many of the young oysters are red-brown in colour, they can be easily mistaken for the latter. To distinguish barnacles from oysters, one must look carefully for the yellow color and elliptical shape of the former. Oizumi (1953) reported the presence of the cypris larvae of barnacles in the oyster basins of Matsushima Bay. These larvae were found in large numbers during the peak spawning period of *C. gigas*, but their setting period differed by nearly one week. Consequently, Koganezawa worked out guidelines for culturists collecting *C. gigas* seedlings in 1959 (Figure 2.95).

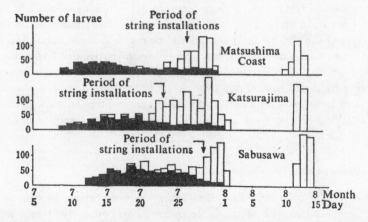

Figure 2.95. Variations in the number of setting larvae of oysters and barnacles (1959).

■—Barnacle;   □—Oyster.

2.4 Control and management

The young oysters which settle develop the characteristic bordering shells within a day, and rapidly grow to 2 to 3 mm in length within 4 to 5 days. They become 10 mm in size by late September, and when transferred to the rearing beds grow to 30 to 50 mm by late December. Collected from winter to the following spring, the seedlings are easily injured during packing or transport to long distances, and are also susceptible to mass mortality. Hence when the temperature declines in early September, they are kept in shallow places in control racks made from bamboo, along with pine twigs, and the strings of seedlings are placed in groups of 3 and 4. By this method, popularly known as floor-raising (*Yuka-age*), the period in which the seed oyster remains in water is shortened, and consequently growth is controlled. At the same time, sunlight and wind strengthen and thicken the shell margins.

The places selected for "floor-raising" have weak tidal currents, and since a number of strings are placed one above the other in places with a mean tidal level, the duration of low tide is longer and the amount of available food less. Furthermore, since the seedlings are exposed to cold air in winter, their growth is nearly arrested, and thus seedlings of 10 mm shell height and strongly resistant to air are obtained. These seedlings are able to withstand long journeys in February to April and can be reared by transplanting them to cultivation beds far away from the seedling collecting grounds. They can even be exported to North America. Furthermore, oysters which are transplanted to rearing beds after growth has been controlled through the floor-raising method during the seedling period, grow rapidly, being capable of withstanding variations in the environment. The growth of seed oysters collected from Sendai Bay is particularly remarkable.

The period of "floor-raising" as well as the selection of location, is very important in oyster culture. When the floor-raising process is initiated, abnormal temperatures due to residual matters and direct sun rays should be avoided. Since a number of strings are piled on the control racks, the flow of sea water is obstructed and hence the seed oysters on the middle strings may not get enough sunlight and air. Therefore the strings must be adjusted now and then to expose the oysters in the middle.

The height of the control racks should also be carefully considered. If it is too low, the strings may touch the sea bottom and be destroyed by drills (*Urosalpinx cinera*). If it is too high, the seedlings might freeze to death when the climate is too cold. To avoid boring by drills, control racks should be positioned in muddy places, but here it is necessary to fix cross bars so that the racks cannot sink. In some places racks are painted with drill-repellent substances. All beds should be carefully cleaned of drills during low tides.

2.5 Exportation and transplantation regulations

Seed oysters are collected after completing the floor-raising process in February or March, and washed with water to remove mud and other sticky substances. Any drills which may be mixed up in the collection are thus washed out. After washing, the oysters are arranged on checking platforms and all extraneous matter, such as eggs of drills, removed. Seedlings are selected on the basis of their condition. The standards of collectors are given in Table 2.35, which also shows the number of young oysters adhering to a single shell. Shells are cut in such a manner that 5 to 6 seedlings can adhere to each piece, and the smaller broken bits of shell are removed by repeated washing with water.

The standards for checking seedlings are presented in Table 2.36. Whole shells are separated from cut pieces and placed in different boxes. After fitting lids to them, the boxes are tied and transferred to racks kept under water in those places where there is no possibility of pollution or drill infestation. Following frequent submersion in water, the cases (about 10,000) are loaded on ships and covered with straw mats to prevent drying during transportation.   As an additional preventive to mass mortality, the straw mats are sprayed with sea water two or three times daily during the

Table 2.35. Types and specifications of collectors

Types of Collectors	Specifications	Size of shell Height cm	Length cm	Number of shells per ren	Number of young oysters (4 to 18 mm) per shell
Oyster	Large	Above 13.5	Above 7.5	55	30
	Medium	Above 9.0	Above 4.5	70	25
	Small	Below 9.0	Below 4.5	100	20
Scallop	Very large		Above 13.5	50	80
	Large		Above 10.5	70	60
	Medium		Above 8.5	100	40
	Small		Below 8.5	130	30
Clam	Large		Above 9.0	100	50
	Small		Below 9.0	130	30
Abalone			Above 9.0	70	30

Table 2.36. Standards for checking whole and cut shell packing, A-grade products

Materials	Whole shell	Cut pieces
Vitality	Outstanding	Outstanding
Size of seedlings	3 to 18 mm	3 to 18 mm
Number of seedlings	More than 10 seedlings per shell More than 12,000 seedlings per case	More than 16,000 per case
Extraneous substances	Absence of extraneous substances such as mud, debris, etc.	Absence of extraneous substances such as mud, debris, etc.

10-day voyage to North America, where the seed oysters are then released into the growing beds and reared.

In exporting seed oysters, the standard is based on the specifications mentioned in the Export Control Act and the export agreement made accordingly. Table 2.36 shows the standard for A-grade products, the only ones exported. Inspection and checking are carried out jointly by the Japanese Ministry of Agriculture and Forestry, and a representative of the American government. However, exporters are expected to do the checking themselves to ensure conformity with the specifications. The Seed Oyster Association, as well as the Fisheries Experiment Stations, also participates in the checking of control racks, management of seed oyster washing at the time of packing, selection of seed oysters, installation of control racks, etc.

Distribution within the country follows more or less the same standards, although the specifications are not so rigid. Materials are transported on their original strings, both overland and by sea and, consequently, the controls are not very satisfactory.

Whether for export or transplantation, the duration of the transport should be as short as possible, and the seed oysters should be maintained in a state of minimum temperature and maximum humidity. It is needless to mention that great care should be exercised in packing and shipping.

(Oizumi)

3. Condition and management of culture and fishing grounds

3.1 Requirements for culture and culture fishing grounds

A good yield of oysters requires a rich supply of plankton feed and proper culture management. Toward this end, various culture methods like the sowing method and the hanging method are being experimented with in Japan. According to the 1967 statistical reports published by the Ministry of Agriculture and Forestry, the Hiroshima Prefecture produced 28,438 tons of oyster meat, the Miyagi Prefecture 3,727 tons, the Iwate Prefecture 1,141 tons, and the Okayama Prefecture 1,098 tons. Hence the total amount of oyster meat in 1967 was 38,037 ton. A major portion of this tonnage was produced by the hanging culture method with rafts (Table 2.37).

Table 2.37. Conditions of *C. gigas* **culture**
(1967 Statistics Published by the Ministry of Agriculture and Forestry, Japan)

Method of culture		Number of enterprises	Number of installations
Hanging culture method	Raft method	3,298	14,974
	Long-line method	1,049	5,927
	Rack method	1,770	3,182 km²
"Hibitate" method		339	3,618 km²
Sowing method		597	2,570 km²

3.1.1 Sowing Culture

The sowing culture method for developing *C. gigas* is the oldest in Japan, and records show that it was utilized as early as the 17th century in the Hiroshima Prefecture.

For this culture, a sea bottom at the time of low tide is selected, and the seedlings either grown without transplant or with it. The season chosen for transplantation is the autumn of the same year or the spring of the following year. According to Kuroda *et al.* (1957), oysters from the age group of one and a half to two years, which show a healthy growth, are collected in Ariake-Kai. Production is said to be 1.5 kg of flesh per 1 m² area.

The fishing grounds selected for sowing culture must be firm since oysters in muddy soil are lost by sinking or by being carried away by the currents. Hence, while sowing is possible in the grounds of the coastal regions of Hokkaido Bay where a mixture of sand and mud still remains firm, grounds with loose mud only are not suitable until strengthened (Figure 2.96).

Kuroda *et al.* (1957) reported that one method for improving the ground to use the sowing culture method for *C. gigas*, is to reinforce it with stones, bamboos, empty

Figure 2.96. Sowing culture of *C. gigas* in Hokkaido Bay (photo by M. Tamura).

shells, etc. This method was adopted in Ariake-Kai in Kyushu. In some places, the culture beds are rectangular, measuring 30 to 60 m in length and 4 to 6 m in width, arranged along the tidal current at intervals of 6 to 10 m. In Ariake-Kai, *C. rivularis* are reared by this method.

The sowing culture method adopted by Japan leaves little room for artificial improvement. However in France, *parcs* (Yonge, 1960), and in America, *dikes* (Galtsoff, 1964), are built around the culture grounds so that at the time of low tide, some sea water still remains in the grounds to prolong the feeding time. The seedlings are also thereby protected from exposure to cold air during winter. Yet another technique is employed in France, namely, culture grounds are located in shallow sea water ponds or *claires* (Yonge, 1960) in which the water temperature is increased to cause an acceleration in the growth of diatoms which, in turn, accelerate the growth of *C. gigas*.

3.1.2 HANGING CULTURE

Since oysters reportedly open their shells to collect food by drawing in and filtering the sea water, the hanging culture method was devised with the idea of increasing the duration of feeding and thereby accelerating growth. This method proved effective not only in the use of surface sea water, but also in preventing possible damage from the sand and mud of the sea bottom and from predators. The method can be divided into three types, namely, the rack method, the raft method, and the long-line method.

1. *Rack method:* This method of culturing *C. gigas* is carried out in shallow waters of 2 to 4 m depth during low tide. Two rows of poles are planted vertically in the sea bottom and a rack prepared by fixing bamboo poles or thin rods crosswise on top. Galvanized wires or tar-painted units to which seedlings adhere, are hung vertically from the crossbars (Figures 2.97 and 2.98).

About 6 to 10 pieces of empty shells are strung at intervals of about 20 cm on 1.5 m string, and 20 such strings (units) are hung vertically for every 3.3 m² area. The units are collected during the following spring and, according to Tamura (1967),

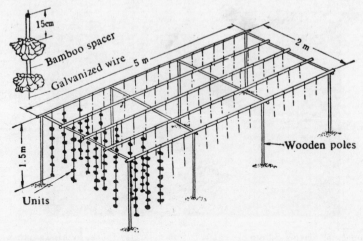

Figure 2.97. Rack culture method of *C. gigas* in the Miyagi Prefecture (Cahn, 1950).

Figure 2.98. Rack culture method of *C. gigas* in the Hiroshima Prefecture (photograph, courtesy of the Hiroshima Fisheries Experiment Station).

an average of 0.48 kg of oyster meat is obtained for every 1 m² of the culture area in Hiroshima Bay.

Since the rack culture of *C. gigas* is used inside the bays where the water is calm, water is easily arrested. To overcome this difficulty, the density of oysters on the strings is limited and the spaces between the racks kept wide. Furthermore, channels made between the racks allow the water to pass freely through them. Ito (unpublished) carried out an investigation on the effect of artificial channels of about 1,200 m length, 20 m width, and 3 m depth at the corners of Matsushima Bay, and found that by increasing the tidal current in the culture grounds, the weight of flesh of *C. gigas* at the time of harvest could be raised to about 20 g, whereas the mean weight of the control group was 15 g. Ito (1960) also tried to increase the sea water current in Matsushima Bay so that *Zostera* growths could be prevented. He removed *Zostera* from an area of 60 m width and about 500 m length, and found that the tidal current in the treated regions was 0.16 to 0.35 cm/second, or about 1.8 times that of the untreated regions. Accordingly, the growth in the treated regions in which the racks were placed at a lower layer was about 1.3 times that in the untreated regions. Hence a significant difference was found in the results of treated and untreated regions. Furthermore, Ito and Imai (1955) measured the weight of the dry meat of oysters (*C. gigas*) caught in the culture racks of Matsushima Bay, and found that it decreased when culture was continued for several years (Table 2.38).

Table 2.38. Effect of continuous cultivation of *C. gigas* by rack cultivation method in Matsushima Bay, Miyagi Prefecture (Ito and Imai, 1955)

Number of years	Cultivation for 2 years		Cultivation for 3 years		Cultivation for 6 years	
Depth	Middle layer	Lower layer	Middle layer	Lower layer	Middle layer	Lower layer
Number of oysters examined	42	33	33	32	28	26
Mean dry weight of meat (g)	0.834	0.713	0.725	0.578	0.754	0.422

When the cultivation is continued in the same bed, the metabolic wastes of the oysters (*C. gigas*) collect at the bottom. In fact, an oyster weighing about 90 g will produce 0.036 g (dry weight) of excreta daily. Oysters on a raft of the size of 60 m² will excrete 0.6 to 1.0 ton (dry weight) of waste. This includes 9.1% of decomposed plant material and 0.8% of total nitrogen. The sediment gradually floats as silt in the culture grounds due to the stirring of the water by wind and waves. This causes a decrease in the amount of dissolved oxygen in the sea water. Hydrogen sulfide is released, which adversely affects the physiological activities of the oyster. As countermeasures, Ito and Imai suggested the removal of the mud by a sand pump, etc., or shifting the culture equipment from the polluted

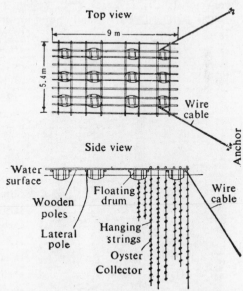

Figure 2.99. Raft culture method in Miyagi Prefecture.

120 to 150 strings; length 9 m; number of collectors (per string) 35 fixed at intervals of 24 to 27 cm (Cahn, 1950).

grounds to other culture grounds, or the rotation of crops.

2. *Raft method:* Wood poles or bamboos are placed parallel to one another and fastened to lateral poles. These are suspended at a depth of 4 m. Shells are strung

Figure 2.100. Raft method of culture of *C. gigas* in Hiroshima Prefecture (photo, courtesy of Hiroshima Fisheries Experiment Station).

at intervals of 10 to 15 cm on rope or galvanized more than wire units of 3 to 6 m length, and hung on the rafts at intervals of 40 to 50 cm. On a raft of 4.5 m × 9.0 m, about 200 strings are usually hung. Since the density of young oysters is closely related to their growth, it is better to limit the number of young oysters to 20 per collector (shell) (Figures 2.99, 2.100, 2.101, 2.102).

The seed oysters are usually hung in the spring following the year of seeding. From October to the next April, they grow to about 10 cm when they are harvested. The amount of harvest depends greatly upon the conditions of the cultivating grounds. According to Sato (1967), 3.5 to 14.8 kg of meat per 1 m^2 of raft is produced in various regions of Japan (Table 2.39).

In some regions, the seedlings of *C. gigas* for cultivation are produced in the same place, and in others the seedlings are transplanted from other regions. The production of seedlings is large in various prefectures in Japan, particularly in Miyagi, Hiroshima, and Kumamoto. According to Imai and Sakai (1961), the quality of *C. gigas* differs according to the place of seedling production. As already explained, the oysters of the southern regions have smaller and thinner shells. Furthermore, the shells have more brown pigment and there is a considerable difference in physiological characteristics such as growth, spawning, etc. When the seedlings of

Figure 2.101. *C. gigas* catch after cultivating by raft method in Hiroshima Prefecture (photo, courtesy of Hiroshima Fisheries Experiment Station).

Table 2.39. Raft culture method of C. gigas and the amount of catch (Sato, 1967)

	Raft			Number of strings				Amount of catch				
	Number of rafts	Size of raft (m²)	Total (m²)	No. of strings per raft	No. of strings per 1 m²	No. of ropes	Rope length (m)	Total (ton)	Raft For one raft (kg)	Raft For 1 m² (kg)	Collector For one string (kg)	Collector For 1 m² (g)
Hiroshima	8,230	164.0	1,990,000	500	3.05	4,115,000	39,092,539	20,042.0	2,433	14.83	4.86	511
Matoya	100	53.0	5,300	100	1.89	10,000	40,000	30.0	299	5.65	3.00	750
Kagami-Ura	1,200	58.1	69,720	130	2.24	156,000	1,092,000	312.0	260	4.48	2.00	375
Shiraishi-Ko	150	20.8	3,120	125	6.02	18,750	75,000	33.6	223	10.70	1.80	448
Kesen-numa	4,200	62.4	262,500	190	2.98	798,000	5,580,000	1,563.3	384	6.15	1.95	284
Kamo-Ko	4,167	38.8	161,500	100	2.58	416,700	1,875,000	552.0	136	3.50	1.88	418

Figure 2.102. *C. gigas* catch after cultivating by raft method in Hiroshima Prefecture (photo, courtesy of Hiroshima Fisheries Experiment Station).

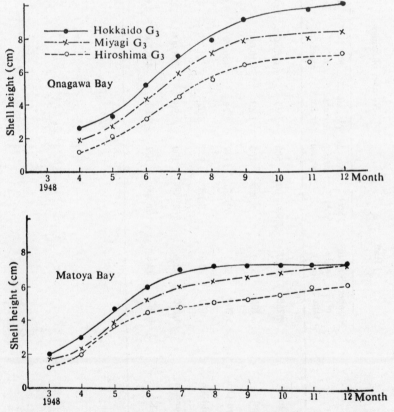

Figure 2.103. Comparison of the growth of *C. gigas* by raft culture method. The seedlings were obtained from various regions and grown in Onagawa Bay and Matoya Bay (Imai and Sakai, 1961).

Hokkaido, Miyagi, and Hiroshima were grown by the hanging culture method in Onagawa Bay of the Miyagi Prefecture and Matoya Bay of the Mie Prefecture, those reared in the northern regions matured earlier than those reared in the southern regions (Figure 2.103).

From these results, it appears that it is possible to select the variety of seedlings according to requirements. On the other hand, efforts are also being made to improve the quality of oyster by hybridization and artificial selection.

The growth and flesh content in *C. gigas* vary according to the season. In an investigation carried out by Sato (1967) in which he reared the seed oysters of Miyagi Prefecture in Matoya Bay of the Mie Prefecture by the hanging method in the January following the year of seeding, it was found that the average shell length after 15 months, i.e. in March of the following year, was 7.5 cm, and the shell height 10.0 cm. In this case, the growth was good during March to June when the water temperature was below 20°C, but in August to September when the water temperature increased, growth was temporarily arrested due to spawning. When the temperature decreased from October onwards, growth resumed (Figure 2.104). According to Mori *et al.*

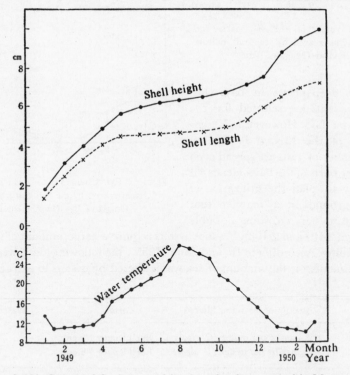

Figure 2.104. Growth of *C. gigas* reared by the raft culture method in Matoya Bay. The seedlings were obtained from the Miyagi Prefecture (Sato, 1967).

(1965), a decrease in the rate of growth of *C. gigas* during the spawning period was observed at Onagawa Bay of the Miyagi Prefecture. The decrease in the weight of the soft parts was particularly significant (Figure 2.105). Hosonaka (1940) also observed a decrease in the weight of soft parts and reported the subsequent decrease in the protein, fat, and glycogen contents (Figure 2.106).

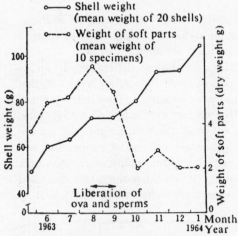

Figure 2.105. Weight of the soft parts and shells of *C. gigas* reared by the raft culture method at Onagawa Bay.

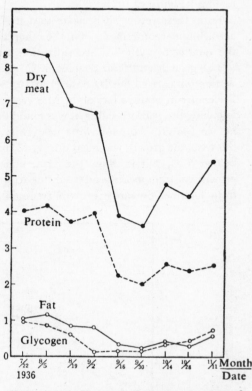

Figure 2.106. Changes in the meat components of *C. gigas* reared by the raft culture method at Onagawa Bay (Hosonaka, 1940).

Relatively deep places in bays with fast tidal currents are selected for the raft culture method. However, in Kesennuma Bay of the Miyagi Prefecture, a large number of rafts are spread even in the corner parts of the bay. Hence, it is necessary to plan the utilization of the fishing grounds in a manner that will obviate oyster crowding (Table 2.40; Figures 2.107 and 2.108), because oysters require a large amount of sea water to filter for feeding. According to Ito and Imai (1955), the following data were obtained from experiments on the amount of sea water filtered by oysters to extract food:

Period	Amount of sea water filtered	Size	Water temperature
June to Sept.	5 liter/hr per oyster *(C. gigas)*	Shell length 1-5 cm	18-23°C
Sept. to Dec.	15 liter/hr per oyster *(C. gigas)*	Shell length 5-8 cm	23-8°C

On the basis of the data supplied by Ito and Imai on the amount of sea water filtered, estimates on the amount of food required by one row of rafts (9 m ×4.5 m) with a density of 250 oysters/m³ were made in the fishing ground of Kesen-numa when the water was flowing in one direction. When the residual amount of food available in the middle part of the row of rafts was measured after the water had

Table 2.40. Details of raft culture of *C. gigas* of Kesen-numa Bay of Miyagi Prefecture (Data collected in December, 1954) (Imai *et al.*, 1957)

Culture ground	Area	Number of rafts	No. of rafts per unit area	No. of strings per raft	Hanging depth	No. of oysters per raft
	Hectare		Rafts/hectare	No. of oysters/raft	m	No. of oysters/m³
Shishiori	30 8	825	26.8	49,728	3.5	292
Matsu-Iwa	63.1	960	15.2	68,376	6.1	231
Hashikami	92.7	563	6.1	57,554	5.6	211
Oshima	12.9	314	24.5	91,113	6.5	289
Kaihama	32.7	410	12.5	81,696	7.1	237
Others	—	1,890	—	—	—	—
Total	—	4,962	—	—	—	—
Mean	—	—	17.0	—	5.8	252

Table 2.41. Residual amount of food present in the raft culture grounds of Kesen-numa Bay, Miyagi Prefecture (Imai *et al.*, 1957)

Raft No. and side		June to September	September to November
Outer side of fishing ground		100%	100%
Passing through :	Raft 1	88.9	77.4
	Raft 2	78.9	60.0
	Raft 3	70.1	46.4
	Raft 4	62.3	36.0
	Raft 5	55.4	27.8
	Raft 6	49.2	21.6
	Raft 7	43.7	16.7
	Raft 8	38.8	12.9
	Raft 9	34.5	10.0
	Raft 10	30.7	7.7
	Raft 11	27.2	6.0
Inner side of fishing ground		24.2	4.6

passed through raft 11, it was found to be 4.6 and 24.2% (Table 2.41 and Figure 2.109). This showed that the growth of oysters on the inner rafts was unlikely to be satisfactory (Table 2.42).

According to Yokota and Sugiwara (1914), the number of spat per shell influences their growth. Tanida and Kikuchi (1957) have studied the relationship between the number of seedlings and the survival rate, and the relationship between the growth rate and density. They found that the maximum number of surviving oysters on scallop shells of 10.5 to 12.0 cm diameter is about 60. They also found that there is a constant functional relationship between the number of seedlings and the survival rate.

Table 2.42. Growth of *C. gigas* in September, 1955 in Kesen-numa of Miyagi Prefecture (Culture carried out by raft method) (Imai *et al.*, 1957)

Raft No. and side	Total weight (g)	Weight of meat (g)
Outer side of fishing ground		
Raft 1	22.5	2.53
Raft 2	23.8	3.45
Raft 3	25.0	4.01
Raft 4	30.9	4.20
Raft 5	27.9	3.16
Raft 6	32.7	4.11
Raft 7	32.0	4.63
Raft 8	27.4	3.17
Raft 9	22.5	2.19
Raft 10	20.3	1.72
Raft 11	25.0	2.76
Inner side of fishing ground		

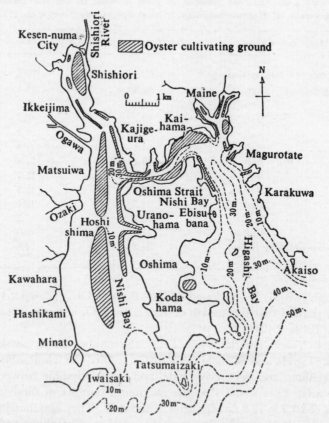

Figure 2.107. Distribution of rafts used for hanging culture of *C. gigas* in Kesen-numa Bay of the Miyagi Prefecture (Imai *et al.*, 1957).

Imai and Kikuchi (unpublished) reared oysters in Onagawa Bay in the Miyagi Prefecture, fixing the seedlings at the rate of 10, 20, 30 and 40 per shell and counted the surviving oysters after 7 months of rearing. They determined correlation between the number of individuals (x) and the mean total weight (y) of each seed oyster per shell (Table 2.43), and expressed the relationship by the following equation:

$$y = -0.64\ x + 72.20.$$

The growth of *C. gigas* is greatly affected by adhering organisms (Figure 2.110). Imai and Fukuchi (unpublished) kept mussels adhering to the seed oyster collectors after hanging at the rate of 0, 10, 20, 30, 40, and 50 for each collector (mean weight of seed oysters of which 19 were kept adhering to each collector, was 0.05 g), and observed the growth rate of *C. gigas* after hanging on July 14, 1955 in Onagawa Bay in the Miyagi Prefecture. The oysters were taken out on December 14 and their growth

Figure 2.108. Distribution of rafts in culture beds of Kesen-numa Bay, Matsuiwa, and Hashikami (photo, courtesy of Kesen-numa Fisheries Experiment Station, Miyagi Prefecture).

Table 2.43. Influence of the number of seed oysters per collector shell on the growth of *C. gigas* (Imai and Fukuchi, unpublished)

Date	Measured values	10 seed oysters/ collector	20 seed oysters/ collector	30 seed oysters/ collector	40 seed oysters/ collector	50 seed oysters/ collector
May 16, 1955 (strings hung)	Mean of total weight (g)	0.05	0.05	0.05	0.05	0.05
	Mean of meat weight (g)	—	—	—	—	—
December 15, 1955 (strings taken out)	Mean of total weight (g)	65.2±9.7	60.4±9.5	52.8±8.9	46.2±8.0	40.3±8.5
	Mean of meat weight (g)	7.15±1.21	6.30±0.85	5.49±1.01	4.6±0.97	4.16±0.82

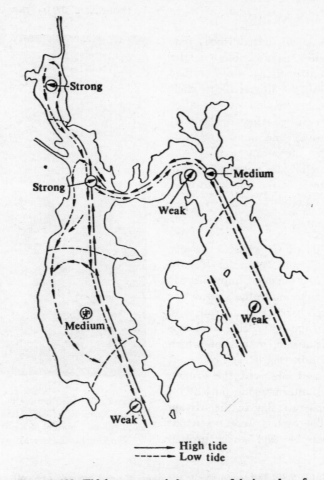

Figure 2.109. Tidal current and the extent of the inner bay of
Kesen-numa Bay (Imai *et al.*, 1957)

Figure 2.110. Adherence of mussels during a raft culture of *C. gigas* in the Hiroshima
Prefecture (photo, courtesy of Hiroshima Fisheries Experiment Station .

Table 2.44. Influence of adhering mussels on the growth of *C. gigas*
(Imai and Fukuchi, unpublished)

Date	Measured values	Number of mussels adhering to each collector					
		0	10	20	30	40	50
July 14, 1955 (strings hung)	Weight ratio of mussels to *C. gigas*	0	0.009	0.028	0.051	0.047	0.233
	Total weight of *C. gigas* (g)	5.45	4.69	4.88	4.57	6.07	6.00
December 14, 1955 (strings taken out)	Weight ratio of mussels to *C. gigas*	0	0.120	0.385	0.742	1.129	3.251
	Total weight of *C. gigas* (g)	62.7	59.2	57.7	51.2	43.4	33.9

checked. A reverse correlation was seen between the total weight ratio (x) of the mussels and *C. gigas*, and the mean total weight (y) per each specimen of *C. gigas* (Table 2.44). The relationship can be expressed by the following equation:

$$y = -7.58\,x + 59.39.$$

The raft method of culture is also used for cultivating the European oyster and the Olympia oyster. Imai *et al.* (1967) produced seedlings from maternal oysters imported from Europe and America, which were strung on strings painted with coal tar. The strings were hung in Mohne Bay of the Miyagi Prefecture in the autumn of the year of the seedlings. After two years, the European oyster showed a growth of more than 10 cm in shell height, and was of a marketable size (Figure 2.111). Even when the European oysters were hung in metal wire nets, they had the same rate of growth (Figure 2.112).

3. *Long-line culture method:* Wooden barrels connected by two parallel lengths of straw rope are floated on the sea surface, and strings of seed oysters are hung from the ropes

Figure 2.111. Rope culture of European oysters on rafts in Maine Bay of the Miyagi Prefecture (photo, courtesy of Oyster Research Center).

Figure 2.112. Wire net culture of European oysters on rafts in Maine Bay of the Miyagi Prefecture (photo, courtesy of Oyster Research Center).

(Figure 2.113). The major advantage of this method is that possible damage to the equipment and the crop is greatly minimized. Hence this method of cultivation is adopted at the mouths of bays or in open seas where the waves are rough, and is popular in the Iwate and Miyagi Prefectures (Figure 2.114). The seed oysters are usually hung in the spring and harvesting begins in October of the same year. The yield rate per unit and the growth of *C. gigas* are more or less the same as in the raft culture method. It is said that the growth is not satisfactory when the culture zones are far removed from the coasts.

(Ito and Sato)

3.2 Management of culture

In Japan as well as in foreign countries, the culture of oysters seems to have started from the sowing method. This method has been greatly modified in Europe and

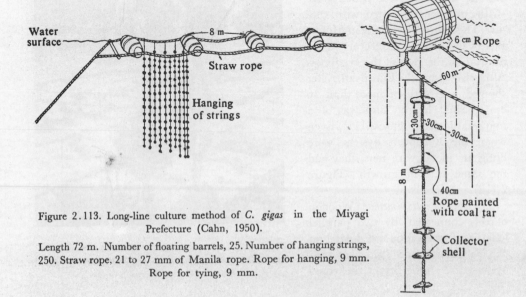

Figure 2.113. Long-line culture method of *C. gigas* in the Miyagi Prefecture (Cahn, 1950).

Length 72 m. Number of floating barrels, 25. Number of hanging strings, 250. Straw rope, 21 to 27 mm of Manila rope. Rope for hanging, 9 mm. Rope for tying, 9 mm.

Figure 2.114. Long-line culture method of *C. gigas* in Kesen-numa Bay of the Miyagi Prefecture (photo, courtesy of Kesen-numa Fisheries).

America, but still remains one of their main methods of culture. In Japan, however, this method is losing its importance because of the long period it involves for culture and harvesting. The hanging culture method, on the other hand, has become more and more popular since it has the advantage of being employable even in the deep waters below the ebb tide line. Hence, the area of culture by this method is rapidly expanding. Among the various types of hanging culture methods, the yield rate per unit area is maximal with the raft type. For example, in the Hiroshima Prefecture, this yield rate is 15 times more than that from the sowing culture method, or that from the rack method (Table 2.45).

In view of the above data, the raft culture method is spreading more rapidly than

Table 2.45. Comparison of yield rates of *C. gigas* meat in 1955 to 1962 in the Hiroshima Prefectures (Tamura, 1967)

Method	Unit area of culture	Yield rate	
Raft (per raft)	18 m×9 m=162 m²	2,552.0 kg	15.74 kg/m²
Rack	99	47.8	0.48
Bamboo pole	99	17.8	0.19
Sowing	99	91.9	0.93

any other method. In the Hiroshima Prefecture, raft culture extended to only 5% of the culture grounds in 1935, but by 1960 comprised 96.6% (Table 2.46).

This trend is expected to persist in the future. However, since the shallow seas around Japan, especially the water inside the gulfs and bays, are becoming more and more polluted, culture by the raft method in the corner parts of the gulfs and bays has become impossible. Hence, culture grounds are gradually shifting from the corner parts of the bays to the mouths and even to regions beyond them. On the Sanriku

Table 2.46.   Changes in the production ratio of cultured oyster meat by various culture methods in the Hiroshima Prefecture (Ogasawara *et al.*, 1962)

Year	Method of culture		
	Hanging (raft type) %	Hanging (rack type) %	Sowing %
1926	0	0	100
1935	5.0	35.0	60.0
1941	4.2	63.0	32.8
1945	4.4	65.6	30.0
1947	9.1	73.3	17.6
1950	20.0	57.3	22.7
1953	51.6	39.9	8.4
1954	61.0	32.5	6.5
1955	66.7	29.3	4.0
1956	75.0	20.0	4.9
1957	81.0	15.2	3.8
1958	83.7	13.0	3.3
1959	95.8	2.7	1.3
1960	96.6	2.2	1.1

coast where the waves are rough, the culture of *C. gigas* is carried out widely in open seas by the long-line hanging method (Figure 2.115).

In the raft type of hanging culture, the spat of *C. gigas* are usually hung in the spring following the year of seeding, and harvested either in the winter of the same year or the following spring. However, in the Hiroshima Prefecture, in order to avoid mass mortality which generally occurs every summer, the spat are hung in the autumn of the year of seeding and harvested the following spring. This method not only avoids the mortality period but also shortens rearing. However, the growth and fattening of the oysters are inferior to those of the two-year-old oysters. Another method of preventing mass mortality is to check the growth of the seedlings by exposing them for 10 to 18 hours daily, and then to rear them by hanging the same in water from the spring of one year to the summer of the following year. Mass mortality in the summer of such growth-checked seedlings is very low and, furthermore, the growth rate after hanging them in water is higher. Experiments carried out in the Hiroshima Prefecture by Ogasawara *et al.* (1962) showed that in the growth-controlled seedlings, the growth rate after hanging was so rapid that by December it was very near the growth rate of normal seedlings and, in fact, after February was much better than that of normal seedlings. This is due to the fact that in growth-controlled seedlings, loss during the spawning period is minimal, and the growth rate after the spawning period is high. Hence this method is expected to gain in popularity.

(Ito and Sato)

3.3 Harvest

According to studies carried out by Tsuchiya *et al.* (1962), the cyclic changes of

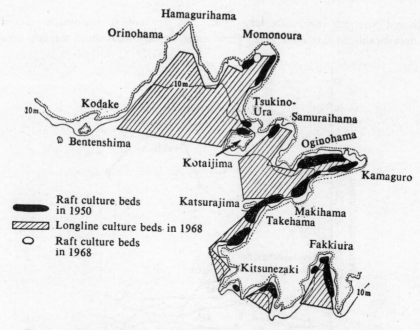

Figure 2.115. Distribution of long-line culture zones in Ogihama Bay, Miyagi Prefecture (data from the Miyagi Fisheries Experiment Station).

protein and glycogen follow opposite courses. Glycogen starts increasing from the late winter and continues to increase up to early summer when the oysters are fat. However, with an increase in water temperature, gonads develop, and there is a corresponding reduction in the glycogen content which is at its minimum in August or the spawning period.   On the other hand, the protein content which is minimal when the glycogen content is maximal, rapidly increases with the development of the gonads and attains its maximum in June to July, or just before spawning, and decreases after spawning (Figure 2.116). The luster of oysters fattened by glycogen differs from that of oysters fattened by protein. Usually the flavour of oysters fattened with glycogen is more delectable than protein-fattened oysters. In summer, if the protein content is too high in the oyster, rapid decomposition occurs; even if these oysters spawn, they become inedible due to thinning and high water content.

The meat content of oysters is greatly influenced by the water temperature, amount of planktons, etc. According to Sato (1967), when there is an abundant growth of planktons in the winter (more than 3 ml as sediment), even the so-called "water oyster" (*Mizugaki*) in which the flesh is transparent, resumes growth in about 10 days, and the digestive diverticulum can be identified. Sakai *et al.* (1964) reported that transplantation at this stage is an effective measure for improving flesh content.

According to Ogasawara (1962), the environmental conditions which accelerate the growth of shell, the fattening of meat, and spawning, differ in various cultivation areas. By effectively combining the most favorable conditions, it is possible to accelerate growth and fattening. When oysters growing in grounds with conditions favorable to growth and fattening are transplanted to beds with less favorable conditions, growth and fattening deteriorate very rapidly. By the same token, oysters transferred from unfavorable to favorable beds, grow more rapidly than those already present.

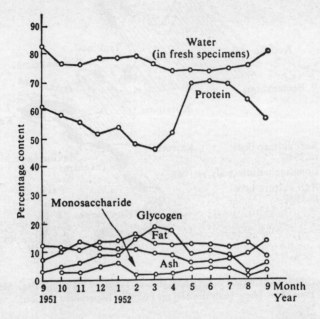

Figure 2.116. Variation in meat content during growth of oysters (Tsuchiya, 1962).

Harvesting is carried out according to the geographical conditions of various growing regions and the condition of the meat. Harvesting takes place from October to January in the Mie Prefecture, from October to May in the Miyagi Prefecture, and from December to June in the Hiroshima Prefecture. However, harvesting is carried out now in the Hiroshima Prefecture even after June, firstly because the meat content is not sufficiently high in the winter, and secondly, because of consumer demand (Figure 2.117).

Since the extent and conditions of cultivation differ greatly in various regions, it

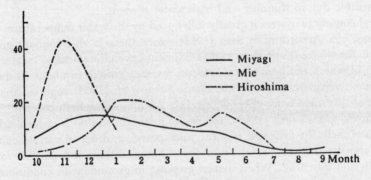

Figure 2.117. Seasonal variations in production.

is very difficult to compare the productivity of these regions. Hence the assessment shown in Tables 2.47 and 2.48 is based on the amount of catch per collector shell in 1 m of collector string. The maximum catch in the gulf area, i.e., 126.8 tons for 1 km², is seen in Kagami-Ura. Kamo-Ko comes next with 113.6 tons, Hiroshima with 99.5 to 115.6 tons, and the other regions produce 50 tons or less. The maximum production of 750 g per 1 m of collector string is found in Matoya, followed by 513 g in

Table 2.47. Production rate per 1 km² area and production rate of meat in various fishing grounds

Regions	Year	Area of bay (km²)	Production of meat (tons)	Production rate per km² (tons)
Hiroshima	38 (1963)	201.46	22,217	110.28
,,	39 (1964)	201.46	20,042	99.48
,,	40 (1965)	201.46	23,295	115.63
Shiraishi-Ko	39 (1964)	4.16	34	8.17
Kagami-Ura	38 (1963)	2.39	303	126.78
Kesen-numa	38 (1963)	28.27	1,386	49.03
,,	39 (1964)	28.27	1,360	48.11
,,	40 (1965)	28.27	939	33.22
Kamo-Ko	38 (1963)	4.86	552	113.58
Matoya*	40 (1965)	12.73	307	24.12
Mushiake	39 (1964)	4.59	177	38.56

*Includes 275 tons of meat from Akoyagai on 5,500 rafts.

Hiroshima, and 500 g or below in other regions. The amount of meat per collector shell is maximum in Matoya at 158 g, followed by Hiroshima with 128 g, Kesen-numa Bay with 124 g, Shiraishi-Ko with 117 g, and the other regions with 100 g or less. The fact that the amount of meat per unit area in Matoya Bay is very high seems to be due to the low density of strings and favorable environmental conditions, particularly the tidal current.

Regarding the relationship between the shape of raft and production rate per unit area, it was found that in all the growing beds, except those of Matoya Bay, the production rate increased when the raft was narrow and long. Hence the shape of the raft requires improvement.

Usually the oysters are shucked and only the meat shipped for consumption (Figures 2.118 and 2.119). However, the practice of shipping the oysters in their shells is presently increasing. The price of these oysters is several times more than the shucked ones, and hence this method of shipping is more profitable. However the shells should be uniform in shape and the meat particularly good. In Matoya Bay, the oysters on the strings are checked individually and only those with uniform shells selected for marketing. To ensure good meat, these oysters are then placed in metal-net baskets and reared by hanging the baskets in water for 15 to 30 days. During this period, any slight damage caused at the tips of the shells while separating them from the collector shells is cured by regeneration, and the product thus becomes fit for

Table 2.48. Yield rate and equipment in various regions (Sato, 1967)

Culture grounds	Number of rafts	Size of rafts (m)	Ratio	Area of raft (m²)	Number of collectors			Length of collector		Number of collector shells		Total (tons)	Yield rate		
					Per raft	Number of strings	Density of strings	Length of one string (m)	Length of rope (km)	Per 1 m rope	Per rope ×1,000		Per collector string (kg)	Per 1 m collector string (g)	Per collector shell (g)
Hiroshima	8,230	18.2×9.1	1:0.50	165.6	500	4,115,000	3.02	9.0-10.0	39,093	36-40	156,370	20,042	4.87	513	128
Kesen-numa	4,200	10.1×7.3	1:0.72	73.7	150	630,000	2.98	6.0-7.6	4,284	20	12,600	1,563	2.48	365	124
Matoya	100	9.1×6.4	1:0.70	58.2	200	10,000	1.89	4.0	40	18-20	190	30	3.00	750	158
Kagami-Ura	1,200	9.1×6.4	1:0.70	58.2	150	180,000	2.24	5.0-6.0	990	20	3,600	312	1.73	315	87
Shiraishi-Ko	150	6.1×3.4	1:0.56	20.7	125	18,750	6.04	4.0	75	15-16	291	34	1.81	453	117
Kamo-Ko	4,167	7.3×4.5	1:0.62	32.8	100	416,700	2.58	4.5	1,875	17-18	7,292	552	1.32	294	76

Figure 2.118. Catch and culling of oysters.

Figure 2.119. Workers shucking oysters.

marketing. These oysters are finally dipped in sterilized sea water and tinned, fetching a high price. But this mode of selling involves a stringent procedure for selection, careful removal of foreign bodies, and furthermore, can only be used in fishing grounds which are capable of producing oysters with healthy meat.

(Sakai)

4. Pests and prevention

The cultivation grounds of oysters are also habitats for organisms of various phyla such as Protozoa, Porifera, Annelida, Echinodermata, and other Mollusca and Protochordata. These organisms compete for the food available in the bed and also obstruct the tidal currents. Hence the growth of flesh in the oyster is delayed. Also, some of the organisms parasitize the oysters and even injure them. In extreme cases, the organisms are found to prey directly on the oysters. The fauna of the fishing grounds thus adversely affect the oyster culture in some way or another.

While the various types of organisms are either competitors for food or are predators, gastropod drills such as *Thais bronni, Thais tumulosa, Ceratostoma burnetti,* etc., are popularly known as "oyster pest" and cause much worry to cultivators. In the culture grounds of Matsushima Bay, Mangoku-Ura, and Ojika Hanto of the Miyagi Prefecture, three species of gastropods, namely, *Purpura clavigera* Kuster (*Ibonishi*), *Rapana thomasiana* Crosse (*Akanishi*), and *Tritonalia japonica* Dunker (*Ouyouraku*) are found as oyster pests. The number of such pests is large, and they injure the oysters by boring through their shells. Sometimes adults and eggs of these pests are found mixed in seed oysters and infest the culture grounds when the seeds are planted (Chapman and Banner, 1949).

The distribution of drills in Matsushima Bay shows that there is an increasing trend in the number of drills, from the low chlorine concentrated regions of the interior parts of the bays to the mouth of the bay with a high chlorine concentration. Of the three types of pests just mentioned, *P. clavigera* shows the widest distribution and the chlorine concentration which controls its existence is found to be in the range of 14.75 to 53.75‰. The chlorine concentration limiting the distribution of *Tritonalia japonica* is 16.25 to 43.75‰. However, the duration of their existence in both the low and high chlorine concentration regions, is related to water temperature; when the temperature is increased, the duration of their existence is shortened. Engle (1953), who studied *Urosalpinx cinerea* Say, which is also a type of drill in the Atlantic coast of North America, reported that the area for existence in the high chlorine concentration zone narrows when the temperature of the water is high.

Depending upon environmental conditions, the density of drill distribution in Matsushima Bay is maximal at the mouth of the bay. The habitat of drills thus coincides with the oyster culture grounds, and is a major factor in oyster damage. Drill damage usually begins in mid-March when the water temperature is about 8°C. As the temperature of the water rises, the degree of damage increases and continues up to December when the water temperature falls to 10°C. According to experiments carried out on drills, the damage by *Purpura clavigera* is at the rate of 5 oysters in 10 days, the damage by *Tritonalia japonica* 2 oysters in 10 days and *Rapana thomasiana* 12 oysters in 10 days (Koganezawa, 1963).

Of the various treatments for eliminating pests in the oyster culture grounds, copper compounds were found to be effective (Engle, 1949; Newcombe, 1941-1942; Lindsay and McMillin, 1963). However, since copper salt (copper sulfate) when sprayed either in the form of a solution or powder was found to have an adverse effect on other organisms, its use was stopped. Glude (1956) tried to control the movement of drills by using copper plates, and claimed the results were good. Imai *et al.* (1958) experimented with both copper plates and paint containing copper, and reported that both could be used in the fishing grounds. Interestingly, while copper salts were found to be very effective against *Tritonalia,* its effectiveness with *Purpura* and *Rapana* was judged doubtful.

The study of the reaction of various chemicals (Loosanoff *et al.,* 1960) and Sevin (1-naphytyl-n-methyl carbamate) (Koganezawa, *et al.,* 1968) on pests, including drills, showed that pests could be more or less completely eliminated by preparing a chemical barrier (Loosanoff *et al.* and Koganezawa *et al.*). When the fishing grounds are used for culturing more than one species, it was found that the effect on prawns, crabs, and abalones was adverse, but the effect on bivalves was good. Hence, drugs used in the control and eradication of the enemies of oysters should be very carefully used.

(Koganezawa)

5. Considerations for the future

The history of oyster culture is very old, but in spite of the fact that research on oysters, and the development and improvement in culture techniques are considerably advanced, the oyster cultivation beds are facing serious problems with water pollution due to the establishment of more industries, especially in the cities, in the field of metallurgy, fish processing, and agricultural pharmacology. The loss of seedling grounds in the Seto Inland Sea, and the mass mortality due to poisonous substances in Hamana-Ko and Matsushima Bay, have resulted in a decrease in the number of culture beds throughout Japan for several species of seaweeds such as *Porphyra, Undaria*, etc. At the same time, since the culture beds are concentrated in limited regions, the same beds are used repetitively, resulting in the loss of quality, another serious national problem. Doubtless the location and development of new fishing grounds is very essential, but this solution is rather facile in view of the fact that there are multiple problems involved related to culture techniques, raw materials, and other fishing industries. Hence, in order to prevent the aging of fishing grounds, a natural cleaning device should be found by which a concentrated cultivation could be carried out, or efforts should be made to clean the fishing grounds by artificial means such as dredging, or by changing the sea water. In order to achieve this, appropriate equipment or that suited to ensuring rich growth, is required; simultaneously, a method for improving the ecological conditions of the culture ground must be provided. Regarding the latter, not only oysters, but several other cultivated plants or organisms, should be considered collectively. As to the growth of shell and the fattening of meat, since the environmental conditions in various culture grounds differ, the favorable characteristics of each should be utilized fully in an effective manner.

Problems related to the improvement of quality and parasite control also exist. *C. gigas* is the major type of oyster cultivated in Japan, and it is well known that various types and strains of this oyster exist, which have specific regional characteristics. While utilizing these special features in a profitable manner, a suitable culture technique should be evolved, and an attempt made to improve quality by hybridization. By introducing better varieties from other countries, culture possibilities could be vastly improved, and in the future many problems related to all these aspects eventually resolved.

Recently the damage to oysters caused by adhering organisms such as mussels, barnacles, and bryozoa, as well as by drills, etc., has become a major problem. Mussels are found throughout Japan, but are particularly concentrated on the Sanriku coast; measures to eliminate them must be developed in the interests of oyster culture.

(Sakai)

Bibliography

1. AKASAKA, Y., 1952. Kesenuma-wan ni okeru tane-kaki saibyo ni tsuite (On the Seed Oyster at Kesen-numa Bay), *Nihon Suisan Gakkai Tohoku Shibu Kaiho (Bulletin of Tohoku Branch of the Japanese Society of Scientific Fisheries)*, **3** (1, 2), 33.

2. AMEMIYA, I., 1921. Kaki hassei shoki ni tsuite no kansatsu oyobi endo to no kankei (On the Early Development of Oysters), *Suigakuho*, **3** (3), 150-181.

3. ——, 1928. Ecological Studies of the Japanese Oyster, *J. Coll. Agr. Imp. Univ. Tokyo*, **9** (5), 333-382.

4. ——, 1928. A Preliminary Note on the Sexuality of a Dioecious Oyster, *Ostrea gigas* Thunberg, *Jap. J. Zool.*, **2** (1), 99-102.

5. ——, 1931. Kaki no hanshoku (Reproduction in *Ostrea gigas*), *Iwanami Koza: Seibutsugaku (Biology)*. Tokyo: Iwanami Publishers.

6. ——, 1933. Jiki ni yoru magaki seishokokusu no seijuku henka (Gaiho) (Seasonal Variations in the Maturation of Gonads of *Ostrea gigas*), *Suigakuho* (Preliminary Report), **5** (4), 341-353.

7. ——, and N. NAKAMURA, 1940. Magaki wa yu-sei senjuku ka? (Is *Ostrea gigas* Protandrous?), *Gakukyoho*, **15** (3), 338-340.

8. ——, M. TAMURA, and H. SENUMA, 1929. Magaki no shiyu-sei ni kansuru kosatsu (On the sexuality of *Ostrea gigas*), *Suigakuho*, **5** (2), 234-257.

9. CAHN, A. R., 1950. Oyster Culture in Japan. Report No. 134, G. H. Q., Nat. Res. Sec., 1-80.

10. CHAPMAN, W. M., and A. H. BANNER, 1949. Contributions to the Life History of the Japanese Oyster Drill, *Tritonaria japonica*, with Notes on other Enemies of the Olympia Oyster, *Ostrea lurida*. State of Washington, Department of Fisheries, Biological Report No. 49A, 167-200.

11. DAVIS, H. C., 1958. Survival and Growth of Clam and Oyster Larvae at Different Salinities, *Biol. Bull.*, **114** (3), 296-307.

12. ENGLE, J. B., 1941. Further Observations on the Oyster Drills of Long Island Sound, with Reference to the Chemical Control of Embryos, Conv. Addr. Wat. Shell Ass., Atlantic City, N. J.

13. ——, 1953. Effect of Delaware River Flow on Oysters in the Natural Seed Beds of Delaware Bay, Rept. U. S. Fish. Wild. Serv., Washington, D. C., 1-26.

14. FUJITA, K., 1930. Magaki no haiyobunka ni tsuite (On the Cleavage of *Ostrea gigas*), *Dozatsu (Japanese Journal of Zoology)*, **43** (513), 488-496.

15. ——, 1933. *Suisan Hanshokugaku: Suisan Gaku Zenshu (Culture of Marine Organisms)*. Tokyo: Koseikaku.

16. ——, 1934. Note on the Japanese Oyster Larva, Proc. 5th Pacific Sci. Congr., B8, 4111-4117.

17. FUJITA, T., 1929. On the Early Development of the Common Japanese Oyster,

Jap. J. Zool., **2** (4), 353-358.

18. FUJITA, T., 1952. Kaki-saibyo jiki no yochihokaki no tanemi (Method of Determining the Seedling Period of *Ostrea gigas*) *Rakusui*, October issue, 10-12.

19. GALTSOFF, P. S., 1964. The American Oyster, *Crassostrea virginica* Gmelin, *Fish. Bull.*, **64**, 1-480.

20. GLUDE, J. B., 1956. Copper, a Possible Barrier of the Oyster Drill, Urosalpinx. Proc. Nat. Shell Ass.

21. HATANAKA, M., 1940. Kakitai kakubu no ippan sosei ni tsuite (On the General Structure of Various Parts of *Ostrea gigas*), *Nihon Suisan Gakkaishi (Bulletin of the Japanese Society of Scientific Fisheries)*, **9** (1), 21-26.

22. Hiroshima Fisheries Experiment Station, 1949. *Suishi-Dayori*, **47,** 1-30.

23. ——, 1950. Preliminary Report on the Seedling Period of Oysters, *Suishi-Dayori*, **9**.

24. ——, 1954. *Suishi-Dayori*, **34,** 435-440.

25. HOPKINS, A. E., 1935. Attachment of Larvae of the Olympia Oyster, *Ostrea lurida*, to Plane Surface, *Ecology*, **16,** (1), 82-87.

26. HORI, J., 1926. Fuchaku jikini tasseru magaki shichu oyobi sono chishi ni tsuite (Study on the Larvae and Spat of *Ostrea gigas*), *Suikoshiho*, **22** (1), 1-10.

27. ——, 1926. Notes on the Full-grown Larva and the Spat of the Japanese Common Oyster, *Ostrea gigas* Thunberg, *J. Fish. Inst. Tokyo*, **22** (1), 1-7.

28. ——, 1928. Magaki Kuchu katsuryoku to ondo (Activity and Temperature of Japanese Oysters in Air), *Suikoshio*, **23** (5), 178-183.

29. ——, 1935. Beikoku taiheiyo engan ni okeru magaki no hanshoku ni no tsuite (Reproduction of *Ostrea gigas* in the Pacific Coast of America), *Suikenshi*, **30** (4), 171-180.

30. ——, 1937. Miyagi-Ken Kaki Suisan Kumiai (Oysters of Miyagi-Ken). Pamphlet. Miyagi Prefecture Fisheries Union.

31. ——, 1942. Kaki no suikashiki yoshoku (Vertical Culture of Oysters), *Kaiyo No Kagaku (Bulletin of Marine Science)*, **2** (5), 337-342.

32. ——, 1957. Kaki shubyo kenkyu no kaiko (On the Study of Oyster Seedlings), *Suisan Shigen*, **3** (7), 32-34.

33. ——, and D. KUSAKABE, 1926. Magaki no jinkoshiiku oyobi tennen ni okeru kaigara shichu no gaiteki ni tsuite (On the Artificial Culture of *Ostrea gigas* and the Enemies of Oyster Larvae), *Suikoshiho*, **22** (3), 77-188.

34. ——, ——, 1926. Preliminary Experiments on Artificial Culture of Oyster Larvae, *J. Fish. Inst. Tokyo*, **22** (3), 47.

35. ——, ——, 1927. Magaki shichu no jinkoshiiku (dai-ni-ho) [Artificial Culture of Oyster Larvae (Report No. 2)], *Suikoshiho*, **23** (3), 112.

36. ——, ——, 1927. On the Artificial Culture of Oyster Larvae (II), *J. Fish. Inst. Tokyo*, **22** (3), 79.

37. IKEMATSU, W., 1949. Kumamoto Suisan Shikenjo Jiho (Annual Report of Kumamoto Fisheries Experiment Station) (1947), 31-34.

38. IMAI, T., 1942. Jinko baiyo ni yoru shubyo no kakuho (Seedlings Through Artificial Culture), *Kaiyo No Kagaku (Bulletin of Marine Science)* **2** (5), 328-331.

39. ——, 1957. Saibyo ni kansuru shomondai (Some Problems Related to the Seedlings of Oysters), Abstract of a Symposium. Suisan Zoshoku Danwakai.

40. IMAI, T., 1967. Mass Production of Mollusks by Means of Rearing the Larvae in Tanks, *Venus, Jap. J. Malacology,* **25** (3, 4), 159-167.

41. ――, and M. HATANAKA, 1949. Mushoku benmochu ni yoru magaki no jinkoshiiku, *Tohoku Daigaku Nogyo Kenkyu Hokoku (Bulletin of the Agriculture Research Laboratory of Tohoku University)* **1** (1), 39-46.

42. ――, and S. SAKAI, 1952. Kaki no ikushugakuteki kenkyu (Dai-goho) [On the Breeding of Oyster (Report No. 5)], *Nihon Suisan Gakkai Tohoku Shibu kaisho (Bulletin of the Tohoku Branch of the Japanese Society of Scientific Fisheries),* **3** (1, 2), 34.

43. ――, and Y. NAKAMURA, 1955. Yoshoku dankai ni okeru shizen to keizai no junkankozo (Matsushima-Wan Tanegaki yoshoku ni tsuite no seibutsugaku to keizaigaku no kyodo kenkyu) (Biological and Economical Aspects of Oyster Culture in Matsushima Bay: Report of the Agriculture Research Laboratory of Tohoku University), *Tohoku Gyogyo, Keizai Kenkyukai.*

44. ――, and M. HATANAKA, 1950. Studies on Marine Non-colored Flagellates, *Monas* sp., Favorite Food of Larvae of Various Animals: I. Preliminary Research on Cultural Requirements, *Sci. Rep. Tohoku Univ.,* 4th Ser. **18** (3), 304-315.

45. ――, and S. SAKAI, 1961. Study on Breeding the Japanese Oyster, *Crassostrea gigas, Tohoku J. Agr. Res.,* **12** (2), 125-171.

46. ――, ――, and R. YUKI, 1949. Kaki shubyo seisan jikken (Experiments on Oyster Seedlings) *Nihon Suisan Gakkaishi (Bulletin of the Japanese Society of Scientific Fisheries),* **15** (1), 48.

47. ――, Z. MATSUTANI, and S. SAKAI, 1948. Kaki no ikushugakuteki kenkyu (Dai-sanpo) [On the Breeding of Oyster (Report No. 3)], *Nihon Suisan Gakkaishi (Bulletin of the Japanese Society of Scientific Fisheries),* **14** (1), 68.

48. ――, T. YOSHIDA, and A. FUJI, 1952. Yushutsu tanegaki no fuchakuki ni kansuru shiken (Experiments on the Hold-fast of the Exported Variety of Seed Oyster), *Nihon Suisan Gakkai Tohoku Shibu kaiho (Bulletin of the Tohoku Branch of the Japanese Society of Scientific Fisheries),* **3** (1, 2), 34.

49. ――, M. HATANAKA, R. SATO, and S. SAKAI, 1951. Ecology of Magoku-Ura Inlet with Special Reference to Seed Oyster Production, *Sci. Rep. Tohoku Univ.,* **1-2,** 137-156.

50. ――, ――, ――, ――, and R. YUKI, 1950. Artificial Breeding of Oyster in Tanks, *Tohoku J. Agr. Res.,* **1** (1), 69-86.

51. ――, S. ITO, K. NAKAMURA, and H. ONODERA, 1957. Kesen-numa Wan kaki yoshokujo no seitaigakuteki kenkyu—kankyojoken to kaki no seisansei (Ecological Studies of Oyster Beds in Kesen-numa Bay with Special Reference to the Environmental Conditions and the Outbreak of Oysters), *Kesen-numa Wan Kaihatsu Kenkyukai,* 1-39.

52. ――, ――, K. SHIRAISHI, and K. SHIBUTANI, 1958. Kaki Yoshokujo ni okeru nishigai kugo no kenkyu. Dai-ippo. Dozai no koka shiken (On the Study of Spiral Shellfish Prevention in Oyster Culturing Beds: Report No. 1), *Nihon Suisan Gakkaishi (Bulletin of the Japanese Society of Scientific Fisheries),* **24** (6, 7), 507-510.

53. ――, T. YOSHIDA, M. ENDO, T. SANO, and S. SATO, 1953. Tanegaki seisan no

tachiba kara mita matsushima-wan no seitai teki tokusei (Ecological Characteristics of Matsushima Bay with Regard to Seed Oyster Production), Proceedings of the Congress of the Japanese Society of Scientific Fisheries.

54. INABA, F., 1963. Studies on the Artificial Parthonogenesis of *Ostrea gigas* Thunberg, *J. Sci. Hiroshima Univ.*, Ser. B., Div. **15** (2), 29-46.

55. ITO, S., 1960. Yoshokujo ni okeru seisan teikagensho to sono taisaku ni kansuru
kenkyu. I. Kaki yoshokujo ni okeru seisan teika gensho to sono taisaku
(Decrease in the Production of Oysters in the Culture Beds and Measures
to Prevent the Same: I. Report of the Investigations Sponsored by the
Ministry of Education, Government of Japan. (Pamphlet, 3-7).

56. ——, and T. IMAI, 1955. Ecology of Oyster Beds : I. On the Decline of Productivity Due to Repeated Cultures, *Tohoku J. Agr. Res.*, **5** (4), 9-26.

57. Kagaku Gijutsucho Shigen Chosakai, 1964. Present Conditions and Future
Trend of Marine Culture Industries at Setonai Kai, 1-176.

58. ——, 1968. Report on Marine Culture in the Northern Part of the Pacific Coast,
1-294.

59. KIKUCHI, S., 1960. Matsushima-wan ni okeru kaki yosai no bunpu ni kansuru
kosatsu (Distribution of Oyster Larvae in Matsushima Bay), Tohoku Suisan
Kenkyujo Kenkyu Hokoku (Bulletin of Tohoku Fisheries Research Laboratory) (16), 118-126.

60. ——, 1961. Matsushima-wan-san kaki no hassei ni oyobosu suion endun no
eikyo (Influence of Temperature and Chlorine Content of Sea Water on the
Outbreak of Oysters in Matsushima Bay), *Ibid.* (19), 154-163.

61. ——, and S. TANIDA, 1951. Tanegaki saibyoki no fuchaku seibutsu ni yoru
osen to kaki chigai no fuchakuritsu (Infection through Organisms Settling on
Seed Oysters and Setting Rate of Spat), *Ibid.* (19), 149-153.

62. ——, H. KANNO, S. SATO, and T. SANO, 1951. Sendai-wan ni okeru kaki yosei
no bunpu ni tsuite (On the Distribution of Oyster Larvae in Kesen-numa
Bay), *Ibid.* (19), 164-170.

63. KINOSHITA, T., 1927. Kaki saibyo no ichijikken (Experiments on Oyster Seedlings), *Suikenshi*, **22** (4), 94-95.

64. KOGANEZAWA, A., 1958. Kaki shubyo no shoki genmo (Early Infection of Oyster
Seedlings), Suisan Zoshoku Danwakai (Report of the Symposium Held under
the Auspices of the Japanese Society of Scientific Fisheries).

65. ——, 1963. Kaki yoshokujo ni okeru ni shigai no seitai-gakuteki kenkyu (Ecological Study of Oyster Culturing Beds) Miyagi Suisan Shikenjo Hokoku
(Report of the Miyagi Prefecture Fisheries Experiment Station) (3), 1-11.

66. ——, 1965. Kaki no saibyo jiki no yochi ni tsuite (On the Method of Determining the Seedling Period of Oysters). Pamphlet. Miyagi Prefecture Fisheries
Experiment Station. (3), 1-11.

67. ——, and N. ISHIDA, 1968. Tanegaki gyojo ni okeru nishigai bojo no kenkyu
(Investigations to Prevent Infection from the Seed Oyster Culture Beds),
Miyagi Suisan Skenjo Hokoku (Report of the Miyagi Prefecture Fisheries
Experiment Station) (4), 1-6.

68. ——, ——, and T. IMAI, 1964. Miyagi-ken ojika hanto kaiiki ni okeru tanegaki seisan no seitaigakuteki chosa (Ecology Studies of Seed Oyster Produc-

tion in the Regions around Ojika Hanto, Miyagi Prefecture), Nissuigakki Taidai Koen Yoshi (Proceedings of the Congress of the Japanese Society of Scientific Fisheries).

69. KORRINGA, P., 1953. Recent Advances in Oysters, *Quart. Rev. Biol.*, **27,** 266-308, 339-365.

70. KURANAKA, M., 1909. Kaki-ran no jusei noryoku to kaisuigendo (Fertilizability of Oyster Eggs), *Suikenshi,* **4** (1), 18-20.

71. KURODA, T., Y. TSUCHIDA, Y. TANIZAWA, and H. UEMOTO, 1957. Senkai yoshoku no riron to jissai (Theories and Practice of Shallow Sea Water Culture), *Gyoson Bunka Kyokai,* 1-24.

72. LINDSAY, E. E., and D. C. McMILLIN, 1950. Control of Japanese Oyster Drills, Stat. Wash. Dept. Fish., Puget Sound Oyster Bull., Ser. 9 (1).

73. LOOSANOFF, V. L., C. L. MAKENZIE Jr., and L. W. SHEARER, 1960. Use of Chemical Barriers to Protect Shellfish Beds from Predators, *Fisheries,* **3,** 1-5.

74. Ministry of Forestry and Agriculture Fisheries Department, 1931. Yoki no chishiki (Study of Culture Methods). Pamphlet.

75. Miyagi Prefecture: Fisheries Experiment Station: Oyster Culture Study Association, 1955. General Account of the Oyster Culture Industries in Miyagi Prefecture, *Suishi-Dayori,* **3,** 47-65.

76. ——, 1957. On the Seed Oysters of Miyagi Prefecture, *Suisan Jiho (Monthly Research of Fisheries),* **9** (12), 52-53.

77. Miyagi Prefecture, 1962. Plan to Improve the Structure of Coastal Fishing Industries in Miyagi Prefecture.

78. Miyagi Prefecture: Agriculture and Fisheries Department, 1961, 1963, 1965. Fisheries in Miyagi: Reports Published in 1961, 1963, and 1965.

79. Miyagi-Ken Tanegaki Gyogyo Kyodo Kumiai, 1947-1967. Report on the Observations Carried Out on Oyster Larvae of Matsushima Bay and Ojika Hanto.

80. ——, 1951. Study on Export Variety of Seed Oysters. Pamphlet.

81. MIYAZAKI, I., 1957. Nimaigai no sanran hassei oyobi shiga no shusei ni tsuite (Spawning of Bivalves: Development and Characteristics of Veliger Larvae), *Suisangaku Shusei, Todai Shuppankai,* 433-443.

82. ——, 1957. *Nimaigai to sono yoshoku (Bivalves and Their Culture).* Tokyo: Inasa Publishers.

83. ——, 1962. Nimaigai no fuyuyogai no shikibetsu ni tsuite (On the Veliger Larvae of Bivalves), *Nihon Suisan Gakkaishi (Bulletin of Japanese Society of Scientific Fisheries),* **28** (10), 955-966.

84. MORI, K., T. IMAI, S. TOYOSHIMA, and K. USUKI, 1965. Matsushima-wan ni okeru kaki no dairyo heishi ni kansuru kenkyu. IV. Seiseijiku oyobi sanran ni tomonau kaki no seiriteki kassei to togenryo no henka (On the Mass Mortaiity of Oysters at Matsushima Bay: IV. Changes in the Physiological Activities and Carbohydrate Content in Oysters during Maturation and Spawning), *Tohoku Suisan Kenkyujo Kenkyu Hokoku (Bulletin of Tohoku Regional Fisheries Research Laboratory)* (25), 49-63.

85. NELSON, T. C., 1924. The Attachment of Oyster Larvae, *Biol. Bull.,* **46** (3), 143-151.

86. NEWCOMBE, C. L., 1941-1942. Preliminary Results of Drill Studies at York-town: Summary of 1941, and Notes on Drill Trapping Experiments Conduct-ed during the Period June 3 to September 18, 1942. Unpublished Report, Virginia Fish. Lab., Va.

87. Nihon Suisan Shigen Kyokai, 1957. Kaki yoshoku no onjin (Pioneers of Oyster Culture), *Suisan Shigen*, **3** (7), 81-86.

88. NUMACHI, K., 1955. Kaki bunrui no kijun to shite no yoseiki ni okeru tane oyobi hinshu no keishitsu ni tsuite (Characteristics of Species and Strains during the Larval Period of Oysters Forming the Basis for the Classification of Oysters), *Nihon Suisan Gakkai Tohoku Shibukaiho (Bulletin of Tohoku Branch of the Japanese Society of Scientific Fisheries)*, **6** (1, 2), 34.

89. ——, 1960. Kaki juseiran no koteki hassei enbun nodo ni taisuru oyagaki shiiku enbun nodo no eikyo ni tsuite (Influence of Chlorine Concentration during the Rearing of Parent Oysters in Relation to the Chlorine Concentration Studied for the Development of 12 Eggs), Nissuigakki Koen Yoshi (Proceed-ings of the Meeting of the Japanese Association of Scientific Fisheries).

90. ——, 1965. Tane no hozon-kaki (Preservation of Seed Oysters). Symposium on Breeding Oysters Held by the Japanese Society of Scientific Fisheries.

91. ——, 1966. *Ibid.*, *Suisan Zoshoku (Aquaculture)*, **13** (3), 156-158.

92. OGASAWARA, Y., U. KOBAYASHI, R. OKAMOTO, A. FURUKAWA, M. HISAOKA, and K. NOGAMI, 1962. Kaki yoshoku ni okeru kyakusei shubyo no shiyo to sono seisanteki igi (On the Utility of Control Seedling in Oyster Culture and Its Effect on the Production Rate), *Naikai-Ku Suisan Kenkyujo Kenkyu Hokoku (Bulletin of Naikai-Ku Regional Fisheries Research Laboratory)* (19), 1-153.

93. OIZUMI, S., 1953. Tanegaki seisan ni kanshite no Matsushima-wan no seitai chosa (Ecological Condition of the Matsushima Bay Studied in Relation to the Production of Seed Oysters), *Seitaigaku Kenkyu*, **13** (3), 183-195.

94. ORTON, J. H., 1937. *Oyster Biology and Oyster Culture*. London: E. Arnold Co.

95. OTA, F., 1959. Kaki Shubyo zosei ni kansuru shiken (Experiments on the Struc-ture of Oyster Seedlings), *Kumamoto Suisan Shikenjo Jiho* (1952).

96. ——, 1959-1960. Taibei yushutsu tanegaki zosei jigyo (Experiments on the Structure of Seed Oysters Meant for Export to America), *Ibid.* (1950-1952), (1953-1958).

97. SAKAI, S., I. HIROZAWA, and Y. TAKAHASHI, 1964. Kesen-numa wan kaki yoshokujo no seitaigakuteki chosa (Ecological Studies Carried Out at the Culture Beds of Kesen-numa Bay), *Miyagi-Ken Suisan Shekenjo, Kesen-numa Bunba Hokoku* (1), 1-46.

98. SASAKI, A., 1966. Magaki ran no kaonshori (Heat Treatment of *C. gigas* Eggs). Thesis submitted to the Fisheries Department of Tohoku University of Agri-culture.

99. SATO, S., H. KANNO, and S. KIKUCHI, 1960. Matsushima-wan no tanegaki saibyoki ni okeru teien bunsui, akashio oyobi kakiyosei ni tsuite (Low Salinity of Water in Matsushima Bay during the Seedling Period of Seed Oysters), *Tohoku Suisan Kenkyujo Kenkyu Hokoku (Bulletin of Tohoku Regional Fisheries Research Laboratory)* (16), 78-117.

100. SATO, S., 1951. Matsushima-wan no kaki no sanranjiki no hendo to suion to no

kankei (Relationship between the Changes in the Spawning Period and Water Temperature at Matsushima Bay), *Nihon Suisan Gakkai Tohoku Shibu Kaiho (Bulletin of Tohoku Branch of the Japanese Society of Scientific Fisheries)* **7** (1, 2), 16.

101. SATO, S., M. HATANAKA, and T. IMAI, 1943. Jinkoteki shubyo seisan ni kansuru Futatsu no jikken (Experiments on the Artificial Production of Seedlings of Oysters), *Nihon Suisan Gakkaishi (Bulletin of the Japanese Society of Scientific Fisheries)*, **11** (5, 6), 217-218.

102. SATO, T., 1948. Magaki yosei no seiiku to hyosōsuion oyobi hiju to no kankei (Relationship between the Growth of Oyster Larvae, Surface Water Temperature and Density), *Suiken Kaiho*, **1** (1), 90-110.

103. ——, 1967. Oyster. Yogyogakukakuron, Tokyo: Koseisha Koseikaku.

104. SCHAEFER, M. B., 1937. Attachment of the Larvae of *O. gigas*, the Japanese Oyster, to Plane Surfaces, *Ecology*, **18** (4), 523-527.

105. ——, 1938. The Rate of Attachment of the Larvae of the Japanese Oyster, *O. gigas*, as Related to the Tidal Periodicity, *Ibid.*, **19** (4), 543-547.

106. SEKI, H., 1937. Hiroshima (Magaki) to Sendai (Magaki) to no soi ni tsuite (A Comparison of the Hiroshima Strain and the Sendai Strain of *C. gigas*), *Suishi Chosa Shiryo*, **4**, 45-50.

107. ——, 1943. Suion josho ni yoru kaki sanran jini sokushin (Acceleration in the Spawning of Oysters by Increasing Water Temperature), *Nihon Suisan Gakkaishi (Bulletin of the Japanese Society of Scientific Fisheries)*, **11** (5, 6), 217.

108. SENO, H., 1938. Honpo ni okeru yoki hattatsu no kaiko (On the Progress of Culture Methods in Japan), *Kagaku (Science)*, **8** (6), 33-38.

109. ——, 1939. Kaki yosei no seitaigakuteki kenkyu (Ecological Studies of Oyster Larvae), *Dobutsu Zasshi (Journal of Zoology)*, **51** (2), 94.

110. ——, 1941. Kaki yochu no bunpitsu suru semento ni tsuite (On the Cement Secretion of Oyster Larvae), *Ibid.*, **53** (2), 87.

111. ——, J. HORI, and D. KUSAKABE, 1926. Magaki-ran no hassei to ondo oyobi hiju to no kankei (Relationship between the Development of Oysters, Temperature, and Density of Water), *Suiko Shiho*, **22** (3), 169-176.

112. ——, ——, 1927. Suikashiki yoki shiken hokoku (Report on a Study of the Vertical Culture Method), *Suiko Shiho*, **22** (4), 211-261.

113. SENO, J., 1950. Magaki hoseisokushin ni kansuru kenkyu (On the Acceleration of Sperm Release in the Oyster, *C. gigas*), *Dobozatsu Zasshi (Journal of Zoology)*, **59** (2, 3), 48-49.

114. *Shallow Water Culture of 60 Species*, 1965. Tokyo: Daisei Publishers.

115. SHIRAI, K., 1939. *Ostrea gigas* Thunberg no tamago no hassei (Development of *Ostrea gigas* Thunberg), *Hakubutsugaku Zasshi*, **37** (66), 1-4.

116. Suisancho Gyogyo Chosei Dai-ni-ka, 1950. Report on the Outcome of the Meeting on Shallow Water Culture Technique Held in 1950 by T. Imai.

117. Suisancho Chosa Kenkyubu Kenku Dai-ni-ka, 1955. Proceedings of the 2nd Oyster Culture Expert Committee Meeting by Muta, Miura, and Imai.

118. TAKATSUKI, S., 1949. *Oyster*. Tokyo: Gihodo.

119. TAMATE, H., K. NUMACHI, K. MORI, O. ICHIKAWA, and T. IMAI, 1965. Matsushima-wan ni okeru kaki no tairyoheishi ni kansuru kenkyu: VI. Byori soshi kigakuteki kenkyu (Study on the Mass Mortality of Oysters at Mat-

sushima Bay: VI. Study on the Pathological Tissues), *Tohoku Suisan Ken-kyujo Kenkyu Hokoku (Bulletin of Tohoku Regional Fisheries Research Laboratory)* (25), 89-104.

120. TAMURA, T., 1956. *Suisan Zoshokugaku.* Tokyo: Kigensha.

121. ——, 1967. *Shallow Water Culture.* Tokyo: Koseisha Koseikaku.

122. ——, 1967. Zankai zoshokuba no kankyo: II. (Environmental Conditions of Shallow Water Culture Beds: II), *Nihon Suisan Shigen Hogi Kyokai*, **15** (2), 96-166.

123. TANIDA, S., and S. SATO, 1953. Suika yoshoku kaki no fuchaku seibutsu ni kansuru kenkyu (On the Study of Organisms Settling on Oysters Grown by the Vertical Culture Method), *Tohoku Suisan Kenkyujo Kenkyu Hokoku (Bulletin of Tohoku Regional Fisheries Research Laboratory)* (2), 56-66.

124. ——, and S. KIKUCHI, 1957. Suika yoshoku kaki no mitsudo koka ni kansuru kenkyu. Dai-ippo. Genbannai no kotaimitsudo koka (The Effect of Crowding on Oysters Grown by the Vertical Culture Method: Report No. 1), *Ibid.* (9), 133-142.

125. TERAO, S., 1928. Kaki no sanran to ondo (Spawning of Oyster and Temperature), *Suikenshi (Journal of Imperial Fisheries Institute)*, **23** (12), 380.

126. TOHOKU SUIKAN, 1952. Kaki no yoshoku (Culture of Oysters), *Tohoku Suiken Gyoshoku*, **2,** 1-23.

127. TSUCHIYA, Y., 1962. *Suisan Kagaku Suisan Gaku Zenshu.* Tokyo: Koseisha, Koseigaku.

128. YASUGI, R., 1938. On the Mode of Cleavage in the Eggs of the Oyster, *Ostrea spinosa* and *O. gigas,* under Experimental Conditions (A Preliminary Note), *Ann. Zool. Japan*, **17** (3), 293-300.

129. ——, 1947. Magaki ran no fukatsu ni oyobosu enrui no eikyo (Influence of Various Salts on the Viability of Oyster Eggs), Dobutsu Zasshi (Journal of Zoology), **57** (6), 75-76.

130. ——, 1947. Magaki juseiran ni kansuru ni san no jikken to kosatsu (tokuni kyokuyo ni tsuite) [Experiments on the Fertilized Eggs of Oysters *(C. gigas)*: A Discussion in Particular on the Polar Lobe], *Dobutsu Zasshi (Journal of Zoology)*, **57** (11), 176-178.

131. YOKOTA, T., 1932. Magaki no seishoku ni tsuite (Gaiyo) (Reproduction in *C. gigas*), *Yoshoku Kaishi*, **2** (1), 7-11.

132. ——, 1935. Kaki shichu no fuchaku ni tsuite (Setting of Oyster Larvae), *Ibid.*, **5** (9, 10), 147-154.

133. ——, 1936. Kaki shichu no fuchaku ni tsuite (Setting of Oyster Larvae), *Ibid.*, **6** (11), 203-206.

134. ——, 1936. Magaki no yurankansenmo ni tsuite [On the Cilia of the Oviduct (Report No. 1)], *Suikenshi*, **31** (2), 61-66.

135. ——, 1936. Mangoku-ura no kaki suion ni tsuite (On the Water Temperature in Summer in Mangoku-Ura), *Ibid*, **31** (21), 697-702.

136. ——, 1936. Kaki shichu no fuchaku ni tsuite (Setting of the Larvae of *Ostrea gigas*), *Dobutsu Zasshi (Journal of Zoology)*, **48** (3), 224-225.

137. ——, 1957. Nimaigai no fuchaku ni tsuite (Setting of Bivalves), *Suisangaku Shusei*: Todai Publishers, 565-568.

138. YOKOTA, T., and K. SUGAWARA, 1914. Ikada shiki suika yoki ni okeru kaki no mitsudo ni tsuite (Crowding of Oysters in Raft-type Vertical Culture), *Suikenshi (Journal of Imperial Fisheries Institute)*, **95,** 28-30.

139. ——, and M. HIROTSUGE, 1950. Magaki chigai no kuchu katsuryoku ni tsuite (On the Activity of Spat in *Ostrea gigas* in Air), *Nihon Suisan Gakkaishi (Bulletin of the Japanese Society of Scientific Fisheries)*, **15** (11), 632-634.

140. YONGE, C. M., 1960. *Oyster*. London: Collins.

141. YOSHIDA, H., 1964. *Kairui Shubyogaku (Seedlings of Mollusk)*. Tokyo: Hokuryu-kan.

PART III
The Evolution of Scallop Culture

An answer to whether or not "through" culture of scallops has been achieved is difficult, but it can be asserted that steady progress has been made in studies relating to artificial seedlings, natural seedlings, the rearing of spat (which are prone to heavy mortality just before the phase when they settle on the sea bottom), management of fishing grounds, and fluctuations in the harvesting rate (resulting from a so-called "abnormal fecundity"), and these studies have resulted in considerable progress in the study of culture techniques from the early stages of the life history of scallops to their harvesting.

Comprehensive data on the culture of scallops have been published by Kinoshita (1936, 1949) and Yamamoto (1964). These publications are still useful in the current investigations on this bivalve and the facts and data given in them ought to stimulate research work on the biological aspects of this organism, information which is essential to the development of a really complete and efficient technique for scallop culture.

The authors received much guidance from several workers in compiling this chapter. Special mention should be made of Prof. Tanaka, Director, and Mr. Ohara of the Hokkaido Fisheries Experiment Station, the research workers of the Abashiri Fisheries Experiment Station, the staff of the Fisheries Department, Iwate Prefecture, and of Mr. Miyazaki, who allowed the authors to use his unpublished data. The authors are also grateful to the Association for the Conservation of Fisheries Resources of Japan (Inc.) for allowing them to use part of their data and charts.

CHAPTER I

Biological Research on Scallops

1. Introduction

A species of mollusk first found in the China Sea by an American warship captained by M. C. Perry, was identified in 1852 to 1854 as a new species of scallop, *Patinopecten (Mizuhopecten) yessoensis* (Jay), and thus a systematic biological study of this organism began (Jay, 1856).

A reference to scallops in Japan dates from a much earlier period, however, in the era of the Sixth Shogun of Tokugawa when a description of the mollusk appeared in *Wakan Sansai Zuei* (1716) edited by Terajima Ryoan, in which the features and characteristics of *Pecten yessoensis* and *Pecten albicans* were combined, and the species referred to as "hotategai" and "itayagai"; the species was said to be abundantly distributed on the "western side of Hokkaido." Some features of the described scallop were identical to those of "itayagai," i.e. *Pecten albicans*. The larger bivalve was reported to be 1 *shaku* (0.303 m) or 2 *shaku* (0.606 m). All the bivalves moved in groups of a few hundred, each with its mouth open, and with one shell positioned like a boat while the other looked like a sail—hence the name "hotategai" (*ho* for sail, *tate* for erected, and *gai* for shellfish). This description of the mollusk's locomotion is very interesting, and some of the facts stated in later Japanese literature are similar to those mentioned in Ryoan's publication.

In 1896, Kishigami studied scallops obtained from Mutsu Bay, with special reference to the fishing conditions inside the bay, the distribution of the animal, and its locomotion. Jay, of course, had already done studies on scallops in the early half of the 1800's, which were followed by Nomura's interest in their anatomical features (1918, 1919, 1922), and by Tamura's detailed physiological studies (1932, 1933), from which data compiled on the physiological aspects were published in 1939 (a, b).

While such studies were in progress, research on culture techniques had begun in Hokkaido, particularly on the Okhotsk coast and in Lake Saroma. Isahaya, Ono, Kinoshita, Imai, Sato, and others, carried out a series of valuable investigations. Because of the research work of Kinoshita and his associates, biological studies of this bivalve and its culture technique progressed rapidly. Indeed, it would not be an exaggeration to state that the work done by Kinoshita and his associates represents a significant phase in the evolution of scallop culture techniques. Tanaka, Wakui, Ohara, and others, pursued several new lines of research on this bivalve on the basis of the data and results published by Kinoshita and his colleagues.

Nishioka and his colleagues also took up the study of this mollusk at the Aomori Fisheries Experiment Station, attached to Tohoku University, in 1940, and Yamamoto later studied the ecological aspects; their findings were of considerable help in the development of a culture method for this animal.

In 1963, Masuda carried out a detailed systematic research on *Pecten albicans*, and published his findings in the same year.

The history of scallop study in other countries begins in England with Fullarton who did work on its biological aspects in 1890. Dakin carried out a study which was primarily concerned with the morphological features of *Pecten maximus* in 1909. Amirthalingam (1928), Mason (1953, 1957, 1958, 1959), and others, studied the physiological and ecological aspects and added much valuable information. In America, Drew (1960) published valuable data on the morphology, ecology, and embryology of scallops. Other reports on these aspects came from Gutsell (1930), Posgay (1953), Sastry (1963), Rathjen (1963), and others. There are recent reports of active research work being undertaken in Canada where Dickie (1955, 1958), Dickie and Medcof (1963), Bourne (1964), Medcof and Bourne (1964), and others, are primarily concerned with the ecological aspects in relation to fishing operations. Olsen (1955) published valuable data on the ecological aspect of fishing the *Notovola meridionalis* found off Tasmania, data which are particularly significant since this is one of the few species of scallop found in the southern hemisphere.

2. Classification and distribution

Hotategai (also known as *umioogi, akitagai* in Japan) belongs to Phylum Mollusca, Class Pelecypoda (Bivalvia), Order Pseudolamellibranchia, Family Pectinidae, and as mentioned earlier was recorded by Jay as *Pecten yessoensis* n. sp. (Figure 3.1). Later it was referred to as a separate group known as *Patinopecten* belonging to the genus *Pecten* (Dall, 1895), but others preferred to classify it as a subgenus (Arnold, 1906). Later, *Pecten caurinus* Gould, belonging to the group *Patinopecten* suggested by Dall, was considered an allied species of *Hotategai*, and classified under the genus *Patinopecten* as either *P. caurinus* or *Patinopecten* (Jay). Recently, Masuda (1963) carried out a comparative study of *Hotategai* and *P. caurinus* and found that the ribs of the right valve of the former were not flat but slightly rounded, that the left valve was either flat or slightly convex, and that the byssus was short and wide. Thus, it differed from *Patinopecten*, and so he placed the species under a separate genus, *Mizuhopecten yessoensis* (Jay). Watanabe(1956)accepted the description given by Masuda and referred to this species as *Patinopecten (Mizuhopecten) yessoensis* (Jay).

This species is found in Chishima Retto, Karafuto, Hokkaido, the northern part of Honshu, and in the Sea of Japan off the northern part of Korea. However, scallops were also found in considerable numbers in the Okiai off the Wakayama Prefecture, which means that they are found on the Pacific coast. This is unusual, but in fact they are found on the northern side of Tokyo Bay on the Pacific coast and on the northern side of Toyama Bay in the Sea of Japan. Fishable stocks are available at Chishima Retto, the Okhotsk coast of Hokkaido, Funka Bay (Uchiura Bay) and Mutsu Bay of Honshu. As will be explained later, seedlings have recently been transplanted from Mutsu Bay and Hokkaido in various coastal regions of Sanriku and propagated by the hanging culture method. Future fishing operations will no doubt extend to the southern regions as well (Figure 3.2).

Pecten (Notovola) albicans (Schröter), a smaller variety found distributed from the southern part of Hokkaido to Kyushu and Korea, like *Hotategai* is also found to give rise at times to a so-called "abnormal fecundity" in western Japan, and the Kyushu and San-in regions. Hence this species attracts the attention of fishery research workers because of its capacity for production.

The allied species of *Hotategai* found in other countries will now be discussed.

Aequipecten irradians (Lamarck) (Bay scallop), which is found from Nova Scotia on the Atlantic coast of North America to Tampa, Florida, has a high yield, i.e. 10,000

Figure 3.1. *Hotategai* recorded as a new species (Jay, 1856).

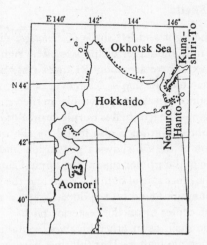

Figure 3.2. Distribution of *Hotategai* in Japan.

tons per year, and is thus generally considered the maximum yielding variety in the world. In fact, however, its yield is slightly lower than that of *Hotategai* of Japan (Yamamoto, 1964). This species is said to differ from the *A. i. irradians* found in Massáchusetts and Maine, and the *A. i. concentricus* found off the coasts of Florida and North Carolina (Sastry, 1963, and others).

Placopecten magellanicus Gmelin (Sea scallop, Digby, Giant or Smooth scallop) is found in the deep waters in Okiai and distributed more toward the northern side of North America than the previous species. Since 1945, fishing operations have progressed rapidly in Canada, and the amount of catch has shown a phenomenal increase through the years. While the coastal catch is about 400 tons, catches of the offshore areas total about 4,500 tons. This species is largely distributed from Nova Scotia to Prince Edward Island and New Brunswick (Dickie, 1955; Dickie and Medcof, 1963; Bourne, 1964).

Patinopecten caurinus (Gould) is a large variety found near Juneau in Alaska. It is also found in large numbers in the offshore areas of Cape Fairweather and Cape Saint Elias (Rathjen, 1963, 1964).

In Europe, 2,000 tons of *Pecten (Chlamys) opercularis* (L.) (Queens, Common scallop, or Zamborina in Spain) are caught in the Firth of Forth and off the Isle of Man. Bantry Bay in Ireland is said to be the center of fishing operations for this species. On the other hand, *Pecten maximus* L. (scallop, Escallop, or Great scallop, known as Vieira in Spain), is widely distributed around England although the yield rate is not very high (Dakin, 1909, and others).

Natovola meridionalis (Tate) (Tasmanian scallop) is found in the southern hemisphere from New Castle (N. S. W.), the southern part of China, Port Lincoln, Albany (S. A.), Tasmania Island, and up through the Bass Strait. The yield rate per year is said to be 200 tons (Olsen, 1955).

3. Morphology

3.1 Shells

The right and left shells of *Hotategai* differ considerably in shape, the right being generally the larger of the two. The right shell is yellowish white in color and very convex. There are about 21 to 24 ribs,[1] and in a transverse section the rib is almost semicircular in shape. Moreover, the growth rings (annual rings) are also distinct. A distinct byssal notch is visible on the ventral side of the cochlear part on the anterior margin. The left shell is purple-brown or orange-yellow in color, and slightly convex; a slight curvature is also found on the inner side of a few specimens. A transverse section of the rib shows a protrusion and, compared to the right shell, neither the shape of the rib nor that of the annual rings is distinct. The byssal notch is absent.

The inner side of both the right and the left shell is white in color, and the hinge and resilium are found on the dorsal edge; a muscular scar is seen inside the shell either at the center or on the posterior part.

The shell, as in other bivalves, contains 94 to 99% calcium carbonate and about 1% organic matter. It is thin but hard, and not so rough as that of *Crassostrea gigas*. As the distribution extends northward, the shell becomes gradually thicker.

Boring *Polydora ciliata* (Johnston), a parasite, is often found in the shells (Figure 3.3). It is found in large numbers on the left shell, and more than 70% of the specimens

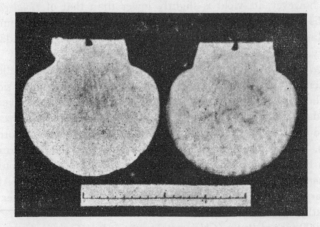

Figure 3.3. Infection of *Polydora ciliata* on the shells, particularly the left one, of *Hotategai*.

[1]According to Kinoshita (1935), the number of radiating ribs of scallops varies in different regions of Hokkaido. The number is characteristic of a particular zone up to the growth of 1 cm of shell length. In Kunashiro-to, 38% of the scallops have the minimum number of ribs, 21, while in Kitami Tokoro, 23% have the maximum, 24.

living on a muddy bottom are infested. According to Kinoshita (1949), the rate of infection by this parasite in Hokkaido increases with age. In 1939, every scallop in Saroma, Bakkai, Shari, Tonbetsu, and Sarufutsu was infected, and 67 to 75% of those in Tokoro. The infected scallops became weak, particularly in the adductor muscles, and lost their luster. The dry meat is also affected in some individuals. Scallops in which the adhering surface of the adductor muscles was infected, died sooner or later, and an infection in the area of the midgut gland caused considerable damage.

3.2 Mantle

Nomura (1922, 1923) carried out detailed studies on the anatomy of *Hotategai*, but unfortunately his studies were incomplete. As mentioned earlier, a few anatomical reports are available on this species from other countries, and these will presently be discussed.

The weight of the soft parts of *Hotategai* is 40 to 50% of the total weight. Except for the adductor muscles, all of the soft parts are covered by a thin transparent mantle (Figure 3.5). The outer margin of this mantle is thick and gives rise to the shell fold, pallial fold, or secretory fold, the ophthal pallial fold (sensory fold), and the velum (Figure 3.4). The shell fold has a periostracal groove and tentacles. Eyes, special to

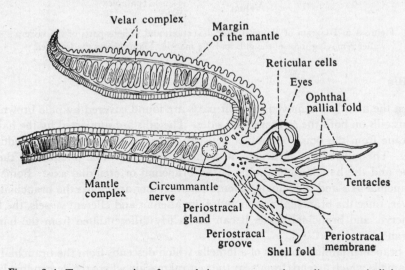

Figure 3.4. Transverse section of a mantle (some parts are shown diagrammatically).

this animal, are found on the ophthal pallial fold, and their number differs from individual to individual, but 20 to 30 are quite common. The tentacles are usually 3 to 4 cm in length and, when the shells are opened gently, can be seen moving; however, the tentacles on the anterior tip of the velum are shorter, about 1 cm. As will be detailed later, the velum plays the important role of holding sea water in the mantle cavity during locomotion (Figure 3.7).

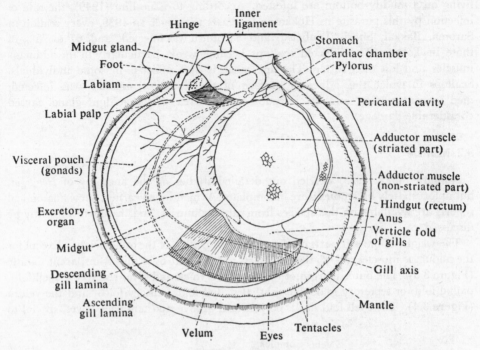

Figure 3.5. Diagram of *Hotategai* after dissection, and various parts of the viscera after removing most of the gills and the mantle. Figure shows the left shell.

3.3 Gills

When the mantle is removed, the soft parts are found covered by pale brown, filamentous gills on both the right and left sides. The gills are supported by the hanging membrane from the point of contact between the visceral pouch and the adductor muscles. The gills consist of a pair of branchial laminae. The juncture of the two laminae and the hanging membrane is the branchial or ctenidial axis.   Sometimes the branchial axis and the hanging membrane together are called the branchial axis. However, since the branchial axis includes the afferent and efferent vessels, the branchial nerve, and branchial tissue, it can be clearly differentiated from the hanging membrane.

The branchial lamina consists of a lamella which descends from the branchial axis, and a lamella which ascends from near the ventral margin where the descending lamella folds back. In this species, the ascending lamella does not reach the branchial axis. Each branchial lamella consists of a row of branchial filaments, which are connected through the interfilamentar junction, while the lamellae are connected through the interlamellar junction (Dakin, 1909; Nomura, 1922; Gutsell, 1930).

3.4 Digestive system

The mouth is in the form of a vertical slit and is enclosed by dorsal and ventral lips, which are covered by labial palps. It leads into the esophagus, which joins a midgut gland or the digestive diverticulum. The esophagus opens through the midgut gland into a large stomach. The latter, largely covered by the midgut gland, consists of several blind tubules which are joined to the digestive diverticulum.

The intestine bends around the center of the visceral pouch, and extending up to the anterior end, takes a reverse course; after passing through the dorsal part of the pericardial cavity, it opens into the mantle. The rectum opens through the anus near the non-striated margin.

A crystalline style, which supplies enzymes to the stomach, lies in the groove between the stomach and the midgut. The anterior edge of the crystalline style rubs against the gastric shield in the stomach and, rotating in the groove, is gradually worn away, while its posterior end continues to grow.

3.5 Foot and adductor muscles

The foot in the adult animal is vestigial and its original function lost. However, a byssal groove is present at the base of the ventral surface of the foot. From this byssal groove, a number of byssal strands are secreted which assist the young mollusk, about 1 cm in shell length, to remain attached to the substratum. Hence the byssus seems to play an important role during the development of the mollusk. The byssal glands are present in the form of functional cellular lumps in the tissues of the foot. In a number of scallops, the byssal glands rapidly degenerate when the animal reaches 1 cm in shell length.

The adductor muscles may be divided into two parts, namely, a large, striated muscle and a small, non-striated muscle which is elliptical in shape. The striated adductor muscle is used when the animal moves quickly and the non-striated one when the animal moves slowly.

The adductor muscles of *Pelecypoda* seem to have both the striated and non-striated muscles. However observations carried out under an electron microscope, show that this is only true of two families of bivalves, namely, *Pectinidae* and *Spondylidae*. In a number of bivalves, the presence of striated muscles has not been confirmed (Kawaguti and Ikemoto, 1958).

3.6 Circulatory system

The pericardial cavity is composed of two auricles and a single muscular ventricle. An anterior and posterior aorta proceed from the ventricle, the former supplying blood to the midgut gland, the visceral pouch, the mantle, etc., and the latter supplying blood to the hindgut, the mantle, the adductor muscles, and the gills.

Like other bivalves, the scallop also has a closed circulatory system, but there are several lacunaries in the arteries and veins.

The blood of the scallop does not seem to mix with sea water, but it has been observed in *P. maximus* (found in Port Erin, Britain) that the falling degree of the freezing point of the blood is $\Delta = -1.905$ to -1.920, which is equal to that of sea water, i.e., $\Delta = -1.910$ (Dakin, 1909).

3.7 Nervous system and sense organs

The central nervous system consists of three ganglia in most *Pelecypoda*. The position of these ganglia is characteristic. The cerebral ganglion is found below the epidermis between the lower lip and the foot, and is a fusion of the right and left nerve cell masses. The pedal ganglion lies between the mouth and the esophagus near the ventral surface. The cerebral and pedal ganglia are very close to each other. The largest of the three ganglia, the visceral or parieto-splanchnic ganglion, is found on the ventral surface of the adductor muscle, entangled in the connective tissue.

Among the sense organs of *Hotategai*, the eyes found at the margin of the mantle are the most significant. Dakin (1909) and Gutsell (1930) have furnished details of their structure. The eye consists of a cornea, a crystalline body, lens, reticular cells, argentea, tapetum, sclerotica, and internal and external branches of the optical nerve. It seems that the structure closely resembles the eyes of vertebrates (Figure 3.4).

According to Bauer (1912), the eyes are sensitive to the strength of light and to moving objects. By sensing the strength of light, they aid the body to move in a suitable direction. At the same time, they are also useful during feeding. Wenrich (1916) reported that the eyes of scallops are more sensitive to a reduction in light strength than to an increase.

3.8 Urinogenital system

The excretory organs are present on either side of the visceral pouch and in close contact with the adductor muscles. They are a pair of slender pale yellow-brown organs shaped like a 3-day moon. The excretory pore opens on the ventral surface into the mantle cavity, while the excretory organs connect with the coelom through the pericardial cavity.

A large part of the visceral pouch is covered with gonads. During the breeding season, the visceral pouch greatly enlarges, attaining ten times its weight in summer when it reaches its maximum dilation. The midgut, which runs through the visceral pouch, is fully covered by the gonads. During the peak period of breeding, the gonads seem to cover even a part of the midgut gland.

Hotategai are usually unisexual. Since the gonads of the male are orange-red in color, and those of the female yellow-white, it is possible to differentiate them easily (Yamamoto, 1943, 1964).

A pair of ducts arise from the gonads and open on both the right and left sides into the excretory organs. The products of the gonads are released into the mantle cavity through the excretory pores.

4. Ecology

4.1 Regional distribution and factors controlling distribution pattern

Regarding the regional distribution of *Hotategai*, Nishioka and Yamamoto (1943) reported the species in Mutsu Bay. According to them, scallops prefer regions in which particles smaller than 100 µ constitute less than 30% of the total mud content.

In other words, scallops live in seas where the bottom is fairly hard with little mud. Their number varies, but distribution is restricted to the coastal zones of up to 40 m depth, and they are not seen in mid-sea.

Kinoshita *et al.* prepared a report on the distribution of scallops in the Tokoro and Nemuro fishing grounds of Hokkaido, after carrying out surveys by diving to study the density of distribution and the environmental factors (1942, 1944a, b, 1949). They categorized density of distribution in the fishing grounds as high, medium, and low. In all three categories, the mud content was below 4%, the gravel 56% in high-density fishing grounds (6 scallops per 1 m² area), and 43% in medium-density grounds. The sand in the medium-density grounds was 22%, but in low-density grounds (1.5 per 1 m²), fine sand constituted 83% of the soil; hence as density of distribution decreased, soil compactness increased. The same trend was observed in the regions of Lake Saroma and in Oshima. In the high-density regions, the silt content was below 25%, and the amount of organic carbon in the range of 0.5 to 1.5% (1966). That *Hotategai* prefer a sea bottom with coarse material was observed long ago. Dakin (1909) reported this fact in relation to the distribution of *P. opercularis* and *P. maximus*; Imai (1941) and Shibuya *et al.* (1942) reported it in relation to *P. magellanicus*; Bourne (1964) recently confirmed it for *P. magellanicus*.

In fishing grounds suited to scallops, it has been found that *Hotategai* keep their ventral margin facing the tidal currents in places where the ground material is mainly formed by such currents. According to Kinoshita *et al.*, when the tidal current is rather strong, it controls the regional distribution of *Hotategai*, and thus they conclude that the ground material is influenced by the strength of the tidal current. That the tidal current plays an important role as a controlling factor in the life of *Hotategai* was reported as early as 1906 by Drew with regard to *P. tenuicostatus*.

In Mutsu Bay, "abnormal fecundity" may occur in consecutive years, or once in several years. At such a time, a dense population of *Hotategai* can be observed in the soft mud bottom of Okiai, at a depth of 50 or 60 m; Kodera (1958) claims that the scallops pile one on top of the other. But since the growth of *Hotategai* resulting from abnormal fecundity is very poor, and since the scallops living in the lower layers become victims of mass mortality when the abnormal condition continues for some time, it seems rather extraordinary that *Hotategai* were seen in the soft mud bottom of Okiai as reported.

Yamamoto (1950) carried out group ecological studies on the conditions of the habitat of scallops. According to him, the bottom-dwelling organisms of Mutsu Bay can be divided into four groups, based on the frequency of the appearance of the dominant species (Figure 3.6). Since *Hotategai* can be included as a member of the coastal group (group No. 4), one would conclude that this region is its habitat. Yet, in some sea basins with 30% or less silt content and a 40 meter depth along the coast, *Hotategai* are not found. For example, in Mutsu Bay and in Okiai of Tsugaru Hanto on the western coast, a species composition of Group 1 was found far away from the species composition of Group 4. However, when young shells of 5 to 7 cm shell length were transplanted and allowed to settle, the behavior after setting showed that most of the transplanted *Hotategai* had migrated toward Okiai. The migration took considerable time but, ultimately, none of the transferred scallops were seen in the original place of transplantation. This was one of the striking phenomena observed by Yamamoto and Edo (1951) during a short period of a few months, and established that

Tsugaru Hanto is not a suitable habitat for *Hotategai*. Therefore the authors decided to carry out a large-scale investigation to improve the culture operations, without wasting time on culturing *Hotategai* through transplantation in this region.

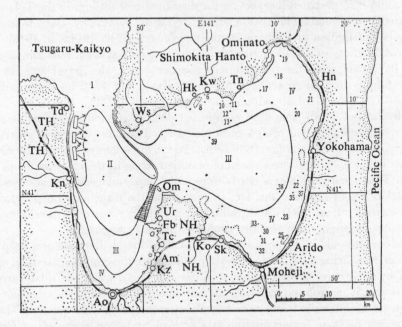

Figure 3.6. Regions of Mutsu Bay divided on the basis of the frequency of appearance of a dominant species of benthic organisms (Yamamoto, 1964).

Region No. 4 (coastal) could be said to be the habitat of *Hotategai* judged from the various biological analyses and observations. It is clear from the figure, however, that the young scallops transplanted to the coastal regions between Tairadate and Kanita of Region No. 1 disappeared, and new fishing grounds could not be formed in these places. Region No. 4 could be further divided into suitable habitats for young scallops, judging from the distribution of 2 or 3 index organisms. Areas surrounded by dotted lines show the suitable regions for young scallops. Areas shown by slanting lines in Futagobana and the western side of Oshima are places where the so-called "abnormal fecundity" phenomenon occurs frequently.

Kw—Kawauchi; Hk—Hinokigawa; Tn—Tsunotagai; Hn—Hamaokunai; Om—Oshima; Ws—Wakinasawa; TH—Tsugaru Hanto; Td—Tairadata; Kn—Kanita; Ur—Urata; Fb—Futagobana; Tc—Tsuchiya; Am—Asamushi; Kz—Kukurizaka; NH—Natsudomari Hanto; Sk—Shimizukawa; Ao—Aomori; Ko—Kominato.

It was thus found that the localized distribution of *Hotategai* is not only controlled by the strength of the tidal currents or the distribution of benthic materials, but also by the environmental complex of the benthic layer. One must assume that biological investigations on benthic organisms would serve as indices for determining the controlling factors. The distribution of the 4 group regions mentioned earlier, correlated well with the distribution of the chlorine content of the benthic layer of water as studied for several years. However, it also correlates with the distribution of benthic materials, except for one or two aspects (Figure 3.6).

4.2 Locomotion and migration

Regarding the migration of *Hotategai*, several interesting points were made above relating to transplantation experiments carried out at the Tsugaru Hanto Okiai of Mutsu Bay. However, the phenomenon of transplanted scallops completely disappearing within one night is not restricted to *Hotategai* only; it has been observed with several mollusks, in particular *Hamaguri* (clam).

In studying the migration of *Hotategai*, the authors first referred to the data published in *Wakan Sansai Zukai*, and obtained the data collected in the Aomori Prefecture in Meiji 28 (1896); while the latter mentioned that studies on *Hotategai* could not be satisfactorily completed in a two-week period, the animal's locomotion was correctly described in this report. The shells open to allow the sea water to enter the mantle cavity and then close gently. The water in the mantle cavity flows toward the sides, the flow being stronger toward the dorsal side. By spurting water from the two corners of their shells, the scallops move backwards by leaps. At times, the water comes out only from the anterior part of the shell, in which case, the scallop moves in a diagonal direction on its posterior side. It has also been observed that the scallop is capable of a rapid expulsion of water by making an outlet in a part of the velum. When this happens, it moves in a leaping manner on its dorsal side (Figure 3.7).

This type of movement is also seen in young scallops of about 1 cm shell length. Since the direction of water expulsion can be changed, and because even in the sea the right shell bulges more than the left, a scallop is capable of various types of motion such as forward and backward movements with its right shell facing downward, and oscillating movements on both the dorsal and ventral sides. By combining the leaping and oscillating movements, the scallop

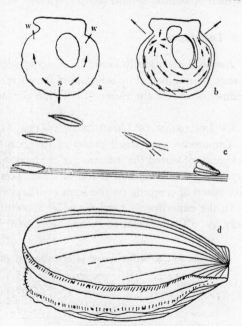

Figure 3.7. Mechanism of the leaping movement of *Hotategai*. Arrows show the direction of water current (Figures *a* and *b* from Yonge, 1936; Figures *c* and *d* from Kishigami, 1895).

can float on the surface of the sea, an interesting phenomenon not seen in any other marine animal.

Needless to say the long-distance, free movement, and the group movement in a single direction seen so commonly among fishes, are rarely observed in scallops. As mentioned earlier, movement in a particular direction and in groups is seen among scallops only when environmental conditions are not suitable.

Between 1928 and 1930, Isahaya carried out a series of studies in Mombetsu Okiai of Okhotsk on scallop migration, and published his field experiments in 1933. Of the

1,843 scallops released in 1928, and the 1,306 released in 1929, 296 and 166 respectively were recaptured. Their average migrating distance in 59 days was 5.8 nautical miles; the maximum distance was 16.9, and the minimum 2.1 nautical miles. All of the specimens had migrated toward Abashiri, i.e., southward. This directional movement was not attributed to the scallops, but rather to the influence of the strong southern current in Mombetsu Okiai. Kinoshita (1936) confirmed Isahaya's observations, among them that there is no fixed directional movement in *Hotategai*. In addition, even within Hokkaido, there are morphological variations in the shells of various regions, and significant regional differences in the age composition of each catch. The growth rate of *Hotategai* shows regional variations just as that of *P. opercularis* in Britain does. As mentioned earlier, there are subspecies of *Aequipecten irradians* found in the Atlantic Ocean of North America, which also show variations in their spawning season (Sastry, 1963).

4.3 Tolerance

Hotategai are usually found in the cold regions of the northern part and along the coasts. A number of investigations are currently underway to determine the range of temperature and the range of chlorine content which scallops can tolerate.

4.3.1 INFLUENCE OF CHANGES IN WATER TEMPERATURE AND SALINITY

Yamamoto used small pieces of the gills from adult and young *Hotategai* (Table 3.1) and measured the influence of changes in water temperature and chlorine content according to the method of Nomura and Tomita (1932, 1933) for measuring the relative speed of creeping on the basis of ciliary movement (1956b, 1957a).

In the experiments carried out in September, when the temperature was lowered only by 5°C, the relative ciliary movement in adult specimens decreased to 16% of

Table 3.1. Relative speed of creeping of gill pieces under various temperatures

Initial temperature	Experimental temperature	Adult specimens 110—140 mm		Young specimens 10—13 mm	
		100% sea water	125% sea water	100% sea water	125% sea water
5° C*	5° C (control)	100%	100%	—	—
,,	10	16.2	17.3	—	—
,,	15	11.0	11.6	—	—
,,	20	10.0	7.9	—	—
20 °C†	5	11.7	9.2	0.0	0.1
,,	10	12.0	11.9	0.8	0.6
,,	15	16.0	16.4	3.4	3.0
,,	20 (control)	100	100	100	100

* and † signify experiments carried out in February and September respectively.

the control, and in young specimens to 3% of the control. When the temperature was decreased by 10°C, the corresponding decrease was to 12% and 0.8%. In the experiments carried out in February, when the temperature was increased by 5°C, the relative speed of ciliary movement decreased to 16.2%, and when the temperature was increased by 10°C, decreased to 11%. While this decrease was sharp in both adult and young specimens, it was more striking in the latter.

The ciliary movement becomes disturbed when the temperature is 23°C and the relative speed of movement drops to 0 to 0.2%. When the temperature is lowered to 5°C the ciliary movement is very slow, and when it is 0°C, stops intermittently. From the foregoing, it can be concluded that the ciliary movements of the gills of *Hotategai* required for normal animal motion, occur at a temperature range of 5 to 23°C (Yamamoto, 1964).

According to Posgay (1953), the optimum temperature or activity of *Aequipecten magellanicus* is 10°C. For the same species, Dickie (1958) claimed that the upper limit of the temperature is between 20 and 23.5°C. This high temperature limit is very similar to that of the Japanese *Pecten*. Now, in the case of oysters, cells of *Isochrysis galbana* and *Monochrysis lutheri*, food items for the larvae of *Crassostrea virginica* and *Mercenaria* found in America, can tolerate water temperature up to 27.5 to 30°C. *Chlorella* sp. can tolerate a temperature up to 33°C. Hence the larvae of bivalves fed on *Chlorella* are able to tolerate and grow at a temperature of 33°C, or about 2.5°C higher than the maximum temperature tolerated by the larvae fed on the other two species. Growth is more certain to continue on a wide range of temperature in sea water with optimum chlorine content, than in sea water with low chlorine content (Davis, 1963).

Regarding tolerance to chlorine concentration, it seems that sea water with a 25% higher chlorine concentration than the normal (18.11 $Cl^o/_{oo}$, pH 8.3 T. B.) is suited to the growth of scallops. The lower limit of chlorine concentration is 12.00 to $13.50^o/_{oo}$ Cl, at which the ciliary movements of the gills rapidly decrease. The tolerance range of chlorine concentration is therefore narrow.

Kinoshita (1949) measured the density of water in the habitats of scallops in Lake Saroma. During the three months of spring, the density was 1.02281 to 1.02455, and the mean 1.02341.

4.3.2 OXYGEN CONTENT

Both the adult and young *Hotategai* become weak when the oxygen content is low. When the oxygen content is 1.5 to 1.7 cc/L, the relative speed of ciliary movement in the gill piece of an adult specimen is 40 to 50%, and after 45 minutes zero (Table 3.2). In young specimens, the effect of a low oxygen content is much more severe and the tolerance poor.

4.3.3 INFLUENCE OF SUSPENDED PARTICLES

Dried soft mud from the sea bottom was added to sea water in amounts of 0.05, 0.10, 0.15, ..., 0.35 g, and the relative ciliary movement in the gill pieces measured (Yamamoto, 1956b). As seen in Figure 3.8, the relative ciliary movement for adults was 50% when 0.1% mud was added, which decreased to 0 to 20% in the young (Figure 3.8). In the gill pieces of spat of shell length 17 to 19 mm, which had just

Table 3.2. Relative ciliary movement in the gill pieces of *Hotategai* at three different stages of growth, and the duration after which creeping stops when the oxygen content is low

Stages of growth, shell length	Relative ciliary movement after 15 minutes	Time of ceasing
11-18 mm	21.7(15.4-28.0)%	20(13-37)*
23-27	43.3(38.7-48.0)	30(22-34)
120	40.0(38.2-41.7)	45(36-53)

*Reliability limit, 95%.

entered bottom-dwelling life, the ciliary movement of the gills stopped when the sea water contained 0.05% mud. When the gill pieces were observed under a microscope, along with the mucus accumulation of mud, particles measuring several hundred

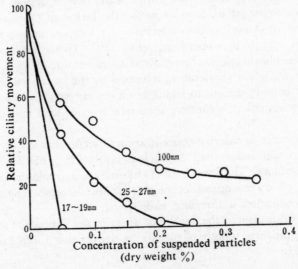

Figure 3.8. Relative ciliary movement of gill pieces of *Hotategai* at different stages of growth (shell length 17 to 19 mm, 25 to 27 mm, and above 100 mm) in sea water with various concentrations of dry suspended particles.

microns were found settled on the gill surface. The cilia had stopped movement at several places and, moreover, the cilia which did move appeared very weak. These results agree with those obtained by Loosanoff (1947, 1948) in *C. virginica*, in which the mud content of 0.3 g/l was found to weaken the ciliary movement by 95%.

4.4 Reproduction and rate of mortality

The presence of a large number of larvae in sea water during the spawning period of *Hotategai* has been reported by Kinoshita (1939, 1940, 1949), Edo (1953), Sato (1954, 1961), Kato (1962), Yamamoto (1950, 1951, 1964), and others. When the number of larvae is large, 1,265 larvae per 100 liters of sea water are found, e.g. in Lake Saroma. Even when the number is small as in Mutsu Bay, 10 to 50 larvae per 100 liters of sea water are found. This does not mean that the larvae are equally distributed throughout the region, as the density of larvae differs. However, the number of planktonic larvae found in Lake Saroma or Mutsu Bay is phenomenally high. For example, if the condition in Mutsu Bay during 1949-1950 is taken as a representative example, 75 trillion larvae would be present in the entire region. Among marine invertebrates which have no parental care in the larval stage, such a condition is said to be rather an abnormal phenomenon. A number of research studies have been done on the fecundity of organisms with this type of spawning habit.

Various methods for determining the spawning capacity of *Hotategai* have been devised. Yamamoto (1950) followed that suggested by Belding (1910) and Hopkins (1937), in which the value is obtained after subtracting the cubic volume of ovary from the cubic volume of an ovary with mature ova of 70 μ in size. The number of immature ova in the ovary is much more than the number of mature ones. Although it has been said that an accurate value could be obtained by taking the mean value of these ova, in the method mentioned, only mature ova of 70 μ in size are taken into account. Needless to say, the ova in the ovary are smaller than the volume equal to the cube of the diameter. Again, the ovary not only contains the ova but also the epithelium of the follicles, oviducts, connective tissues, some muscle tissues, etc. Hence the volume occupied by these structures in the ovary inflates the value obtained by the calculation of the ovarian volume. The results of the calculation of the follicles and spermatocytes are given in Table 3.3. It is clear from this Table that in the first year

Table 3.3. Number of follicles and spermatocytes in *Hotategai*

Age	Sex	Shell length	Total weight (including shell)	Weight of gonads	Number of follicles and spermatocytes
2	Female	122 mm	211g*	39g*	$11,440 \times 10^4$
2	,,	108	162	29	8,448
2	,,	111	191	37	11,000
3	,,	137	294	43	12,584
3	,,	127	282	40	11,704
3	,,	131	299	45	13,200
3	,,	126	288	44	11,088
4	,,	148	367	53	15,488
4	,,	156	352	59	17,248
4	,,	138	330	52	15,224
4	,,	142	341	61	17,864
5	,,	148	359	60	17,600
5	,,	149	381	62	18,128
6	,,	151	392	59	17,248
6	,,	152	420	64	18,744
2	Male	121	199	23	$3,700 \times 10^9$
3	,,	148	368	52	8,307
4	,,	156	381	56	8,928

*Wet weight.

the mother scallops spawn 100,000,000 eggs, and in the fourth year 160,100,000 eggs.

When the spawning is stimulated by heat treatment, it is possible to obtain approximately 10 to 30 million eggs in one spawning. From this fact, it appears that the figures mentioned in the Table are not only acceptable, but within a reasonable limit. The fact that the number of eggs is very large is also reported for *C. virginica* found in America, in which 115 million eggs are relased in one spawning cycle (Galtsoff, 1947).

However, the number of young scallops of 6 to 10 mm shell length setting in Mutsu Bay rarely reaches one billion. In fact, the figure 892,510,000 in 1968, is the maximum obtained thus far. Usually the figure is much lower than this. In Lake Saroma, the maximum figure of 10 mm scallops is said to be 500,000,000 (Table 3.4).

Table 3.4. Number of seedlings produced in Lake Saroma
(Kinoshita, 1949; Ohara and Maru, 1970)

Year	10,000 Unit	Year	10,000 Unit	Year	10,000 Unit
1936	3,200	1941	21,700	1966	17,580
1937	8,000	1942	22,700	1967	44,920
1938	5,900	1943	38,200	1968	17,297
1939	15,500	1944	35,200	1969	4,963
1940	22,200	1945	52,500		

Although the figures in Table 3.4 are very large, compared to the number of follicles mentioned earlier, and to the spawning rate, one can easily visualize the drastic reduction in the number during the stages of development up to setting. Indeed the mother scallop does not release all the ova from the ovary, but only one-third or one-fourth. In some cases, a large portion of the ova in the ovaries are not released but gradually degenerate within the body (Yamamoto, 1951c). However, when the conditions of spawning are suitable, the scallop releases more than 10 million eggs at a time. Moreover, after a few days, it spawns again at the same rate. The same phenomenon has also been reported for *C. virginica*; spawning occurs 4 times in an individual, and a total of 500,000,000 eggs are released in a single spawning (Galtsoff, 1947). In view of this fact, it can be said that the extent of spawning, loss or large-scale mortality during the different stages of growth of planktonic larvae in the first 30 to 40 days, changes in the environment, etc., greatly influence the stock of *Hotategai*.

After leading a sessile existence for nearly two months, the scallop then begins a bottom-dwelling life. In Mutsu Bay, a large-scale mortality is often experienced at this stage. From late July, when the scallop begins bottom-dwelling life, to late October, a three-month survey was done by divers on the fate and behavior of young scallops. It was found that only about 5 to 10% of the scallops had survived by late September. Although this survey was carried out at Arito Okiai in Mutsu Bay, the facts established have also been verified in other regions of the bay. Usually, during the three months subsequent to beginning bottom-dwelling life, only about 1% of the scallops survive (Yamamoto and Edo, unpublished). Thereafter, during the period of growth up to a shell length of 2.5 to 3.0 cm, the mortality rate is not very high.

Thus it is found that the mortality rate is high during the period of growth from embryo to planktonic larvae, and soon after the beginning of bottom-dwelling life. Hence, the mortality rate during these two periods greatly affects the yield rate of scallops, and constitutes a survival factor in organisms capable of producing a large number of eggs.

5. Physiology

5.1 Food and digestion

It is well known that the gills are the organs of feeding in scallops. When the sea water enters the mantle, it passes over the surface of the ctenidia whereby the food substances in the sea water are covered by mucus, and this viscous mass is drawn toward the labial palp where the actual feed is selected. Larger particles and unfavorable substances are eliminated. Sawano has reported that the starch particles of rice are completely eliminated. Yet, Yamamoto discovered the buccal organ of cuttlefish in the stomach of a *Hotategai*. *Mytilus californianus* selectively rejects flagellates (Verwey, 1952). Those substances rejected by the labial palp are imbedded in the mucus, which carries them to the byssal groove from where they are eliminated.

The feed, after the elimination of the unwanted substances, is carried into the stomach. The stomach content is a mixture of different substances. In earlier investigations, the contents were said to be mainly biological organisms which could be easily identified. Kinoshita (1935) reported the presence of 44 species of diatoms, 19 species of protozoa, crustaceans, spores of green algae and seaweeds, as well as larvae of echinoderms. The main components of *Hotategai* feed are diatoms and protozoa. *Hotategai* is not very efficient in selecting its feed, selection being based on the size of the planktonic organisms, morphology, and presence or absence of movement. Gutsell (1930) reported that microscopic algae and microscopic fauna from the sea bottom were found in the stomach of *Hotategai*. Besides these, a large amount of detritus was found in the stomach, but it is doubtful that this was utilized as food, a fact strongly supported by other authors (Martin, 1923; Hunt, 1923; Savage, 1925; Yonge, 1926). The authors also studied the contents of the alimentary canal of *Hotategai*, and found that 80 to 95% consisted of detritus. Besides detritus, they found diatoms, protozoa, small pieces of polychaetes, amphipods, etc. Regarding the role of detritus as feed, Savilov (1957), Verwey (1952), Zobell and Feltham (1938) made a detailed study and found that *Pelecypoda* are capable of digesting this substance (Yamamoto, 1967).

In the coastal regions, which are the habitats of *Hotategai*, and in the benthic layer of sea water in the bays, a large amount of detritus is found suspended in water (Table 3.5). In a unit area the weight of detritus is several times more than the dry weight of plankton in the same unit area. There can be little room for doubt that scallops which take in large amounts of detritus by filtering sea water do ultimately digest it (Yamamoto, 1964). This type of sea water filtration could be said to be directly related to feeding. However, data on the filtering capacity of *Hotategai* are very few. Yamamoto and Kanno (1963), Cole and Hepper (1954) used Neutral Red (N. R.) and found that the rates of filtration at night in juvenile scallops with

Table 3.5. Comparison of the amount of planktons and suspended organic particles in Mutsu Bay (Dry weight /10 liters)

Survey points	Water depth	Surface layer		Benthic layer		Remarks
		Amount of plankton*	Suspended organic particles	Amount of plankton*	Suspended organic particles	
Asamushi Oki	27 m	27.3 mg	43.1 mg	53.0 mg	72.5 mg	Millipore filter R. 1.2 μ
,,	,,	23.0	13.0	43.1	21.2	R. 5.0 μ
Futagobana Oki	50 m	39.5	207.0	37.5	507.0	R. 1.2 μ
Oshima Oki	50 m	46.8	231.0	19.0	393.0	R. 1.2 μ
,,		42.0	561.0	22.0	154.0	R. 1.2 μ

*Collected by using gauze NBC No. 25.

4.2 to 6.6 cm shell length, and spat of 1.0 to 2.5 cm shell length, are 10 cc/g/hr (16.3 to 18.3°C) and 16 cc/g/hr (14.5 to 17.9°C) (Figure 3.9). According to the N. R. method, the absolute value seems to be smaller than that found by the direct method. The

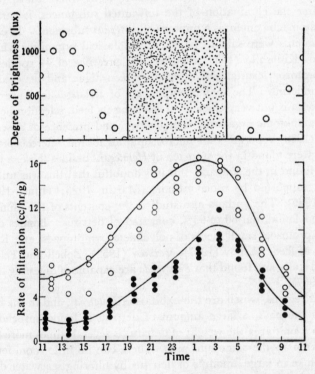

Figure 3.9. Changes during the day in the amount of filtered water in *Hotategai* (Yamamoto, 1964).

●—juvenile scallops, shell length 4.2 to 6.6 cm; ○—spat, shell length 1.0 to 2.5 cm. Changes in the degree of brightness during the day are shown in the upper part of the figure. The temperature of the water was 16.3 to18.3°C for the juvenile scallops and 14.5 to 17.9°C for the spat.

correct value can be obtained by multiplying the value with coefficients of 16 to 20 (Jørgensen, 1949). The results obtained by Jørgensen for *Mytilus,* and by Chipman and Hopkins (1954) for *Pecten irradians* by the direct method, are more or less the same (Yamamoto, 1964).

Recent studies on digestion are few, but several were done in the early 20th century. Enzymes capable of reducing the starch in the crystalline body into sugar, were detected by Nelson (1918, 1925), Allen (1909), and others. Dakin (1909) reported the presence of proteolytic enzymes, amylaze, and lipaze in the digestive gland. Yonge pointed out the importance of intracellular digestion in the midgut gland. Regarding the intracellular digestion of lamellidans aided by phagocytes, there have been a number of controversial reports in the postwar period. Mansour (1946) reported that animal planktons taken in as food in the stomach of *Tridacne* are quickly digested by the enzymes. By removing the phagocytes in *Tridacne* and *Pinctada,* Mansour nee Bek (1946, 1948) found fat-breaking and proteolytic enzymes present in the stomach. While the source of these enzymes was not confirmed, she suggested that the enzymes were probably secreted by the digestive diverticulum (midgut gland) in the absence of any influence of phagocytes. On the other hand, Nelson (1947), Loosanoff and Engle (1947) and Cerruti (1941), used *Crassostrea, Tapes, Mytilus,* and *Venus* as experimental animals and found that *Chlorella, Coscinodiscus, Nitzschia, Pleurosigma,* etc., which were taken in as food, were not assimilated and were eventually excreted without change (Korringa, 1952). This discovery drew attention to the role of phagocytes in the assimilation of food, and to the fact that some scallops cannot assimilate *Chlorella* and the diatoms, *Coscinodiscus* and *Nitzschia,* due to the presence of certain food organisms.

5.2 Growth

The growth of shellfish is usually considered in relation to the changes in the shell length. Soon after fertilization, scallop eggs grow into veliger larvae with a type D shell in 5 to 7 days at 7 to 9.5°C. The shell length at this stage is $72 \times 58 \mu$, and increases to 120μ in 15 to 16 days after fertilization, at which stage the bulging at the umbo begins. The embryo now passes into the second stage. In *Akoyagai (Pinctada martensii)* the shell of the D-type embryo consists mostly of dahllite, but in the second stage mostly of calcite (Watanabe and Yuki, 1952; Watanabe, 1956). When the shell length reaches about 300μ, the larvae settle down and begin a sessile life. While free swimming, the water temperature is 8 to 14°C, but after the larvae attain a shell length of 300μ in 30 to 40 days, growth continues at the rate of 100 to 150μ per day for 40 to 60 days under a temperature of 15 to 20°C. In other words, during the sessile period of life, the growth rate is unusually high. When the two-month sessile existence is completed, the larvae lose their byssal filaments and enter a bottom-dwelling life. The shell length at this stage is 6 to 10 mm.

When *Hotategai* enter bottom-dwelling life, their growth slows down for a short period which, in Mutsu Bay comes up in summer when the temperature is high. As autumn approaches, growth resumes rapidly and continues up to the spring of the next year when a shell length of 5 to 6 cm is attained. The growth rate and weight increase of scallops are shown in Figure 3.10. Regarding the 2-year-old, fully mature scallops, growth seems to increase at a lower rate in the third year and the body weight shows a lower rate of increase after the fourth year. On the other hand, scallops growing in the Hokkaido region continue to develop until the end of autumn with growth slightly retarded in winter. The average size of a 1-year-old is 1.9 cm and of a 2-year-old 6.2 cm. Thus there is a difference or delay in growth by one year in the Hokkaido scallops compared to the growth of those in Mutsu Bay (Figure 3.10).

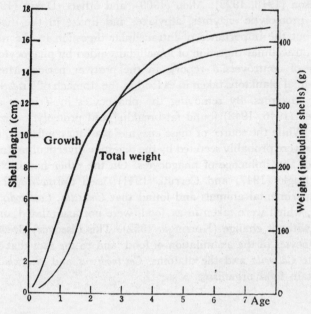

Figure 3.10. Growth of *Hotategai* (Yamamoto, 1964). Shell length and wet weight show a reciprocal logarithmic phase up to the third and fourth years. The growth rate during this period is high, after which it enters a stationary phase.

It is clear from the foregoing that while in Mutsu Bay the growth of scallops is checked by the unfavorable environmental conditions in summer after spawning, in Hokkaido growth is checked by environmental conditions in winter. This is very interesting in view of what has been discussed about resistance to temperature. It should be pointed out that the retardation of the growth rate in summer in Mutsu Bay is, in fact, related to spawning; this means that the gonads of one-year-old *Hotategai* are such that it is very difficult to differentiate the sexes with the naked eye. Even examining the tissues under a microscope, reveals little difference between the two, and hence it is not possible to state definitively whether the gonads can display reproductive activity. One-year-old *Hotategai* do not show any adverse change in their growth rate even in summer, and their annual rings are not very distinct. Hence the two-year-old scallops are known as the "biologically smallest form"* at both Hokkaido and Mutsu Bay, and scallops of more than two years of age are known as juveniles.

Concerning the life span of the scallop, as far as the authors know, it seems to

---

* In Japanese this term is used to signify the smallest size of the organism at which it begins to show reproductive activity—Translator.

range between 10 and 12 years. However, in about 7 or 8 years, most scallops are either caught by fishermen or succumb to natural death.

5.3 Respiration

Although gills are usually considered to be the organs of respiration, Dakin (1909) and Gutsell (1930) have pointed out that the mantle also plays a very important role as one of the respiratory organs, since the capillaries are distributed throughout it.

According to the investigations on respiration in *Hotategai*, the oxygen consumption is more or less constant in adult scallops up to that point when the oxygen content decreases from normal to 1 or 0.4 cc/liter. Oxygen consumption in spat is somewhat constant, up to 2.5 to 1.0 cc/liter. When the content decreases further, the amount of oxygen consumed by the adult as well as the spat decreases by about 1/10 (Figure 3.11) (Tamura, 1939 a, b; Yamamoto *et al.*, 1950). The critical point of oxygen

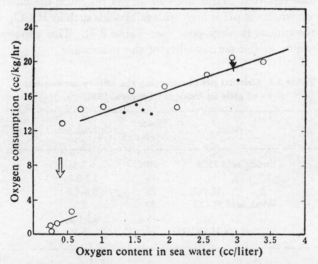

Figure 3.11. Relationship between the consumption rate of oxygen and oxygen content in sea water. Adult animals used in experiment. Water temperature shown by ●; ○ refers to noon temperature, 14°C, 12.3 to 13.4°C (● according to Tamura, 1939; ○ according to Yamamoto *et al.*, 1950).

content of 1 cc/liter for *Hotategai* is very close to that of *C. gigas,* but much higher than in the short-necked clam (*Pophia phillipinarum*). It is interesting to note that scallops live in tidal currents with their ventral side facing the current, and the critical point is high for these animals especially during the bottom-dwelling period. Nevertheless, bottom dwelling seems to be an environmental condition which scallops require.

The amount of oxygen consumed by adult animals and spat as shown in Table 3.6 reveals that for spat the amount is almost 10 times more than the weight of their soft parts[2] (Yamamoto *et al.*, 1950).

[2]Value of Q_{10} in the temperature range of 10 to 23°C of the scallop is 1.8 to 2.1 (Yamamoto, unpublished).

Table 3.6. Consumption of oxygen by spat (Yamamoto *et al.,* **1950)**

Water temperature	Oxygen tension in sea water	Consumption of oxygen	Remarks
	4.56 cc/1	0.344 cc/g/hr	Shell length 0.8 to 1.2 cm.
	4.22	0.340	Experiments carried out on groups
	3.74	0.316	of 40 scallops.
22.5–23.0°C	3.18	0.292	Respiratory chamber closed.
	2.78	0.228	Winkler method.
	2.52	0.018	
	2.43	0.016	

The ciliary movement of the gills in this animal stops due to a decrease in the oxygen content in sea water, rather than the influence of CO_3^{--} in the water (Nomura, 1932). Nomura also reported that a decrease in pH obstructs the ciliary movement of gills, and likewise when the pH is high with mild acids such as H_2CO_3, CH_3COOH, etc., the ciliary movement is very poor (see Table 3.7). This problem should be studied with reference to the permeability of the membrane.

Table 3.7. Critical pH obstructing the ciliary movement of gills in *Hotategai* **(Nomura, 1932)**

Acids	Water temperature	Critical pH
Strong acid HCl	20°C	3.8
,, ,, ,,	13	$3.7–3.8_5$
,, ,, H_3PO_4	20	3.8–4.8
Weak acid H_2CO_3	20	5.5
,, ,, ,,	13	$6.1–6.3_5$
,, ,, CH_3COOH	20	5.4_5

5.4 Sex and maturation

The sexes are generally separate in *Hotategai*. The gonads of the females are orange-red in color 4 to 5 months before the spawning period; the gonads of the males are yellow-white. Rarely, hermaphrodite specimens are found in which two different conditions are observed: In the first, the basal half of the gonads represent the male part and the anterior half the female part, and each acquires its characteristic coloration; in the second type, the male and female gonads are mixed and form a mozaic (Figure 3.12). *Hotategai* are also capable of changing their sexual characteristics. Among the European varieties, many are hermaphrodites such as *Pecten (Chlamys) opercularis* found in Britain, *P. maximus, P. varius,* and *P. glaber* found in the English channel, and *Aequipecten irradians* and *P. gibbus* found in America. In *P. tenuicostatus* and *Placopecten magellanicus,* however, the sexes are separate.

Spawning in the southern part of Hokkaido occurs between early April to mid-May, and in the northern part from either early or mid-May to mid-June. In

Nemuro and Kunashiri, spawning takes place between early June to mid-July (Kinoshita, 1949). The spawning period in Mutsu Bay is from March to mid-May,

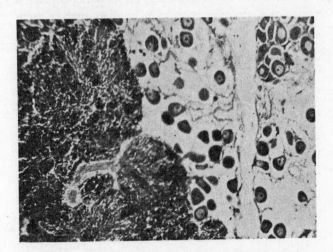

Figure 3.12. Gonads of a hermaphrodite showing the mozaic arrangement of male (left) and female (right) gonads.

with maximum spawning occurring in late March to late April. The critical temperature of spawning is 8.0°C to 8.5°C, but Kinoshita indicates that for coastal waters 9.0°C is the critical temperature. However, spawning in Mutsu Bay begins in early March when the water temperature is only 4 to 5°C, considerably lower than the recognised critical temperatures just given. Such variations occur in several other organisms as well and would make an interesting subject for further investigations.

5.5 Chemical composition

The chemical composition of this bivalve is given in Table 3.8, which establishes that the scallop is rich in protein and poor in fat. Hence its food value is very high. The protein content comprises 35% of the soft parts (wet weight).

Compared to that of vertebrates, the protein content of the meat in mollusks is poor in lysine and rich in alginine (Otani, 1934). However, the meat of *Hotategai* is comparatively rich in lysine (Sekine and Kokutsuno, 1920). Furthermore, compared to beef which is only 0.007%, *Hotategai* is also rich in succinic acid at 0.370%. This means that it contains 10 times more succinic acid than abalones, and about 3 times more than arkshell and clams. It is this percentage of succinic acid which makes the meat of scallops very tasty (Aoki, 1932).

The fat content of scallops is given in Table 3.9. Compared to other mollusks, which are close to this variety, the unsaturated fatty acid content is higher, and so the food value of scallops is subsequently greater.

Mason (1959) pointed out that in *Pecten maximus* most of the carbohydrates are in the form of glycogen. In the British varieties, there is not much change in the chemical components of the scallops throughout the year. It is only after spawning

Table 3.8. Chemical composition of *Hotategai* **compared with** *Magaki* **(***C. gigas***) and**
Ubagai (Mactra Sachalinensis)

(According to the standard food chart prepared by the Nutrition Association of Japan, 1965)

	Calories (cal)	Water content (g)	Protein (g)	Fat (g)	Carbo-hydrate (g)	Ash conten (g)	Calcium (g)	Na (mg)
1	2	3	4	5	6	7	8	9
Hotategai meat (raw)	106	74.3	20.8	0.8	2.4	1.8	11	—
Magaki whole body (raw)	96	79.6	10.0	3.6	5.1	1.7	40	380
Ubagai whole body (raw)	98	76.1	17.8	1.3	2.4	2.4	24	—

(Contd.)

	P (mg)	Fe (mg)	A (effective) (I.U.)	Carotin (I.U.)	B_1 (mg)	B_2 (mg)	Nico-tinic acid (mg)	C (mg)
					Vitamins			
1	10	11	12	13	14	15	16	17
Hotategai meat (raw)	76	0.4	8	24	0.04	0.10	1.4	3
Magaki whole body (raw)	140	8	100	90	0.30	0.20	1.2	5
Ubagai whole body (raw)	480	13	—	—	—	—	—	—

Table 3.9. Fatty acid content of *Hotategai* **and a few allied varieties**
(Weight % for the whole acid content is given)

(Data source the same as Table 3.8)

Species	Saturated fatty acids			Unsaturated fatty acids				
	C_{14}	C_{16}	C_{18}	C_{14}	C_{16}	C_{18}	C_{20}	C_{22}
Hotategai[1]	6	10	6	1	18 (2.0)	47 (3.8)	12 (4.0)	—
Akazaragai *Chlamys nipponensis*[1]	11	9	7	3	15 (1.8)	20 (2.4)	24 (4.4)	11 (4.5)
Placopecten magellanicus[2]	4	35	tr	—	16 (1.6)	18 (1.0)	14 (3.1)	9 (2.7)

[1]According to Igarashi, H., *et al.*, 1961.

[2]According to Brockehoff, H., *et al.*, 1963.

in the spring that glycogen shows a decrease value of 2.8 wt. %. It is at its maximum value of 4.6 wt. % soon after spawning in the autumn. On the other hand, the chemical components of the gonads show considerable variation in the course of the year in relation to spawning. The protein content for one month before spawning is 17.3%, i.e. twice that in the period following spawning. The fat content shows similar changes, being 3 times greater after spawning, while the glycogen content is 4 times more in the pre-spawning period than it is in the post-spawning period.

The meat of scallops is sometimes found to be orange-red in color. Obata *et al.* (1950) report that this coloration is due to a carotinoid pigment, which turns brown when exposed to sunlight for some time. Occasionally, scallops are found which have lost their color and become pale.

(Yamamoto)

CHAPTER II.

Techniques of Scallop Culture

1. History and problems of scallop culture

1.1 Fluctuations in harvest and the so-called "Abnormal Fecundity"

It has long been known that the harvest of scallops is subject to sharp fluctuations, which are particularly remarkable in Mutsu Bay. Back in 1896, Kishigami wrote:

"About 34 or 35 years ago, scallop production in Noheji Bay was high, but the harvest gradually became poor in the years to follow, until finally there were almost no scallop to harvest. In the early years of the Meiji era (1867), scallops were highly fecund for two or three years at Kawauchi Muramae-oki but thereafter diminished in this region also. In Meiji 23 (1890), groups of spat, thought to consist of two-year-olds, were found at a depth of 78 *hiro* (one *hiro*=1.818 m) in Okiai off the coast of Noheji-mura. Here also scallops became scarce from 1891 until a stage of 'no scallop' was reached... Following the discovery in the middle of 1885 of small scallops (the size of a zeni[1]) in large numbers in the sea off Aomori, the use of fishing implements harmful to scallops was prohibited until 1888. Thus scallops were allowed to grow and harvested only when fully grown. The growing area was extended in 1889 and the capital invested reached 300,000 yen. However the yield began decreasing from 1890, and in 1891-1892 adult specimens were rarely caught." Calculating the yield from the 1889 price of scallops, production was roughly 10,000 tons.

Even after the periods covered above, there were fluctuations in scallop harvests (Figure 3.13), and various reasons for such fluctuations have been put forward. While Kishigami attributes it to the mode of harvesting, Nishioka finds an explanation in nature, i.e. the fluctuations follow a rhythmic pattern, each cycle covering a period of 18 years.

As may be noted from Figure 3.13, scallop harvest at times suddenly increased in Mutsu Bay resulting in a yield exceeding 10,000 tons for two to three years, and hereafter dropped to 200 to 300 tons in the next few years. Interestingly, from 1950 the level of the harvest did not go below 1,000 tons, a fact attributed to better management and improvements in the method of culture.

The severe fluctuations in the harvest of scallops from year to year have been attributed to what are called additional resources, and the difference in the additional resources in particular years. One phenomenon which may be described as such an explosive additional resource, is the "abnormal fecundity" of scallops in certain years.

---

[1]*Zeni*: an old Japanese coin about 1 inch in diameter.

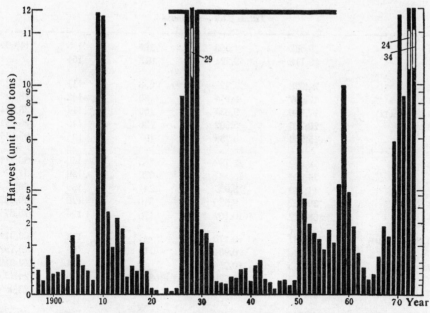

Figure 3.13. Fluctuations in the harvesting of scallops in Mutsu Bay.

Fluctuations in scallop production were studied in Hokkaido (Table 3.10). Similar fluctuations have also been observed in Bakagai of Awajikaya, Itayagai of Hokuriku, San-in and Kyushu, and even in the harvest of *Aequipecten irradians* of the Pacific coast of North America where the yield fluctuated between 16,000 and 300 tons. In Canada, *Placopecten magellanicus* of the Atlantic coast fluctuated in yield from 840 tons to 16.4 tons (Gutsell, 1930; Dickie, 1955).   England, too, has reported fluctuations ranging from 1,300 tons to 200 tons on the yield of *Chlamys opercularis* (Dakin, 1909).

Table 3.10. Scallop harvest in Hokkaido (Unit=ton)

(Data based on reports by Yamamoto, 1964, and Wakui)

Year	Okhotsk Coast	Sea regions of Nemuro	Sea of Japan	Pacific Coast	Total harvest
1910 (Meiji 43)	11,783	295	179	6	12,263
1911	8,972	55	794	24	9,845
1912	12,997	323	432	24	13,776
1913 (Taisho 2)	11,174	747	514	112	12,547
1914	18,179	1,588	1,233	105	22,105
1915	37,359	7,793	244	158	45,554
1916	5,693	1,557	926	169	8,347
1917	1,873	—	1,202	609	3,684
1918	2,738	1,028	1,299	182	5,247
1919	1,225	1,262	10,732	303	13,522
1920	8,660	364	4,910	190	14,124
1921 (Taisho 10)	9,166	1,069	1,619	7	11,861
1922	13,629	1,137	609	234	15,609

Table 3.10—*Continued*

1923	9,582	1,583	424	105	11,694
1924	16,116	9,221	943	164	26,444
1925	24,029	11,727	658	111	36,525
1926	17,997	9,914	69	142	28,122
1927 (Showa 2)	17,979	9,831	150	111	28,071
1928	20,726	5,502	226	110	26,564
1929	15,903	8,739	797	105	25,544
1930	8,042	22,792	5,285	100	36,219
1931	14,904	15,181	1,029	156	31,270
1932	11,140	8,057	1,241	135	20,573
1933	30,922	9,025	319	126	40,392
1934	68,272	10,102	176	124	78,674
1935 (Showa 10)	59,587	8,270	41	114	68,012
1936	35,914	8,984	115	182	45,192
1937	31,675	10,954	47	164	42,840
1938	13,803	10,529	328	78	24,738
1939	18,195	4,716	3	141	23,055
1940	41,345	5,184	6	38	46,573
1941	13,956	3,007	3	54	17,020
1942	58,533	4,174	3	65	62,775
1943	39,378	4,109	—	25	43,512
1944	14,324	1,863	—	2	16,189
1945 (Showa 20)	918	84	—	6	1,008
1946	8,279	1,604	—	5	9,888
1947	7,216	1,845	—	126	9,187
1948	5,602	1,572	15	187	7,376
1949	13,650	279	—	142	14,071
1950	9,759	10	1	208	9,978
1951	5,296	16	—	190	5,502
1952	6,502	2,707	—	277	9,486
1953	9,226	1,895	1	198	11,320
1954	13,134	3,245	—	403	16,782
1955 (Showa 30)	10,115	4,198	—	242	14,555
1956	7,215	3,754	—	547	11,621
1957	11,838	2,920	—	499	15,257
1958	12,912	1,599	—	406	14,917
1959	9,017	1,633	—	891	11,541
1960	7,651	1,642	—	805	10,098
1961	7,789	864	—	380	9,033

One could conceivably attribute the abnormally high reproductive potential of scallops, which results under certain circumstances, to the phenomenon of "abnormal fecundity." As already mentioned in the sections on ecology, reproduction, and mortality in the previous chapter, a single, full-grown scallop (four-year-old) is capable of producing 160,000,000 eggs. If all the eggs of this one scallop were released and fertilized, and developed and grew to an adult stage, about 50,000 tons of fully grown scallops would result in about 4 years. But to expect all the eggs of one scallop to be viable, whether the scallop is grown in Hokkaido or Mutsu Bay is to stretch the imagination beyond reasonable limits. However, it is possible that under certain conditions, perhaps considered abnormal, which would provide some sort of environmental resistance to egg damage, a large number of eggs might prove viable.

Environmental resistance manifests itself in three ways: (a) variations in the rate of spawning, (b) variations in the rate of setting spat, and (c) variations in the rate of surviving spat during setting. All three types have been discussed under the section entitled "Ecology." A point to note is that when the survival rate is high, there is rarely any variation noticeable in the survival rate of benthic spat. This probably results in "abnormal fecundity" and leads to, and can even be called, the "abnormal survival" phenomenon.

1.2 Mechanism of "Abnormal Fecundity"

The phenomenon of "abnormal fecundity" occurs often in Mutsu Bay. The why and how of such an occurrence have not yet been sufficiently investigated in order to answer all the queries it engenders. Kishigami in 1896 gave the mode of harvesting as the reason for variations in production. Nishioka (1943) proposed the rhythmic cycle theory. Both explanations are regarded as insufficient by the present authors because of their views on the effect which additional resources might have. Furthermore, Nishioka's theory of the rhythmic cycle cannot satisfactorily explain the several gaps in "abnormal fecundity" after 1945. As mentioned earlier, the authors prefer to think of "abnormal fecundity" as an "abnormal survival" phenomenon. Investigations on this aspect are currently in progress and some useful data have already been collected (1956a, 1957a, 1957b, 1960, 1964).

"Abnormal fecundity" occurs in waters 35 to 50 m deep in the open seas off Futago and Oshima in Mutsu Bay (1927, 1948, 1956, 1965). It was observed in waters of 35 to 40 m deep in Yokohama Okiai and Tsunotogai (1927), as well as at a depth of 50 to 70 m to the west of Wakinozawa and Oshima (1956, 1965). These waters are not normal habitats of scallops and cannot properly be considered growing beds. The floor of the sea is composed of soft mud mixed with shells and sometimes gives off hydrogen sulfide. Plankton is plentiful, particularly zooplankton, at the center of the eddies, and the planktonic larvae of scallops can also be seen in large numbers. At the bottom of the sea dwell *Asabellides sibirica* Wirén and large *Obelia plana* (Figure 3.14), and scallop spat are frequently found attached to the hydrocaulis of *Obelia plana* and the tube-like burrows of polychaetes.

Since the center of the eddies is quite deep, the flow of water frequently stops in summer, which gives rise to new layers and results in a decreased percentage of dissolved oxygen on the bottom layer. Moreover, the decomposed products of planktons, etc., convert to organic suspended particles (ditoritas) and are carried to these water basins. At times they settle on the sea bottom. Scallop spat entering a benthic life under such environmental conditions, become victims of mass mortality. This seems to occur more or less regularly and the authors discovered in July, 1959. a large number of shells belonging to spat which had died the previous summer along

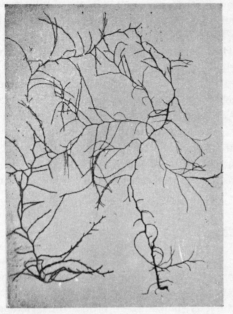 

(A) *Obelia plana*   (B) *Asabellides sibirica*

Figure 3.14. Natural substrata on which scallop spat settle.

with various other organisms (Table 3.11).

In some years the formation of new layers of various organisms and dead spat in summer is not so noticeable and the accumulation of suspended organic particles is not significant due to their dispersion by underwater currents. In such years, the survival rate of spat settling on the sea bottom is high, resulting in the so-called "abnormal survival" phenomenon.

1.3 Problems relating to fluctuations in stocks and cultivation

The problems relating to the culture of scallops are threefold: (a) to increase the spawning rate, (b) to increase the number of setting spat and devise more efficient methods for collecting them, and (c) to increase the survival rate of setting spat i.e. reduce the mortality rate.

Experiments to increase the resources for the harvest by supplementing the naturally produced seedlings with artificially grown ones, are presently underway in Mutsu Bay and the results thus far seem to point to this method as a practical proposition.

The problem of variation in the number of setting spat can be overcome by more efficient use of natural seeding, and considerable technological progress has been made along this line in Lake Saroma of Hokkaido and Mutsu Bay. Hence rapid developments in this regard may be expected in the near future.

Efforts to improve the survival rate of spat by artificially controlling the number which settle to the bottom of the sea, have yielded promising results. The area of

Table 3.11. Benthic organisms in regions where ' abnormal fecundity"
occurred in 1956 (from Yamamoto, 1960)

Scientific name	Japanese name	Number of organisms/m²
August 16, 1957		
Pinnixa ruthbuni	Rasuban-mame-gani	22
Episiphon makiyamai	Rosoku-tsuno-gai	7
Pectinaria hyperborea	Umi-isagomushi	3
Thyasira tokunagai	Hanashigai	2
Dead shells		
Hiatella orientalis	Kinumatoigai	
Balanus rostratus	Minefujitsubo	
Episiphon makiyamai	Rosoku-tsuno-gai	
Raeta pulchella	Chiyonohanagai	
Others		
July 2, 1959		
Prionospio pinniata	Yotsubanesupio	24
Telepsavus costarum	Tsubasagokai	9
Episiphon makiyamai	Rosoku-tsuno-gai	12
Anisocorbula venusta	Kuchibenide	8
Thyasira tokunagai	Hanashigai	10
Lucinoma annulata	Tsukigaimodoki	6
Echinocardium cordatum	Okamebunbuku	4
Pectinaria hyperborea	Umi-isagomushi	2
Ringicula niinoi	Ni-inom meurashima	2
Pinnixa ruthbuni	Rasuban-name-gani	2
Raeta pulchella	Chiyonohanagai	2
Cavernularia obesa	Umishaboten	1
Dead Shells		
Patinopecten yessoensis (spat)	Spat of Hotategai	1,960
Arca boucardi	Koberutofunegai	18
Thyasira tokunagai	Hanashigai	16
Ringicula niinoi	Ni-inomameurashima	8
Balanus rostratus	Minefujitsubo	5
Others		

operation is increasing every year, and a significant rate of increase in the harvest is expected soon.

Of the problems relating to stock fluctuation and cultivation, reducing the mortality rate must be given priority since this will lead to more effective methods for increasing the yield.

(Yamamoto)

2. Seedling production and through culture

The raft-type hanging culture method was recently introduced into the coastal regions of Sanriku, and has gained popularity with each year. Availability of seed-lings is, of course, very important to this type of culture, and, at present, these are being procured from the natural sources of the southern part of Hokkaido and Mutsu Bay of the Aomori Prefecture. However, since the rate of development of spat fluctuates greatly if left to nature, availability of seedlings cannot be maintained satisfactorily. Therefore, developing a supply of seedlings produced through arti-ficial means is essential to the establishment of a planned cultivation of scallops as a small-scale industry.

Producing artificial scallop seedlings is a comparatively recent endeavor, so funda-mental research on this line is limited. Kinoshita (1943) succeeded in inducing artificial spawning in the scallop by simultaneously increasing the temperature and alkalinity of sea water during the spawning period. Yamamoto (1951) improved

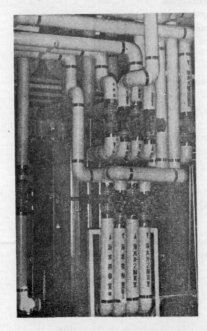

Figure 3.15A. Aquaculture center of Aomori Prefecture (Hiranaimachi, Higashi Tsugaru-Gun, Aomori Prefecture).

Upper left—Aquaculture center. *Upper right*—artificial seedlings of scallop.
Lower left—titan heat exchanger. *Lower right*—temperature control panel.

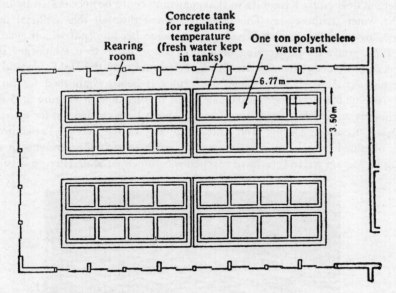

Figure 3.15B. Plan of rearing room (front view) (Aquaculture Center of Aomori Prefecture).

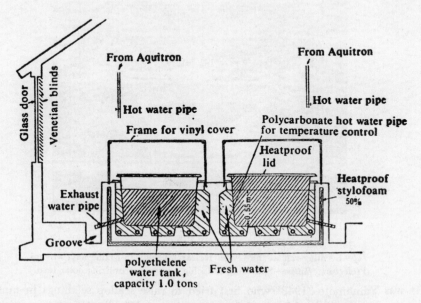

Figure 3.15C. Plan of rearing room (side view) (Aquaculture Center of Aomori Prefecture).

the method devised by Kinoshita so that spawning could be accelerated by increasing just the water temperature. The larvae obtained through this artificial induction of spawning were fed on *Protomonad* sp., and some became spat which even settled. After 1951, study on the rearing of scallop larvae came to a temporary standstill but has recently been resumed at Aomori Prefecture, and the Hokkaido and Miyagi Prefecture. At Aomori, an aquaculture center is being established for an annual culture of 900,000 seedlings measuring 3 cm in shell length (Figure 3.15 A, B, C). Experiments on artificial seedlings of scallop have been carried out for some time at the Oyster Research Center (Inc.) at Maine Bay of the Miyagi Prefecture. The results obtained have been encouraging, so much so that the products of artificial seeding in Miyagi Prefecture have reached a commercial scale (Figure 3.16).

Figure 3.16.

Top—Seedling raft at the Oyster Research Center in Maine Bay, Miyagi Prefecture. *Bottom*—Spawning inducing equipment utilizing solar heat.

It was Yamamoto (1948) who first tried to raise scallop seedlings in tanks. By using a concrete tank of $3 \times 2 \times 1.3$ m capacity, and feeding the larvae with non-shelled flagellates belonging to the genus *Protomonad*, which had been cultured in

Erdschreiber fluid of Föyn, he was able to obtain about 20 to 30 spat. On the basis of this experiment, Yamamoto drew attention to four major problems in developing an artificial seedling technique for scallops. The first problem concerned the maintenance of fertilized eggs; by studying the results of previous investigations, Yamamoto found that it was possible to obtain fertilized eggs by induced spawning through a sudden increase in temperature. The Oyster Research Center then considered the feasibility of setting up an apparatus to induce spawning by using solar heat, and finally designed one which is now in operation (Figure 16, Bottom).

The second problem was feeding the larvae. Yamamoto found that *Monochrysis lutheri* was a very effective food, and Umebayashi reported that *Chaetoceros calcitrans* grown in isolated culture was also effective. Since the latter grows easily on a culture medium, it can be used on a large scale. Imai *et al.* (1950) made an isolated culture of *Monas* sp. and used this as food for oyster larvae; it seems that the same may be fed to scallop larvae. However, since these organisms have a saprophytic characteristic, care must be exercised not to allow any mixing of the culture fluid with the water in which the larvae are kept. The Aquaculture Research Center at Mutsu Bay (1965) and the Oyster Research Center (1965, 1966) recently obtained good results by using a mixed diet of *Mc. lutheri* and *Ch. calcitrans*. This is an interesting point and should be kept in mind during further studies on the nutritional requirements of this bivalve.

The third problem which Yamamoto faced concerned the environmental conditions of the rearing water. The major defects in his rearing experiments were the lack of aeration and the absence of water circulation facilities. The rearing technique of larvae has gradually advanced, and now the adjustment of temperature, aeration, and circulation of the water are commonly done (Figure 3.17A). The rearing temperature of scallop larvae is higher than the temperature of spawning in natural conditions and should be in the range of 13 to 15°C. When the rearing is carried out in this temperature range, it is possible to shorten the period of setting (about 40 days) by about 10 days. In the investigations carried out on oysters, spawning was controlled by keeping the mother oyster under a low temperature (6°C), and allowing

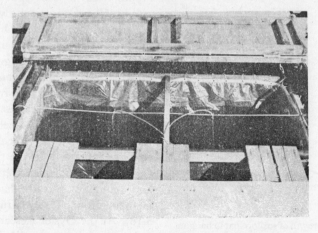

Figure 3.17A. Rearing tank of planktonic larvae of scallop.

it to spawn only after bringing the temperature of the sea water to 12°C. Doing this made it possible to rear the larvae with greater efficiency.

The fourth problem pointed out by Yamamoto related to the equipment and media for rearing the larvae. Clean sea water must be available in abundance and the supply of mother scallop must be uninterrupted. Maine Bay in which the Oyster Research Center is located, has all these necessary conditions and the Aquaculture Center of Mutsu Bay is also ideally suited.

Figure 3.17B. Rearing of spat in running water.

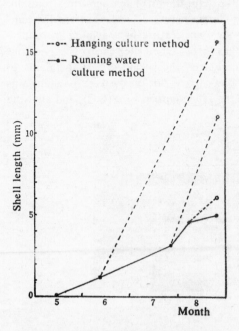

Figure 3.18. Growth of scallop spat in hanging culture and running water culture compared.

Thus, the various problems involved in the rearing of scallop larvae are gradually being solved. At present the major problem in the field is to find a suitable method of rearing spat obtained from seedlings produced on a large scale. Yamamoto (1955) tried to rear spat in tanks and ponds. He was particularly interested in the good results obtained by rearing spat in natural sea water, and in the high rate of survival of the spat thus grown. In 1958, Onodera *et al.*, tried the hanging culture method of Akoyagai for rearing scallop spat and obtained good results. This method has also brought excellent results in the through culture of scallop. Rearing spat on a large scale by the running water culture method is not very economical and, compared to the hanging culture method, growth is rather inferior (Figure 3.18). At the Oyster Research Center, in order to minimize the disadvantages, the method was modified in the following way: the running water rearing of spat (Figure 3.17B) was stopped 1 to 2 weeks after setting, and the seedlings were reared in sea water after covering them with collecting nets (Figure 3.19).

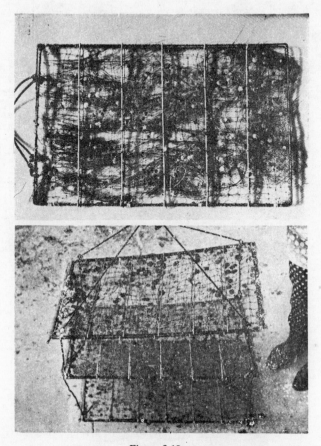

Figure 3.19.

Top—Seedling collector and spat after setting (S. L. 1.0 to 1.5 cm);
Bottom—Hanging culture of spat (S. L. 1.5 to 2.0 cm).

Table 3.12. Relationship between hanging depth and growth of spat

Hanging depth (m)	Shell length (mm) ± standard deviation				Increase in shell length during period of experiment (mm)
	October 30 1965	November 25	December 11	January 13 1966	
1.2	20.10±1.47	22.63±2.32	22.70±1.86	24.35±2.96	3.73
2.5	20.62±1.76	26.79±2.82	28.85±1.88	30.59±2.97	9.97
4.0	21.30±1.46	26.35±1.94	28.25±2.68	32.60±3.32	11.30
5.2	20.35±2.02	26.01±2.78	27.73±3.04	31.80±1.21	11.55
6.7	20.78±2.00	26.56±2.51	29.29±3.52	33.00±1.14	11.22
8.5	21.76±2.32	27.51±2.01	29.82±2.57	34.30±2.71	12.54

Figure 3.20A. Interim culture (circular nets made of nylon threads).

Figure 3.20B. Interim culture (open-close net).

Figure 3.20C. Raft hanging culture: *Left*—one-year-old; *Right*—two-year-old scallop.

The hanging culture of scallops consists of the following steps. Seedlings of 3 cm shell length are reared in pearl-shaped nets (Figure 3.20) until they attain 4 to 6 cm in shell length; at this time they are removed and a tiny hole, 1 to 2 mm in diameter, is bored at the ear part. Nylon thread No. 30 to 40 is passed through this hole and the scallop tied to a rope painted with tar. Further rearing proceeds in the same way as in the raft hanging culture of *C. gigas (Magaki)* (Figure 3.20C). In view of the fact that scallops grow well in deep waters (Table 3.12, Figure 3.21C) and that

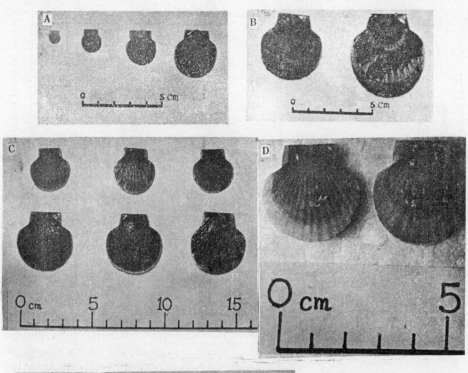

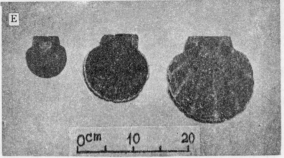

Figure 3.21. (A) Setting spat. (B) Bottom-dwelling spat. (C) *Top row* —spat grown by hanging at 1.2 m depth; *Bottom row*—spat grown by hanging at 8.5 m depth. (D) Annual rings formed by high temperature in summer. (E) *From left*—1-year-old; 2-year-old; 3-year-old scallops.

growth is temporarily arrested in summer (August to September) when the water temperature increases (above 20°C), it is better to hang the strings deep in the water (Figures 3.21D, 3.22, and 3.23). Hence it is possible to carry on a simultaneous culture of *C. gigas* and O. edulis which grow well in shallow waters, and scallops which

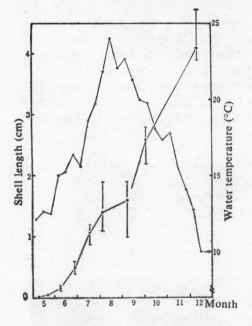

Figure 3.22. Growth of spat at Maine Bay.

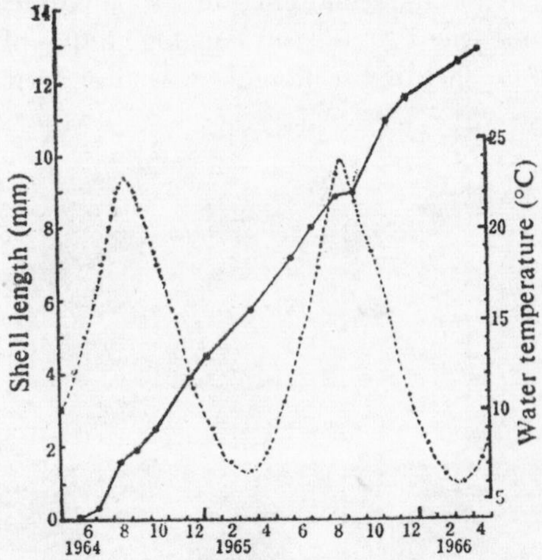

Figure 3.23. Growth of scallops at Maine Bay.

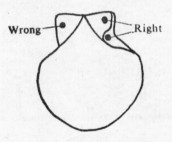

Figure 3.24. Position of hole in scallop for rearing by hanging culture method.

grow well in deep waters, by culturing them one above the other (Figure 3.25).

The hole in the ear part of the scallop must be on the anterior side of the ear (Figure 3.24). If the hole is made on the posterior side, the mantle might be injured and consequently its growth in this part arrested; subsequently, when the animal grows, the mantle may easily slough off which may ultimately result in the animal's death (Figure 3.26).

(Nishikawa)

3. Natural seeding and ecology of fishing grounds

3.1 Evolution of seedling collection methods

Collecting natural seedlings of scallop was first tried by Kinoshita in Lake Saroma in 1935, and again in the following year. Various methods were tried such as placing barriers (fixed wooden frames) in shallow waters along the edge of the lake, hanging scallop shells in strings on them, and allowing spat to set on these shells (Figure 3.27). Further investigations to improve this method were made whereby it was found that the deeper places of Akagawa Oki in Lake Saroma were more suitable for collecting seedlings. Later rafts made of wood and steel tubes were used (Tanaka,

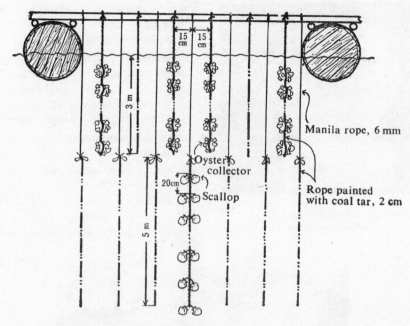

Figure 3.25. Simultaneous culture of scallop and European oyster.

Figure 3.26A. Scallop (2-year-old) grown by raft hanging culture method.

Figure 3.26B. Effect of bad positioning of hold on the development of scallop. *Left*—condition of shell inside.

1963). At present, the long-line method is widely used and instead of shells, sheets of hemp, hyzex, etc., are tied to ropes as seedling collectors (Figure 3.28). As shown in Figure 3.29, baskets made of polyethelene nets are now widely used for collecting seedlings (Ohara and Shinobu). Spat setting on Piccard trap nets has been used for a very long time at Mutsu Bay. It was only after 1953 that scientific seedling equipment was introduced in the field. At first, based on the studies of Nishioka and Yamamoto (after 1943), fence nets with twigs scattered on them were used for collection. Later the long-line hanging method was adopted (Figure 3.29) in which leaves and twigs were entangled on the ropes to serve as collec-

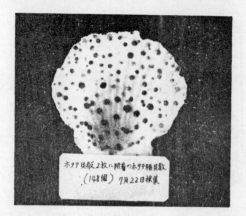

Figure 3.27. Scallop spat setting on scallop shell (Ohara).

tors. Nowadays, instead of leaves, "HYZEX" film is widely used. In 1965, Kofuji (member of the Okunai Fisheries Association) devised an easy and profitable hanging method for setting spat, an onion-shaped polyethelene bag which made it impossible for the collected spat to be carried away. This method has become common in a very short time.

Since 1954, seedlings have been collected in the regions around Date in Funka Bay, and recently collection has spread to the neighborhood of Date itself. Seedlings were initially collected

Figure 3.28. Scallop spat grown from seedlings collected in polyethelene plates. (Ishihara, 1963)

by hand nets, just as fence nets were first used in Mutsu Bay. Soon, however, the hanging method employing leaves and twigs was introduced. Ishihara (1963) was the first to use seedling collectors made of Hyzex film, and subsequent to his success, this method was adopted in other regions. The following year (1964),

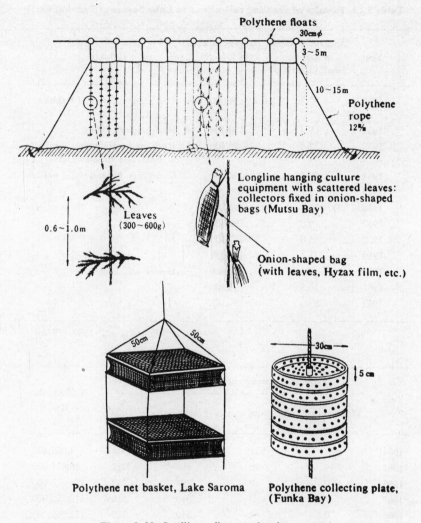

Figure 3.29. Seedling collectors of various types.

Ishihara evolved a method of collecting seedlings by Hyzex plates (Figure 3.29) and growing them in the plates up to the stage of interim culture.

3.2 Results of seedling collections and certain problems involved

Several investigators (Kinoshita, 1940; Yamamoto, 1951; Tanaka, 1963; Ito, 1967) have referred to the enormous fluctuations in the amount of setting spat when scallops are reared from natural seedlings. Table 3.13 shows the results of seedling collection in Lake Saroma, revealing that considerable fluctuation in the amount of setting spat per shell collector occurred, with an average range of 0.05 to 58.6. Even now in Mutsu Bay, the average number of spat setting per 360 g of seedling

Table 3.13. Results of seedling collections in Lake Saroma (Tanaka, 1963)

Year	Average number of spat per shell of scallop	Average number of spat per raft	Remarks
1951	0.3	4,800	Collection in original habitat, rack type
1952	9.4	150,400	Rack and raft types
1953	2.4	38,400	Rack and raft types; heavy losses due to silt
1954	39.6	633,600	Collection by raft type only
1955	12.7	203,200	
1956	20.6	332,800	
1957	9.8	156,800	
1958	58.6	937,600	
1959	31.2	499,200	
1960	0.4	6,400	Heavy loss due to silt
1961	10.3	164,800	
1962	0.05	768	

(Ohara et al., 1967, 1970)

Year	Type and number of seedling collectors				Total number of spat
	Wooden raft	Steel-tube raft	Long-line	Total	
1964	217	331	33	581	4,066,400
1965	14	82	186	282	20,871,600
1966	0	0	293	293	175,800,000
1967	0	0	310	310	449,203,000
1968	0	0	310	310	172,968,000
1969	0	0	320	320	49,628,000

collectors in the form of leaves and twigs, is on the wide range of 27 to 1,836 (Table 3.14). In fact, it is very difficult to find a single water basin in which a uniform state of spat setting occurs every year. Even in the Date regions of Funka Bay, the production rate of scallops cannot be said to be uniform (Table 3.15).

As shown in Table 3.13, the collection of seedlings in Lake Saroma has greatly increased. The same trend also seems to have started in Mutsu Bay. Although there are various reasons for such improvement, one of them is the more efficient method of seedling collection, and another is the use of modern materials in the making of the collectors. Seedling collection can be further improved by studying the development of scallops, their setting conditions, and the ecological conditions of the

Table 3.14. Setting condition of spat in Mutsu Bay (Ito *et al.*, 1967; Sasagi *et al.*, 1970)

	1960	1961	1962	1963	1964	1965	1966	1967	Mean
Wakinosawa						35	27	380	147
Kawauchi					80	8	4	272	91
Yokohama	7	3,100	37	1		72	27	30	468
Noheji	14	2,970	13	4	88	22	37	39	399
Shimizu Kawa			17		145	8	51	309	106
Kominato		1,005	14	8	65		2		219
Urata	58	2,890	159	15	246	14	277	312	496
Moura		1,760	150	8	147	62	266	659	436
Tsuchiya	18	1,540	122	38	91	32	141	270	282
Kukurizaka	132	2,500			438		663		933
Nonai	97	1,300	117	48	100	59	72	202	250
Zodo	116	1,910	34	56	126	34			259
Okunai				88	761	63	209	627	349
Ushirogota	47				241	61	137		122
Yomogita					518	12	105	474	277
Mean	61	2,028	74	29	234	37	104	353	365

Note: Number of setting spat per 360 g of twigs and leaves.

fishing grounds. Once the collection method of seedlings has been stabilized, the loss in scallop production should be greatly reduced.

3.3 Spawning and factors contributing to its fluctuation

It is well known that a properly matured female scallop contains more than 100 million eggs (Yamamoto, 1964). Hence, the reproductive potential of the scallop is extremely high. If a major portion of these eggs could be fertilized and then successfully reared, a large number of spat could be obtained. The first and major problem in achieving this desirable goal relates to spawning.

Yamamoto (1943) carried out a histological research on the gonadal tissues of scallops in Mutsu Bay. He found that the gonadal cells mature by the end of February and continue to remain in that state up to mid- or late April. From May onwards the ova begin to degenerate and the gonads to contract. Given this fact, Yamamoto expressed the view that the ova which remain in the ovary and degenerate, are ultimately absorbed by the body tissues. When scallops in such a condition are transferred to a tank of sea water at 8 to 12°C, spawning is induced by the sudden temperature change. As shown in Table 3.16, when this treatment is administered between the beginning of March and the middle of April, spawning is easily induced, but if the treatment is delayed to May, even though the gonads appear quite large in size, spawning cannot be induced (Ito *et al.*, 1968). The results obtained by Ito *et al.* seem to support the views of Yamamoto.

Table 3.15. Results of seedling collections in regions of Date in Funka Bay (Ishihara, 1963)

Year	Type of equipment used for seedling collection	Number of collectors	Mean number of spat per string	Total number of spat
1954	Hand net	3		
1955	Hand net	2		
1956	Hand net	2	5 per 30 cm 10 per shell	240,000
1957	Hand net Hanging method using scallop shell collectors	1 1	4 per 30 cm 1,707 per string	365,900
1958	Hand net Hanging method using scallop shell collectors	1 2	3 per 30 cm 785 per string	326,600
1959	Hanging method using twigs and leaves	2	2,250 per string	225,000
1960	Hanging method using twigs and leaves	9	1,485 per string	668,300
1961	Hanging method using twigs and leaves	20	6,750 per string	13,500,000
1962	Hanging method using twigs and leaves	20	4,030 per string	8,060,000
1963	Hand net Hanging method using twigs and leaves Hanging method using hyzex film	5 5 9	8 per 30 cm 17,850 per string 18,623 per string	9,864,600

Table 3.16. Seasonal variations in the induction rate of spawning, and the weight of gonads in scallops of Mutsu Bay (Ito _et al._, 1968)

Date of survey	$\dfrac{\text{Weight of gonad}}{\text{Weight of meat}} \times 100$		$\dfrac{\text{Number of eggs}}{\text{Number of scallops stimulated}} \times 100$	
	♂	♀	♂	♀
	%	%	%	%
1965, March 3	32.7	32.9	25	17
March 12	31.5	30.8	82	33
March 23	34.7	34.3	67	17
April 1	32.0	36.0	100	20
April 7	32.3	32.5	100	71
April 17	34.3	36.6	100	90
April 29	21.9	36.1	90	50
May 11	17.5	15.4	14	0
May 28	8.9	8.8	0	0

According to Wakui (1967), the scallops of Lake Saroma mature by March; however, spawning takes place much later and is usually in late May to late June when the temperature of the water rises to 9°C (Kinoshita, 1951). Scallops of Mutsu Bay spawn from late March to early May when the water temperature is 8.0 to 8.5°C or more (Figure 3.30) (Yamamoto, 1964). According to the results of the investigations on seasonal changes in the gonadal tissues of scallops in Mutsu Bay the gonadal tissue

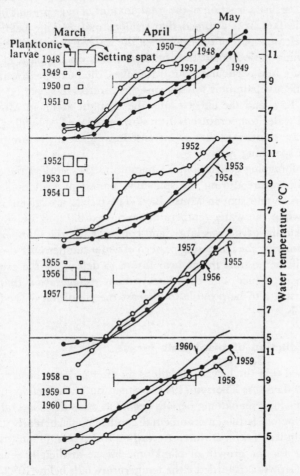

Figure 3.30. Relationship between an increase in water temperature
and the variation in number of planktonic larvae and setting spat
in Mutsu Bay (Yamamoto, 1964).

shrinks somewhat during late March to early April when the temperature of the water on the bottom layer exceeds 5°C. The gonads show a temporary recovery from early April to mid-April, but again diminish in size during mid-April to late April, and in late June shrink rapidly. While every year the number of planktonic larvae assumed to be derived as a result of the first gonadal release is always large, the number of planktonic larvae derived as a result of the second gonadal release fluctuates from year to year (Ito *et al.*, 1967; Tsubata *et al.*, 1968).

A study of the spawning behavior of female scallops in which spawning is induced by artificial means, shows that the ova are released vigorously in a few minutes by opening and closing the shells two or three successive times. Once the release of ova ceases, spawning cannot be artificially induced by heat. The number of eggs resulting from induced spawning is only about 10,000,000 per scallop (Yamamoto, 1964; Ito *et al.*, 1968), or just a fraction of the total amount of eggs present in the body.

In view of this fact, it is assumed that scallops mature long before the period of spawning and release a part of the eggs when the water temperature exceeds a certain level; the spawning period could thus be determined on the basis of the increase in water temperature in a particular year. Moreover, the increase in water temperature serving as an effective stimulus for spawning is limited to a period from March to April in Mutsu Bay, and the effect cannot be brought about in May. In the years when the rise in water temperature is very slow or delayed, spawning does not occur in the normal way; this means that an increase in water temperature plays an important role in the spawning rate.

To verify the foregoing conclusion, Yamamoto (1951) studied the correlation between water temperature during the spawning period of scallops and the number of planktonic larvae and spat in Mutsu Bay. The results are given in Figure 3.30. In those years when the water temperature exceeded 8.0 to 8.5°C in mid-April, the number of planktonic larvae was large, and consequently the number of spat.

By carefully studying the increasing pattern of water temperature during the spawning period, it will be possible in the near future to determine the exact condition of larval development; hence, one can state without exaggeration that on the basis of these data, it will also be possible to forecast the condition of seedling collections for the next crop.

3.4 Natural reduction of planktonic larvae

It is well known that the larvae of scallops set after passing through the planktonic stage for 30 to 40 days. As Thorson (1947) pointed out, a reduction in the number of larvae during this long planktonic period, occurs as a result of several factors.

One of these factors is the environmental condition, particularly the influence of sudden changes in water temperature. According to Ito *et al.* (1968), the temperature of water suitable for the growth of planktonic larvae artificially reared, is 15°C or slightly higher. Growth is retarded if the temperature falls below 10°C, and the larvae die overnight if the water temperature is suddenly dropped to 2 to 3°C. In Mutsu Bay the water temperature at the time of larval formation is 6 to 15°C and, according to the results of a Marine Survey, water basins with a lower water temperature (by 2 to 3°C) are sometimes very close to the larval habitats. Planktonic larvae which enter these water areas either suffer growth retardation, or become victims of mass mortality. According to the results of an investigation by Kinoshita (1940) on the relationship between the rising pattern of water temperature during the period of seedling collection in Lake Saroma, and the setting condition of spat, setting is good in those years in which the critical temperature for spawning occurs early. However, according to the survey carried out by the authors at Mutsu Bay planktonic larvae

hatching from eggs spawned too early, are highly susceptible to low water temperature, and suffer a high mortality during their planktonic life.

Figure 3.31 shows the periodic distribution of planktonic larvae in Mutsu Bay
(Ito *et al.*, 1967). Within a short period, planktonic larvae are widely spread through-

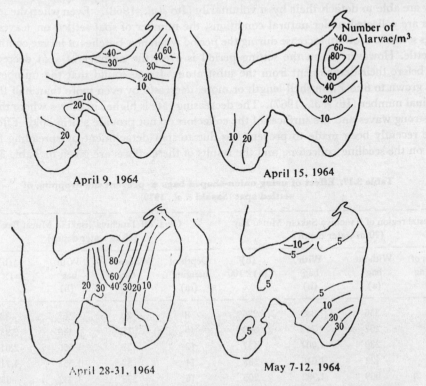

Figure 3.31. Distribution condition of planktonic larvae in Mutsu Bay (Ito, 1967).

out the bay, due to the fact that water in the bay is moved by wind. Sometimes the
wind is so strong that a bottle thrown in the bay crosses from one side to the other
within a few days (Tsubata *et al.*, 1968).   That planktonic larvae may move rather
rapidly into areas of low water temperature is therefore not surprising, and that
efforts to collect seedlings which have been carried away from the collecting equipment should fail, is predictable. This drifting of the planktonic larvae obviously
accounts for part of the fluctuation in the amount of setting spat shown in Table
3.14, and should be studied so that the right type of equipment for gathering seedlings
could eventually be installed.

3.5 Survival rate of settled spat

At Mutsu Bay, scallop spat continue to live a sessile life by attaching to other
materials with the help of fine byssi until they attain a size of 6 to 10 mm in mid-July
to early August. At Lake Saroma, however, spat lead a sessile life until they attain a
size of 15 mm shell length in early September to late September. When spat reach

these sizes, they lose their byssi and drop to the bottom to begin the bottom-dwelling phase of life. Spat reared by artificial means, even those which are only about 3 mm in shell length, are found to change their places of setting. It would seem that either their capacity to attach themselves to other material at this stage is very weak, or that they are able to detach their byssi voluntarily (Ito *et al.*, 1968). Even when the seedlings are collected under natural conditions, the number of spat settled on leaves or twigs seems to show an increase during the period when new batches of larvae continue to settle. However, when the setting period is over, the number of spat decreases just before their detachment from the substratum. It was found that the number of spat grown to 6 to 7 mm shell length or more, decreases by even more than half their original number (Ito *et al.*, 1967). The decreasing rate is higher in places where there are strong waves and the surfaces of the collectors do not provide a firm hold. Efforts have recently been made to prevent loss due to spat detachment by spreading fine nets on the seedling collectors, and the results of these efforts are given in Table 3.17

Table 3.17. Effect of using onion-shaped bags to prevent the dropping of settled spat (Sasaki *et al.*, 1970)

Coastal region of Tsuchiya Saki in Mutsu Bay (20 m water depth)				Okiai off Tsuchiya Jisaki in Mutsu Bay (25 m water depth)			
Depth of hanging (m)	Without bag (a)	With bag (b)	(b)/(a)×100	Depth of hanging (m)	Without bag (a)	With bag (b)	(b)/(a)×100
3	356	839	236%	8	20	65	325%
5	207	745	360	10	15	440	2,933
7	230	1,007	437	12	25	506	2,012
9	422	1,616	383	14	33	1,553	4,710
11	969	1,982	205	16	538	1,932	359
15	2,688	2,016	75	18	548	3,184	582
17	1,556	2,716	174	20	1,863	3,584	192
19	1,763	2,746	156	22	1,058	4,170	395
Mean	1,024	1,708	166	Mean	513	1,929	376

Note: Number of spat per 1 kg of twigs and leaves.

(Sasaki, 1970). It is clear from the Table that the number of spat when onion-shaped bags are used is almost double the number of spat when such bags are not used. The effect of using these bags is most striking when they are attached to the collectors near the upper layers of water where the spat is easily affected by strong waves. The use of nets to prevent the detachment of spat should play a large role in increasing the efficiency of natural seedling collecting.

At Mutsu Bay, an outbreak of *Asterias amurensi* L. is found to occur between June and July, which causes much damage to the spat. *Mytilus edulis* L. start setting from early summer and cause further damage, particularly to the byssi. Some simple and suitable measures are needed to prevent these damages.

(Ito)

4. Culture and management

4.1 Interim culture of spat

When the spat of Mutsu Bay attain a shell length of 8 to 10 mm between mid-July and early August, they detach from the substrata and settle on the sea floor. Kodera (1953), Yamamoto and Kato (1965), Yamamoto and Tsubata (unpublished), and several others, tried to transfer spat which had dropped from the substrata, to fishing grounds for further rearing. For some unknown reasons, mortality in all the experiments was heavy, and to this day no one has succeeded in transferring spat directly to the fishing grounds. Yamamoto (1950) confirmed through physiological experiments that the spat are too weak at this stage to withstand environmental conditions such as silt, low oxygen content, high temperature, etc. It seems that spat cannot survive in places other than those where the environmental conditions in the summer, particularly on the sea floor, are exceptionally good.

For this reason spat should be transferred to the fishing grounds for interim culture until they become healthy. One procedure tried by Yamamoto and Sato (1952),

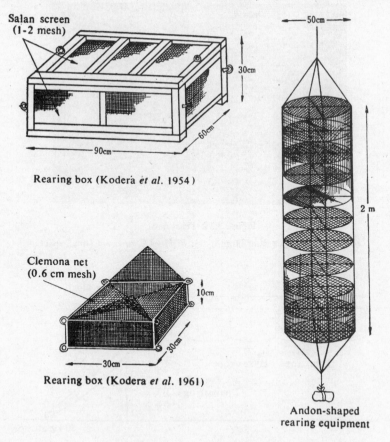

Salan screen
(1-2 mesh)

30cm

90cm

60cm

Rearing box (Kodera *et al.* 1954)

Clemona net
(0.6 cm mesh)

10cm

30cm

30cm

Rearing box (Kodera *et al.* 1961)

50cm

2 m

Andon-shaped
rearing equipment

Figure 3.32. Rearing box and rearing baskets.

Nozawa (1952, 1953), and Takeda (1966) was to rear the spat in tanks, but the survival rate was not good. On the other hand, Kodera (1954) reared spat in boxes hung vertically in sea water until December when the shell length reached 3 cm. The survival rate of these spat when transferred to the fishing grounds was more than 50%, and thus the foundation for interim culture as practiced today was laid.

At Lake Saroma, spat start detaching from the substrata from September onwards. According to Tanaka *et al.* (1958), even when spat are transferred to the fishing grounds they do not die, but interim culture is necessary because transportation involves long-distance travel.

4.1.1 EQUIPMENT AND ITS MODIFICATIONS

Initially, a rearing box spread with Salan net (Figure 3.32), devised by Kodera *et al.* (1954), was hung near the surface of the sea water. Eventually the box was replaced with rearing baskets spread with Clemona which, in 1959, were replaced with the pearl nets in use at the present time (Figure 3.33). Strings of pearl nets, consisting of 5 to 7 nets, are hung at intervals of 1 to 1.5 m, and the entire conglomerate tied to

Figure 3.33. Pearl nets.
Left—6-mesh net (used for large spat); *Right*—1.5 mesh net (small spat).

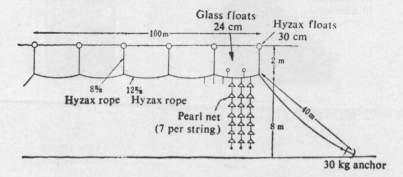

Figure 3.34. Interim culture equipment in Mutsu Bay.

ropes at a depth of 10 to 20 m in coastal regions. The number of spat per pearl net is initially 300 to 500 (August), which gradually decreases to approximately 70 to 150 in December.

The rearing of spat is carried out in Lake Saroma by using either a rearing basket of hyzex (Figure 3.35) or rearing nets of Clemona, which are hung at a relatively

Figure 3.35. Rearing basket (Kodera).

shallow depth (Ohara and Shinobu) (Figure 3.36). In addition to the hyzex basket, in Funka Bay the andon-shaped* rearing equipment shown in Figure 3.32 is used for rearing spat and hyzex plates are used for collecting the seedlings for interim culture (Ishihara, 1964).

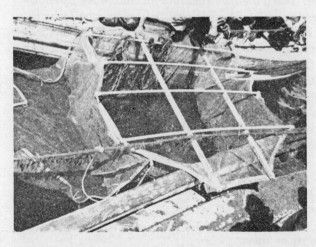

Figure 3.36. Rearing nets (Kodera).

4.1.2 NUMBER OF SPAT PER UNIT AREA AND SURVIVAL RATE

The survival rates of spat during interim culture vary per unit area in various re-

---

* An Andon is a particular style of Japanese lantern, i.e., an accordian-like cylinder-Translator.

gions as shown in Table 3.18, 3.19, and 3.20. When the number collected in each basket is low, the spat grow to 3.5 to 4.0 cm in length by December, and their survival rate is as high as 85 to 90%. But when the number collected is high, the growth of spat during interim culture is poor, and the survival rate is somewhat low.

Table 3.18. Growth of scallop spat during interim culture using various equipment (Mutsu Bay) (Kodera et al., 1961)

Type of equipment	Size of equipment (cm)	Number of spat involved	Survival rate (%)	Average weight (dry) (g)	Average shell length (cm)
Rearing box	Width Length Height 90 ×60× 30	3,000	57.8	0.70±0.27	2.06±0.32
		5,000	73.6	0.47±0.23	1.87±0.38
		7,000	51.6	0.41±0.30	1.69±0.51
Rearing basket	Width Length Height 30 ×30× 10	400	61.5	0.88±0.25	2.40±0.26
		400	94.3	0.99±0.38	2.44±0.35
		500	79.8	0.72±0.34	2.21±0.40
		500	86.0	0.86±0.36	2.38±0.37
		700	75.4	0.64±0.30	2.10±0.28
		700	88.4	0.61±0.25	2.04±0.33
		1,000	79.7	0.80±0.24	2.23±0.23
		1,000	82.9	0.58±0.29	2.00±0.39

Note: Rearing started on August 4, 1959; evaluation carried out on November 19, 1959.

The number of spat per collector suited for interim culture depends upon the type of rearing culture to be used later. When the spat after interim culture are reared by hanging culture, those which are large already grow well, a phenomenon to be explained later, and the period of hanging culture is shortened. Hence it appears that during the interim culture of spat which are to be used for hanging culture, the number of spat collectors should be low (less than 100 per pearl net), and the spat grown, as far as possible, to a large size. On the other hand, when the interim culture is followed up by releasing the spat in fishing grounds, a larger number of smaller spat should be used, provided these are healthy. This method is not only economical, but also reduces the labor involved. According to Kodera (1961), spat of average size, 2 cm or more in shell length, could be used, but Iwagishi (1964) recommends spat which are 1.7 to 2.1 cm in shell length. Akahoshi et al. (1970) released spat of about 2 cm in shell length in early November, and found that the survival rate was in no way inferior to those spat released in December, despite the fact that the period of interim culture for the former was shortened.

4.1.3 WATER DEPTH AND GROWTH OF SPAT

Usually the hanging depth of spat at the time of interim culture is 2 to 10 m. Nishikawa (unpublished) recently carried out a survey in Maine Bay, which established that growth is poor when the hanging depth is very shallow, and growth improves with a depth of a few meters (Table 3.21).

Table 3.19. Growth of scallop spat during interim culture using various equipment (Funka Bay) (Ishihara, 1964)

Types of rearing equipment	Size of equipment (cm)	Number of spat involved	Name of rearing places	Date of investigation	Average shell length (mm)	Average weight (g)	Remarks
Net-spread box	Width Length Height 30 ×30× 15	100	Tomiura Machi, Daigan	November 5, 1964	27	1.7	Uneven growth
		80	Tomiura Machi, Daigan	November 10, 1964	28	3.3	Uneven growth
		200-150	Date-Machi, Usu (mouth of Usu Bay)	November 16, 1964	35	4.5	Spat collected in Hyzex plate were used
		300	Date-Machi, Usu, Noyano-Ma (A)	November, 16 1964	29	2.0	Uneven growth
Andon-shaped rearing equipment	Diameter Height 50 × 20	300	Date-Machi Jiyukyu, Noyano-Ma (B)	November 16, 1964	29	2.1	Uneven growth
		400	Date-Machi Usu (mouth of Usu Bay)	November, 16 1964	31	3.0	Uneven growth
		100	Date-Machi Higashi-hama Jisaki	November 10, 1964	30	3.1	Uneven growth
		100-80	Date-Machi Higashi-hama Jisaki	November 10, 1964	33	4.9	Average
Hyzex plate	Diameter Height 30 × 5	30	Date-Machi Higashi-hama Jisaki	November 10. 1964	29	4.0	Average

Table 3.20. Relationship between the number of spat per unit area and growth (Nishikawa, unpublished)

Spat: artificial seedling. *Place of interim culture:* Maine Bay (depth 7 m) off Karakuwa Machi, Motoyoshi-gun, Miyagi Prefecture. *Rearing method:* interim culture in pearl net (40 × 40 cm, 1 cm mesh) at various densities.

Number of spat per pearl net	Shell length (mm)				Growth of shell during the period of experiment
	October 30, 1965	November 25, 1965	December 11, 1965	January 13, 1966	
5	23.54±2.30	31.34±2.90	35.36±1.00	41.80±4.00	18.26
10	23.25±2.21	30.15±2.05	33.24±1.96	37.68±2.21	15.43
20	23.71±1.80	28.90±2.23	31.31±2.25	33.86±2.21	10.15
30	23.38±1.90	27.72±1.95	29.39±2.70	32.22±2.48	8.84
40	23.44±2.72	26.12±2.34	28.24±2.52	30.54±2.95	7.10

Table 3.21. Hanging depth and growth of spat (Nishikawa, unpublished)

Spat: artificial seedling. *Place of interim culture:* Maine Bay, Karakuwa Machi, Motoyoshi-gun, Miyagi Prefecture. *Rearing method:* interim culture in pearl net (40 × 40 cm, 1 cm mesh).

Hanging depth (m)	Shell length (mm)				Growth of shell during the period of experiment
	October 30, 1965	November 25, 1965	December 11, 1965	January 13, 1966	
1.2	20.62±1.76	22.63±2.32	22.70±1.86	24.35±2.96	3.73
2.5	23.10±1.47	26.79±2.82	28.85±1.81	30.59±2.97	9.97
4.0	21.30±1.46	26.35±1.94	28.25±2.68	32.60±3.32	11.30
5.2	20.35±2.02	26.01±2.78	27.73±3.04	31.80±1.21	11.55
6.7	20.78±2.00	26.56±2.51	29.29±3.52	33.00±1.14	12.22
8.5	21.76±2.32	27.51±2.01	29.82±2.57	34.30±2.71	12.54

4.1.4 FACTORS CAUSING MASS MORTALITY DURING REARING

It is often found that scallop spat die in pairs locked together by their shells. This type of death is particularly high when the spat are small, about 1.5 cm in shell length. "Pair deaths" are also high when the number of spat per basket is high or when the baskets are taken out of the water too often, or are otherwise disturbed.

In July to August, *Asterias amurensis* L. often appear in large numbers in the baskets used for interim culture, and cause damage to the spat. Since the damage caused by this starfish is very severe, the baskets should be examined every now and then and every *Asterias* eliminated.

(Ito)

4.2 Planting and management

The spat setting on the collectors usually lose their byssi between the end of July to the end of August, when they enter the stage of bottom dwelling. Spat at this stage

(0.6 to 1.0 cm shell length) are either released in suitable places carefully pre-selected, or are reared by interim culture up to 3 cm in shell length and then released in the fishing grounds.

The former method is usually known as natural planting, since the spat are released without artificial rearing.

When a large number of juvenile scallops form as a result of "abnormal fecundity," they often collect in groups in soft mud at a depth of 50 to 60 m, and suffer a poor growth rate after a year. Hence, one measure for coping with "abnormal fecundity," which has proven successful, is to transplant the juvenile scallops (4 to 7 cm shell length) to suitable regions.

At present the transplanting of juvenile scallops is carried out in three stages of growth, in which case the actual ecological conditions and the technique to be applied need not be studied, since the environmental conditions of coastal regions and inner bays are never the same, and scallops show various reactions to environmental changes. While in some cases the groups migrate in search of better places, in others the transplanted groups die, being unable to resist the adverse environmental conditions.

4.2.1 GROUPS OF BENTHIC ORGANISMS AND THE HABITATS OF JUVENILE AND MATURE SCALLOPS

1. *Group variation:* Nishioka and Yamamoto (1943) studied the relationship of scallop distribution to the bottom soil qualities and benthic organisms at 87 selected points in Mutsu Bay. For six months the scallops showed a restricted distribution along the coastal regions at a depth of 6 to 30 m, and a close relationship between their distribution and the bottom soil characteristics was apparent. The scallops preferred sea beds with sand and pebbles mixed with less than 30% mud particles of less than 0.1 mm in diameter. However, the bottom characteristics were not the only factors to control distribution, as a number of other effective conditions existed. Yamamoto classified the benthic organisms collected from 24 points in Mutsu Bay into four groups (see Figure 3.6).

Group I exists near the mouth of bays where the bottom is mainly composed of sand, and consists of *Dentalium makiyamai, Chama reflex, Turcina coreensis,* etc., with the first as the dominant species.

Groups II and III, mainly annelids and the *Dentalium* sp., are found in the water basins at the western and eastern parts respectively of the bay where the bottom consists entirely of compact mud.

Group IV is found along the coast of the bay and consists of *Ophiura sarsii,* annelids, *Echinocardium cordatum,* various types of mollusks, *Styela clava,* etc., and about 0.7% scallops. The bottom characteristics differ in various regions and contain sand, gravel, or mud mixed with small pieces of shell. The water movement also differs in each region (Yamamoto, 1950c; Yamamoto and Habe, 1958, 1959, 1962).

It is very difficult to identify the actual environmental factors which control benthic organisms. In fact, a number of environmental factors are collectively involved in giving rise to such groups. A graph of the chlorine content of the water in the bottom layer, prepared on the basis of the data of Ocean Surveys carried out six times in two years, 1947 and 1948, shows that the distribution of benthic groups is more or less similar to the distribution of the chlorine content. The chlorine concentration gradually decreases from the regions of Group I to the regions of Group IV, i.e. from above 32.60S $^0/_{00}$ to 32.40 to 32.60, 32.00 to 32.40, and less than 32.00. However, one cannot say that the chlorine concentration of sea water plays a major role in controlling group distribution; the mixing of land water and open sea water with the water of the bays, and physical and chemical conditions, seem to play important roles as well (Yamamoto, 1950c).

2. *Suitable habitats judged on the basis of juvenile scallop transplant:* It was mentioned that Group IV of the benthic organisms is located in the place which is the habitat of scallops at that stage of growth. This is due not only to the collection of scallops in this region but to various other factors. In 1948, the temperature of sea water in the middle of April when scallops spawned, showed a rapid increase, and a large number of planktonic larvae and setting spat were observed. The survival rate of the setting spat was high and a large number of juvenile scallops were seen in Futagobana Okiai and Arito-oki in Noheji Bay. The following year, about 100 million one-year-old scallops measuring 4 to 6 cm in shell length had migrated to various points along the coast of Mutsu Bay.

Balanus trigonus were found attached to the shells of some of the scallops from Futagobana Okiai. A competition for food between the scallops and the *Balanus* ensued and the growth of the scallops was very poor. Consequently, a suspicion arose that the growth rate of juvenile scallops growing at the bottom of the deep waters in Futagobana Okiai and Noheji Bay, which were suddenly transplanted to the coastal regions of about 10 to 20 m depth, would be impaired rather than improved. Hence the behavior and fate of these transplanted scallops were studied carefully and, as suspected, growth was arrested—but only temporarily (Figure 3.37). The greater part of the transplanted scallops in the coastal regions resumed normal growth, and their dispersal as well as mortality was appreciably low. It was also noted that a larger part of the *Balanus* sticking to the shells degenerated or died.

Figure 3.37. Rings formed due to transplanting (Yamamoto, 1964). The transplantation was done after one year and a few months of hatching, and the juvenile scallops collected one year after the transplantation (at Noheji).

However, the juvenile scallops transplanted to areas between Tairadata and Mushida either dispersed or degenerated, and efforts to prepare a scallop fishing region in Okiai off their capes completely failed. Thus, the chances of juvenile scallops transplanted to the regions of Group I settling to the bottom and resuming a normal life did not seem encouraging. On the other hand, the juvenile scallops settled in the regions of Group IV. In other words, one can assert that the regions of Group IV are growing beds for juvenile scallops (Yamamoto and Ewatari, 1951; Yamamoto, 1951b).

In view of the foregoing, a detailed survey of the regions to which juvenile scallops are to be transplanted appears mandatory. While it is safe enough to transplant juvenile scallops to regions in which scallops are one of the biological organisms present, several factors such as the affinity of the scallop to the biological group in the region, the environmental conditions of the region, etc., should be carefully studied when transplanting to a region where the scallop will be a new member in the animal community. In such a study it might seem that the endurance range of the transplanted animal should form an index, but with scallops this index alone is not enough; the affinity of the animal to the biological group present in the region seems to be a very important factor.

4.2.2 RE-CLASSIFICATION OF THE REGIONS OF GROUP IV, AND THE SUITABLE HABITATS OF SPAT

It is often reported that the survival rate of spat soon after their entering a bottom-dwelling life is very low. At times, however, the survival rate is moderate, or comparatively high, resulting in the so-called "abnormal fecundity" phenomenon. Arito-oki off Noheji Bay and Horogan in Lake Saroma of Hokkaido are regions in which abnormal fecundity occurs, and the characteristics of their water basins, using the benthic organisms as an index, was decided upon as a place for studying this pheno-

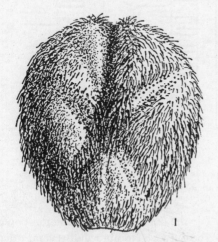 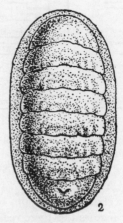

Figure 3.38. Animals constituting the index for the survival rate of bottom-dwelling spat of scallops.

1—*Echinocardium cordatum*; 2—*Lepidopleurus assimilis*.

menon. The floor of the sea in Arito-oki is mainly composed of gravel mixed with a little mud, the depth is a little over 20 m, and the tidal current is favorable. Regarding the biological organisms present in this region, besides the dominant species already referred to in connection with Group IV, there are large numbers of *Lepidopleurus assimilis, Gammarus* sp., *Echinocardium cordatum, Pectinaria hyperborea,* and annelids. However, the first two varieties show a relatively low frequency of appearance (Figure 3.38).

Kawauchi Hinokigawa-oki off Mutsu Bay also has the same conditions as those of Arito-oki, and bottom-dwelling spat are found every year. Hamaokunai-oki on the other hand, has the lowest survival rate in Mutsu Bay. In this region, with a water depth of less than 20 m, the floor of the sea is a mixture of soft mud and shell pieces, and a large number of *Echinocardium cordatum* and *Pectinaria hyperborea,* and some *Theora lubrica* and *Asahikinutaregai,* etc., are found.

In order to study the biological index species which are assumed to be closely related to the survival rate of 1 cm spat, Group IV can be further re-classified (Figure 3.39). From this Figure, it seems that the survival rate of scallops is high or low in the water regions in which a large number of *Lepidopleurus assimilis* and *Gammarus* sp. live, and that these areas are sources of scallop supply to neighboring regions.

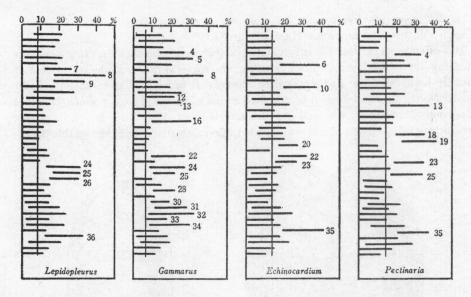

Figure 3.39. Reliability limit (60%) of the appearance of *Lepidopleurus, Gammarus, Echinocardium,* and *Pectinaria* at various points. The survey points have been given in Figure 3.6 (Yamamoto, 1956b).

To better classify the regions of Group IV, the oxidation and reduction potentials of the floor material in these regions were measured. Nomura and several other investigators have studied this point (Nomura and Kagawa, 1949; Nomura, 1952; Nomura *et al.,* 1955), but the ecological conditions were not always clear. The apparatus used for the measurement of oxidation and reduction is shown in Figure 3.40,

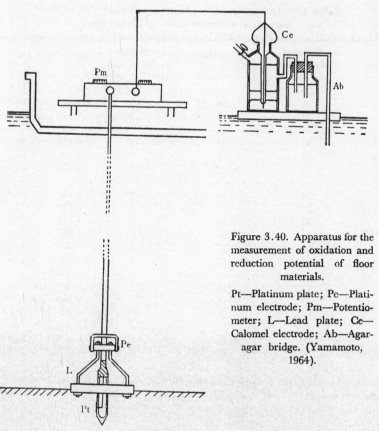

Figure 3.40. Apparatus for the measurement of oxidation and reduction potential of floor materials.

Pt—Platinum plate; Pe—Platinum electrode; Pm—Potentiometer; L—Lead plate; Ce—Calomel electrode; Ab—Agar-agar bridge. (Yamamoto, 1964).

and the data obtained are given in Table 3.22. According to these data, it seems that 20.0 to 22.5 of rH value is suitable for a 1 cm spat. The suitable rH value at point 20 (Hamaokunai) and 38 (Yakohama-oki, Group III) was lower, however, 16.8 to 18.0, although higher than that usually observed in the investigations of floor materials, and the value at Onagawa Bay was lower still, 14 to 15. The rH value suitable to the life of scallops is quite high, particularly for the 1 cm spat, and the range of variation is narrow.

4.2.3 NATURAL PLANTING OF SPAT

Needless to say, the reason for re-classifying the regions of Group IV, is to devise a method by which a 5 to 20% survival rate of spat released to a suitable region could be obtained without doing an interim culture. It should be mentioned here that spat collected are released from the end of July to the end of August. However, the investigations have not progressed very far and, in fact, there have been occasional failures. The success attained in 1952 at Arito-oki in Noheji Bay deserves mention however.

Spat of 1 cm length, which had settled in the Piccard trap nets at Arito-oki, were transferred to such points as were considered suitable for their growth in August. The nets were lowered to the bottom of the sea and a survey of the survival rate done two months later. Although the survival rate was only 18% of the original number,

Table 3.22. Oxidation and reduction potential (rH) of floor materials along the coast of Mutsu Bay

Date of investigation	Points of observation	Water depth	pH of water[1] at the bottom	Eh (mV)	rH[2]	Group
July 11, 1955	32	20 m	7.9	+95	19.0	IV
	36	10	8.2	+291	26.0	,,
	28	20	7.9	+132	20.3	,,
July 12, 1955	27	25	7.9	+138	20.5	IV
	24	18	7.9	+136	20.4	,,
	25	22	7.8	+177	21.6	IV—1
July 13, 1955	24	18	7.9	+132	20.4	IV—1
	25	22	7.8	+122	19.7	,,
	37	20	7.8	+113	19.3	IV
	35	35	7.8	+93	18.6	,,
	38	45	7.8	+35	16.8	III
	20	17	7.8	+72	18.0	IV
July 14, 1955	7	16	7.9	+84	18.6	IV
	8	18.5	7.9	+198	22.4	IV—1
	8′	23.5	7.8	+136	20.1	,,
	8″	38	7.8	+160	21.0	,,
	16	10	7.9	+ca 435[3]	30.3	IV
	15	10	7.9	+ca 180[3]	21.9	,,

[1]Values taken by measuring the water sample after filtering floor materials.
[2]rH value corresponding to pH at various places.
[3]Value measured before complete stabilization (because of sunset).

it was not clear what percentage of those lost had died, and what percentage had been carried away by the currents. Although it looks small, the survival rate of 18% at Arito-oki is considered fairly high. Later experiments at Yokohama-oki, Tsuchiya (Nishihama), etc., were not successful, and the reasons for these failures could not be pinpointed. Unless the number of spat used is very large, it is difficult to confirm the results. Also, since harmful organisms such as flatfish, *Asteria amurensis*, etc., were present, these must be taken into consideration. Recently (1966), spat were released just before entering their bottom-dwelling life in the Horogan region of Lake Saroma— an area which seems to correspond to Mutsu Bay where *Lepidopleurus*, *Gammarus*, etc., are the dominant species—and allowed to winter there before being transferred to the open sea. These spat had been subjected to interim culture before their release in Horogan, and 18,000,000 to 19,000,000 were transplanted to Okhotsk in May to June of the following year (1967). The survival rate was examined in the summer of 1968 and found to be an astounding 76% (Okezukuri and Tanaka, 1969).

4.2.4 TRANSPLANTATION AND MANAGEMENT

The mode of transplantation of scallops in Hokkaido differs from that in Mutsu Bay. In Hokkaido, 1 cm spat transplanted to various regions for some years have reportedly done very well (Kinoshita, 1949). Transplantation of 1 cm spat from

Lake Saroma to Mutsu Bay, first done in 1943, was not successful according to Yamamoto, who carried out a detailed investigation on the mode of transplantation, and recommended that the latter be improved (1950). Nevertheless, "abnormal fecundity" has been observed at Mutsu Bay at intervals of ten or more years and, as a result, 100 to 1,000 million one-year-old spat seen. Such unusual larval explosions pose problems for transplantation while making transplantation imperative. A large number of 3 cm spat can be obtained, some of which are utilized for hanging culture, but the majority are released to the bottom of the sea.

As mentioned earlier, scallops are transplanted at three stages of growth, i.e. 1 cm spat, 3 cm after interim culture, and 5 cm (one-year-old) resulting from "abnormal fecundity." Naturally, the management of transplantation for these stages differs.

When transporting 1 and 3 cm spat, a rise in temperature and desiccation should be avoided. Regarding temperature, Kinoshita (1937) used 1 cm spat from Hokkaido and reported that there was no ill effect during transportation in August, which took almost 30 hours, as the temperature inside the boxes was maintained at 6 to 7°C with ice. According to Tanaka, Kohara, and Wakui (1962), the mortality rate would be high after 15 hours of transportation if the temperature were about 20°C. Regarding desiccation, it has long been known that spat have a weak resistance to abnormal conditions. To avoid desiccation therefore, the spat must be covered with seaweeds or eelgrass before transportation, a precaution which ought to be observed when transporting juvenile and adult scallops also.

As said before, the suitability of the region of transplantation is also an important factor. At Mutsu Bay, the area is restricted to the regions of Group IV. The special characteristics of the transplanting region for 1 cm spat still need to be studied, and an effective method for removing the enemies of scallops such as starfishes, etc., devised. On the other hand, a problem of density arises in the transplanting of one-year-old spat. According to the experience of the authors, a maximum of 5 to 6 scallops, and a minimum of 1 to 2 scallops, should be allowed for 1 m². Growth is retarded when density exceeds 5 to 6 scallops per square meter, and overcrowding may also lead to diseases resulting in abnormal mortality. Although the actual nature of these diseases is not always clear, abnormal mortality does occur due to adverse environmental conditions, and in Canada *Placopecten* were seen which had been infected with *Trichodina* (Medcof and Bourne, 1964).

A detailed discussion on the various problems of managing the sources of scallops will be given later. However, an important point should be mentioned here, namely, that when 3 cm spat are transplanted to regions of Group IV, the harvest expectation is 50%, but with one-year-old spat, estimated at 80 to 90%. Hence a planned production on the basis of the transport pattern can be made on the basis of the yield rate. In fact, the need for planned production should be strongly stressed, and this point will be discussed under the section on "Management of Sources."

(Yamamoto)

4.3 Hanging culture

Hanging culture of scallops was first started by Tanida and Sato at Asamushi-Jisaki in Mutsu Bay in 1956, and the scallops hung by boring holes in the ear part of the shells. The results of this experiment are not conclusive since the time chosen was not ideal and the collector shells were carried away by currents (Tanida and Shinobu). In 1958, Shibui and Yamamoto tried the hanging culture method at Yamada Bay of the Iwate Prefecture. They used oyster culturing rafts and tried the ear-hole method. The growth and survival rate of the scallops was exceptionally good and in no way inferior to that of oyster culture (Shibui and Shinobu).

A lull in hanging culture followed the 1958 experiment, but the method was actively adopted in 1963 with a view to industrialization, and cultures carried out at Iwate, Aomori, and Miyagi Prefectures, as well as Lake Saroma, Funka Bay, etc. Since then hanging culture has been done with various types of nets—pearl net, open-close net, pocket net, andon baskets, etc.—and the merits and defects of each type compared. The hanging culture will very soon be used on a commercial scale.

The suitability of hanging culture in various regions from Tohoku to Kanto, Chubu, and Hokuriku is currently under study through pilot experiments.

Hanging culture for scallops is a comparatively recent introduction, and hence requires much improvement both with regard to implements and technique. A series of investigations should be done to establish the culture technique specifically suited to each environment.

4.3.1 CULTURE EQUIPMENT AND CULTURE METHOD

1. *Method:* The ear-hole technique, devised in 1956, was described earlier. After boring the hole, either with a needle or an electric drill, nylon tags of about 1 mm thickness (No. 8 to No. 12) are passed through it and, as shown in Figures 3.41 and 3.42, the shells attached to straw rope or synthetic fiber rope painted with tar, and the

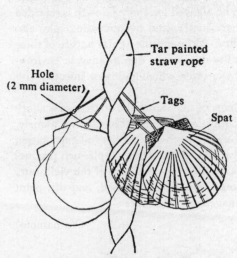

Figure 3.41. Ear-hole technique
(Hatayama unpublished).

Figure 3.42. Scallops in the course of hanging culture by ear-hole technique
(Maine Jisaki, Kesen-numa Bay).

ropes hung in water. Usually, 4 to 5 shells are kept in one place at intervals of 25 to 30 cm. One of the advantages of this method is the low cost. However, there are also disadvantages such as the labor involved in the preparation, poor growth, and a low survival rate during rough waves.

Not surprisingly, the basket method has become more popular since this operation is easier, even though slightly more costly, than the ear-hole technique.

Pearl nets and open-close nets were the first to be tried. While pearl nets have a limited capacity for shells, the open-close nets involve much labor in plucking and fixing the scallops. Hence these two nets have almost disappeared.

Pocket nets are now widely used in the coastal regions of the Iwate Prefecture and Lake Saroma (Figures 3.43 and 3.45). Hyzex net of 2 to 3 cm mesh is stitched into several pockets and attached to the side of the main hyzex net. The scallops are placed in the pockets, the pocket openings tied with nylon yarn, and the net hung in the water. The fixing and removal of the scallops in this net are easy, and changes in shell shape during culture are few. Moreover, the growth and survival rate of the

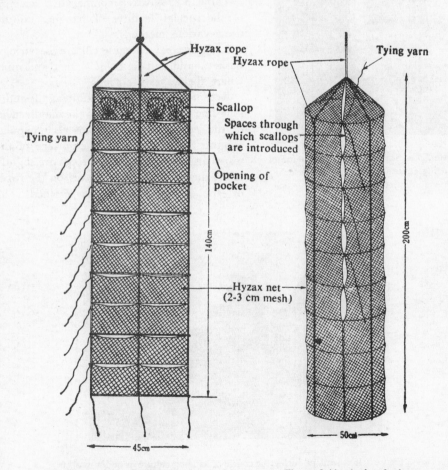

Figure 3.43. Pocket net.   Figure 3.44. Andon basket.

scallops is good even in places with rough waves. The expense is however slightly higher.

At Mutsu Bay, andon baskets are mainly used (Figures 3.44 and 3.46). The baskets are round in shape and made of hyzex net of 2 cm mesh with 10 slits spaced 20 cm apart, through which the scallops are placed inside, and the baskets then hung in the water by a nylon rope. It is very easy to introduce or collect the scallops by this method, and they rarely show any change in their shape during the course of culture. Moreover, other organisms find it difficult to settle on the scallops inside the baskets. But though this method has all these advantages, the growth and survival rate of the scallops in places with rough waves are not as good as when pocket nets are used.

Table 3.23 gives a comparative account of the special features of hanging culture using various methods.

2. *Equipment:* Hanging culture uses strings, baskets, and rafts in calm waters, and ropes where the waves are rough.

Recently in Mutsu Bay, dipping the entire long-line implement down to the middle layer of the sea water has become very popular (Figure 3.47). Floats on the surface of the water are few and, since the ropes are slightly loose, the effect of rough waves on the ropes or the hanging baskets is lessened. The

Figure 3.45. Scallops during the course of hanging culture in pocket nets (Otsuchi Bay) (Sasaki, 1966).

Figure 3.46. Scallops during the course of hanging culture in andon baskets (Okunai Jisaki, Mutsu Bay) (Kofuji).

Table 3.23. Special features of hanging culture methods for scallops (Ito, unpublished)

Method	Material	Specifications	Unit cost of material	No. of scallops	Cost of material per scallop	Durability	Special features
Ear-hole	Nylon tags, coal-tar painted ropes	Length of chain, 8 m	About 60 yen	About 96	About 0.63 yen	About 1 year	Involves much labor. Culture becomes difficult in places with rough waves. Ear part changes shape.
	Pearl net	45 cm × 45 cm	47 yen	4	11.8 yen	2-3 years	Capacity for holding scallops small.
Hanging culture using baskets	5-stage open-close net	45 cm × 70 cm	115 yen	20	5.75 yen	2-3 years	Involves much labor in introducing and plucking scallops. Shells often change in shape.
	10-stage pocket net	45 cm × 140 cm	280 yen	40	7.00 yen	2-3 years	Involves some labor in introducing and plucking scallops. Shells sometimes change in shape.
	10-stage andon basket	50 cm × 200 cm	680 yen	100	6.80 yen	2-3 years	Other organisms do not easily settle on shells. Growth of scallops inferior in places with very rough waves.

growth of scallops and the survival rate are also better since shaking is minimized.

4.3.2 GROWTH OF HANGING SCALLOPS

1. *Effect of hanging method:* The results of rearing scallops by the ear-hole method and pearl nets at Maine Jisaki and Kesen-numa Bay (Hatayama, unpublished), are given in Figure 3.48. It is clear that in places like Maine Jisaki where the water is calm, the growth of scallops by the ear-hole method is good.

On the other hand, Sasaki (1967) reared scallops by the ear-hole method and also in pearl and pocket nets in Otsuchi Bay where the waves are very rough. The results given in Table 3.24 show that pocket nets gave much better results than the ear-hole method.

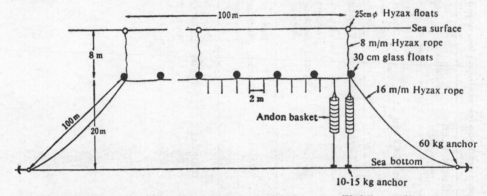

Figure 3.47. Long-line culture in middle layer of the sea water (Kofuji, unpublished).

Akahoshi *et al.* (1968) compared the results of rearing scallops by the ear-hole method with those obtained by using open-close nets at Mutsu Bay where the waves are also rough. The ear-hole method failed as can be seen in Figure 3.49.

Regarding the merits and defects of pocket nets and andon baskets, Akahoshi

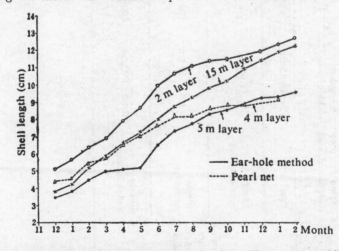

Figure 3.48. Depth of hanging and growth of scallops (Hatayama, unpublished).

Table 3.24. Growth of scallops by different hanging methods (Otsuchi Bay)
(Sasaki, 1967)

Hanging method	Mean weight of scallops (g)				Increase in weight in 3 months (g)
	May	June	July	August	
Ear-hole	20.1	35.0	42.5	63.4	43.3
Pearl net	26.0	39.5	55.1	72.5	46.4
Pocket net	26.9	39.2	61.6	79.3	52.4

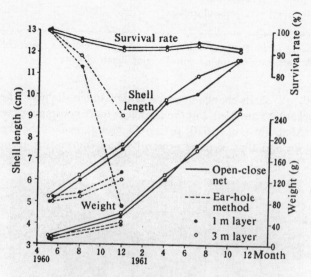

Figure 3.49. Growth and survival rate of scallops by various
hanging culture methods (Akahoshi *et al.*, 1970).

et al. (1970) are now making several studies. In places with slightly rough waves, the results of rearing do not differ much, but in places with very rough waves, the survival rate of pocket nets seems to be slightly better.

Ito *et al.* (1970) compared the results of rearing scallops by the ordinary long-line method with those obtained by the modified long-line method, in which the ropes are fixed at the middle layer of the sea so that the baskets and the scallops are not affected by waves (Table 3.25). The growth with the modified method was much better than with the ordinary.

In view of the above results, it is clear that a hanging culture method suited to one region is not suited to all. In other words, the best method can be selected after considering the environmental conditions of the region, particularly the condition of the waves.

Other points which are clear from these results are (1) the growth of the ear-part of the shell is affected on the side through which the tag passes in the ear-hole method

Table 3.25. Growth of scallops in ordinary and modified long-line methods (Mutsu Bay) (Ito *et al.*, 1970)

Long-line method type	Age of scallop	Increase in shell length in 4 months (cm)	Increase in weight in 4 months (g)
Ordinary	Scallops of current year	0.64	1.80
	2-year-old scallops	0.43	19.30
Modified (middle layer)	Scallops of current year	1.11	3.52
	2-year-old scallops	0.68	31.30

Note: Scallops of current year from pearl net hanging culture; 2-year-old scallops from pocket net hanging culture with hanging depth of 8 m.

(see Figure 3.50); (2) shells change in shape as they touch the nets when either open-close nets or pocket nets are used, but the change is more conspicuous with the former.

Figure 3.51 (Akahoshi *et al.*, 1970) presents a comparison of the shell morphology of natural scallops and that of those grown by hanging culture in open-close nets.

Figure 3.50. Scallops grown by ear-hole method of hanging culture
(Noheji Jisaki, Mutsu Bay) (Akahoshi *et al.*).

The shell width of scallops grown by hanging culture is more than that of naturally grown ones. According to Akahoshi *et al.* (1970), there is no difference in growth between the vertical and horizontal open-close nets.

2. *Influence of hanging depth:* Sasaki (1967) compared the growth of scallops cultured by the hanging method at various depths from 2 m to 11 m in Otsuchi Bay. The results shown in Figure 3.52 indicate that except in 2 m depth (the reasons for the good growth at this level are not clear), growth improves in deeper waters.

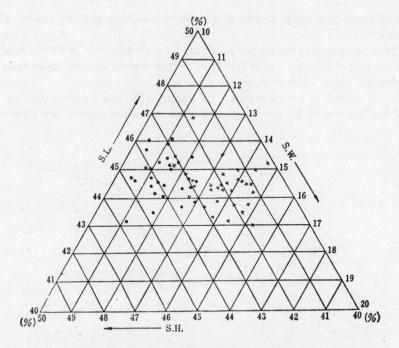

Figure 3.51. Comparison of shell length (S. L.), shell height (S. H.), and shell width (S. W.) in naturally grown scallops and cultured ones.

●—naturally grown; ×—cultured (hanging culture) (Akahoshi *et al.*, 1970).

A comparison of the growth of scallops at various depths in Maine-Jisaki, Kesen-numa Bay (Hatayama, unpublished), has been seen in Figure 3.48. Growth was

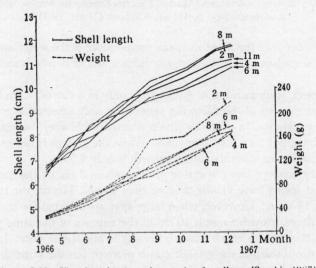

Figure 3.52. Hanging depth and growth of scallops (Sasaki, 1967).

best at 15 m depth (but again growth at 2 m depth was fairly good for reasons which are not clear).

This type of comparative data is not available for scallops grown in Mutsu Bay, but it has been reported that growth is better in deeper waters, and in most cases the hanging depth is about 10 m or more.

3. *Growth of scallops in various regions:* In Figure 3.53, comparative data on the growth of scallops reared by the hanging method in various regions are presented.

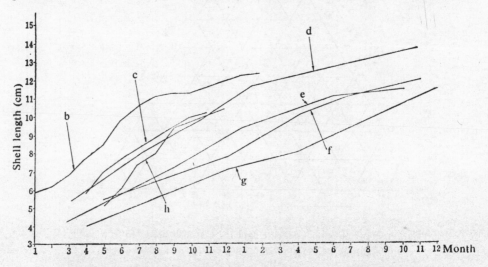

Figure 3.53. Growth of scallops reared by hanging culture in various regions.

b—Maine, Kesen-numa Bay: ear-hole method, 5 m layer (Hatayama, unpublished); c—Obunato Bay: 15 m layer (Akazaki, unpublished); d—Mutsu Bay: andon baskets, selected spat (Kofuji, unpublished); e—Noheji, Mutsu Bay: open-close net (Kanno, unpublished); f—Tsuchiya, Mutsu Bay: open-close net, 3 m layer (Akahoshi *et al.* in press); g—Lake Saroma: pocket net (Abashiri Fisheries Experiment Station, unpublished); h—Otsuchi Bay: pocket net, 6 m layer (Sasaki, 1967).

The relationship between the monthly mean temperature of the water and the monthly growth rate of scallops reared by hanging culture at 3 m depth, is shown in Figure 3.54.

At Lake Saroma, scallops hung in spring grow only to a size of 7 cm by the end of the year, and at the end of the following year have only attained a growth slightly exceeding 10 cm. Hence, in order to obtain a marketable size, the scallops have to be reared (by hanging) for about two years. While the growth is better at high temperature, the maximum monthly rate of growth is only 0.5 cm.

At Mutsu Bay, scallops hung in spring grow to about 8 cm by the end of the year, and by the next spring have grown to more than 10 cm. Hence, the rearing period requires about 1½ years. However, when large spat are selected for transplanting in the hanging culture, growth exceeds 10 cm in the autumn of the same year, and the growth period required to reach marketable size is about one year. In the year in which the culture is started, the growth rate is more or less the same throughout the year, with a monthly mean rate of 0.5 to 0.6 cm, but the following year the growth

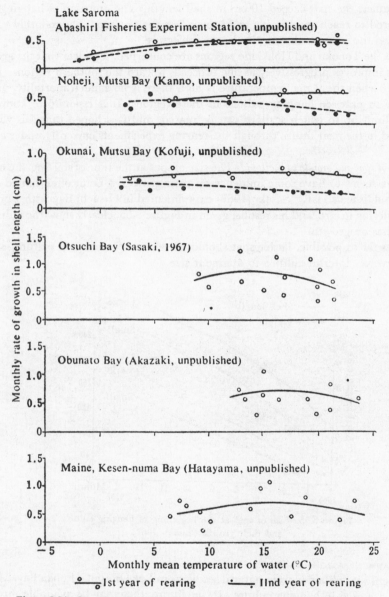

Figure 3.54. Growth rate of scallops and water temperature in various regions.

rate at a high temperature is slightly slower.

At Otsuchi Bay and Obunato Bay, scallops hung in spring grow rapidly, and in the autumn of the same year their shell length exceeds 10 cm. They need to be reared by hanging for just about one year. The growth rate is at its maximum when the water temperature is around 15°C and, at times, the monthly growth rate reaches 1.0 cm. At other times, the rate is maintained at 0.5 cm.

At Kesen-numa Bay, spat used for hanging culture are large in winter. A little

before summer, the spat exceed 10 cm in shell length. The duration of hanging culture required to reach marketable size is even less than a year. The monthly rate is
more or less the same as that in the Iwate Prefecture.

As far as the Tohoku and Hokkaido regions are concerned, it is clear that the growth
of scallops improves progressively as one proceeds toward the southern regions. Even
in summer when the water temperature is high there is no sudden mortality, and it
is possible to practice hanging culture in these regions. It is essential to know the
southern limit up to which scallops can be reared, and it is hoped that this will be
established in the near future through the rearing experiments currently under consideration.

4. *Size of spat and growth of scallops:* The size of spat at the time of starting the hanging culture seems to have some influence on their growth, a point often hinted at in
the data published so far. Sasaki (1967) experimented in Otsuchi Bay with 2 groups
of spat differing in size, and his results, given in Figure 3.55, clearly show the influence
that size has on growth.

Thus, as far as possible, larger spat should be selected for hanging culture, which
usually involves interim culture to augment size.

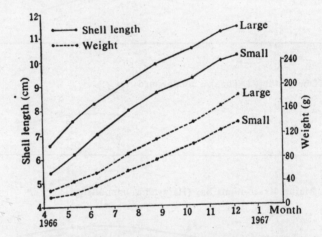

Figure 3.55. Size of spat at the beginning of hanging culture
and their growth (Sasaki, 1967).

4.3.3 TRANSPORTATION OF SPAT

At present spat obtained from natural seedlings in Lake Saroma, Funka Bay, Mutsu
Bay, etc., are used in hanging culture. In the future, they may be available through
artificial seedlings procured from the Oyster Research Center and other sources. In
any case, spat usually have to travel a considerable distance to the rearing places.

According to the experiments of Tamura (1956), it is difficult during transportation to prevent mass mortality among small sized spat of late August to early September, even though sufficient moisture is provided, if the temperature outside is about
20°C and the transportation time more than 15 hours. When spat are priorly subjected to interim culture, the mortality rate can be reduced to about 10%, even after
20 hours of travel, by providing sufficient moisture, if the outside temperature is

10 to 15°C. When the temperature is low, say 6 to 8°C, mortality after 50 hours of travel, can be held to 30%.

Since scallop spat cannot withstand long exposure to air of high temperature, they are usually subjected to interim culture at the place of seeding, and then transported from the end of October to December-March.

As shown in Figure 3.56 (A), spat should be packed in boxes and then transported as early as possible by truck or freight train. The Akazaki Fisheries Association of the

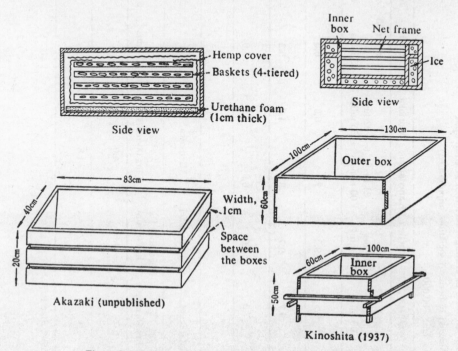

Figure 3.56. Boxes used in the transportation of scallop spat.

Iwate Prefecture uses special boxes for transporting the spat at the end of November from Funka Bay near Abuta and Mutsu Bay near Okunai, and mortality during transportation is less than 5% (Table 3.26). In winter, it is also possible to transport spat for short distances in fish boxes covered with wet hemp.

Kinoshita (1937) reported that it was possible to maintain the temperature inside the boxes at 6 to 7°C during transportation, by packing ice around the 0.5 to 0.7 cm spat at the end of August, and that by doing so transportation time could extend to almost 30 hours [Figure 3.56 (B)].

4.3.4 Care During Hanging Culture

When hanging culture is continued for some time, various types of organisms settle on the growing scallops and on the baskets, such as mussels, barnacles, serpula, hydrozoa, diatoms, marine algae, etc. Of these, barnacles and mussels are the most difficult to eliminate.

The influence of these organisms on the growth and survival rate of scallops has

Table 3.26. Results of transportation of scallop spat

(Akazaki Fisheries Association, unpublished)

a) Abuta→Akazaki, 2-ton truck, spat 2-3 cm, 700,000

	Abuta	Hakodate	Oma	Ohata	Noheji	Kanedaitsu	Morioka	Otomo	Akazaki	Time taken
Time	November 24 8 hrs 40 mins	12 hrs 10 mins	14 hrs 40 mins	17 hrs 00 mins	19 hrs 00 mins	21 hrs 00 mins	22 hrs 55 mins	November 25 1 hrs 05 mins	2 hrs 40 mins	18 hrs
Middle layers on truck left		8.0 (°C)	10.0	9.5	9.5	7.6	6.3	6.0	5.3	
Middle layers on truck right		8.4 (°C)	9.8	9.7	9.7	8.2	4.8	5.3	5.0	
Outside temperature	6.2 (°C)	10.0	10.2	7.0	7.0	6.0	6.5	3.0	3.5	

b) Okunai→Akazaki

	Okunai	Noheji	Kitafukuoka	Senboku machi	Mizusawa	Setamai	Akazaki	Time taken
Time	November 30 19 hrs 00 mins	20 hrs 30 mins	22 hrs 45 mins	December 1 1 hr 00 mins	3 hrs 00 mins	5 hrs 00 mins	6 hrs 00 mins	11 hrs
Temperature inside truck	5.8 (°C)	6.7	6.0	4.8	4.0	3.4	3.2	
Outside temperature	5.6 (°C)	6.2	4.8	4.2	1.8	2.2	3.8	

Survival rate after transportation was more than 95%.

not yet been detailed, but it is known that growth is greatly affected when such organisms are present in large numbers. It is also commonly known that the sale value of scallops decreases if these organisms are present in even small numbers.

Since scallops are not resistant to exposure to air, high temperature, and fresh water, these organisms cannot be eliminated by exposure, hot-water dipping, rain-water immersion, etc., as is done with oysters. Instead, such organisms are usually scraped off by hand with a wire brush, or mechanically cleansed with the blunt edge of small knives. Nowadays, an electric rotating brush is also employed. Nevertheless, the operation is difficult in those places where the organisms are abundant.

Care must be taken in the removal operation to complete it within a short time as otherwise the scallops become weak through exposure to air and direct sunlight. Also, if the new shell happens to be injured during cleaning, growth is very much affected thereafter. Thus, the removal operation must be done quickly and the temperature not allowed to rise during it. Just what length of time the removal operation can safely take, and how often it can be safely done, can only be determined when fuller details are available.

One way to prevent the settling of these harmful organisms is to paint the scallops with chemicals. But since the application of such chemicals is difficult, and their subsequent cumulative effect on the scallop is known to be adverse, this method is still very much in an experimental stage. Another preventive measure, since most such harmful organisms settle in the summer, might be to culture scallops in places where these organisms are less in number. Mention should also be made about increasing the use of andon baskets, since it is known that the number of such organisms is comparatively lower when this method of culturing is employed.

(Ito)

5. Pests and prevention

Pests which cause damage to scallops consist of those which merely bore through the shell, and predators which live on the scallop inside.

5.1 Boring pests

Scallops are often found whose shells have been damaged by a number of holes along the line of growth, in the fishing grounds of Hokkaido. These holes are usually made by the annelid, *Polydora ciliata*. Table 3.27 shows the percentage of bored scallops, and the various degrees of boring; according to this Table more than 60% of the scallops on the Okhotsk coast are affected (Kinoshita *et al.*, 1959), particularly old scallops living in the muddy bottom. Usually the cavities made by *Polydora* just touch the surface of the pearl layer of the shell, but sometimes the cavities go deeper and even reach the inner surface of the shell (Kinoshita, 1951). Although scallops do not die because of these holes, the shells become brittle and, when the muscular scar is particularly affected, the muscles do not develop well because of the continuous irritation. As a result, the commercial value of the animal greatly decreases (Kinoshita, 1951).

Polydora ciliata has caused damage to scallop production in Mutsu Bay for many

Table 3.27. Damage caused by *Polydora* **in scallops of the Okhotsk coast**
(Kinoshita *et al.,* **1959)**

Fishing ground	Water depth (m)	Number of specimens examined	Condition of boring (%)				
			None	Some	Considerable	Severe	Shell damaged
Omuke	30.3	1,890	17.1	53.1	18.7	9.8	1.3
Benten-shima	31.8	857	28.8	36.2	24.5	10.0	0.2
Ipponmatsu	30.3	675	5.0	28.0	26.4	28.0	12.6
Yubetsu	33.3	325	32.0	31.4	22.2	12.3	3.0
Shoryu	—	326	38.0	31.3	20.9	9.8	0

years, but recently the damage has become a major problem (Kanno, 1970[2]; Takahashi and Akahoshi, 1970[3]). Usually, when *Polydora* bore a well-grown scallop, the damage is not very severe even when the number of *Polydora* is large. But, if *Polydora* settle on small scallops soon after their release, the margin of the shells are eaten away and high mortality results. Takahashi and Akahoshi (1970) have even reported instances of an entire population perishing as a result of this pestilence.

To prevent the damage caused by this pest, the ecology of *Polydora* in Mutsu Bay was studied and the following points came to light.

According to the survey done by Takahashi (1970), the life history of *Polydora* involves setting during the winter from January to March, followed by a gradual multiplication of the organism in October to November of the same year. When the water temperature is around 15°C, *Polydora* spawn for the first time; in the following year the ova form at the end of the spring and the second spawning occurs in autumn. Since a small number of *Polydora* are found to bore scallop shells in July to August, they seem to spawn to some extent even in the beginning of summer. A study of the *Polydora ciliata* which settled on *Akoyagai* of Toribane Bay of the Mie Prefecture, revealed two infestations, namely, a large group which settled in July to August, and a smaller group which settled in November to December. The first group derived from the spawning in June, and the second group from the spawning in October to November (Mizumoto, 1966[4]). Eggs are laid in a capsule inside shell cavities, embryonic development occurs inside the egg capsule, the larvae (with four segments) hatch and, after living a planktonic life for some time (attaining 20 segments), settle on the scallops. In Mutsu Bay, the larvae settle sometime between January and March. Initially they live in shallow cavities on the surface of the shells, then in mud ducts formed on the shells, and finally, enter the cavities made in the shells.

[2]Kanno, 1970. Survey of the Damage of Scallops Caused by the Outbreak of *Polydora* on Karibasawa-Jisaki, *Bulletin of Mutsu Bay Fisheries Culture Research Center,* Aomori Prefecture, No. 11: 309-324.

[3]Takahashi and Akahoshi, 1970. Survey of the Damage to Transplanted Scallop Spat Caused by *Polydora* in Kominato-Jisaki. Data presented by the Aquaculture Center of Aomori Prefecture, S. 45, No. 4: 1-24 (Photocopy).

[4]Mizumoto, S., 1966. Study of the Damage to *Akoyagai Shells:* II. Seasonal Variations in Polydora ciliata (Johnston) found in *Akoyogai, Bulletin of National Pearl Research Center,* 11: 1368-1377.

The relationship between the floor material and the rate of setting shows that *Polydora* prefer muddy bottoms, or a mixture of sand and mud, to gravel. Takahashi *et al.* (unpublished) studied the relationship between the composition of bottom material and the setting rate of *Polydora* in the fishing grounds of Mutsu Bay. The results showed that the setting rate of *Polydora* is very high in places where fine mud of 125 μ, or less, makes up more than 10% of the bottom material. When the percentage of fine mud decreases, the setting rate of *Polydora* also shows rapid decrease. The mud ducts made by *Polydora* are reported to contain 66 to 76% fine mud particles of 125 μ, or less (Takahashi, unpublished). The young *Polydora* which settle on the shells evidently require such fine particles of mud in the making of their mud ducts.

There are various countermeasures to prevent the damage caused by *Polydora*, but it seems that the method which would best prevent its outbreak would be shrewd management of the fishing grounds on the basis of the ecological features of the pest. Takahashi *et al.* (1970) have suggested that places with fine mud, or a mixture of sand and mud, should either be avoided, or the release of scallops should be so planned that it does not correspond to the setting period of *Polydora*. It has also been suggested that when scallops are found to be affected by a large number of *Polydora*, they should be collected before the spawning period of the pest, thus checking its multiplication.

(Ito)

5.2 Predators

That predators include several varieties of starfish is undisputable since scallop spat have been found in the alimentary canals of *Asterias rubens, Asteria glacilis, Asterias forbesi, Solaster papposus, Palmipes placenta, Asteropecten irregularis, Luidia ciliaris,* etc. (Hunt, 1925; Galtsoff and Loosanoff, 1939; Hancock, 1955; Tamura *et al.*, 1954).

By examining the contents of the alimentary canals of various types of bottom-dwelling fish and starfish, the food relationship between them and bottom-dwelling organisms was determined and the results are shown in Figure 3.57. The thick lines in the chart represent the primary food and the thin ones the secondary food of the animals. The first and important characteristic feature of the relationship of these animals to food is that there is a severe competition among starfish and bottom-dwelling fish for polychaetes, isopods, amphipods, small prawns, crabs, and small shellfish. The second important feature is that scallops come second in the order of preference as food only for *Asterias*, particularly *Asterias rubens* (Figure 3.58), as barnacles, mussels, mesogastropods, etc., constitute their main food. Oysters and scallops as well as their spat, are rarely eaten by *Asterias rubens* (Hancock, 1955).

A study of the food taken by *Asterias* in natural environment in relation to the exact position of scallops, showed that these do not always constitute the main food of the starfish under natural conditions.

The distribution of scallops and their spat, as well as *Asterias*, with special reference to the concentrated distribution of spat on the Kitami coast and the eastern shores of Shiritoko-Hanto of Hokkaido was observed continuously for some time, and the results are summarized in Figure 3.59 (Tamura *et al.*, 1964). It is clear from the Figure that even in places with a high density of spat, the number of starfish does not always increase. From observation of the habitats of *Asterias rubens, Crepidula,* and oysters

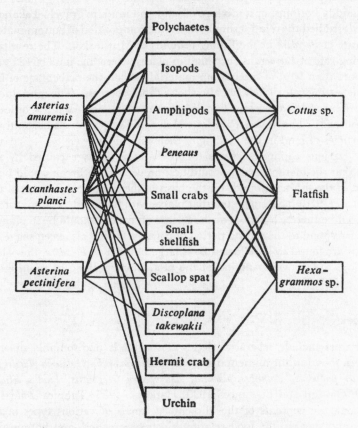

Asterias Small bottom-dwelling animols Bottam-dwelling fish

Figure 3.57. Food relationship among bottom-dwelling animals in scallop fishing grounds (Hamai and Kinoshita, 1963).

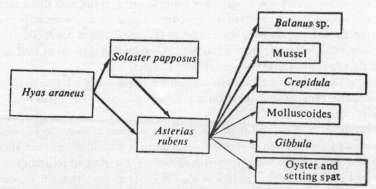

Figure 3.58. Food relationship of *solaster* and *Asterias* to bottom-dwelling animals (Hancock, 1955).

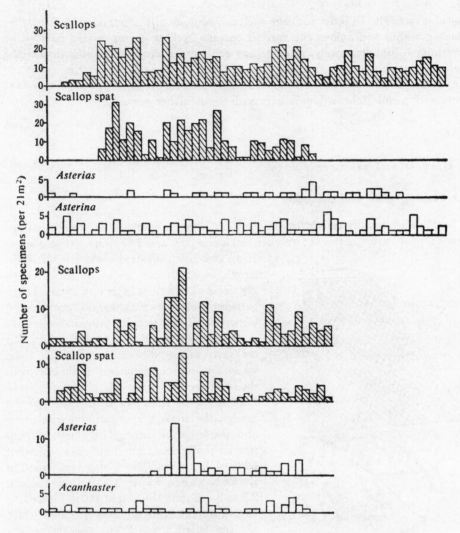

Figure 3.59. Distribution of scallops, their spat, and *Asterias* (Tamura *et al.*, 1964).

in the Crouch River, it became clear that the distribution of *Asterias rubens* was concentrated in those places where the density of *Crepidula*, their main food, was high (Hancock, 1955).

From these facts it is clear that starfish must be prevented from entering the fishing grounds, especially when scallops are being reared in baskets during interim culture and heavy damage could be caused. Although it would seem that the damage by starfish cannot be avoided altogether in places with a large number of spat in a narrow area, the damage in natural environment is not all that severe.

Among the predators which live on small bottom-dwelling animals, the removal of starfish is possible only after the removal of all the other predators; therefore it is difficult to estimate the changes in the bottom-dwelling groups resulting from the removal of starfish. In order to make such an estimation, the quantitative food relationship within and among the various bottom-dwelling groups would have to be determined. Simultaneously, whether any changes in the size and structure of the scallop community occurred should be established, so that the quantitative changes in the starfish group would become clear. This aspect is very important for fishing operations and useful information is expected from further research.

(Fuji)

6. Harvest and resource control

6.1 Fishing gear and methods

Scallops are usually caught by dredge nets, commonly known as "hasshyaku" in Japan; the structure of this net is shown in Figures 3.60 and 3.61. The dredge consists of an iron rod 8 shaku (1 shaku=0.303 m) in length, to which teeth about 50 cm in length are soldered to form a large iron comb. Large hair-pin hooks known as "kaijiki" are soldered to two of the teeth to help maintain uniformity in the movement of the dredge, when it is dragged along the fishing ground. The scallops which normally live partly buried in the beds, are thus uprooted. A pair of nets in the form of an open bag are attached to the dredge, the upper net consisting of iron wires, and the lower of twine. The length of both nets is usually about 3 m, but varies somewhat with each trawler. The fishing vessel used in the Okhotsk Sea is 30 to 40 HP, the net length 2.3 to 2.4 m, and the weight 150 to 160 kg. At Mutsu Bay, the fishing vessel is about 7 HP, the net length 1.5 to 2.5 m, and the weight 40 to 50 kg. At Hokkaido, the comb of the dredge is known as "otafuku" (meaning "mumps") and consists of two rows of crossed teeth.

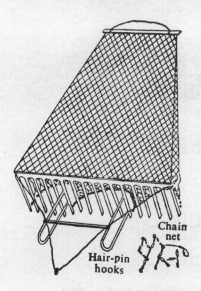

Figure 3.90. Structure of scallop dredge net (beam trawl) (Ito, 1964).

In the past when mechanistic fishing vessels were not yet known, two types of net operations were used, the forward winding method (*zensoho*) and the tidal method (*chodaseho*). Nowadays, trawling operations are carried out by powered vessels by a method commonly known among Japanese fishermen as "gangarahiki". In this method a pair of dredge nets (beam trawls) tied to ropes are let down into the sea, one from each side of the vessel, dragged, and after 20 to 30 minutes drawn back into the vessel by rotating drums.

Other fishing gear include spoon nets (Figure 3.62) of 30 cm diameter, fitted with a 4 m handle, scissors, etc., which were used in methods known as view fishing and dive fishing. These methods are not so common any more.

Figure 3.61. Scallop beam trawl (Kawauchi Mutsu Bay).

Figure 3.62. Spoon net (Kawauchi, Mutsu Bay).

6.2 Limit imposed on various species for stock management

Since scallops which lead a sessile life in the coastal regions are very easy to harvest, indiscriminate catching occurs unless proper management exists. But management of resources and the implementation of culture measures, should not be merely "hunting" operations but "cultivating" ones. This aspect has been subject to study for a long time, and methods have gradually improved; nevertheless there is still much room for further improvement.

6.2.1 LIMIT IMPOSED ON FISHING BOATS AND GEAR

At Mutsu Bay the size of fishing vessels is restricted to 2 tons and 7 HP (or with an internal combustion engine, 5 HP). However, there is no limit on the number of such boats. At Hokkaido the size of the fishing vessel has not been restricted and hence boats of 30 to 40 HP are employed; however, the number of fishing boats has been restricted by the cooperative Societies which,. by reducing the cost of operation, have increased the per capita profit.

There is no specific limitation at Hokkaido on the spaces between the combs of beam trawls, but usually, the gap is 11.5 to 13.5 cm. According to the Marine Fishing Regulations in Mutsu Bay, the space between two combs should be 12 cm or more. Regarding the size of the mesh in the nets, Hokkaido imposes no restriction, but the size is usually 15 cm, while Mutsu Bay limits the size to 12 cm or more.

6.2.2 REGULATION SHELL LENGTH

At Hokkaido the regulation of shell length is 82 mm or more. However, in view of the fact that this length sometimes includes scallops which are not two years old, at Okhotsk the limit has been raised to 106 mm or more, and in Lake Saroma to 91 mm or more, while in Mutsu Bay the shell length must equal 10 cm or more.

Scallops are considered biologically mature when they are one year old, but only a few follicles have formed at that age and spawning is also ineffective. Effective spawning occurs only when the scallop has completed two full years. Since more than half the number of two-year-old scallops measure 10 cm or more in shell length at Mutsu Bay, this restriction on shell length ought to be reconsidered.

6.2.3 PROHIBITED PERIOD OF FISHING

According to Regulations, fishing is prohibited during January to June in Shitani and Abashiri of Hokkaido, and in other regions from March to July. The restriction is intended to prevent catching during the spawning period and, in fact, in Shitani and Abashiri this period is further shortened to July 11 to September 10 by a self-imposed regulation. Fishing is prohibited in March to April and August to October in Mutsu Bay. The prohibition in the spring is to protect the period of spawning, and the prohibition in summer to autumn to protect the young spat which formed in that year. The reason for the latter prohibition is that the young spat are very weak against the suspended silt raised by the net operations. Hence the fisheries unions further shorten the fishing period in Mutsu Bay.

6.2.4 LIMIT ON THE AMOUNT OF HARVEST

According to Ito (1964), the optimum amount of harvest on the Okhotsk coast is calculated by the method given below and has been adhered to since 1953. As shown in Table 3.28, resources have already begun a gradual recovery due to the rigid imposition of this restriction on harvest.

The results of surveying the resources soon after fishing had ceased in the previous year, were applied to the following equation, and the optimum amount of harvest for the year thus calculated.

$$(W'_{(o)} - C_{(t)sw} + R'_{2s2w2} = W^{\cdot}_{(o)}$$

$$\text{Now, } SW = \frac{W'_{(o)}}{W_{(t)}} = \frac{R_2 S_2 W_2 + R_3 S_3 W_3 + \ldots R_t S_t W_t}{R_2 + R_3 + \ldots R_t},$$

$W_{(t)}$ ...Resources after fishing for the current year.

$R_2 + R_3 + \ldots \ldots Rt$ Amount of age groups from 2-year-old to t-year-old.

$W'_{(o)}$ ...Amount of resources before fishing in the following year.

S ...Survival rate.

W ...Increasing rate of weight.

$W''_{(o)}$ ...Amount of resources before fishing in the second year after the current one.

R_2 ...Number of 2-year-old groups expected in the following year.

$C_{(t)}$ ...Amount of harvest.

When the amount of resources is intended to be maintained at the same level, the calculation should be on the basis: $W'_{(o)} = W''_{(o)}$. When the amount of resources is intended to be increased, then $W''_{(o)}$ should be greater than $W'_{(o)}$. However, since there is some deviation in the numerical values judged from the reliability of the survey of resources, a production plan within this range was projected. The amount arrived at on the basis of the production plan made by each of the fishing unions is strictly imposed, after taking into consideration the production aim of each fishing operation and the HP of the fishing vessels.

In fishing grounds where reproduction is due to the natural development of scallop spat, the optimum harvest should be modified according to the capacity for reproduction and the resources maintained.

Table 3.28. Transition in the harvest and amount of resources in various fishing regions along the Okhotsk coast (Ito, 1964)

Fishing ground	1958	1959	1960	1961	1962	1963
Abashiri	?	3,750 ton	1,520 ton	2,700 ton	500 ton	?
	(56%)	(53%)	(42%)	(41%)	(39%)	
Tokoro	12,188 ton	12,750	8,100	8,500	7,300	5,000 ton
	(55)	(55)	(47)		(22)	
Yubetsu	3,375	2,250	1,700	2,200	4,300	3, 00
	(60)				(21)	
Bunbetsu	3,375	2,625	4,700	7,400	15,000	14,000
	(55)				(20)	
Sunaryu	1,687	1,500	2,100	3,100	3,500	3,700
Osutake	938	563	970	700	700	?
Total	21,563	23,438	19,090	24,600	31,300	26,500
Index	100.0	108.7	88.6	114.3	145.7	123.2

Note: Parenthetical figures=Harvest rate. Other figures=Estimated resources before fishing.

6.3 Positive conservation policies

6.3.1 RELEASE OF SPAT

In fishing grounds where there is no natural formation of spat, or where spat formation is poor, it is very essential to transplant them from other regions. Since the natural formation of spat is usually very poor in Mutsu Bay, resource management should begin such transplantation. It must be emphasized however that the places of transplantation should be changed each year to avoid damage to the spat since, as mentioned earlier, the larvae and spat have a low tolerance for the floating silt raised during harvesting with beam trawls.

Scallop growth and the rate of weight increase was seen in Figure 3.10 (Yamamoto, 1964). The weight is high from the first to the third year and shows a decrease after the third year. Hence it is profitable to harvest scallops during the third year. Again, since the scallops are thin after spawning up to late summer (the yield rate of meat is also poor), harvesting should not be carried out during this time.

6.3.2 MEASURES AGAINST PESTS

As explained earlier, there are quite a few animals which live on scallops, but *Asterias amurensis* L. is their worst enemy. In Mutsu Bay, scallop spat are usually released only after the *Asterias* have been cleaned from the beds and the surroundings by beam trawls. At Hokkaido, *Asterias* are removed while the scallop culture is on, and sometimes a fixed amount of *Asterias* are removed at regular intervals; consequently the number of removed *Asterias* increases every year (Table 3.29), but the amount

Table 3.29. Transition in the amount of *Asterias* removed from scallop beds

(From data compiled by the Fisheries Department of Hokkaido)

Year	All of Hokkaido	Abashiri
	kg	kg
1931	902,243	722,095
1932	1,304,309	672,421
1933	1,769,157	1,160,841
1934	2,015,339	1,232,289
1935	2,083,292	1,545,732

removed nevertheless covers only 35% of the total amount estimated as present in the beds (Table 3.30) (Ito, 1964). Hence, a more effective method of removing this animal should be devised.

Table 3.30. Estimated amount of *Asterias* **inhabiting the scallop growing beds in Abashiri**

(From data compiled by the Fisheries Department of Hokkaido)

Fishing bed	Number of *Asterias* for m²	Area of scallop growing bed	Estimated number of *Asterias* inhabiting the bed
Abashiri	0.036	102,400,000 m²	3,686,400
Tokoro	0.036	198,400,000	7,142,400
Yubetsu	0.075	96,000,000	7,200,000
Bunbetsu	0.067	240,000,000	16,080,000
Kambe	0.041	91,200,000	3,739,200
Osutake	0.036	66,000,000	2,376,000
Total	0.049 (mean)	794,000,000	40,224,000

(Ito)

7. Prospects for the future

Trends in the study of scallop culture and its present position have been discussed in detail. The question now is whether scallops may also enter the era of "through culture." While all the steps devised in planned production through artificial management are possible, some problems still exist for which solutions must be found before through culture for scallops can be successful. These are production through natural seedlings, interim culture, transplanting combined with production through induced spawning, artificial seeding, and hanging culture. In the first instance, the number of spat obtained is unusually high. Again, technical development of natural seedlings is also yielding spat at a steady rate. If only the variations in the number of artificial

Table 3.31. Number of natural scallop seedlings in Mutsu Bay (1968)

(Data obtained from Report No. 9 of Mutsu Bay Fisheries Experiment Station and from the Aomori Aquaculture Center, unpublished)

Year	Number of instruments used for collecting seedlings	Total number of spat
		(10,000)
1964	83	147.4
1965	79	21.0
1966	65	105.8
1967	228	10,213.6
1968	683	48,430.0

Note: The total number of spat was calculated in the following manner: On the basis of the survey of scallop setting conditions in early July, it was estimated that spat of 2 mm shell length, or below, were 0%; 2 to 5 mm, 50%; 5 to 10 mm, 90%; 10 mm and above, 100%. The total was calculated by taking these percentages into account.

seedlings could be predicted before the period of collections, and if the technique for growing healthy spat from natural seedlings could be further improved, it would then be possible to achieve a planned production similar to that based on artificial seeding through hanging culture.

7.1 Artificial and natural seeding

It has recently become possible to obtain a regular supply of artificial seedlings from the Aquaculture Center in the Aomori Prefecture. The artificial seedlings are pre-

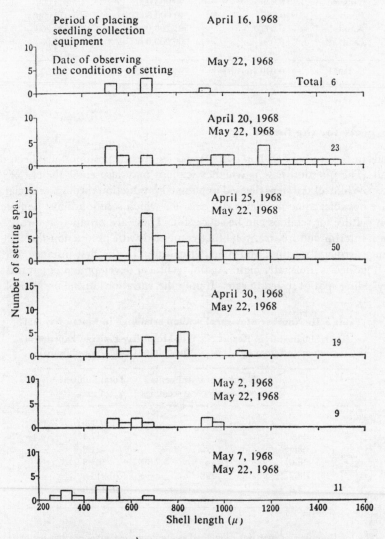

Figure 3.63. Period of placing equipment for seedling collection, and the condition of spat seedlings (Kanno, unpublished).

Collection of seedlings is poor if the placement of the equipment is too early or too late.

pared in Mutsu Bay (Figure 3.15). The capacity of the preparing equipment is about 1,000,000 artificial seedlings. However the total number of natural seedlings in Mutsu Bay (Table 3.31) has now reached 500,000,000 to 1,000,000,000. Hence the number of artificial seedlings is not comparable with that of natural seedlings. For this reason, plans for artificial seedlings have generally declined except at the Oyster Research Center (Inc.) at Maine in the Miyagi Prefecture. Encouraged by their efforts, Iwate, Akita, and Yamanashi Prefectures are now sponsoring artificial seedlings, and consequently hanging culture is progressing. On the other hand, natural seeding has shown great progress in Hokkaido, and even in Lake Saroma where there was a sharp temporary decrease in the number of natural seedlings, the total number has now reached 150,000,000 (Tables 3.4 and 3.13).

Detailed investigations were carried out to stabilize natural seeding and, according to Kanno *et al.*, various factors (shown in Figure 3.63) play significant roles during the placing of collecting equipment.

Under present circumstances, it will be interesting to find a suitable way of combining or adjusting the artificial seedling and the natural seedling, it seems that of all the factors suggested the cost of production of seedlings must be given greater importance.

7.2 Technical improvement of interim culture

As already indicated, a high mortality rate is seen in Mutsu Bay among the settled spat during their changeover to bottom-dwelling life from summer to early autumn, and at Hokkaido, mass mortality is observed during wintering. Hence, interim culture is essential and, at present, is carried out with pearl nets and andon baskets (Figures 3.32 and 3.35). However, the capacity of this equipment, as well as the technique involved, needs further improvement.

The natural release of seedlings, which has often ended in failure, should also be studied so that a successful method could be evolved, and applied not only to scallops but to other shellfish as well.

7.3 Hanging culture and transplantation of 3 cm scallop spat

According to the data already presented, it is clear that spat of 3 cm shell length are relatively strong against changes in the environment and their survival rate high. Moreover, when 3 cm spat are transplanted, the harvest never goes down below 50%, and when hanging culture intervenes, the harvest rate is even higher. The other advantage of hanging culture is that a marketable scallop size is reached within a short time. One problem besets hanging culture, however, the high cost of operation. Furthermore the technique for augmenting the survival rate after transplantation, could also be further improved, and should be studied in places like Lake Saroma and Mutsu Bay in particular where natural seedlings can be obtained.

7.4 Abnormal fecundity

One of the most significant problems in scallop culture relates to "abnormal fecundity," which also occurs in several other mollusks. When the larvae resulting from "abnormal fecundity" are preserved, an unusually high density results which leads to mass mortality and inferior growth in the survivors. The number of scallops caught, of course, increases, but the catch includes quite a few small scallops. A "normal" catch occurs only after 1 to 2 years of growth. "Abnormal fecundity" therefore increases cost of production, resulting in a loss of profit to management. In order to solve this problem, either new fishing beds should be prepared in which the larvae resulting from "abnormal fecundity" could be transplanted and grown, or the harvest should be carried out according to a yearly plan. In fact, this approach to the problem has already been put into effect at Mutsu Bay, providing a model experiment from which suitable solutions may eventually be evolved.

(Yamamoto)

Bibliography

1. Abashiri Fisheries Experiment Station, 1965. Abashiri suishi (1965) Hotategai zoshoku ni kansuru kisokenkyu [Fundamental Research of Scallop Culture, Annual Report (1965) of Abashiri Fisheries Experiment Station, Hokkaido], Hokkaido Abashiri Suishi Showa 40-nen, Jigyo Seisekicho, 85-107.

2. AKAHOSHI and SASAKI [sic], 1970. Hotategai chikukai no hayakihobyo shiken (Early Seedling Experiment of Scallop), Report published by Mutsu Bay Fisheries Culture Research Center, Aomori Prefecture, 11, 272-274.

3. ——, ——, KANO, and TAKEDA [sic], 1970. Suika Yoshoku shiken [Culture of Scallop with Special Reference to Hanging Culture (Study Done in 1967)], Ibid., 11, 275-290.

4. ——, ——, ITO, KANO, SATO, and YOKOYAMA [sic], 1970. Hotategai no zoshoku ni kansuru kenkyu, suika yoshokushiken [On the Culture of Scallop with Special Reference to Hanging Culture (Study Done in 1966)], Ibid., 11, 24-38.

5. AKAHOSI, S., ITO, and H. KANNO, 1968. Hotategai no zoshoku ni kansuru kenkyu. Suika yoshoku shiken (Study on Scallop Culture with Special Reference to Hanging Culture Experiment), Ibid., 10, 61-74.

6. AMIRTHALINGAM, C., 1928. On Lunar Periodicity in the Reproduction of *Pecten opercularis* near Plymouth in 1927-1928, *J. Mar. Biol. Assoc. N. S.*, 15, 605-641.

7. Aomori Prefecture: Fisheries Experiment Center, 1930. Mutsu-Wan nai hotategai chosa (Study of Scallops in Mutsu Bay), Report of Aomori Fisheries Experiment Station (1929, 1930), 32-38.

8. BAUER, V., 1912. Zur Kenntnis der Lebensweise von *Pecten jacobaeus* L. im besonderen uber die Funktion der Auge, Zool. Jahrb. Abt. allg. *Zool. Physiol.* 33, 127-150.

9. BAYLISS, L. E., E. BOYLAND, and A. D. RITCHIE, 1930. The Adductor Muscle of *Pecten, Proc. Roy. Soc. Lond.*, B. **CVI**, 363-376.

10. BELDING, D. L., 1910. A Report on the Scallop Fishery of Massachusetts. The Common Wealth of Massachusetts, Boston.

11. BOURNE, N., 1964. Scallops and Offshore Fishery of the Maritimes, *Fish. Res. Bd. Canada, Bull.*, No. 145, Ottawa, 60 p.

12. BROCKEHOFF, H., R. G. ACKMAN, and R. J. HOYLE, 1963. Component Fatty Acids of Some Marine Invertebrates, *Arch. Biochem. Biophys.*, **100**, 9-12.

13. BUDDENBROCK, W. V., 1911. Untersuchungen uber die Schwimmbewegungen und die Statocysten der Gattung *Pecten, Sitz. Heidelberger Akad. Wiss. Jahrg.* 1911, 28 Abhandl.

14. ——, 1918. Die Statocysten von *Pecten,* ihre Histologie und Physiologie, *Zool. Jahrb. Abt. allg. Zool. Physiol.*, **35**, 301-356.

15. CHIPMAN, W. A., and J. G. HOPKINS, 1954. Water Filtration by the Bay Scallop, *Pecten irradians,* as Observed with the Use of Radioactive Plankton, *Ibid.*, 107, 80-91.

16. COE, W. R., 1945. Development of the Reproductive System and Variations in Sexuality in *Pecten* and other Pelecypod Mollusks, *Trans. Conn. Acad. Arts and Sci.*, **36**, 673-700.

17. COLE, H. A., and HEPPER, B. T., 1954. The Use of Neutral Red Solution for the Comparative Study of Filtration Rates of Lamellibranches, *J. du Cons.* **20**, 197-203.

18. DAKIN, W. J., 1909. *L. M. B. C. Memoirs XVII Pecten.* London.

19. ——, 1928. The Eyes of *Pecten spondylus, Amussium* and Allied Lamellibranches, with a Short Discussion on Their Evolution, *Proc. Roy. Soc. London,* B. **C111**, 337-354.

20. DAVIS, H. C., 1963. The Effect of Salinity on the Temperature Tolerance of Eggs and Larvae of Some Lamellibranch Mollusks, *Proc. XVI Intern. Cong. Zool.,* Washington, D. C., 226.

21. ——, and R. R. GUILLARD, 1958. Relative Value of Ten Genera of Micro-organisms as Food for Oyster and Clam Larvae, *U. S. Fish and Wildl. Serv. Bull.,* 136, **58**, 293-304.

22. ——, and A. D. ANSELL, 1962. Survival and Growth of Larvae of the European Oyster, *O. edulis,* at Lowered Salinities, *Biol. Bull.* **122**, 33-39.

23. DICKIE, L. M., 1955. Fluctuation and Abundance of the Giant Scallop, *Placo-pecten magellanicus* (Gmelin) in the Digby Area of the Bay of Fundy, *J. Fish. Bd. Canada,* **12**, 707-857.

24. ——, 1958. Effects of High Temperature on the Survival of the Giant Scallop, *Ibid.,* **15** (6), 1189-1211.

25. ——, and J. C. MEDCOF, 1963. Causes of Mass Mortality of Scallop, *Placo-pecten magellanicus,* in the Southwestern Gulf of the St. Lawrence, *Ibid.,* **20**, 451-482.

26. DREW, G. A., 1960. The Habits, Anatomy, and Embryology of the Giant Scallop (*Pecten tenuis costatus* Mighels), *Univ. Maine Stud.,* No. 6, 1-71.

27. EWATARI, T., 1953. Fuyuchigai shutsugen jokyo chosa. Fuchaku chigai no chosa (Hotategai ni Kansuru Kenkyu) (Study on Planktonic and Settling Spat as Related to Scallops), *Mutsu Wan Suiken Jigyo Hokoku* (Bulletin of Mutsu Bay Fisheries Research Center), 1951, 1952, 54-58.

28. FULLARTON, J. H., 1890. On the Development of the Common Scallop, *Pecten opercularis.* 8th Ann. Rep. Fish Bd. Scotland for 1889, **111**, 290-298.

29. GALTSOFF, P. S., 1930. The Fecundity of the Oyster, *Science* **72**, 97-98.

30. ——, 1947. Egg Number in *Ostrea virginica, Ibid.,* **196**, 342.

31. ——, and V. L. LOOSANOFF, 1939. Natural History and Method of Controlling the Starfish (*Asterias forbesi* Desor), *Bull. U. S. Bur. Fish.,* **49**, 75-132.

32. GUTSELL, J. S., 1930. Natural History of the Bay Scallop, *Ibid.,* **46** (Doc. No. 1100), 569-632.

33. HAMAI, I., and T. KINOSHITA, 1963. Wazumi hanto kaiiki no gyorui bunpujokei ni tsuite (On the Distribution of Fishes in Wazumi Hanto). Data published by Hokkaido Development Center (Hokkaido Kaihatsu Kyoku Chosa Shiryo), 315-3-206.

34. HANCOCK, D. A., 1955. The Feeding Behavior of Starfish on Essex Oyster Beds, *J. Mar. Biol. Assoc.* **34**, 313-331.

35. HANEHARA, M., 1953-1954. *Nihon Gyogyo Keizaishi (Fisheries Economics of Japan)*, Vol. 12. Tokyo: Iwanami Publication.

36. HATANAKA, SATO, and IMAI [sic], 1943. Asari, hamaguri no shichu no jinko-shiiku ni tsuite (Artificial Rearing of the Trochophores of the Short-necked Clam and Clams), *Nihon Suisan Gakkaishi (Bulletin of the Japanese Society of Scientific Fisheries)*, **11**, 218.

37. Hirano Sangyo Keizai Kenkyusho, 1950. Aomori-ken sogo shiryo dai 20-shu. Honken no hotate gyogyo keizaishi (A History of Scallop Fishing Operations in Aomori Prefecture). Tokyo.

38. HIRANO and OSHIMA [sic], 1963. Kaisan dobutsu yosei no shiiku to sono jiryo ni tsuite (On the Rearing and Feeding of the Larvae of Marine Animals), *Nihon Suisan Gakkaishi (Bulletin of the Japanese Society of Scientific Fisheries)*, **29**, 282-297.

39. Hiroshima Prefecture Fisheries Experiment Center, 1952. Sogai jinkosaibyo no chukan ikusei senkai zoshoku koka chosa hokokusho (Interim Culture of Artificial Seedlings), *Seikai Zoshoku Koka Chosa Hokokucho (Bulletin of Shallow Sea Water Culture)*.

40. Hokkaido Fisheries Experiment Center of Hiroshima Prefecture, 1953. Showa 27-nendo zenkai zoshoku shiaen iji chosa (Survey on the Sources of Shallow Sea Water Culture Carried Out in 1952), *Senkai Zoshoku Koka Chosa Hokokusho (Bulletin of Shallow Sea Water Survey)*.

41. ——, 1954. Showa 28-nende hogosuimen kanri jigyo chosa hokokusho (Survey of Sea Surface Protection Management).

42. HOPKINS, A. E., 1937. Experimental Observations on Spawning, Larval Development and Setting in the Olympia Oyster, *Ostrea lurida, Bull. U. S. Bur. Fish.*, **28**, 439-503.

43. HUNT, O. D., 1925. The Foods of the Bottom Fauna of the Plymouth Fishing Grounds, *J. Mar. Biol. Assoc.*, **13**, 560-598.

44. IGARASHI, H., K. ZAMA, and K. TAKAMA, 1961. Kairui shishitsu (Fat of Mollusks), *Hokudai Suisan Kenkyu Tho (Bulletin of Hokkaido University Fisheries Research)*, **12**, 196-200.

45. IMAI, HATANAKA, and SATO [sic], 1942. Asari, hamaguri shichu no jinko shiiku ni tsuite (Artificial Rearing of Larvae of Short-necked Clam and Clams), *Nihon Suisan Gakkaishi (Bulletin of the Society of Scientific Fisheries)*, **11**, 79.

46. IMAI, S., 1941. Kaisen to teishitsu ni tsuite (On the Nature of the Sea Bottom). Hokusuishi junpo (Fortnightly Report of Hokkaido Fisheries Experiment Station), 510.

47. IMAI, T., and S. SAKAI, 1961. Study of Breeding of Japanese Oyster (*Crassostrea gigas*), *Tohoku J. Agr. Res.*, **1**, 69-86.

48. ——, and M. HATANAKA, 1949. On the Artificial Propagation of the Japanese Oyster, *Ostrea gigas* Thun. bu Non-colored Naked Flagellates, *Bull. Inst. Agric. Res. Tohoku Univ.*, **1** (1), 33-46.

49. ——, ——, 1950. Studies on Marine Non-colored Flagellate, *Monas* sp., Favorite Food of Larvae of Various Marine Animals: I. Preliminary Research on Culture Requirements, *Sci. Rep. Tohoku Univ.* Ser. IV (*Biol.*) 18, 304-315.

50. ISAHAYA, T., 1933. Hotate no ido chosa (Study on the Migration of Scallops),

Hokusuishi Junpo (Fortnightly Report of Hokkaido Fisheries Experiment Station), 204, 5-8.

51. ISAHAYA, T., M. SAKUMA, and Y. HIRANO, 1934. Hotate no fuchaku seikatsu jidai (Study on the Sessile Life of Scallops), *Ibid.*, 233, 2-3.

52. ISHIHARA, S., 1963. Haizekkusufirumu shiyo ni yoru hotategai gaikai saibyo shiken hokokusho (Experimental Studies on the Collection of Seedlings of Scallops in Open Seas by Using Hyzex Film), *Bosuikairyo Shiryo*, No. 1, 1-30.

53. ——, 1964. Obon-gata (H. Z-Sei) ni yoru hotategai gaikai saibyo ken ihusei shiken chukan hokokusho (Experimental Study on the Rearing of Seedlings of Scallops Collected in the Open Seas by Plates Made of Hyzex), *Ibid.*, No. 2, 1-12.

54. ITO, SASAKI, KANO, AKAHOSHI, SATO, YOKOYAMA, TAKEDA, OGAWA, AOYAMA, ZUBO, and NAGATANI [sic], 1970. Zankai gyoba kaisatsu jigyo chosa (Tsuchiya chisaki ni okeru choryu, suishidsu to hotategai no scicho narabi ni taiha yoshoku chitetsu no kaisatsu shiken) [Investigations Carried Out by Shallow Water Culture Development Group (On the Tidal Current and Water Quality of Tsuchiya Chisaki and Growth of Scallop)], Report published by Mutsu Bay Fisheries Culture Research Center, Aomori Prefecture, 11, 451-462.

55. ITO, S., 1964. Ohotsuku-kai engan ni okeru-hotate-gai gyogyo (Fishing of Scallops in Okhotsk Coast), Suishan Zoshoku Gyocho 7 (Association for the Protection of Marine Resources).

56. ITO, S., KANNO, and AKAHOSHI [sic], 1965. Hotategai no jinko saibyo shiken (Study on the Artificial Seedlings of Scallops). Data published by Mutsu Bay Fisheries Culture Centre, S. 40, No. 1).

57. ——, F. TSUBATA, K. TAKEDA, H. CHIBA, and Y. HASE, 1967. Hotategai no zoshoku ni kansuru kenkyu (1) tennen saibyo shiken (Study of the Culture of Scallops: I. Experiments with Natural Seedlings). Monthly Report of Mutsu Bay Fisheries Research Center, 9, 1-22.

58. ——, H. KANNO, and S. AKAHOSHI, 1968. Hotategai no zoshoku ni kansuru kenkyu (Study on the Rearing of Scallops: II. Experiments with Artificial Seedlings), *Ibid.*, 10, 1-9.

59. IWAGISHI, S., 1964. Showa 39-nendo hotategai saibyo shiken jigyo kekka hokoku oyobi ikusei shiken chukan hokokucho (Report of the Results of Experiments on the Seedlings of Scallops, and Interim Report on Rearing Experiments for the year 1964). 37 p.

60. JAY, J. C., 1856. Report on the Shells Collected by the Japan Expedition under the Command of Commodore M. C. Perry, U. S. Navy, together with a List of Japanese Shells, *Narrative of the Expedition of an American Squadron in the Years 1852, 1853, and 1854 under the Command of Commodore M. C. Perry, United States Navy, by Order of the Government of the United States*, Vol. II, 291-297.

61. JØRGENSEN, C. B., 1949. The Rate of Feeding of *Mytilus* in Different Kinds of Suspension, *J. Mar. Biol. Assoc.*, 28, 333-344.

62. KATO, T., 1962. Fuyu chigai chosa (Hotategai zoshoku ni tsuite no chosa kenkyu) [Study on the Planktonic Spat (Relation to the Culture of Scallops)], Monthly Report of Mutsu Bay Fisheries Research Centre (1960), 6, 4-8.

63. KAWAGUTI, S., and N. IKEMOTO, 1958. Electron Microscopy on the Adductor Muscle of the Scallop, *Pecten albicans, Biol. J. Okayama Univ.*, **4** (3-4), 191-206.

64. KINOSHITA, T., 1934. Hotategai jinko-saibyo jono ichi shiryo (On the Artificial Seedlings of Scallops), Hokusuishi Junpo (Fortnightly Report of Hokkaido Fisheries Experiment Station), **230**, 280-281.

65. ——, 1934. Hotategai no sanran to ondoto no kankei (Relationship between the Spawning of Scallops and Temperature), *Ibid.*, **233**, 3-8.

66. ——, 1940. Hotategai no saibyoritsu ni oyobosu sanran jiki no suion no eikyo ni tsuite (Influence of Water Temperature on the Spawning Period Studied in Relation to the Number of Seedlings Collected), *Nihon Suisan Gakkaishi (Bulletin of the Japanese Society of Scientific Fisheries)*, **9**, (1).

67. ——, 1940. Hotategai jinko zoshoku no sonogo (Artificial Rearing of Scallops and Its Recent Progress), Hokushuishi Junpo (Fortnightly Report of Hokkaido Fisheries Experiment Station), **455**, 5-10.

68. ——, and HIRANO [sic], 1935. Hokkaido-san hotategai no shokuji ni tsuite (On the Rearing of Scallops of Hokkaido), *Ibid.*, **272**, 690-693.

69. ——, ——, 1935. Hotategai saibyo shiken (Experiments on Scallop Seedlings), *Ibid.*, **273**, 1-8.

70. ——, ——, 1935. Hokkaido-san hotategai no kara no hoshanoku no chihoteki heni (Regional Variations in the Number of Radical Lines on the Shells of Hokkaido Scallops), *Ibid.*, **290**, 10-12.

71. ——, ——, 1935. Hokkaido-san hotategai *Pecten (Patinopecten) yessoensis* Jay no shokuji ni tsuite [Rearing of Hokkaido Scallop, *Pecten (Patinopecten) yessoensis* Jay], *Dozatsu (Zoological Magazine)*, **47**, 1-8.

72. ——, ——, 1935. Hokkaido hotategai no kara no hosharoku no chihoteki heni (Regional Variations in the Number of Radial Lines on the Shells of Hokkaido Scallops), *Venus* **5** (4), 223-229.

73. ——, ——, 1936. Hotategai no chishiki (Study of Scallops). Pamphlet published by Hokkaido Fisheries Experiment Station, No. 1, 62 p.

74. ——, ——, 1949. Hotategai no zoshoku ni kansuru kenkyu (Study on the Culture of Scallops). Sapporo: Hoppo Publishers, 106 p.

75. ——, ——, 1951. Hokkaido senkai zoshoku gaiteki seibutsuhen (Harmful Organisms during Shallow Water Culture in Hokkaido). Collection No. 7 edited by Suisan Kagaku Gyocho. Samonji Book Publishers, Otaru, 64 p.

76. ——, and SHIMIZU [sic], 1945. Hotategai no jinko saibyo zoshoku ni okeru saibyojiki (Period of Seedling Collections in the Artificial Rearing of Scallops) Hokusuishi Geppo (Monthly Report of Hokkaido Fisheries Experiment Station), **1** (12, 13), 467-470.

77. ——, and Y. NAKAJIMA (1940). Saroma-ko ni okeru hotategai oyobi arushu no makigai shichu no shutsugen shocho ni tsuite (Rate of Emergence of Scallop Larvae and Larvae of other Mollusks in Saroma Lake), *Suisan Kenkyushi (Journal of Fisheries Research)*, **35** (1).

78. ——, ——, 1940. Saroma-ko ni okeru hotategai oyobi arushu no makigai no zonchu no shutsugen shocho ni tsuite (Study of Scallops and Allied Mollusks of Saroma Lake) Hokushuishi Junpo (Fortnightly Report of Hokkaido Fisheries Experiment Station), **456**, 8-10.

79. KINOSHITA, T., and SAKUMA [sic]. 1935. Hotategai no sanranchosa (Study on the Spawning of Scallops of the Northern Regions), *Ibtd.*, **272**. 688-690.
80. ——, and S. SHIBUYA, 1944. Saroma-ko hyomen ni okeru hotategai shichu to fukusokurui shichu no kotaisu no chuna narabi koko henka (yoho) [Variations in the Number of Scallop Larvae and Larvae of Gastropods during Daytime, Nighttime, and Tidal Variations (Preliminary Report)], *Ibid.*, **10** (1).
81. ——, and SHIMIZU [sic], 1947. Sensui kekka kara mita hotategai no shui to kaitei ni okeru peta-ami no unko ni tsuite (Habitat of Scallops and Dragnet Operation on the Sea Bottom), *Ibid.*, **4** (2), 19-21.
82. ——, SHIBUYA, and SHIMIZU [sic], 1943. Hotategai *Pecten (Patinopecten) yessoensis* Jay no sanran yuhatsu ni kansuru shiken (Yoho) [Experiments to Induce Spawning in *Pecten (Patinopecten) yessoensis* Jay (Preliminary Report)], *Nissui-kaishi (Bulletin of the Japanese Society of Scientific Fisheries)*, **11**, 168-170.
83. ——, ——, ——, 1944. Hotategai gyobya sensui chosa kekka (Yoho) [Submarine Survey of Scallop Beds (Preliminary Report)], Hokusuishi Geppo (Monthly Report of Hokkaido Fisheries Experiment Station), **1** (6).
84. ——, ——, ——, 1944. Kaisengyogyo ni okeru Saroma-ko yoranjo igi (Dai-ippo) [Survey of Saroma Lake Where Coastal Fishing Operations are Done (Report No. 1)], *Ibid.*, **1** (8, 9).
85. ——, ——, ——, 1944. Hotategyoja no kosei yoso ni kansuru kenkyu. Dai-niho. Nemuro kaigan narabi Nemuro kinkai no hotategai gyogo no sensui chosa (Some Essential Features of Scallop Fishing Beds: Report No. 2. Study of the Characteristics of Scallop Beds in Nemuro Sea Shores and the Sea Region near Nemuro), *Ibid.*, **1** (1), 49-54.
86. ——, K. TAMOTO, and S. HARADA, 1959. Showa 32-nen Kitami Mombetsu hotategai gyojo ni okeru byogai mondai ni tsuite (Study on Some Diseases Breaking Out in Scallop Fishing Beds of Bunbetsu in 1957), *Ibid.*, **16**, 101-105.
87. KISHIGAMI, K., 1896. Mutsu-wan hotategai chosa (Survey on Scallops of Mutsu Bay).   Report on the Fisheries Survey Carried Out in 1895, 43-49.
88. KODERA, S., SATO, and KATO [sic], 1961. Shiiku-hako, shiiku-kago ni yoru chigai no ikusei (Rearing of Spat in Boxes and Baskets), Mutsu Wan Suisan Kenkyu Gyomu Hokoku (Report of Mutsu Bay Fisheries Research Center), No. 5 (1957-1959), 21-25.
89. ——, ——, HASEGAWA, and SASAKI [sic], 1958. Shiiku-hako ni yoru chigai ikusei ni tsuite no kenkyu (Hotategai zoshoku ni tsuite no chosa kenkyu) [Study on the Rearing of Spat in Boxes (Related to the Rearing of Scallops)], *Ibid.*, No. 4 (1954-1956), 96-107.
90. KORRINGA, P., 1949. More Light upon the Problem of the Oyster's Nutrition?, *Bijdr. Cierk.*, **28**, 238-248.
91. ——, 1952. Recent Advances in Oyster Biology, *Quart. Rev. Biol.*, **27**, 266-308, 339-365.
92. LOOSANOFF, V. L., and J. B. ENGLE, 1947. Feeding of Oysters in Relation to Density of Microorganisms, *Science*, **105**, 259-261 [sic].
93. ——, and F. D. TOMMERS, 1948. Effect of Suspended Silt and Other Substances on Rate of Feeding of Oysters, *Ibid.*, 107, 69-70.
94. MARSHALL, N., 1960. Studies on the Niantic River, Connecticut, with Special

Reference to Bay Scallop, *Pecten irradians, Limnol. Oceanogr.*, **5**, 86-105.

95. MASON, J., 1953. Investigations on the Scallop *Pecten maximus* (L.) in Manx Waters. Ph. D. Thesis, Univ. Liverpool.

96. ——, 1957. The Age and Growth of the Scallop *Pecten maximus* (L.) in Manx Waters, *J. Mar. Biol. Ass., U. K.*, **36**, 473-492.

97. ——, 1958. Breeding of the Scallop *Pecten maximus* (L.) in Manx Waters, *Ibid.*, **37**, 653-671.

98. ——, 1959. The Food Value of the Scallop *Pecten maximus* (L.) from Manx Inshore Waters, *Rep. Mar. Biol. Sta. Port Erin*, 71, 42-52.

99. MASUDA, K., 1963. The So-called *Patinopecten* of Japan, *Trans. Proc. Palaeont. Soc. Japan*, N. S., No. 52, 145-153.

100. MEDCOF, J. C., and N. BOURNE, 1964. Causes of Mortality of the Sea Scallop, *Placopecten magellanicus, Proc. Nat. Shellfish Assoc.*, **53**, 33-50.

101. MIHARU, H., and T. TOYOJIMA, 1944. Kitami engan ni okeru hotategai no nenreisosei ni tsuite (Composition of Age Groups of Scallops in Hokumi Coast), Hokusuishi Geppo (Monthly Report of Hokkaido Fisheries Experiment Station), **1** (6), 327-333.

102. ——, 1940. Nemurokai-wan ni okeru hotategai no sanranki to (Laba) no shutsugen jokyo ni tsuite (On the Spawning Period and Conditions of Larval Appearance of Scallops in Nemurokai-Bay), Hokusuishi Junho (Fortnightly Report of Hokkaido Fisheries Experiment Station), 445, 12-16.

103. ——, 1941. Nemurokai-wan ni okeru *Pecten (Patinopecten) yessoensis* Jay no sanranki to larva no shusugengata ni tsuite [Spawning Period and Larval Appearance of *Pecten (Patinopecten) yessoensis* Jay], *Suisangakuzatsu (Journal of Fisheries)*, No. 49, 72-78.

104. MORITA, M., KANNO, CHIBA, TSUBATA, KANNO, and YAMAMOTO [sic], 1963. Kuroshio-go ni yoru Mutsu-wan sensui chosa hotategai, akagai narabini gyosho no kansatsu (Submarine Survey of Mutsu Bay in Relation to Black Tides: Some Observations on the Condition of Scallop and Ark-shell), Aomori Prefecture, 33 p.

105. NISHIKAWA, N., T. SUZUKI, and T. IMAI, 1965. Hotategai no jinko saibyo to sono kanzenyoshoku. Dai-ippo. Mutsu-wan-san Hotategai no Miyagi ken Maine Wan e no ishoku shiken (Artificial Seedlings and Through Culture of Scallops: Report No. 1. Transplanting Experiment of Mutsu Bay Scallop into Bune Bay of Miyagi Prefecture), *Nihon Suisan Gakkai Tohoku Shibu Kaiho (Bulletin of Tohoku Branch of the Japanese Society of Scientific Fisheries)*, No. 18, 4-8.

106. ——, A. CHIBA, T. SUZUKI, and T. IMAI, 1966. Hotategai no kanzen yoshoku dai-sanpo (Through Culture of Scallop), Report No. 3, Proceedings of the 1965 Conference of Japanese Society of Scientific Fisheries.

107. NOMURA, S., 1917-1918. Hotategai no kaibo (Anatomy of Scallops), *Dozatsu (Zoological Magazine)*, **30**, 43-44; **31**, 45-48.

108. ——, 1922. Hotategai no era no kozo oyobi sono keitogakuteki igi (yoho) [Structure of Gills in Scallops (Preliminary Report)], *Ibid.*, 34, 436-445, Pl. 4.

109. ——, 1932. Studies on the Physiology of Ciliary Movement: I. Effect of

Hydrogen Ion Concentration upon the Ciliary Movement of the Gill of *Pecten, Sci. Rep. Tohoku Imp. Univ.,* Ser. IV (*Biol.*), **7**, 15-42.

110. NOMURA, S., 1933. Studies on the Physiology of Ciliary Movement: II. Intracellular Oxidation-reduction Potential Limiting the Ciliary Movement, *Protoplasma,* **20**, 85-89.

111. OBATA, KONISHI, and ISHIDA [sic], 1950. Hotategai-bashina no daidai-iro shikiso to hotategai-bashira nijiru no aminotai chisso (Color Pigments and Amino-nitrogen in Scallops), Hokusuishi Geppo (Monthly Report of Hokkaido Fisheries Experiment Station), **7** (3), 35-37.

112. OHARA K., and K. MARU, 1967. Saroma-ko no hotategai yoshoku no genkei ni tsuite (Scallop culture in Saroma Lake), *Ibid.,* 24, 469-482.

113. ——, ——, 1970. Showa 44-nendo saroma-ko ni okeru hotategai saibyo kekka ni tsuite (Scallop Seedlings in Saroma Lake Studied in 1969), *Ibid.,* **27** (1), 1-11.

114. OLSEN, A. M., 1955. Underwater Studies on the Tasmanian Commerical Scallop, *Notovola meridionalis* (Tate) (*Lamellibranchiata: Pectinidae*), *Austr. J. Mar. Freshw. Res.,* **6**, 392-409.

115. OKEZUKURI, H., and S. TANAKA, 1969. Hotategai chigai no gaikai horyu koka ni tsuite. Ichi-nenkai shubyo no dairyo horyu kei ko (Effect of Releasing Scallop Spat in the Open Sea: Large-scale Transplantation of 2-year-old and 1-year-old Seedlings), Hokusuishi Geppo (Monthly Report of Hokkaido Fisheries Experiment Station), **26**, 638-655.

116. OSHIMA, WATANABE, SATAKE, SHIOSAWA, OHARA, and MARU [sic], 1966. Hokkaido Saroma-ko no seitagakuteki kenkyu-keiseishi to teishitsu ni tsuite (Ecological Studies of Saroma Lake, History of Its Formation and Floor Materials), *Hokusuishi Hokoku (Bulletin of Hokkaido Suisan Fisheries Experiment Station),* No. 6, 1-32.

117. OTANI and FUJIKAWA [sic], 1934. *Nantaido-butsu no kagaku (Chemistry of Mollusks).* Tokyo: Koseikaku, 208 p.

118. OYAMA, K., 1952. Pecten-rui no koseitaigakuteki kenkyu (Sono ichi) [Ecological Studies of *Pecten* (Report No. 1)] *Shigenkeniho,* 25, 24-30.

119. POSGAY, A., 1953. Sea Scallop Investigations. 6th Rep. Invest. Shellfish, Massachusetts, 8-24.

120. RATHJEN, W. F., *et al.,* 1963. Alaska Exploratory Cruise 63-1, m/v John R. Manning, May 20 to June 14, 1963. U. S. Dept. Interior Fish and Wildl. Serv. Bur. Comm. Fish. Cruise Rep.

121. ——, and RIVERS, J. B., 1964. Gulf of Alaska Scallop Explorations, 1963, *Comm. Fish. Rev., Washington D. C.* **26** (3), 1-7.

122. SASAKI, AKAHOSHI, and HASE [sic], 1970. Hotategai no zoshoku ni kansuru kenkyu fuchaku chigai no chosa [On the Culture of Scallops (Study on Setting Spat)] Report published by Mutsu Bay Fisheries Culture Research Center, Aomori Prefecture, 11, 17-21.

123. ——, ——, and TAKEDA, 1970. *Ibid., Ibid.,* 11, 264-268.

124. SASAKI, H., 1945. Kitami engan no hotategai gyogo ni tsuite (Fishing Conditions of Scallops in Kitami Coast), Hokusuishi Geppo (Monthly Report of Hokkaido Fisheries Experiment Station), **2** (7, 8, 9), 101-102.

125. SASTRY, A. N., 1963. Reproduction of the Bay Scallop, *Aequipecten irradians* Lamarck: Influence of Temperature on Maturation and Spawning, *Biol. Bull.*, **125**, 146-153.

126. SATO, S., and S. SHIBUYA, 1940. Soya shicho kannai ni okeru hoyo hotategai chosa. Sono ichi [Survey on the Scallops Released in Soya (Report No. 1)], Hokusuishi Junpo (Fortnightly Report of Hokkaido Fisheries Experiment Station), **447**, 10-12. Report No. 2, *Ibid.*, **448**, 10-16.

127. SATO, S., 1954. Fuyu chigai shutsugenryo no chosa. Chigai fuchoku jokyo no chosa (hotategai ni kansuru chosa kenkyu) [Number of Planktonic Spat and Conditions of Setting (Study on Scallops)], *Bulletin of Mutsu Bay Fisheries Research Center*, No. 3 (1953), 29-39.

128. ——, 1958. *Ibid.*, No. 4 (1954-1956), 85-89.

129. ——, 1961. *Ibid.*, No. 5 (1957-1959), 10-21.

130. SAVILOV, A. I., 1957. Biological Aspect of the Bottom Fauna Groupings of the North Okhotsk Sea, *Trans. Inst. Oceanogr.* Vol. XX of *Marine Biology* edited by B. N. Nikitin, USSR *Acad. Sci.* Press, Moscow (1957), published by *The American Institute of Biological Science*, Washington, D.C. (1959) 67-136.

131. SEKINE and KAKUTSUNO [sic], 1920. *Suisan koshusho shiken hokoku (Bulletin of Fisheries Training Center)* **16** (1) (as cited by Otani and Fujikawa, 1934).

132. SHIBUYA, S., SHIMIZU, and KINOSHITA [sic], 1942. Hotategai gyojo sensui chosa kekka (Survey of Fishing Beds of Scallops), Hokusuishi Jigyo Junpo (Fortnightly Report of Hokkaido Fisheries Experiment Station), 546.

133. TAKEDA, CHIBA, and FUDUSHIMA [sic], 1966. Hotategai no zoshoku ni kansuru kenkyu fuchaku chigai no chosa: Rikujosuiso shiiku shiken [On the Culture of Scallops: Experimental Tank Culture)], Report published by Mutsu Bay Fisheries Culture Research Center, Aomori Prefecture, 8, 14-20.

134. TAMURA, T., 1939a. Kakushu kaisan kairui no kokyu ni oyobosu gaii henka no eikyo: II. Sansoshohiryo to ondo to no kankei narabini chuyo no eikyo (Influence of Changes in the Environment on the Respiration of Various Marine Mollusks: II. Relationship between Oxygen Uptake and Temperature), *Suisan Zatsu (Journal of Fisheries)*, **43**, 1-10.

135. ——, 1939b. *Ibid.*, III. Kankyosui no yozonsanso ga sanso shohi ni oyobosu eikyo (Influence of Dissolved Oxygen in Water on Oxygen Uptake), *Ibid.*, **44**, 64-72.

136. ——, and A. FUJI, 1954. Hitode no shokusei (Food Habits of *Asterias*). Hokusuishi Geppo (Monthly Report of Hokkaido Fisheries Experiment Station), **11**, 17-21.

137. ——, ——, 1964. Mombetsu-Tokoro okiai no hotategai no seisoku bunpu ni tsuite (Distribution of Scallops in Jono Okiai), Hokkaido Kaihatsu Kyoku Chosa Shiryo (Data published by Hokkaido Development Survey), 306-3-196.

138. ——, ——, M. TANAKA, and A. OHARA, 1956. Hotate chigai no kuchu roshutsu jikan to heishiritsu to no kankei (Relationship between the Duration of Air Exposure and Mortality Rate of Scallop Spat), Hokusuishi Geppo (Monthly Report of Hokkaido Fisheries Experiment Station), **13** (8), 27-34.

139. TANAKA, OHARA, and WAKUI [sic] (Scallop Research Group), 1962. Saikin no hotategai chosa kara erareta chiken ni tsuite (Ideas Derived from Recent

Survey of Scallops), *Ibid.*, **19** (11), 28-39.

140. TANAKA, S., 1963. Hokkaido ni okeru hotategai gyogyo to kenkyujo no shomon-dai (Scallop Fishing and Some Related Problems in Hokkaido) Gyogyo Shigen, Zoshokuhen (Proceedings of the Symposium Held under the Autumn Conference of Japanese Society of Scientific Fisheries) (1963), 142-150.

141. THORSON, G., 1947. Reproduction and Larval Development of Danish Marine Bottom Invertebrates, with Special Reference to the Plankton Larvae in the Sound (Oeresund). Medd. Komm. Danm. Fiskriog Havund. Ser. Plankton, Bd. 4, Nr. 1, 1-523.

142. TSUBATA, ITO, KANNO, and AKAHOSHI [sic], 1968. Hyoryubin ni yoru Mutsu-wan nai choryu chosa (Study on Tidal Currents in Mutsu Bay), Report published by Mutsu Bay Fisheries Culture Research Center, Aomori Prefecture, **10**, 210-216.

143. ——, ——, ——, ——, HASE, and SASAKI [sic], 1968. Fuyu yosei no chosa (Study of Planktonic Larvae), *Ibid.*, **10**, 9-14.

144. TUBB, J. A., 1946. The Tasmanian Scallop (*Pecten medius* Lamarck): I. First Report on Tagging Experiment, *J. Counc. Sci. Ind. Res.* **19**, 202-211.

145. TURNER, H., and J. H. HANKS, 1960. Experimental Stimulation of Gameto-genesis in *Hydroides dianthus* and *Pecten irradians* during the Winter, *Biol. Bull.*, **119**, 145-152.

146. VAN DAM, L., 1954. Active and Standard Metabolism in Scallop, *Ibid.*, **107**, 192-202.

147. VERWEY, J., 1952. On the Ecology of Distribution of Cockle and Mussel in the Dutch Waddensea, Their Role in Sedimentation, and the Source of Their Food Supply, with a Short Review of the Feeding Behavior of Bivalve Mollusks, *Arch. Neerl. Zool.*, **10**, 172-239.

148. WAKUI, T. and A. OHARA, 1967. Saroma-ko ni okeru Hotategai *Patinopecten yessoensis* (Jay) seishoku sono shunenhenka ni tsuite [Changes in the Gonads of *Patinopecten yessoensis* (Jay) of Saroma Lake], *Hokusuiken Hokoku (Bulletin of Hokkaido Fisheries Experiment Station)*, **32**, 15-22.

149. WATANABE, T., 1956. Akoyagai, *Pinctada martensii* (Dunker) no dai ichigen kara, chigai kaigara oyobi shinju no denshisen Kaiseki [Electron Analysis of Spat Shell and Pearl of *Pinctada martensii* (Dunker)], *Kagaku (Science)*, **26** (7), 359-360.

150. ——, and R. YUKI, 1952. Akoyagai no fuyushichu yori fuchaku chigai ni itaru kan no kaigara kobutshu seibun ni tsuite (Yoshi) (Mineral Components of the Shell of Pearl Oysters from Planktonic Larval Stage to Setting Spat), *Dozatsu (Zoological Magazine)*, **61**, 118.

151. WENRICH, D. H., 1916. Notes on the Reaction of Bivalve Mollusks to Changes in Light Intensity: Image Formation in *Pecten*, *J. Anim. Behav.* **4**, 297-318 (Cited from Gutsell, 1930).

152. YAMAMOTO, G., 1943. Hotategai *Patinopecten yessoensis* Jay no seishoku saibo keisei narabini seishobu jiki (Gametogenesis and the Reproduction Period of *Patinopecten yessoensis* Jay), *Nihon Suisan Gakkaishi (Bulletin of the Japanese Society of Scientific Fisheries)* **12**, 21-26.

153. ——, 1949. Hotategai yosei no shiiku narabini fuchaku chikukai ni tsuite (Rear-

ing of *Patinopecten* Larvae and Study of Setting Spat), Proceedings of the 19th Meeting of the Japanese Zoological Association.

154. YAMAMOTO, G., 1950b. Ecological Note on the Spawning Cycle of the Scallop, *Pecten yessoensis* Jay, in Mutsu Bay, *Sci. Rep. Tohoku Univ.* Ser. IV (*Biol.*), **18**, 477-481.

155. ———, 1950c. Benthic Communities in Mutsu Bay, *Ibid.*, **18**, 482-487.

156. ———, 1951a. Induction of Spawning in the Scallop, *Pecten yessoensis* Jay, *Ibid.*, **19**, 7-10.

157. ———, 1951b. Ecological Note on Transplantation of the Scallop, *Pecten yessoensis* Jay, in Mutsu Bay, with Special Reference to the Succession of the Benthic Communities, *Ibid.*, **19**, 11-16.

158. ———, 1951c. On Acceleration of Maturation and Ovulation of the Ovarian Eggs in Vitro in the Scallop, *Pecten yessoensis* Jay, *Ibid.*, **19**, 161-166.

159. ———, 1951d. Mutsu-wan-san hotategai no zoshoku ni kansuru kenkyu-V (On the Reproduction of the Scallop of Mutsu Bay, II), Report of the Aomori Prefecture Fisheries Resources Survey, Report No. 2, 29-40.

160. ———, 1951e. Mutsu-wan-san hotategai no sanran no hendo (Changes in the Spawning of Scallops in Mutsu Bay), *Nihon Suisan Gakkaishi (Bulletin of the Japanese Society of Scientific Fisheries)*, **17**, 53-56.

161. ———, 1952. Further Study on the Ecology of Spawning in the Scallop, in Relation to Lunar Phases, Temperature, and Plankton, *Sci. Rep. Tohoku Univ.*, IV Ser. (*Biol.*), **19**, 247-254.

162. ———, 1953. Mutsu-wan-san hotategai no zoshoku ni kansuru kenkyu-III (On the Reproduction of Scallop in Mutsu Bay, III), Report of the Aomori Prefecture Fisheries Resources Survey, Report No. 3, 4-13.

163. ———, 1953a. Ecology of the Scallop, *Pecten yessoensis* Jay, *Sci. Rep. Tohoku Univ.*, IV Ser. (*Biol.*), **20**, 11-32.

164. ———, 1954. Saikin no zoshoku jigyo to kenkyu men no shinpo (Recent Progress in the Study of Fisheries Culture), *Suisan Kagaku*, **15**, 1-5.

165. ———, 1955. On Rearing Scallop Spat in Tanks and Pools, *Bull. Mar. Biol. Stat. Asamushi, Tohoku Univ.*, **7**, 69-73.

166. ———, 1956a. Habitats of Spat of the Scallop, *Pecten yessoensis* Jay, Which Turned to Bottom Life, *Sci. Rep. Tohoku Univ.*, IV Ser. (*Biol.*), **22**, 149-156.

167. ———, 1956b. Shuju no seicho dankai no hotategai no kankyo ni taisuru teikosei ni tsuite, tokuni era-senmo, undo ni taisuru kendaku fudei sanso ketsubo nado no eikyo (Resistance of Scallops in Various Stages of Growth to Environmental Conditions: Study on the Influence of Suspended Silt and Low Oxygen on the Ciliary Movement of the Gills), *Nihon Seitai Kaishi (Journal of Japanese Ecology)*, **5**, 172-175.

168. ———, 1957a. Tolerance of Scallop Spat to Suspended Silt, Low Oxygen Tension, High and Low Salinities, and Sudden Temperature Changes, *Sci., Rep. Tohoku Univ.*, IV Ser. (*Biol.*), **23**, 73-82.

169. ———, 1957b. Mutsu-wan-san hotategai no zoshoku ni kansuru kenkyu no saikin no shinpo (Recent Progress in the Culture of Scallops in Mutsu Bay), Aomori Prefecture, 10 p.

170. ———, 1960. Hotate gyojo ni okeru seisan teika gensho Monbusho kaken sogo

kenkyu (6042). Yoshokuja ni okeru seisan teika gensho to sono taisaku ni kansuru kenkyu [Decline in the Production of Scallops in Various Fishing Grounds: Study of Measures to Prevent the Decline of Production in Various Culture Centers under the Sponsorship of the Ministry of Education, Government of Japan (6042)]. Project led by Prof. Imai, 21-24.

171. YAMAMOTO, G., 1963. Kaiyo ni okeru busshitsu junkan (yoshi) (Circulation of Various Substances in the Sea), *Nihon Seitai Tohoku Chiku Kaiho*, **16**, 2-4.

172. ——, 1964. Mutsu-wan ni okeru hotategai zoshoku (Scallop Culture in Mutsu Bay). Data published by the Association for the Protection of Marine Resources, *Suisan Zoyoshoku Gyosho*, **6**, 77 p.

173. ——, 1967. Food Relations of Dominant Animals in Marine Benthic Communities in Mutsu Bay, *Sci. Rep. Tohoku Univ.*, IV Ser. *(Biol.)*, **33**, 519-526.

174. ——, and EWATARJ [sic], 1950. Noheji-wan ni okeru hotate chigai no ishoku ni tsuite (Transplantation of Scallops from Nohenji Bay). Data published by the Survey Group in Aomori Prefecture, No. 10, 56-60.

175. ——, ——, 1951. Mutsu-wan-san hotategai no ishoku ni kansuru ichi kosatsu (On the Transplantation of Scallops from Mutsu Bay), *Aomori Ken Suisan Shikenjo and Suisan Joho (Bulletin of Fisheries Experiment Station in Aomori Prefecture)*, No. 3, 85-87.

176. ——, and T. HABE, 1958. Fauna of Shell-bearing Mollusks in Mutsu Bay, Lamellibranchia, I, *Bull. Mar. Biol. Stat. Asamushi, Tohoku Univ.*, **9**, 1-20, Pls. 1-5.

177. ——, ——, 1959. *Ibid.*, II, *Ibid.*, 9, 85-121, Pls. 6-14.

178. ——, ——, 1962. *Ibid.*, Scaphopoda and Gastropoda I, *Ibid.*, **11**, 1-20, Pls. 1-3.

179. ——, and KANNO, 1963. Hotategai yogai rosuiryo no nisshu henka (Daily Changes in the Amount of Water Filtered by Scallop Larvae), Proceedings of the Autumn Meeting of the Japanese Society of Scientific Fisheries. (Otaru).

180. ——, *et al.*, 1950a. Mutsu-wan-san hotategai no zoshoku ni kansuru kenkyu (Study of the Culture of *Patinopecten* of Mutsu Bay), Report of the Aomori Prefecture Fisheries Resources Survey, Report No. 1, 145-167.

181. YONGE, C.M., 1936. The Evolution of the Swimming Habit in Lamellibranchia, *Mem. Mus. Roy. D'hist. Nat. de Belgique*, Deuxieme Serie, Fasc. 3, 77-100.

182. ——, 1946. Digestion of Animals by Lamellibranches, *Nature*, London, **157**, 729.

183. ZAMA, K., M. HATANO, and H. IGARASHI, 1960. Nantai dobutsu rinshishitsu (Phopholipid in Mollusks), *Nihon Suisan Gaikkaishi (Bulletin of the Japanese Society of Scientific Fisheries)*, **26**, 917-920.

184. ZOBELL, C. E., and C. B. FELTHAM, 1938. Bacteria as Food for Certain Marine Invertebrates, *J. Mar. Res.*, **1**, 312-327.

PART IV

The Evolution of Abalone Culture

CHAPTER I

Biological Research on Abalones

1. Introduction

Edible mollusks are popular as raw and canned food. One such mollusk, the abalone, has gained popularity in Japan throughout the years, and delicacies such as *noshi-awabi* (in which the meat is sliced, stretched, and dried), *meiho* (in which the meat is cooked and then dried), *kaiho* (in which the meat is baked and dried), etc., are longstanding favourites. The shells of abalones are widely used in Japan for ornamental purposes.

The scientific study of the abalone began in Japan many years ago and scholars and scientists like Matsuhara (1882, 1883, 1885), Kishigami (1894, 1895), Uchimura (1881), and others, are renowned for their investigations. Tago (1931), Uchida and Yamamoto (1942), and others, studied their distribution, while Kinoshita studied the spawning period, growth, food habits, etc., of *Ezo-awabi (Haliotis discus hannai Ino)* (1927, 1934, 1936, 1937, 1947, 1949, 1950). Okamura and Seno (1918, 1920) and Ueda and Okada (1939, 1941) studied the marine algae which serve as food for the abalone. There are also reports on the classification, embryology, food habits, catches, etc., by Murayama (1935) and Iino (1936, 1937, 1941, 1943, 1946, 1947, 1952).

Management measures taken in the prewar years relating to abalone cultivation, were confined to imposing restrictions on the period of harvesting, the banning of fishing in certain regions, the protection of abalone spat, the methods of fishing, extension of the habitats of abalones by the stone-scattering method, etc. The transplanting of mother abalones was also tried during these years. In the postwar period, fishing operations in the coastal regions were taken up, and after introducing several measures, the problems involved in the selective cultivation of high-grade abalones of better food value were studied. The international regulations on fishing operations in the far seas have, since 1960, brought about several corresponding changes in the conditions governing fishing in Japan and "from mere hunting to cultivation" became the ruling slogan. This led to active culture along the coast in the various prefectures. Plans already exist for adopting the technique of seedling culture of abalone to a larger, more commercial scale. Experiments toward this end are being carried out at Iwate, Miyagi, Ibaraki, Chiba, Kanagawa, Shizuoka, Mie, Tokushima, Yamaguchi, and Akita Prefectures, and the results obtained to date are very encouraging.

Abalones are expensive and take a long time to ready for market. Seedlings are released in natural fishing grounds and reared on natural feed. It is necessary therefore to study the oceanography, ecology, and resources in the places where abalones can be gathered. These aspects are under study by Sakai (1959, 1960, 1962), Uno

(1967), and the Shizuoka Fisheries Experiment Station (1968), as well as other Fisheries Experiment Stations, Fisheries Research Centers, and University Faculties of Fisheries.

(Iino)

2. Classification and distribution of the genus Haliotis

There are about 100 species of abalones in the world. They are found in both the northern and the southern hemispheres, but the larger varieties exist in temperate regions, while the smaller ones live in tropic and arctic regions. Abalones are commercially important in Japan, Korea, the Pacific coast of North America, South Africa, South Australia, and New Zealand.

2.1 Species found in Japanese waters and their distribution

The following 10 species of *Haliotis* are found in Japan:

1. *Haliotis (Haliotis) asinina* Linne	*Mimigai*
2. *H. (Euhaliotis)* *gigantea* Gmelin	*Madaka*
3. *H. („) sieboldii* Reeve	*Megai*
4. *H. („) discus* Reeve	*Kuro*
5. *H. („) discus hannai* Ino	*Ezo-awabi*
6. *H. (Sanhaliotis) diversicolor* Reeve	*Fukutokobushi*
7. *H. („) diversicolor supertexta* Lische	*Tokobushi*
8. *H. („) varia* Linné	*Ibo-anago*
9. *H. („) crebrisculpta* Sowerby	*Chirimen-anago*
10. *H. (Padollus) ovina* Gmelin	*Ma-anago*

Among the above-named species, *H. asinina (Mimigai)*, *H. varia (Ibo-anago)*, and *H. ovina (Ma-anago)* are very small in size and few in number. They are also low in value and found in the tropic and subtropic regions.

H. diversicolor supertexta (Tokobushi) and *H. diversicolor (Fukutokobushi)* differ from common abalones by having 6 to 9 inhalant siphons on the dorsal shell (3 to 5 in common abalone) and a smaller shell size. The distribution of the two species is also restricted to warm-current regions. Compared to *H. diversicolor supertexta,* the distribution of *H. diversicolor* is greater in regions with a higher water temperature. The abalones of Hachijojima and Taneshijima are mostly *H. diversicolor*. Sometimes these are used for canning, but they are not as commercially important as common abalones. The following are the four important species:

1. *Madaka (H. gigantea)*: Among abalones this species is the maximum in size with a shell length that extends to 25 cm. The shell is very convex and 4 to 5 inhalant siphons protrude prominently from it. The sole of the foot is mostly light brown in color, and the meat of the foot soft, and suited for *meiho;* the upper surface of the foot is wrinkled. *Madaka* is found in deep waters, sometimes in waters of even 50 m in depth, and have been located in the Pacific Ocean south of Ibaraki Prefecture and in the coastal regions of the Sea of Japan southwest of Hokkaido.

2. *Megai (H. sieboldii)*: The shell is round, about 17 cm in size, and thin, but the

ventral surface is flat and wide. Regularly radiating lines and a few rough patches mark the back of the shell. The inhalant siphon is short. The sole of the foot is light brown in color and the meat is soft. The upper surface of the foot protrudes. The geographical distribution is more or less the same as that of *H. gigantea (Madaka)*, but it dwells in considerably shallower waters.

3. *Kuro (H. discus)*: The shell is elliptical, the cavity narrow, and old shells 20 cm in size. Compared to other species, the umbo lies at a higher level and the surface of the shell is smooth. The inhalant siphon is longer than in *Megai* but shorter than in *Madaka*. The sole of the foot is black in color with a blue edging, and the meat is somewhat hard. The protuberance on the upper foot is complex and slimy, and suited for *kaiho*. The geographical distribution is more or less the same as that of the foregoing species, but the vertical distribution is even shallower.

4. *Ezo-awabi (H. discus hannai)*: The shell is very small, not exceeding 13 to 14 cm, and thin and elliptical. The inner labium is narrow, and the dorsal surface very wrinkled. The number of inhalant siphons is 3 to 5. In most cases the sole of the foot has a black margin. Some soles are yellow in color. The upper foot protuberance is similar to that of *H. discus*. Distribution is limited to shallow seas; in Okushiri-To it extends to 1 to 3 m (Nakajima *et al.*, 1953) and in Miyagi Prefecture, 3 to 5 m. Harvesting this species is easily done by boat fishing.

In the past, this species was considered the same as the *Haliotis kamtschatkana* of the Pacific coast of North America. Morphological differences were eventually noted however, and the two species distinguished. In comparing the ecological and morphological characteristics of *H. discus* and *H. discus hannai*, the latter were once considered a regional strain of the former. Compared to *H. discus*, *H. discus hannai* is comparatively smaller in size, and its shell thinner and slightly longer; wrinkles on the shell are characteristic of this species, and its distribution is restricted to shallow waters. *H. discus hannai* exhibits all these differentiating characteristics in such places as Hokkaido and Tohoku where the existing environmental conditions are similar to those of Higashi Nagashima Cape, south of Kyushu. Nevertheless, the fact that *H. discus hannai* of the northern region gradually changes to *H. discus* when transplanted to the southern regions proves that these are not two entirely different species. Hence, Iino grouped them together in the subspecies *H. discus hannai* after studying a number of specimens. The subspecies was named after Dr. Hanna of the California Academy of Sciences in San Francisco who supplied a number of specimens to Dr. Iino.

This species is not found in the eastern part of Hokkaido, which always has cold currents, but only along the Pacific coast north of Ibaraki Prefecture, and the coast of the Japan Sea off Hokkaido. The northern limit of natural distribution in Hokkaido is Shiya Strait and the southern limit is Ikuyama Jiku. The distribution on the northwest side extends to Okushiri-To.

(Iino)

2.2 Species found in other countries and their distribution

The following is a brief account of the regional distribution of abalone species throughout the world.

A single specimen of *H. pulcherrima* was collected from the deep offshore waters of Florida in 1859, and again in 1913. Another smaller specimen was also found in Rio de Janeiro, Brazil. There are no other reports of this species being found in the Atlantic Ocean. On the Pacific coast, however, distribution ranges from Sitka, Alaska, to Cape San Lucas of Baja California, Mexico. It is not found in South American waters, and so far only one very small specimen of *H. dalli* has been reported as found off Galapagos Island, a survey group from Japan recently visited these places but were unable to obtain specimens.

The following are the commercially important species:
1. *H. refescens* Swainson (Red Abalone)
2. *H. fulgens* Philippi (Green Abalone)
3. *H. corrugata* Gray (Pink Abalone)

The following species are often found with the aforementioned varieties:
4. *H. kamtschatkana* Jonas (Pinto Abalone)
5. *H. cracherodii* Leach (Black Abalone)
6. *H. walallensis* Stearns (Flat Abalone)
7. *H. sorenseni* Bartsch (White Abalone)
8. *H. assimilis* Dall (Threaded Abalone).

Except for *H. sorenseni*, these species are small, but in Japan they are still marketable as food, and are available in large quantities in shallow waters. In size, *H. sorenseni* is similar to the *Madaka* (*H. gigantea*) of Japan but its distribution is limited to 80 to 100 feet depths, and the quantity is small.

Abalones are distributed in the coastal regions of Africa from Tanzania, Mozambique, Natal, Madagascar, the Cape of Good Hope, the Gold Coast, and in the coastal regions of Islands such as Kanaliya, Madeira, Azores, etc. However, of all the species distributed in these regions, only *H. midae* of Cape Town is utilized on a commercial scale. All the other species are small in size and quantity. In Europe, *H. tuberculata* and *H. lamellosa* are found, but are not utilized because of their small size. However, the embryology, physiology, morphology, etc., of *H. tuberculata* have been studied in detail.

In the Indian Ocean, abalones are found from the Arabian Sea and the Persian Gulf to the Bay of Bengal around Andaman and Nicobar Islands, as well as near Sri Lanka (Ceylon). They are also found in the coastal regions of Cambodia, Thailand, Malaysia, Indonesia, Borneo, Celama, the Philippines, Taiwan, Malacca, etc., but their size is small.

Korea ranks second to Japan in abalone interest. *H. discus hannai* is the most important species, commercially speaking, and is widely distributed from Petropavlovsk to Yamato Hanto of China after skirting the western coast of Korea. In Korea, Tago (1931) reported that *H. discus hannai* is found on the eastern coast and the warm-water strains such as *Madaka, Megai, Kuro* in other regions. According to Uchida and Yamamoto (1942), *H. discus hannai* is distributed in February up to the isothermal line of 12°C of water temperature, at a depth of 25 m from the northern part of the Korean coast and Chiju Do toward both the east and west. On the southern part, *Kuro, Megai,* and *Madaka* are distributed in warm currents. For purposes of transplantation, *H. discus hannai* is at present available only in Hokkaido. But if and when proper conditions are created, it should be possible to transplant specimens from

Korea to warm regions on the western part of Honshu and Kyushu. This would lead to a renewed effort in producing better abalones.

In the coastal regions of Australia and New Zealand, abalones are found in Melanesia and Polynesia but their size is small. Large sized abalones are found in 30° to 50°S of Australia, New Zealand, and Tasmania. The harvesting and export of abalones from these regions are gradually increasing. The following are the important species of these regions:

Australia:
1. *Notohaliotis ruber* (Leach)
   Red-Ear-Shell or Black-Lip Abalone
2. *Schismotis laevigata* (Dcmevan)
   Mutton-Fish or Green-Lip Abalone

New Zealand:
3. *Haliotis iris* Martyn.

According to FAO statistics (1965), the total world production of abalone is 16,040 tons, with Mexico producing 7,800 tons, Japan 4,300 tons, America 1,800 tons, South Africa 1,700 tons, and Australia 400 tons.

(Iino)

3. Ecology

Abalones live in rocky belts where the waves are very rough. Their shells are strong and suited to such environmental conditions. The sexes are separate and the larvae which hatch from the eggs through external fertilization, lead a planktonic existence for some time before settling to the bottom of the sea where they enter a sessile life. According to Iino (1952), the difference in the vertical distribution of mature abalones might be related to the difference in the capacity of the planktonic larvae of various species to move away from light. The spat of *H. discus hannai* live in comparatively shallow waters and move toward deeper waters as they grow into adults, but during spawning the mother abalones move back toward shallow waters (Kishigami, 1894; Iino, 1952). On the other hand, Uno (1967) compared the variations in the accumulation rate by $A = \sqrt{\Sigma \alpha^2 / n}$ (α = deviation from mean density, n = number of regions observed) of *H. sieboldii* in their habitats, and found that the rate was at its highest peak at the time of spawning. Hence he concluded that not only do abalones move up to shallow waters, but they also congregate at particular places at the time of spawning.

Abalones are nocturnal, remaining attached to stones or in spaces between rocks during the day. At night they creep around in search of food at which time nets are used to collect them in the Hokkaido and Tohoku regions. In some parts of Kagoshima Prefecture, abalones are even caught with fishing rods from boats at night. According to experiments carried out in tanks, the creeping speed of an *H. discus* of 11.6 cm shell length is 85.2 m per night, and the creeping speed of an *H. gigantea* of 11.2 cm shell length is 35.4 m per night (Kaneda, 1953; Uno, 1967). However, that abalones prefer to remain in one place, particularly in rocky belts rich in food, is apparent from the fact that they seldom migrate to distant places. Very few reports are available on the migration of abalones under natural conditions. Observations in Okushiri-To (Saito, 1965) indicated that abalones moved 100 to 160 m in 10 months. Another report mentions that abalones in the Ibaraki Prefecture migrated to places about 180 m distant in one year. The activity of abalones differs greatly according to the species and its size. *H. discus* is much more active than *H. sieboldii*, and *H. gigantea* and larger abalones move more slowly than the smaller ones (Iino, 1952). In the aged, movement is still more sluggish. The holdfast of abalones plucked from a rocky surface has a distinct scar, which Japanese fishermen call "meshiro" or "tori-ato." Whether abalones remain fixed to rocks so firmly as to be scarred when plucked and, this being so, how do they obtain their food while attached, are important points to be explored. Do they live on seaweed brought near them by water movements, or do they detach themselves to search for food at night and then return to their original place of fixation? Not only are these queries interesting from an ecological point of view, they are extremely significant to abalone culturists.

The spawning period of abalones differs according to species, region, sea conditions, etc. The spawning period of *H. discus hannai* differs according to the place (Table 4.1). While the spawning period in Hokkaido is between late July and late November, the peak period is between late August and late September. The spawning period on the Sanriku coast is between late July and late October, and the peak between early September and mid-October. In either case, however, the water temperature is around 20°C (Kinoshita, 1950). The spawning period of the strains in warm currents is between mid-October and late December, and the peak occurs in November. Spawning commences when the mean temperature is 20°C.

Creeping larvae reportedly find the following acceptable as feed: *Platymonas* (Shiraishi and Imai, 1964), *Amphora* sp. (Fisheries Experiment Station of Iwate Prefecture, 1966), *Navicula* sp., *Cocconeis* sp., *Melosira* sp. (Shibui *et al.*, 1966), *Navicula* sp., *Nitzschia* sp., and other adhering diatoms (Kikuchi, 1963), *Chaetoceros simplex* (Umebayashi, 1961), etc. Kikuchi (1964) has pointed out that the planktonic larvae enter into the creeping stage after 3 to 4 days, and hence it is necessary to maintain the culture of these diatoms so that feed is available to the larvae. Sakai, on the other hand, found that *Melosira, Navicula*, etc., are good feed for adult abalones (Sakai, 1962).

When the spat grow to the shell length of 4 to 5 mm, they are able to eat small and soft algae. At this stage, plumules of *Laminaria japonica (Konbu)* or *Undaria pinnatifida (Wakame)*, as well as *Ulva pertusa (Ana-aosa)*, and diatoms, are said to be effective feed (Fisheries Experiment Station of Iwate Prefecture, 1966). Kikuchi *et al.* (1967) carried out feeding experiments with 20 types of seaweed on spat of 3 cm shell length and found that *Undaria pinnatifida, Eisenia bicyclis, Codium fragile, Aalymenia accuminata, Ulva pertusa*, etc., were acceptable.

Regarding the relationship between feed and growth, adult abalones show greater selectivity than the younger (Table 4.2). When various types of seaweeds were individually provided, the order of preference was brown algae followed by green, then red. Feed was selected in the same order when the seaweed were mixed.

The color and pattern of shells differ according to the type of feed taken. When the abalones were fed on brown algae, green algae, and diatoms, their shells developed a bluegreen edge, and when fed on red algae, their shells were brown. The physio-

Table 4.1. Spawning period of *H. discus hannai* (*Ezo-awabi*)

Regions	Spawning period	Peak of spawning period	Water temperature (°C)	Reference
Yakishiri-To (Hokkaido)	Late August to Early October	Mid-September	20	Ono (1932)
Mashige "	Early August to Mid-October	Mid-September	"	"
Shiotani "	Early August to Late October	Early and Mid-September	"	"
Karo "	Late July to Mid-October	Late August to Early September	"	"
Okushiri-To	Mid-September to Late November	Late September	"	"
Kube "	Mid-September to Mid-October	Late September	"	"
Fukuyama "	Early August to Late October	Late August to Mid September	"	"
Shiriya (Aomori)	Mid-August to Mid-October		17-24	Takahashi (1965)
Furuhama (Iwate)	Late July to Early October	Late September to Early October	18.5-20.5	Aonuma (1953)
Taro (Iwate)	Early August to Mid-October	Late September to Early October	18.0-20.0	"
Oshima (Miyagi)	Early September to Mid-October	Mid-September to Mid-October	16-22	Onodera (1957)
Kazu "	Early September to Mid-October	Mid-September to Mid-October	16-22	"
Onagawa "	Late August to Mid-October	Early September to Early October	17-22	Sakai (1960)

Table 4.2. Feed and growth of *H. discus hannai* **fed on various types of seaweeds**
(Sakai, 1962)

Type of feed	Feeding rate per day	Growth rate	Increase in weight	Selectivity	Conversion rate	Remarks
Undaria pinnatifida	17.6%	9.2%	32.1%		5.07%	Mean water temperature, 20.3°C.
Ulva pertusa	12.4	4.9	8.5		2.29	Period of rearing, 31 days
Covipopeltis affinis	7.6	2.2	3.6		1.50	(July to August, 1958)
Undaria pinnatifida	12.3			73.0		
		4.6	30.7		4.90	
Ulva pertusa	4.7			27.0		
Undaria pinnatifida	13.3			84.2		
		7.8	29.6		5.28	
Covipopeltis affinis	2.5			15.8		
Ulva pertusa	8.7			81.7		
		4.1	6.8		1.98	
Covipopeltis affinis	1.9			18.3		
Undaria pinnatifida	22.4	11.2	40.7	—	4.87	
Calcareous algae						

logical functions are not clear but in future aquacultures, shell color might reveal the environmental conditions of the habitats and make it possible to assess the suitability of the abalone for release.

The feeding activity and growth of *H. discus hannai* are closely related to temperature. When the temperature is below 7°C, movement is very sluggish and the abalones almost cease to feed. When the abalones were grown by feeding with *Undaria pinnatifida,* as shown in Table 4.3, they consumed an amount equal to 7.5% of their body weight in one day at 10.9°C. When the temperature was 20.3°C, they consumed 17.6% of their body weight. The increase in their body weight in one month was 12.8% and 32.1% respectively. Hence, the water temperature suited for the growth of *H. discus hannai* is 15 to 20°C. The movement of *H. discus hannai* is rapid for 2 to 3 hours after sunset and 2 to 3 hours before sunrise, during which time they feed voraciously (Sakai, 1962).

Table 4.3. Relationship between temperature, feeding rate, and growth
(Sakai, 1962)

Water temperature, °C	Feeding rate per day (%)	Growth rate per month (%)	Increase in weight per month (%)	Conversion rate (%)
10.9	7.5	2.1	12.8	4.7
14.7	—	2.6	23.9	—
16.5	14.7	4.7	27.0	3.9
20.3	17.6	9.2	32.1	5.1

Abalone growth differs in various regions. Since annual rings form during the spawning period when growth stops temporarily, the age of abalones is calculated by counting these rings which become prominent when the shells are heated (Sakai, 1960; Takayama, 1940). According to the data (Table 4.4), abalone growth slows down as the latitude increases, and in Hokkaido, it takes 5 to 7 years for the abalones to reach the harvestable shell length prescribed under Fisheries Regulations (7.5 cm), in the Miyagi Prefecture 4 to 5 years (9 cm), the Ibaraki Prefecture 4 to 6 years (11 cm), and the Mie Prefecture 4 to 5 years (10.6 cm).

Table 4.4. Growth rate of *H. discus hannai* **in cm**

Region	Age, years								Reference
	1	2	3	4	5	6	7	8	
Reibun-To (Hokkaido)	2.8	3.9	5.3	6.4	7.3	7.9	8.4	8.6	Saito (1963)
Tenun-To (Hokkaido)	2.0	3.3	5.4	6.0	7.5	8.1	9.0	9.6	Fisheries Experiment Station, Hok-kaido (1927)
Okushiri-To (Hokkaido)	1.6	2.9	4.1	5.4	6.4	7.2	7.8	8.0	Nakajima *et al.* (1953)
Onagawa-Machi (Miyagi)	2.9	4.7	6.7	8.4	9.5	10.1	—	—	Sakai (1962)
Ojika-Machi (Miyagi)	2.8	4.8	7.3	9.1	10.9	12.2	—	—	Sakai (1962)
Ibaraki Prefecture	3.1	5.7	8.3	9.7	10.6	11.6	12.5	—	Fujimoto *et al.* (1964)

Seasonal variations in the growth of abalones are related to the intake of feed. In *H. discus* the feed intake decreases sharply during the spawning period, and when spawning is completed in December, increases with appetite, being maximum during April to June, and decreasing to minimum in October to November. Growth shows a more or less corresponding pattern.

Corpulence increases in *H. discus hannai* from February to July, and following spawning a gradual thinning occurs (Figure 4.1). In September to October, the abalone is extremely lean. The seasonal variations of corpulence are related to the maturation of the gonads and spawning (Sakai, 1962).

(Sakai and Iino)

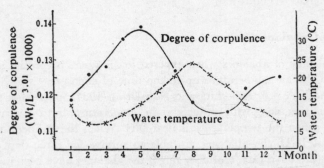

Figure 4.1. Seasonal variations in the degree of corpulence in *H. discus hannai* (Sakai, 1962).

4. Physiology

4.1 Environmental conditions, oxygen consumption, and feeding rate

Tamura (1939) found that the oxygen consumption of *H. discus hannai* differs greatly according to water temperature and the period of the day, i.e. daytime and nighttime (Table 4.5). While the oxygen consumption is 16.8 to 46.6 ml/l at 13 to 14°C

Table 4.5. Oxygen consumption of *H. discus hannai*
(Tamura, 1939)

Temperature	Date	Number of specimens examined	Shell length (cm)	Number of measurements	Oxygen consumption (ml/kg/h)		Daytime/ nighttime
					Mean	Range	
5.5°C	February 27-29	5	5.5-6.6	11	16.42	12.15-20.32	Daytime
13.0	November 10	6	4.5-6.5	4	18.28	16.80-20.00	Daytime
13.0	November 10	6	4.5-6.5	1	46.60	46.60	Nighttime
14.0	November 9	6	4.5-6.5	1	23.60	23.60	Daytime
14.0	November 9	6	4.5-6.5	2	33.60	33.10-34.50	Nighttime
22.0	August 9-10	10	5.5-6.5	6	57.41	52.20-61.80	Daytime
22.0	August 9-10	10	5.5-6.5	1	63.80	63.80	Nighttime
23.0	August 11-12	10	5.5-6.5	5	63.92	57.30-78.00	Daytime
23.0	August 11-12	10	5.5-6.5	4	74.27	65.20-85.30	Nighttime

of water temperature, it is 52.2 to 85.3 ml/l at 22 to 23°C, or almost twice the amount of oxygen consumed at 13 to 14°C. Since oxygen consumption is always more at night, it seems closely related to the activity of the abalone. According to Sano *et al.* (1962), oxygen consumption increases with a temperature increase of sea water up to 24°C, but slows down when the temperature exceeds 26°C. There is no appreciable change in oxygen consumption as long as the chlorine content of the sea water is above 14°/$_{oo}$. When this content is below 13°/$_{oo}$, oxygen consumption decreases rapidly. It also decreases when the ammonia content increases, the decrease being rapid when that content is 10 μ g-atoms/l (Figures 4.2, 4.3, and 4.4).

The feeding rate also decreases when the concentration of nitrogen, in the form of ammonia, increases (Figure 4.5).

4.2 Vitality of abalones exposed to air

The survival time of abalones when exposed to air varies according to the living conditions and the air temperature up to the time of exposure.

As shown in Table 4.6, Kinoshita and Nagakawa (1934) found that the survival time in air decreases along with an increase in temperature, but the survival time is also shortened when the temperature is near 0°C. When the temperature is maintained at about 10°C, the abalones are able to survive for three days. According to Iino (1941), there is a high mortality rate if the abalones have been injured in the part between the head and the fleshy body.

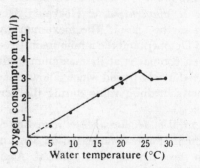

Figure 4.2. Oxygen consumption and water temperature (Sano *et al.*, 1962).

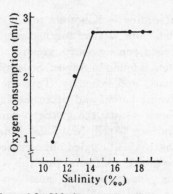

Figure 4.3. Chlorine content and oxygen consumption (Sano *et al.*, 1962).

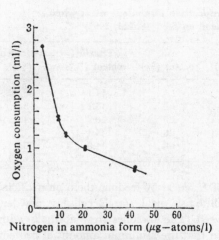

Figure 4.4. Oxygen consumption at various concentrations of nitrogen in ammonia form (Sano *et al.*, 1962).

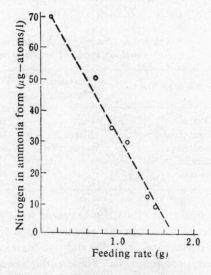

Figure 4.5. Feeding rate at various concentrations of nitrogen in the form of ammonia (Sano *et al.*, 1962).

Table 4.6. Vitality and weight decrease of abalones when exposed to air (Kinoshita and Nakagawa, 1934)

Mean temperature, °C	27.3	20.7	10.6	5.8	−0.78
Mean survival time	16	35	94	140	120
Weight decrease (%)	15.2	15.9	13.0	12.7	7.7

4.3 Seasonal changes in chemical composition

Abalones are very rich in protein and their glycogen content is almost as good as the oyster's.

According to Kinoshita *et al.* (1950), there are a few seasonal changes in the ash content and total nitrogen content of the *H. discus hannai* of Hokkaido. However, glycogen content varies greatly according to the season. The maximum glycogen content is found in August, September, and October, and the minimum in December, January, and February. Thus the glycogen content is at its maximum during the spawning period and decreases during the thinning period which follows spawning. In other words, the glycogen content shows medium value during the period of corpulence (April, May, and June).

Sakai (1962) carried out analyses of the meat of *H. discus hannai* reared by feeding on seaweed, and found that the water content of the abalones fed on *Sargassum fulvellum (Hondawara)* was much less than in those fed on other seaweed. Again, the meat stored much more glycogen when the abalones were fed on *Latimaria* sp. *(Hosomekonbu)* or *Undaria* sp. *(Wakame)* than when they were fed on *Akahada* or *Sargassum fulvellum*. However, there was no appreciable change in the protein content (Table 4.7).

Table 4.7. Changes in the chemical composition of abalone meat when various types of seaweed were used as feed (Sakai, 1962)

Type of feed	Duration of rearing	Water content (%)	Ash content (%)	Protein content (%)	Glycogen (%)
Wakame		72.6	2.4	14.1	3.7
Hosomekonbu	March to	76.0	1.9	13.1	3.3
Hondawara	August, 1954	84.0	1.6	10.2	1.0
Akahada		77.6	1.8	14.1	1.7
Ana-aosa		76.7	1.8	14.0	3.1

Iino *et al.* (1963, 1964) reared some spat of *H. discus* by feeding them an artificial diet of gel of calcium algenate mixed with white fish meal, starch, cellulose powder, and vitamins. The diet did not contain seaweed. The abalones grew normally with green colored shells. These authors concluded that the minimum amount of protein required for the normal growth of spat (*H. discus*) is about 20%.

4.4 Heart beat and water temperature

According to Iino (1952), the heart beat of abalones is 20 to 30 min at 10 to 20°C of water temperature; the pulse increases with a temperature rise up to 29°C, but decreases when the temperature exceeds 30°C. In *H. sieboldii* and *H. gigantea*, the heart beat stops at 3 to 4°C, but in *H. discus*, it does not cease even when the temperature drops to −1.6°C.

(Sakai and Iino)

CHAPTER II

Technique of Abalone Culture

1. Introduction

Abalones have long been recognized in Japan as one of the more valuable marine products and the history of its culture begins many years ago with the efforts of Genzo Uchimura. Soon after his graduation from the Sapporo Agricultural School, he joined the Hokkaido Development Section of the Government, and as early as 1881, the Sapporo Fisheries Culture Center published his paper on the color, morphology, and maturation of the gonads of abalones, as well as their age of maturity and spawning behavior.

In 1882, Matsuhara described the ova and sperms of the abalone by means of diagrams, and Hayagi published a paper on their spawning. Attempts to culture abalones were made in the same year, but more sophisticated experiments were done a few years later by Kishigami in Shirohama of the Chiba Prefecture and in Kawana of the Shizuoka Prefecture. His reports, published in 1894 and 1895, contained details of the reproductive system, the gonadal structure, the spawning, and the feed of abalones. Unfortunately none of his efforts to induce artificial fertilization succeeded.

Rearing of abalone larvae obtained through artificial fertilization was first achieved in 1935 by Saburo Murayama. He utilized *H. gigantea* and reared the larvae for six weeks to a shell length of 1 mm in Misaki of the Kanagawa Prefecture. Some years later, Iino (1952) experimented with *H. discus* in Kominato of the Chiba Prefecture, and succeeded in rearing larvae hatched through artificial fertilization for 13 months and a shell length of nearly 2 cm. Thus Murayama is lauded for making the breakthrough in artificial abalone culture, but Iino is credited with being the first to rear abalone larvae through the early phase of their life history. Iino was also able to rear the larvae of *H. sieboldii* from artificial fertilization for 20 days to a shell length of 0.42 mm; a comparison of the development and ecology of the two types of larvae in the early stage of their growth was thereby possible. Murayama and Iino worked together to rear larvae produced by artificial fertilization in which sperm suspension fluid was used and spawning stimulated by heat treatment. Kishigami, on the other hand tried to induce artificial fertilization by slitting the gonads as is done in sea urchins, etc., and had his method proven effective so that 100% fertilization could be achieved, a spectacular revolution in abalone culture would have resulted. However considerable study of the cytology of the gonads, size-wise and period-wise, and of the environmental conditions of abalones needs to be done before Kishigami's method can be considered. Such studies would also be helpful in improving the technique of inducing spawning by artificial stimulation.

The successful experiments of Murayama encouraged some investigators to try induced spawning in natural media such as pools, tanks, etc. Yaita *et al.* (1939) introduced 100 mature male and 100 mature female abalones into a large pool of about 100 *sho* (330.5 m²) located in a rocky belt in the Hyogo Prefecture. Several carefully thought through precautions were taken to facilitate the collection of seedlings and the feeding of the larvae. Various types of collectors were used, e.g. bamboo bundles, each of one *shaku* (0.303 m) diameter and consisting of bamboo poles 4 to 5 *shaku* in length and 2 to 3 *sun* (1 *sun*=3.03 cm) in diameter, and rows of square concrete plates measuring 8 *sun* in length and 3 *sun* (1 *sun* = 3.03 mm) thick and piled one above the other. He also took care to cultivate seaweed for feeding. But in spite of all these precautions, his experiments did not succeed because the pool became fouled through continuous use.

Rearing seedlings collected from natural conditions was also considered by early investigators. Reports of abalone spat settling on materials floating on the sea surface date back to 1882 near Nagasaki Prefecture, and from various other places such as Oshikiami of the Tokushima Prefecture, Amishiro, Usami, and Ito of the Shizuoka Prefecture, Shinzuru of the Kanagawa Prefecture, and Wadaura-Oki of the Chiba Prefecture (Iino, 1952). Other countries have made the same observation, for example, Boutan (1899) reported a collection of spat (*Haliotis tuberculata*) of 1.5 mm shell length taken from buoys in ports. These facts indicate the possibility of collecting seedlings from natural conditions. Hence in the Fisheries Experiment Stations of the Chiba Prefecture, the Tokushima Prefecture, etc., a collection of seedlings from natural conditions was considered, and investigators tried floating seedling collectors. While no spat were collected in the Chiba Prefecture, a large number were gathered in Tokushima. However, the method of collecting spat from floating materials is not yet adequately developed. Since abalones are found in large numbers in rough seas, the structure and the floating of collecting implements need extra care and, further, the spat thus obtained need special handling since they settle on the substrata not by byssi or an adhesive secretion, but by the suction mechanism of their feet. In order to determine the suitable period for setting up the seedling collecting implements, as well as suitable regions and depth, the distribution of planktonic larvae of abalones in natural conditions, and the ecology of the environment during the migration of larvae to the sea floor, should be carefully studied.

Culture techniques of feed organisms for abalones, and the various equipment necessary for abalone culture, have been greatly improved, and hence artificial seedling production is now possible. But advancement in abalone culture can only come about through improving the preservation of the mother abalone, induced spawning, rearing of larvae, transplanting and rearing of spat, etc.

(Iino)

2. Role of artificial seedling

Several scientists since the Meiji era (1868-1905) have tried to produce artificial seedlings of the abalone, but Murayama (1935) and Iino (1952) provided the method by which it is now possible to release thousands of seedlings through artificial production. However, some unsolved problems with regard to the rearing of these seedlings still remain and until these difficulties are overcome, artificial seedling production cannot reach commercial proportions. One of these problems is the assessment of the recovery in relation to the amount of seedlings initially released.   Hence it seems more appropriate to discuss the role of artificial seedlings only after describing the transplantation of natural seedlings, a process which has been practiced for a very long time.   It began sometime in 1935 when seedlings from Hokkaido and Sanriku were transplanted to other regions (Iino, 1966). The number of seedlings was gradually increased as was the transport distance. At present, 70% of the spat (about 1,500,000) of 3 to 6 cm shell length produced in Okushiri-To in Hokkaido, are utilized in Hokkaido, while the balance (30%) is transplanted to other regions (Figure 4.6).

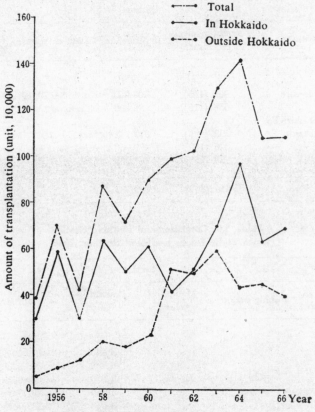

Figure 4.6. Transplantation of abalone.

Seedlings from the Sanriku regions are now transplanted in other regions; the southern limits of the receiving regions are the Mie Prefecture on the Pacific coast, and the Fukushima Prefecture on the Sea of Japan (Saito, 1965). Although the results of transplantation are good in Akita, Fukushima, Ibaraki Chiba, part of the Kanagawa Prefecture, and in Tokyo, they are less than good in other prefectures. Hence, it is not yet certain whether transplantation in regions other than those mentioned here will be effective (Iino, 1966). The effect of transplantation experiments carried out at Shogata Saki were published by Akita Fisheries Experiment Station (1966), and the successful transplantations are given in Tables 4.8 and 4.9. The effect of transplantation was first seen after 3 years (1964) and in the following year (1965) production had increased by about 4 times, and the capital by about 7. Morover the per capita income of fishermen had also increased by about 6 times. The results of the transplanting experiments carried out in Shibane Jisaki of Joge-Shima in the Kanagawa Prefecture, show that the catch represented 33% of the seedlings released (Inoue, 1965).

Table 4.8. Results of transplanting experiments on abalone
(Akita Fisheries Experiment Station, 1966)

Year	Destination	Section			
		Amount involved in experiment	Expenditure	Date of release	Number of marked seedlings
		kg (boxes)	yen		
1962	Miyagi Prefecture	900 (155)	408,000	November 20, 1962	100
1963	,, ,,	656 (115)	190,200	November 23, 1963	100
1963	Minami Okushiri-To Hokkaido	552 (95)	317,732	May 17, 1963	100
1964	,, ,,	1,026 (180)	479,960	April 30, 1964	200
1965	,, ,,	2,280 (263)	707,229	May 8, 1965	200
Total		5,414 (808)	2,099,121		700

Table 4.9. Conditions of abalone catch
(Akita Fisheries Experiment Station, 1966)

Year	Number of fishermen	Fishing period	Method of fishing	Amount of catch	Capital	Per capita income
				kg	yen	yen
1960	98			1,940	813,000	8,900
1961	97			1,730	865,000	8,900
1962	98	July 1 to	Diving	1,970	1,078,000	11,000
1963	106	August 31		1,860	1,367,000	16,200
1964	110			4,510	3,383,000	30,700
1965	118			7,990	6,192,000	51,600

According to the report published by Kyoto Fisheries Experiment Station (1966), the results of transplanting *H. discus hannai* differed in various fishing beds. The mean survival rate after 2 to 3 years was 15% but, in those fishing beds where mass mortality was endemic, revival by transplantation was not possible. The average annual growth rate was 1 cm, but in places where growth was rapid, shell length doubled after two years and the weight increased about 5.5 times. Although a large number of *H. discus hannai* seedlings from Okushiri-To were transplanted in Yamagata

Prefecture, details regarding the survival rate, growth, migration, and production are not available. The reasons for doubting the benefits of transplantation are (1) large-scale mortality after the consignment arrives, and (2) loss through dispersal and predators after transplantation. Nevertheless, transplantation has improved through progress made in transporting techniques, especially in the use of new materials, but the rearing of transplanted spat still requires much attention. The conditions of seedlings produced by artificial means are more or less the same as those just given for natural seedlings; to improve the results of transplantation of the former, the basic facts should be well understood.

Table 4.10 presents the conditions of seedlings reported from various experiment stations in 1966. Even after excluding those cases in which the number of larvae and the number of eggs are not clearly known, it is found that the total number of seedlings is nearly 200,000,000. Even if only 1% of the total survived by the time the seedlings were 2 to 3 cm in size, the number produced by artificial means would be impressive. This fact should bring about a number of changes in the transplantation of abalone seedlings which, until recently, depended upon natural sources. According to statistics released by the Ministry of Agriculture and Forestry up to 1962, the production of abalone in the whole of Japan is in the range of 3,600 to 4,900 tons or, transposed into number of individuals, 24,000,000 to 30,000,000 abalones. To increase the production of abalones through artificial seedlings therefore, requires a larger number of seedlings produced with greater efficiency. The results of artificial seeding done by the Oyster Research Center (1964) are given in Figure 4.7.

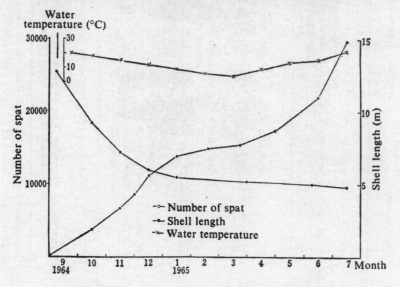

Figure 4.7. Number of abalone spat and growth.

It is clear from the figure that the survival rate of abalones from the planktonic larval stage to a 15 mm spat is 5.8%. This being so, the price of 1 seedling would be 5 to 6 yen. Bearing this fact in mind, and the economic aspect of rearing after transplantation, it seems that the following conditions should be satisfied in the near future:

Table 4.10. Information on the collection of abalone eggs, 1966
(Tokaiku Fisheries Experiment Station, 1966)

Name of place	Type of abalone	Date of egg collection	Method of egg collection	Number of eggs/larvae	Number of spat	Remarks
1	2	3	4	5	6	7
Seed supply section of Akita Fisheries Experiment Station	H. discus	November 15	Increase in temperature during ebb tide (17.0°C-21.3°C)	(Planktonic larvae) 4,800,000	Shell length, 2-4 mm 800,000	January 10, 1967
Ibaraki Fisheries Experiment Station	H. discus	September 19	Natural spawning	(Collection of eggs) 140,000	Shell length, 8 mm 500	
		November 7	Ebb tide	1,200,000	Shell length, 3 mm 4,000	January 10, 1967
		November 9	Increase in temperature during ebb tide	500,000	Shell length, 3 mm 1,000	
Chiba Fisheries Experiment Station	H. discus	November 3 November 6 November 8 November 27		(Planktonic larvae) 4,820,000 178,000 267,000 1,377,000	In several cases, number of spat was not surveyed	January 24, 1967
	H. discus	December 14	Increase in temperature			
Tokaiku Fisheries Experiment Station	H. gigantea	December 5 to December 12	(Eggs during this period obtained from Kanagawa Fisheries Experiment Station)		Shell length, 1.5 mm 10,000	(January 3) Degree of maturation of mother abalones (late October) was poor as compared to previous year. First lot of seedlings collected mid-December.
	Cross between H. discus (♂)/ H. gigantea (♀)	January 11, 17, and 24	Rate of fertilization. about 60%		Shell length, 0.5 mm 10,000	

	Date	Tide	Number	Shell length	Remarks
			(Planktonic larvae)	In several cases, no. of spat not surveyed	(January 15) Larvae were kept in water tanks (90×90×230 cm) capable of holding 1,000,000-2,000,000 larvae. Just before hatching, 60 plates (4.5×4.5 cm) were hung in each tank. To date, 1,000 plates of seedlings collected.
H. discus	November 4	Ebb tide	4,700,000	Shell length, 2-3 mm	
	November 25	Ebb tide	3,500,000	Shell length, 1.5-2 mm	
	December 7	Ebb tide	500,000	Shell length, 0.8-1.2 mm	
Kanagawa Fisheries Experiment Station			(Planktonic larvae)		
H. gigantea	November 11	Ebb tide	5,000,000	Shell length, 2-3 mm	
	November 20	Ebb tide	3,000,000	Shell length, 1.5-2.5 mm	
	November 25	Ebb tide	2,000,000	—	
	December 21	Ebb tide	6,000,000	Shell length, 0.5-0.8 mm	
			(Collection of eggs)	In several cases, no. of spat not surveyed	(January 17) A few seedlings obtained from eggs collected after December 5. Method of seedling collection the same as culture method.
H. discus	October 31	Ebb tide	20,000,000	Shell length, 5-9 mm	
	December 19	Ebb tide	28,000,000	—	
	January 19, 1967	Ebb tide	18,000,000	—	
Kanagawa Fisheries Experiment Station			(Collection of eggs)		
H. gigantea	November 15	Ebb tide	39,000,000	Shell length, 2-3 mm	
	December 5	Ebb tide	40,000,000	—	
	December 6	Ebb tide	Not known	—	
	January 17	Ebb tide	3,600,000	—	

1	2	3	4	5	6	7
Izu branch of Shizuoka Fisheries Experiment Station	H. discus	November 22 December 5 December 7 December 13	—	—	Shell length 1.2 mm 10,000	(January 11) (January 25) Maturation of naturally grown abalones (in the year of study) was earlier than the usual peak period (late October). Many mature specimens seen at end of fishing season (September 15).
Mie Fisheries Experiment Station	H. discus	December 7 December 14	—	130,000 700,000	Shell length, 2 mm 10,000	
		September 27		—	Shell length, 5 mm 1,000	(January 9) Eggs collected on September 27 at Ishikagami fresh water culture farm.
Toribanashi Fisheries Experiment Station	H. discus	November 30 December 15	(Larvae obtained from Mie Fisheries Experiment Station)		Shell length, 2 mm 2,000 Shell length, 0.5 mm 3,000	At present, 20,000 spat of 8 mm shell length being reared.

Station	Species	Dates		Eggs / Shell length	Number	Remarks
Kyoto Fisheries Experiment Station	H. discus	November 1	(January 13)	(Collection of eggs) 1,200,000	Number of spat not surveyed	Rearing by increasing temperature in three-ton water tank.
		November 10				
Yamaguchi Open Sea Fisheries Experiment Station	H. discus	October 24	January 5	Shell length, 2-3 mm 30,000		
		November 18		Shell length, 1-2 mm 30,000		
		December 11		Shell length, 0.5 mm 30,000		
Kamiura Center of Setonaikai Culture Farm	H. discus	November 8	(January 17)	Shell length, 0.5-1 mm 150,000		Seedling collection poor in November.
		November 19				
		November 24				
		December 8				
		December 10				
		December 17				

$$\frac{B \cdot C}{100} > A,$$

where A—value of seedling;
       B—catch rate;
       C—value of abalone at the time of catch.

If it is assumed that the price of an adult abalone is 100 yen, and if B is greater than A, and again the expenditure for the production of a seedling is 10 yen, the catch rate should be increased by more than 10%. However, from the time spat are released to the time they grow to a shell length of marketable size, the abalone spawn once or twice. When such a secondary production is expected (Akita Fisheries Experiment Station, 1966), even when the rate of catch is very low, the effort of transplantation might not be a total loss. The results would be much more effective if a through culture were carried out by using spat of artificial seedlings, and culturing done in ponds or by hanging culture depending upon the region chosen. At Maine Bay of the Miyagi Prefecture, hanging culture of abalone was carried out by keeping the seedlings in polyethelene containers and feeding them *Laminaria japonica (Kombu)*. It took three complete years for the abalones to reach a size of 11 to 12 cm, but the results were in no way inferior to those obtained by the Oyster Culture Center in the same bay.

Considering the progress in the development of the materials used in culturing, that this could become a profitable industry soon appears possible, especially since new varieties can be obtained by artificial seeding. The significance of the role of artificial seeding, related to the selection of species from among the 100 known ones, and the selection of a suitable region, should prompt new investigations since the culture technique has already been perfected. Moreover, since obtaining a sufficient number of spat has been difficult, physiological and ecological studies are incomplete. As research in artificial seeding progresses, many facts related to the early feeding and movement of spat will also come to light. When they do, artificial seeding will inevitably play a considerable role in the management of fishing grounds.

(Shibui)

3. Natural seed stock and fishing ground requirements and management

Research on the natural seed stock of abalones is not progressing at the same rate as is that on oysters and scallops, and the reasons are several. First of all, abalones live in shallow waters in open seas where the waves are rough and the salinity is high, while oysters live in quiet inland bays. Hence the habitat of abalones deters research since the equipment necessary for collecting seedlings cannot be protected. However, experiments have been done on natural seedlings of abalones from spat which settle on the buoys or floating bamboos in the Tokushima and Chiba Prefectures. Noguchi's efforts to collect these seedlings in 1953 were unsuccessful, but from 1958 to 1961 they have been obtained by using barriers and rafts (Table 4.11; Kodake and Tanimoto, 1965). Between November 26, 1958 and May 11, 1959, 48 seedlings of *H. discus*, measuring 10 to 25 mm in shell length, were collected by strands attached to wooden frames floated on the surface of the Pacific Ocean in Tokushima Prefecture. In 1960 and 1961, 41 and 39 seedlings were obtained from

Table 4.11. Data on natural seed stock
(Kodake and Tanimoto, 1965)

Year	Date of placement	Date of collection	Size of spat	Place	Number of seedlings	Remarks
1958	November 26	May 11	10-25 mm	Kinoki	48	A number of spat dropped while gathering collectors
1960	November 29 December 9	May 10	8-12 mm	Kinoki	41	,,
1961	October 24	March 28	7.5-16 mm	Abe	39	Many shells broken and many spat dropped due to wind or rough waves
	November 25	March 29	5.5-13 mm	Bennoki		

different places, and Kodake and his colleagues pointed out that since the possibilities for seedling collection existed, the environmental conditions of spat should be carefully studied in order to increase the intake. This would involve research in three areas: (1) determining the period of maturation and spawning of mother abalones, (2) an ecological study of planktonic larvae (the period of larval appearance, floating layer, daily activities, etc.), and (3) forecasting the period when the larvae would enter a sessile life and the setting layer. Equipment which can withstand strong waves in open seas has to be designed before research along the lines indicated by Kodake can be initiated.

Okushiri-To in southwest Hokkaido is a good place for the collection of natural seedlings, and the conditions for seedling development there are described below.

3.1 Topography

Iino (1966) reported that in selecting a place to settle, the planktonic larvae of abalones are influenced by topographic conditions in places with tidal and coastal currents. Even scallop larvae are influenced by such factors and select places with mild currents and eddies (Yamamoto, 1964). This likewise applies to the abalone off Okushiri-To (Saito, 1965). A large number are found in Inebojiku on the northern edge of the Island, Aobyo-jiku on the southern end, and Mutsuga-To 2 km to the south of Aobyo-jiku. Of these, Mutsuga-To has the maximum water depth of 20 m and the tidal current is very fast, but Nakajima et al. (1953) state that abalones *(H. discus hannai)* live even at a water depth of 20 m. Abalones are also found in large numbers along the coastal regions and on rocky beds in Okiai.

3.2 Quality of the substratum

In Okushiri-To observations revealed that more abalones lived on a rocky bed than on a bed of rolling stones. On 1 m² of rolling stones 6.1 to 8.8 abalones were found as compared to 19.2 to 25.3 counted for the same area of a rocky bed. In other words the density of abalones on a rocky bed is 2 to 3 times greater than on a bed of rolling stones (Table 4.12).

Table 4.12. Comparison of abalone density on various substrata
(per 1 m²) (Saito, 1965)

Year	Rocky bed	Rolling stones	Remarks
1961	22.9	6.1	Values taken are the mean of
1962	25.3	8.8	12 survey points for rocky
1963	19.2	8.7	beds and beds of rolling
1964	23.8	6.2	stones.

3.3 Water depth

A large number of young abalones, 1 to 2 years old, are found at a depth of 0.5 to 1 m. These abalones are found either in between the rocks or hidden in holes. In rolling stone beds, most abalones are attached to the sides of the stones. As mentioned already under "Topography," when the planktonic larvae enter the phase of setting, spat are found in large numbers in a water depth of 0.5 to 1 m in places where the current is mild and eddies form easily.

3.4 Stock management of natural seedlings

When abalone spat are transplanted, it is necessary to carefully study the relationship between the natural seedling and the number of spat transplanted. In order to assess to what extent the seed stock can be transplanted, Nakajima (1953, 1958) carried out a survey of the resources in the fishing grounds. The number of spat to be transplanted was determined on the basis of the study of changes in the number of seed stock, composition of shell length, comparison of age composition, conditions of young abalones, feeding and maturation conditions, shell length composition of transplanted spat, daily reports on catch by fishermen, and changes in abalones due to aging. The transplanting operation was carried out in such a way that the resources of the supplying regions were not affected. The amount of transplantation done since 1955 has been shown in Figure 4.6, which reveals that transplantation was maximum in 1964 with 1,450,000. This amount decreased in the following years to about 1,100,000.

(Saito)

4. Transplanting: fishing ground requirements and management

4.1 Method of transplanting

One of the measures to overcome the decrease in the production of abalones is to transplant them from places in which they do not grow well to those in which they do. For example, while large numbers of *H. discus hannai* are found in Okushiri-To of Hokkaido, their growth in that area is poor (7.5 cm in 7 years), but on transplantation to the western coast of Hokkaido or to Funka Bay, growth is rapid and the yield rate quite good (Kinoshita, 1950). In fact, Kinoshita has reported that the *H. discus hannai* of Okushiri-To took 4 years to grow from 5 cm to 7.5 cm in shell length. Iino (1952), on the other hand, reported that it took only 2 years for abalones in Shimugawa Bay of the Miyagi Prefecture to attain the same length. It is known that abalones generally take 2 years from embryonic development to reach a 5 cm shell length. Hence Kinoshita suggested that Okushiri-To be used for developing seed stock which would then be transplanted to regions more suitable for growth; Iino went a step further suggesting that the best method of culturing abalones lay in developing seedlings in cold regions and transplanting them to warmer waters in the southern regions for further growth.

Hokkaido, Iwate, and Miyagi Prefectures have always been sources for natural abalone seed stock but recently production has decreased, so much so that artificial seeding and transplantation are now employed.

Seed stock is transported either by truck or by ship depending upon the place and the distance. In overland transportation the seed stock is exposed to air as abalones have a high tolerance to such exposure when the temperature is low (see Table 4.6). Nevertheless, the period of transportation should be kept as short as possible, and consideration given to temperature and moisture conditions. It should also be remembered that abalones are very weak and tired after spawning, i.e. in November to February, and hence have very little tolerance for land transportation during that time; furthermore, freezing is apt to occur in those months. In addition, their ability to adjust to a new environment is considerably lessened. March and May then are the best months for transplantation, since the temperature then is neither too high nor too low and the abalone is physically in a much healthier condition.

Various types of containers are used for seedling transportation. However, certain requirements are observed for all containers: (1) the size should be convenient for transportation, (2) the moisture in the container should be adequately maintained, (3) the area for the attachment of seedlings should be wide, (4) the abalones should be separated, and (5) the plates on which the abalones are arranged should be released into sea water without disturbing the plates or the seedlings. Figure 4.8 shows the boxes in current use for transportation from Okushiri-To. Transportation to Ibaraki, Chiba, Shizuoka and Aomori Prefectures is done by sea, and that from Aomori by freight train. However transplantation done in early May to Shirahama in the Shizuoka Prefecture (temperature of air 9.0 to 25.0°C, room temperature, 15.5 to 18.00°C) which took 57 hours, resulted in a sudden mortality of 20.5%.

Seedlings transported from Onagawa, Miyagi Prefecture to Kamoi in the Kanagawa Prefecture by truck in December, 1952, were packed in containers with compartments of 33 cm dia-

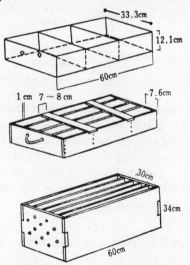

Figure 4.8. Boxes used for transportation of abalones (Kinoshita, 1950). *Top*—Okushiri-To; *Bottom*—Hokkaido Fisheries Experiment Station.

meter and 30 cm height; the containers looked like a popular Japanese salt-and-pepper holder, and had a capacity for 15 kg of seedlings. Since 1959, seed stock transported from Maeami in the Miyagi Prefecture to the Ibaraki Prefecture is allowed to adapt itself to tank conditions for two days and then sent by truck, and the resultant mortality is only an approximate 10%.

Experiments on transportation of abalone seedlings in fish tanks on ships are being done at the Mie Fisheries Experiment Station. Abalones caught at Shizukawa Bay of the Miyagi Prefecture were transplanted to the Mie Prefecture. The voyage took 76 hours and the survival rate was very high at 94.4% or 7,040 abalones (Miyamura, 1954).

After transportation, if the abalones are healthy and active, it is better to release them in the fishing grounds as quickly as possible. However, if the animals are weak, they should be kept in tanks until totally recovered.

There are several reports from the Ibaraki, Chiba and Kanagawa Prefectures on the growth of *H. discus hannai* after transplantation. The results of transplanting from Hakuhama of the Miyagi Prefecture to Yokosuga-Shi are shown in Table 4.13,

Table 4.13. Results of transplanting *H. discus hannai* **from Hakuhama of the Miyagi Prefecture to Yokosugashi of the Kanagawa Prefecture (Iino, 1966)**

	Measurements	December 12, 1952—September 9, 1953 (about 9 months)					Survey in January 1958
		1	2	3	4	Mean	
Before transplantation	Shell length	5.40 cm	6.22	6.24	7.25	6.28	6.55
	Weight	20 g	35	40	55	37.5	40
After transplantation	Shell length	8.53 cm	9.24	9.53	10.15	9.36	14.53
	Weight	77 g	108	105	165	113.8	487
Rate of increase	Shell length	58.0%	48.5	52.7	40.0	49.0	122.1
	Weight	285.0%	208.6	162.5	200.0	203.5	1117.5

which reveals that growth doubled within 9 months, and increased about 11 times in a little over 5 years. According to Iino (1952), if the regions to which the abalones are transplanted have warm currents, *H. discus hannai* grows to a large size and its morphology resembles that of *H. discus*.

The quality of the water, the tidal currents, and the quality of both the rocks and the seaweed are some of the important environmental conditions to be considered when transplanting.

4.2 Fishing ground: requirements and management

The fishing grounds of abalones are the rocky belts in the open seas where the waves are rough and seaweed grows abundantly, especially *Undaria pinnatifida* and *Laminaria japonica*. The exact quality of the rocks is not yet known but from experience it seems that those which have a smooth surface and are relatively soft are better suited than those which are hard and have a rough surface.

Some of the regulations imposed by the Fisheries Departments of the metropolis and the prefectural administration to conserve abalones, pertain to the size of the mollusk at the time of catching and the length of the fishing period. Various unions impose restrictions which relate to the method of fishing, prohibition of certain regions, growth of algae which form the diet of abalones, the amount of total harvest, etc.

The permissible shell length differs in various regions; while it is 7.5 cm for Hokkaido, it is 9.0 cm in Aomori, Iwate, and Miyagi Prefectures. The ultimate aim of these restrictions is to preserve the second generation by allowing the parent abalones to spawn at least once, if not more than once. Even from the biologically minimum size shown in Table 4.14, this means of preserving the progeny seems to be essential. From the commercial point of view, since the weight trebles the length, these regulations must be strictly observed.

Table 4.14. Biologically minimum size of *H. discus hannai*

Region	Shell length (cm)	Age	Reference
Hokkaido	7.5	4	Hokkaido Fisheries Experiment Station (1927)
Okushiri-To	5.0	4	Nakashima *et al.* (1953)
Iwate Prefecture	4.5		Hirose (1953)
Miyagi Prefecture	6.0	3	Sakai (1959)

Fishing is prohibited between July 16 and September 15 in Hokkaido, between August 1 and October 31 in the Aomori and Miyagi Prefectures, and between April 1 and October 31 in the Iwate Prefecture. Fishing is also prohibited in all these areas during spawning and the subsequent recovery period (see Table 4.1).

The harvesting of *H. discus hannai* is done with various implements such as fishing rods, spoon nets, scissors, dredges, nets, long-line ropes, etc. In many regions har-

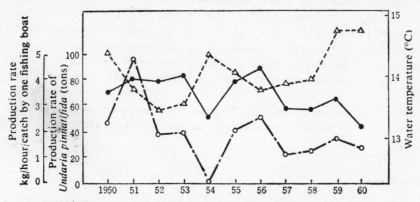

Figure 4.9. Relationship among unit amount of catch (kg/h) of *H. discus hannai* (△), water temperature (°C), and production of *Undaria pinnatifida* (○) in Enoshima, Miyagi Prefecture.

vesting by diving with or without diving gear is prohibited in order to conserve the resources.

The abalone stock, however, cannot be maintained by these protective measures only. According to Sakai (1962), the production rate and growth of *H. discus hannai* fluctuates greatly each year, and the fluctuations are directly related to the production rate of seaweed, particularly *Undaria pinnatifida*. It is said that production of *H. discus hannai* is greatly influenced by the production of feed algae such as the brown *Undaria pinnatifida* and *Laminaria japonica* (Figure 4.9). The increase in the production of seaweed will cause an increase in the growth of abalones, accelerate their growth, and shorten their cycle of reproduction. Attempts are being made in several places to bring about an increase in the production of algae by rock blasting, stone scattering, etc., but more effective methods for increasing the production should be devised so that the abalone crop can also be improved.

(Sakai)

5. Pond culture and management

Nakamura (1940), who cultured abalones in ponds, said : "If the problem of seed stock supply is considered separately, it seems that the rearing of abalones in ponds is technically possible. Moreover, if there is a regular supply of healthy seed stock, pond culture will also play an important role in production." Since the method of transportation is now improved and the duration of transportation considerably shortened, and as artificial seedling techniques have made substantial progress, it seems that through culture from seedling production to harvesting will be possible in the near future on the basis of a careful study and understanding of all its aspects.

5.1 Objectives

Abalone culture in ponds seeks two major objectives: first, to store abalones at a time when the prices are down and to place them on the market when the peak price period occurs, and second, to establish a through culture by rearing spat grown from artificial seedlings in various regions. Since storing time is relatively short and the abalones are usually already of marketable size, the most important consideration in this objective is to prevent any reduction in the yield rate. With regard to the second objective, seedlings must be grown as quickly as possible to attain permissible size. An additional aim is the culturing of pearls, as has been tried by Komatsu of Onagawa Machi.

To achieve these objectives, more efficient collection equipment, faster location of beds, etc., are necessary, depending upon the availability of feed and management efficiency.

5.2 Seedlings

When abalones are used as seedlings, harvesting below the size of 9 to 12 cm shell length (mostly 10 cm) is prohibited in various prefectures south of Chubu on Honshu Island, according to the Marine Fishing Regulations. There are some cases where regulations on the fishing period and the size of catch are not strictly imposed when seedlings are collected for preparing seed abalones. In fact, it is better not to enforce fishing regulations in such places. Hence, the seedlings which can be used for rearing are generally more than 10 cm in shell length. But these very abalones can be sold at high prices. In places south of Chubu, harvesting is carried out on a large scale between April and August and fishing is prohibited during September to December. Hence the price of abalones is low in the summer and high in the autumn. Consequently, abalones collected in natural conditions are stored in large amounts and marketed in the autumn and winter when the prices rise. How to equalize the price of abalones throughout the year is currently under consideration.

Artificial seedlings of *H. discus hannai,* and natural seedlings are considered suitable for pond culture. Seedlings should be carefully checked however and only those without injury selected. Symptoms of ulceration due to injuries should be treated with chloremphenicol or sulfisoxazol, drugs which not only cure the injuries, but also decrease mortality (Takayanagi *et al.,* 1966).

5.3 Ponds

The following are some of the types of ponds built and the methods followed for the rearing of abalones.

1. Ponds or tanks are made in shallow places along the sea coast, in creeks, or rocky beds (Figure 4.10) (sometimes the ponds have to be covered)—Onagawa, Miyagi Prefecture; Daimon, Aomori Prefecture; Truruhara, Shirohama, Kominato, Nidoiwa, Wada, etc., Chiba Prefecture.

Figure 4.10. Abalone culture pond in Onagawa, Miyagi Prefecture (Komatsu).

2. Ponds are made on land (Figure 4.11)—Maeami, Enoshima, Shizukawa, Tenguwa, Miyagi Prefecture; Hirota and Oshimo, Iwate Prefecture; Senkura Machi, Chiba Prefecture; etc.

3. Vessels containing abalones are hung by rafts (Figure 4.12)—Tenguwa, Miyagi Prefecture; Obunewata, Iwate Prefecture; etc.

Figure 4.11. Abalone culture beds in Gyosaki, Miyagi Prefecture
(Karakuwa-Machi Fisheries Association).

Figure 4.12. Through culture at Maine Bay, Miyagi Prefecture
(Oyster Research Center).

4. Baskets made of expanded metal, etc., are placed on the floor of the sea—Senkura-Machi, Chiba Prefecture.

Since abalones live in rocky beds in open seas, the ponds in which they are to be cultured are built in places with rocks that are directly in contact with rough waves from the open sea. On the other hand, these ponds are easily affected by typhoons which damage the culture and collecting equipment in them. Hence, matters like topography, conditions of the sea, weather, water quality, biological organisms, etc., have to be carefully considered in relation to the ecological conditions of the animal. In other words, the places where culture ponds are built should conform to the following conditions:

    a. Be constructed in a rocky belt with no sand or silt in the vicinity.

    b. Be safe from damage by fire.

    c. Receive pure sea water uncontaminated by river or polluted water.

d. Receive sufficient circulation of sea water through waves, tides, and coastal currents.

e. Be accessible for easy transportation of the seedlings.

f. Be able to grow sufficient seaweed as feed.

The flow of sea water through the equipment placed on the surface of the sea can be well maintained by fixing water gates (mud tubes, fume tubes, and concrete windows or rectangular shape). Steps must also be taken to prevent the collection of sand or silt at the bottom of the pond. Masuda (1966) determined the direction and position of water-ways on the basis of the direction of waves, and tried to improve the rate of water circulation through the collecting equipment by making its flow uniform. When the water flow is weak, resistant plates are placed at the mouth of the water path and, to accelerate the water flow, draining paths are made. If the area of abalone setting is too narrow, or if a large part of the area is too exposed to light, abalones crowd in dark places; this leads to a shortage of oxygen for them and a blocking of their inhalant siphons due to congestion, a condition which often results in mass mortality. Since abalones take in oxygen dissolved in water, the amount of dissolved oxygen varies according to the consumption of oxygen by the abalone, and the amount of water introduced into the pond. The amount of oxygen in the sea water is usually 5 to 6 ml/l. At the time of rearing, it should be at least 3 ml/l. Hence the relationship between the amount of rearing and the necessary amount of water introduced into the pond can be expressed by the following equation:

$$V = \frac{E \cdot N \cdot T}{C_1 - C},$$

where V—necessary amount of introduced water l/day ;

E—amount of oxygen consumption (at t°C) ml/kg/h ;

N—amount of abalones, kg ;

T—24 hours ;

C_1—amount of dissolved oxygen in water introduced, ml/l ;

C—minimum amount of dissolved oxygen, ml/l.

Table 4.15. Amount of oxygen consumed by *H. discus hannai* (Tamura, 1939)

Temperature	Date	Number of specimens	Shell length	Number of times measured	Amount of oxygen consumed/kg·h Mean	Range	Daytime Nighttime
5.5°C	February 27 to 29	5	5.5-6.6 cm	11	16.42 ml	12.15-20.32 ml	Daytime
13.0	November 10	6	4.5-6.5	4	18.28	16.80-20.00	Daytime
13.0	November 10	6	4.5-6.5	1	46.6	46.6	Nighttime
14.0	November 9	6	4.5-6.5	1	23.6	23.6	Daytime
14.0	November 9	6	4.5-6.5	2	33.6	33.10-34.50	Nighttime
22.0	August 9 to 10	10	5.5-6.5	6	57.41	52.20-61.80	Daytime
22.0	August 9 to 10	10	5.5-6.5	1	63.80	63.8	Nighttime
23.0	August 11 to 12	10	5.5-6.5	5	63.92	57.30-78.00	Daytime
23.0	August 11 to 12	10	5.5-6.5	4	74.27	65.20-85.30	Nighttime

The amount of oxygen consumed by abalones varies according to the water temperature or size of the specimen (Table 4.15). It is said that the density of abalones at the time of catch should not exceed 110 kg/3.3 m² (Kuroda, 1957).

5.4 Feed

When brown algae was supplied for feed, the monthly growth rate was very high, 1.4 to 9.2%. This was followed by 2.7 to 4.9% when green algae was given; red algae showed a poor rate of 1.4 to 3.9% (Sakai, 1962). In the Tohoku region *Undaria pinnatifida, Laminaria, Eisenia bicyclis, Ecklonia cava,* etc., are usually standard feed and the growth rate is shown in Table 4.16. However the supply and type of feed change

Table 4.16. Standard feed
(Kuroda *et al.*, 1957)

Abalone shell length	Number of abalone totaling 3.75 kg weight	Amount of feed per day
9 cm	35	225 g
10	25	206
11	20	188
12	10	56

according to the seasons; for example, *Undaria pinnatifida* and *Laminaria* are supplied in small amounts in the summer since these two seaweed decompose in 3 or 4 days. In winter, on the other hand, the feed remains fresh for a long time, and the feed-intake is also low; hence, feed is supplied at intervals of 7 to 10 days. But since the growth of abalones is rapid in the spring and the autumn, feed should be given at intervals of 4 to 5 days and the remnants removed. In September, 1962, the seed stock of *H. discus hannai,* obtained from artificial seedlings, were kept in polyethelene packets covered over by netron nets. These were hung in rafts and reared by supplying *Laminaria.* The results are shown in Figure 4.13. In 3½ years the abalones attained

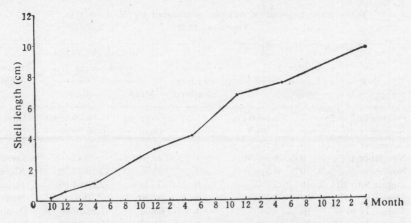

Figure 4.13. Growth of *H. discus hannai* (Oyster Research Center).

the permissible shell length of 9 cm.

Plans for rearing abalones should include a study of the various data already available on the subject, a survey of the various equipment, the types of abalones, the seasonal variations, and past experience. Furthermore, after ascertaining the natural feed available, the development of an artificial diet capable of accelerating abalone growth should also be considered.

5.5 Growth and catch

The results of abalone pond culture by Nakamura (1940) are shown in Tables 4.17

Table 4.17. Rate of growth in pond culture of abalones of Iwate Prefecture transplanted to Chiba Prefecture (I)*

Group	Number of abalones examined	Shell-length range	Shell length		Shell width		Weight	
			Mean shell length before experiment	Mean rate of increase in shell length	Mean shell width before experiment	Mean rate of increase in shell width	Mean weight before experiment	Mean rate of increase in weight
I	7	-89 mm	85 mm	5.4%	57 mm	5.6%	79 g	28.5%
II	13	90-94	92	3.9	63	4.9	100	30.7
III	21	95-99	97	2.9	65	3.3	109	25.7
IV	26	100-104	102	1.4	70	1.9	134	19.0
V	15	105-109	107	1.6	72	1.8	150	19.0
VI	31	110-114	112	1.4	76	2.0	175	16.6
VII	24	115-119	117	0.9	80	1.6	194	23.2
VIII	12	120-124	122	0.0	84	0.5	227	10.4
IX	13	125-129	126	0.4	86	0.9	241	12.9
X	5	130-	132	0.3	90	1.5	311	3.8

*Data prepared by dividing the groups on the basis of the shell length (for every 5 mm) of abalones transplanted on February 26 and March 6, 1938, and removed from the ponds on May 31, 1938.

and 4.18. In this culture, *H. discus hannai* were transplanted from the Iwate Prefecture to the Kominato, Chiba Prefecture.

If *H. discus hannai* are transplanted sometime in November from southern regions and grown up to May or June of the following year, it is better to release seed abalones of 10 cm shell length so that the animals attain 12 cm at the time of marketing.

The density of the abalones released in ponds can be judged on the basis of oxygen consumption by the animals, and the extent of water circulation possible in the pond. In ponds in which rocks are placed, the water level changes during ebb tide and high tide, and unlike those in which water is circulated by pumps (known as the land method), water circulates through tides and waves. Hence it is possible to determine the appropriate amount of water in the pond by taking the water level at high tide and ebb tide as the standard. Care should be exercised during the calm days of the summer when the current of the sea water is slow, that the oxygen does not become deficient due to high water temperature.

When abalones are grown in ponds on land, they often tend to crowd. *H. discus*

Table 4.18. Rate of growth in pond culture of abalones of Iwate Prefecture transplanted to Chiba Prefecture (II)*

Group	Number of abalones examined	Shell-length range	Shell length		Shell width		Weight	
			Mean shell length before experiment	Mean rate of increase in shell length	Mean shell width before experiment	Mean rate of increase in shell width	Mean weight before experiment	Mean rate of increase in weight
I	9	-89 mm	87 mm	10.1%	60 mm	10.4%	85 g	47.5%
II	27	90-94	92	5.2	65	5.0	95	34.1
III	32	95-99	97	4.3	66	3.8	112	26.9
IV	42	100-104	102	2.9	70	2.3	125	27.2
V	26	105-109	107	2.4	72	2.1	147	20.4
VI	23	110-114	112	1.6	77	1.5	172	16.9
VII	8	115-119	117	1.5	79	1.8	195	21.4
VIII	17	120-124	122	0.9	84	0.9	211	18.4
IX	6	125-129	127	0.7	86	0.2	238	13.6
X	7	130-134	132	1.5	90	2.2	263	22.7
XI	6	135-139	137	0.4	95	0.2	312	6.0
XII	2	140-	146	1.1	97	0.0	365	19.2

*Data prepared by dividing the groups on the basis of the shell length (for every 5 mm) of abalones transplanted on December 26, 1938 and January 1, 1939 and removed from the ponds on May 27, 1939.

crowd more than *H. sieboldi* and *H. gigantea*. If U-shaped tubes are introduced to increase the surface area for the setting of the abalones, heavy mortality sometimes results from overcrowding and insufficient oxygen. Hence one should be extremely watchful at the time of crowding, and remove the dead abalones as early as possible. Particular care must be taken to supplement the oxygen inside the U-shaped tubes, and to clean the bottom of the ponds regularly to remove feces and remnants of feed.

To remove the abalones, ebb tide is carefully forecasted, the water drained from the pond, and the abalones taken out. Both the water inlet and outlet are avoided. After the abalones have been removed, the ponds are cleaned of all mud and sediment.

The shipment of cultured abalones is done according to the variations in the market price. Harvesting is greatly influenced by weather and sea conditions, and the price varies according to availability. Hence, it is better to release the cultured abalones for marketing when weather conditions continue to be poor, thus raising the prices, or when the harvesting of abalones is prohibited during October.

(Oba and Shibui)

6. Pests and prevention

One of the major factors affecting the growth of abalones is pests. Octopuses are a major predator in Japan and Table 4.19 shows the extent of the damage caused by them in Tokushima Prefecture. Other predators include several varieties of fish: Rays (*Ei*), *Chimaera* (*Same*), *Sparus macro cephalus* (*Kurodai*), *Gymnothorax kidako* (*Utsubo*), *Semicossycus reticulatus* (*Kandai*), *Lateolabrax japonicus* (*Suzuki*), *Gronistius zonatus* (*Takanohadai*), *Halichoenes* sp. (*Kyusen*), *Ditnema* sp. (*Umitanago*), flatfish (*Kurogashira-*

Table 4.19. Abalone catch and Octopus consumption in Tokushima Prefecture (Iino, 1966)

Year	Abalone catch (kan)	Octopus catch (kan)*	Remarks
1920	889	—	
1921	978	177	Sudden increase in octopus catch
1922	950	126	Decrease in abalone catch
1923	1,040	338	
1924	926	353	
1925	995	2,314	
1926	356	1,944	
1927	335	2,704	
1928	755	507	

*Kan=3.75 kg.

garei), *Cottus* sp. (*Onikajiki*), *Micnostomus stellen* (*Babagarei*), etc. In Hokkaido, Oku-shiri-To is well known for the production of seed abalones, and according to Kino-shita (1950), from the stomach of one *Ditnema* sp. (*Umitanago*), 71 seed abalones were removed. In some specimens nearly 200 abalone spat were found in the stomach. Considering the harvest rate of *Ditnema* sp. in Okushiri-To, an estimated 1,500,000 abalone spat are consumed by these fish. Figure 4.14 presents shells of abalones found in the stomachs of *Ditnema*, which were collected by the Oyster Research Center. One report stated that in the stomach of just one *Ditnema* measuring 22 cm in length, 136 abalone spat were found. Kinoshita (1950) reports that a considerable number of abalone spat were found in the stomachs of octopuses and starfishes. Tokuhisa (1913) reported that *Morphysa iwamushi*, concealed in the mud on the surface of rocks to which

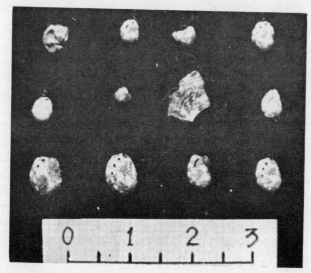

Figure 4.14. Abalone spat collected from the stomach of *Ditnema* sp. (*Umitanago*) (Oyster Research Center).

abalones remain attached, also eat the animals. Crabs have also been reported as abalone predators (Cox, 1962). The authors found that young abalones often fall prey to sea urchins.

To safeguard abalones from predators, studies of the ecological, geographical, and seasonal characteristics of their supplementary organisms must be made of the latter. On the basis of such studies, firm measures against these enemies can be established. A careful search for these predatory organisms is essential in the regions where fishing is prohibited, so that efforts to remove them can be put into effect before the abalones are damaged. Instances in which a large number of spat were eaten by sea urchins and other predators within a very short time have occurred even in those regions where spat grow well every year, i.e. regions well suited to the production of abalone seedlings. If this destruction is predictable, it is better to transplant the sea urchins to other places or harvest them. It is especially important to ascertain the relationship between the fishing ground and predators during the time of seedling release, while considering the size of the seed abalones and the period of their release. For example, in those grounds where the predators are nocturnal, it is better to release the abalones during the day; where the predators are diurnal, abalones are better released at night. Again, in those grounds where a large number of octopuses, starfishes, etc., live, these should be removed to some other areas where *Mytilus* have been previously provided, before releasing the abalones.

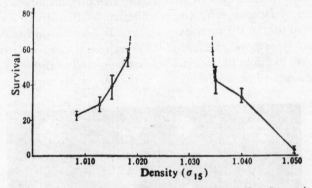

Figure 4.15. Survival period of *Haliotis discus hannai* (8 to 9 mm shell length of spat) (Oyster Research Center).

Besides the biological organisms there are also physical and chemical factors which can damage abalones. Young spat are often found in sea zones in which fresh water is sometimes mixed. Since spat are frequently damaged by swirling sand and mud, extra care is needed when ports, etc., are under construction. On suspicion that a zone might be "polluted" with fresh water, transplanting the young abalones to safer regions is the best solution. There is also the problem of drainage water from various industries, a problem which is likely to become more and more acute with industrial progress.

(Shibui)

7. Harvest

Fishing equipment and fishing methods vary according to the weather and sea conditions and the topography of the sea floor. In regions with warm water currents, collection is made directly from the sea floor, but in the Tohoku and Hokkaido regions, collection is mainly made with fishing rods.

Even with these traditional fishing methods, there are several problems, relating mostly to the pollution of resources. Hence a series of modifications of both the implements and the methods have been tried.

The helmet method of diving gear was introduced in Japan as early as the Meiji era, but since the efficiency of this method is limited, today it is applied only in special conditions where the water is very deep.

Along with the development of the artificial seeding technique, and progress in the various steps of culture methods, it is necessary to devise a harvesting method which ensures a higher yield of abalone. The various types of implements developed recently, their improvement, and the modifications of fishing methods are currently under study.

The present technique of artificial seeding is not adequate for commercial purposes, and hence abalone production depends primarily on natural fecundity. To increase this yield rate, proper regulations for abalone fishing must be stipulated. In some areas abalone resources have greatly decreased due to overfishing, while others have improved much due to the strict imposition of fishing regulations.

In Ibaraki Prefecture, diving gear was introduced in Meiji 12 for abalone collection, and the method was very widely used for some time; as a result, resources were gradually depleted. In Meiji 25, although the area for this method was restricted, depletion was still rapid and so stringent restrictions were imposed in Taisho 15, during which time resources showed some recovery. In Showa 4, even fewer areas were permitted to use diving gear, and the harvesting rate per catch increased. Interestingly, however, when the area permitting diving gear was expanded by Showa 9, the total amount of catch showed a remarkable increment (Iino, 1966). When regulations concerning permissible shell length for harvesting were imposed, abalone culture improved in the Iwate and Hokkaido Prefectures (Iino, 1966).

Restrictions to protect the resources of almost all the fishing grounds exist today, which take into consideration the various characteristics of the fisheries and past experience with exploitation.

To stabilize and maintain higher rates of production through scientific management of resources, the relationship between the spawning rate, replenishment, growth rate, natural and accidental mortality, and methods of controlling resource changes, should be understood. It is now possible to make a quantitative assessment of abalone resources. Inoue (1965) estimated the fishing rate and the number of living abalones in Joga-Shima of the Kanagawa Prefecture; on the basis of this estimate he forecasted the size of the catch by standard methods. Resources can be quantitatively estimated by Delury's method also. But though abalone resources can be mathematically assessed, their management has not yet entered an effective stage. However, present experiments with artificial seeding techniques will, hopefully, correct the latter problem.

Abalones caught by plucking with needles or wires sometimes suffer back injuries, which result in a high mortality when the animals are stored. Needless to say, the first step toward minimizing such losses is to catch the animals without injuring them. Despite efforts to prevent injuries however, about 30% of the abalone catch dies when plucking is done with needles and 5% when plucking is done with wires. Death occurs in storage due to the fact that infection sets in in the injured parts and sometimes spreads to the uninjured animals causing mass mortality.

To prevent or cure such infections, various antibiotic drugs are being experimented with. According to Takayanagi (1966) any infection can be forestalled by washing the injured parts with Syazin solution. A mass treatment method for injured abalones is currently under study.

(Kikuchi)

8. Future trends

Abalone production used to depend solely on natural fecundity, and workers used to blindly try to increase it by experimenting with various methods of improving the natural productive structure. Hence, though some improvement in the local conditions of the fishing grounds resulted, an appreciable increment in abalone production did not occur. Efforts were also made to rear young abalone under artificial environmental conditions, the spat being transferred back to the fishing grounds. These efforts proved successful and the rate of fecundity exceeded that of the natural conditions several times. Measures for controlling abalone resources were likewise carefully studied, as were various culture methods for improving productivity in natural conditions.

At present, improvement is directed toward higher production by determining the rate of spat collection transplanted from natural fishing grounds and their growth rate. A seeding technique economically suited to large-scale production is also being investigated. Culture techniques are expected to show considerable progress on the basis of improvement in the seeding technique, and an increase in production rates is expected through seedling transplantation. The transplantation of abalone seedlings from natural surroundings has been carried out in various regions, but the yield rate of the transplanted abalones is not yet clear. Hence it is necessary to know the actual conditions of transplanted abalones to compare them with those of artificially grown spat which will later be transplanted. The size of the spat at the time of release, the suitable density of them, the period selected for their release, the various conditions of the releasing ground, and the effect of transplantation in terms of production, also require further study.

A review of the seeding techniques shows that little progress has been made in this area, and consequently its impression on productivity is rather negligible. In other words, the collection of eggs has not reached that stage where results can be forecasted. Although the number of eggs collected can be increased by using a larger number of mother abalones, the changes following the spawning period affect seedling development and, consequently, production. Hence, the physiological mechanisms involved in maturation and spawning must be understood in order to forecast seed production.

Another problem is the unpredictable survival rate of planktonic larvae up through the stage of spat formation with 4 to 5 mm shell length. Losses occur mainly during bottom-dwelling life and the first inhalant stage. The factors operating at this time are not clearly understood, but their influence on planned seedling collection is considered paramount.

Transplanted spat should be grown as quickly as possible to the size recommended for release in the fishing ground. The effectiveness of rearing implements and the adequacy and nutritional value of diet should be studied therefore, before pilot programs in large-scale production are put into effect.

As already mentioned, there are several problems involved in devising a perfect abalone culture technique under natural conditions. However, the results obtained thus far by artificial seeding techniques raise hopes of an annual production of about 10,000,000 abalones through the large-scale release of artificial seedlings in natural fishing grounds.

(Kikuchi)

Bibliography

1. Akita Fisheries Experiment Station, 1965. Report on the Results of Experiments for the Year 1965.
2. ———, 1966. Report on the Index Survey for the Year 1965.
3. AONUMA, G., 1953. Iwate-Ken san ezo-awabi *Haliotis kamtschatkana* no seitai ni tsuite [Ecology of *Haliotis kamtschatkana* Produced in the Iwate Prefecture (II)], Iwate Suishi (Iwate Fisheries Experiment Station).
4. BOUTAN, L., 1899. La cause principale de l'asymetrie des Mollusques Gastreopodes, Arch. Zool. Exper. (3), T. VII, 207-331.
5. Chiba Fisheries Experiment Station, 1966. Chiba ken ni okeru chikuyoshoku no genkyo (Culture Technique in Chiba Prefecture), 1-7.
6. ———, 1966. Awabi no kaitei chikuyoshoku gijutsu kenkyu (Investigations on Sea-bottom Culture Technique of Abalones), Report on the Investigations Carried out for the Year 1966, 1-34.
7. Cox, K. W., 1962. California Abalones, Family *Haliotidae, Fish Bull.* (118), Dept. of Fish and Game, The Resources Agency of California.
8. FUJIMOTO, T., and S. YAMAGUCHI, 1964. Ezo-awabi no ishoku seicho koka ni tsuite (dai-ippo) [On the Effect of Transplantation of Ezoawabi (Report No. 1)], *Ibaraki Suishi Hokoku (Bulletin of Ibaraki Fisheries Experiment Station for the Year 1962)*, 33-39.
9. HIROSE, T., 1953. Iwate ken san ezo-awabi *Haliotis kamtschatkana* no seitai ni tsuite [Ecology of *Haliotis kamtschatkana* of Iwate Prefecture: Report on the Abalone Survey of Iwate Prefecture (I)], Iwate Suishi (Iwate Fisheries Experiment Station).
10. IINO, T., 1936. Awabi-zoku bunruijo no ichitokucho ni tsuite (Classification of *Haliotis*), *Nisuikaishi (Bulletin of the Japanese Society of Scientific Fisheries)*, 5 (3), 175-176.
11. ———, 1937. Awabi no hanshoku hogo ni kansuru jirei (On the Fecundity of Abalones), *Yoshoku Kaishi (Bulletin of the Culture Society)*, 7 (7-8), 145-148.
12. ———, 1941. Sokkin ni yoru Awabi no sonshodo ni tsuite (Injuries of Abalones Due to Metal Wires), *Suisan Kenkyushi (Journal of Fisheries Experiments)*, 36 (10), 171-172.
13. ———, 1943. Awabi no setsuji to seicho (Diet and Growth of Abalones), *Ibid.*, 11 (5-6), 171-174.
14. ———, 1946. *Awabi no zoshoku (Fecundity of Abalones)*. Kasumigaseki Publications.
15. ———, 1947. Karihara so tekisei saishuryo kettei ni kansuru kenkyu (On the Study of Yield Rate of Algae), *Nissuikaishi (Bulletin of the Japanese Society of Scientific Fisheries)*, 13 (3), 117-119.
16. ———, 1952. Hosan awabi-zoku no zoshoku ni kansuru seibutsugakuteki kenkyu (Biological Investigations on the Fecundity of Japanese Abalones), *Tokai-Ku Suiken kenkyu Hokoku (Bulletin of Tokai-Ku Fisheries Experiment Station)*, (5), 1-102.

17. IINO, T., 1966. Awabi to sono zoyoshoku (Abalones and Their Culture), Suisan Zoyoshoku Gyosho, **11** (Association for the Production of Japanese Fisheries Resources).

18. INOUE, M., 1965. Hyoshiki horyu kara mita awabi-zoku no ishoku ni kansuru ichi, ni (Transplantation of Abalones), *Suisan Zoshoku (Aquaculture)*, Special issue (5), 23-31.

19. Iwate Fisheries Experiment Station, 1966. Awabi shubyo seisan gijutsu Kenkyu hokokusho (Report on the Study of the Seeding Techniques of Abalones) (for the years 1963-1965).

20. Izu Branch of Shizuoka Fisheries Experiment Station, 1958. Awabi saiho no 2, 3 no jiretsu (Discussion on the Collection of Abalones), Izu Bunjo Dayori, No. 17.

21. KANADA, H., 1952. Awabi-zoku nishu no nisshuki katsudo (Daily Activities of 2 Types of Abalones) (Unpublished).

22. KANNO, H., and S. KIKUCHI, 1963. On the rearing of *Anadara broughtonii* and *Haliotis discus hannai)*, *Bull. Mar. Biol. St. Asamushi, Tohoku Univ.*, **11** (3), 71-76.

23. KIKUCHI, S., 1963. Ezo-awabi no tanku saibyo ni tsuite (On the Tank Seeding of Ezo-awabi), *Suisan zoshoku (Aquaculture)*, Special issue (2), 15-18.

24. ——, 1964. Awabi no yoshoku ni kansuru Kenkyu (On the Culture of Abalones). Data from the Peking Symposium held in 1964.

25. ——, Y. SAKURAI, M. SASAKI, and T. ITO, 1967. Kaiso 20-shu no awabi chigai ni taisuru shiryo koka (Effect of 20 Types of Seaweeds Given to Abalone Spat), *Tohoku Suiken Kenkyu Hokoku (Bulletin of Tohoku Fisheries Experiment Station)*, (27), 93-100.

26. KINOSHITA, T., 1927. Awabi ni kansuru kenkyu (Sonoichi) seichodo ni tsuite [Study on Abalones (Report No. 1) on the Growth of Abalones], Hokusuishi Junpo (Fortnightly Report of Tohoku Fisheries Experiment Station), (12).

27. ——, 1934. Ezo-awabi no seicho ni tsuite (On the Growth of Ezo-awabi), *Rakusaikai, Kaishi*, **29** (11), 1-7.

28. ——, 1950. *Awabi no chishiki to sono zoshoku (Reproduction of Abalones)*. Hoppo Publishers.

29. ——, and K. NAKAGAWA, 1934. Awabi no kuchu katsuryoku to ondo (Activity of Abalone in Air and Temperature), Hokusuishi Jigyojunpo (Fortnightly Report of Tohoku Fisheries Experiment Station), 258.

30. ——, ——, 1936. Fukuyamasan awabi no jikitekisohi ni tsuite (Periodical Fattening of Fukuyama Abalones), *Ibid.*, (334).

31. ——, ——, 1937. Senkai zoshoku no gaiteki seibutsu (Pests in Shallow-sea Culture). Pamphlet published by Hokkaido Fisheries Experiment Station 4.

32. ——, ——, 1974. Hokkaido-san awabi no zosan ni kansuru kenkyu (dai-ippo). [Study on the Production of Hokkaido Abalones (Report No. 1)], Hokusuishi junpo (Fortnightly Report of Hokkaido Fisheries Experiment Station), 4 (3).

33. ——, ——, 1949. Hokkaido-san ezo-awabi no seicho ni kansuru kosatsu (On the Growth of Hokkaido Abalones), *Hokusuishi Kenkyu Hokoku (Bulletin of Hokkaido Fisheries Experiment Station)*, (1), 2-6.

34. ——, ——, 1950. Funka-wan nai yuju ni ishoku shita ezo-awabi no hanshoku ni tsuite (On the Reproduction of Transplanted Ezo-awabi in Funka-Bay),

Ibid., (6) 21-27.

35. KISHIGAMI, K., 1894. Awabi kenkyu dai-ippo (First Report on the Study of Abalones), *Suisan Chosa Hokoku (Fisheries Survey),* 3 (1-2).

36. ——, 1895. Awabi kenkyu dai-niho (Second Report on Abalones), *Ibid.,* 4 (2), 1-16.

37. KURODA, T., Y. TUSHIDA, Y. TANIZAWA, and H. UEMOTO, 1957. *Senkai zoshoku no riron to jissai (Theory and Practice of Shallow-sea Water Culture).* Gyoson Bunka Kyokai.

38. Kyoto Fisheries Experiment Station, 1966. Report on the Index Survey for the Year 1966.

39. MATSUBARA, S., 1882. Sekketsu myo no sanran o kaibojutsu ni yute setsumeisu (Study of Spawning through Anatomy), *Dainihon Suisankai Hokoku* (10), 2.

40. ——, 1883. Sekketsumyo no kenkyu ni kansuru fukumeisho (On the Study of Abalones), *Ibid.,* (10), 20.

41. ——, 1885. Abalone Survey, *Ibid.,* (44), 26.

42. Miyagi Prefectural Fisheries Experiment Station, 1964. Gyogyo Gijutsu Shunrenkai Shiryo (for the Year 1964).

43. MIYAMURA, M., 1954. Ezo-awabi no ishoku shiken (Transplant Experiments on Ezo-awabi), Mie Suishi jiho (Report of the Mie Fisheries Experiment Station), (176), 55-69.

44. MURAYAMA, S., 1935. On the Development of the Japanese Abalone, *Haliotis gigantea, J. Coll. Agr. Tokyo Imp. Univ.,* 8 (3), 227-232.

45. MASUDA, T., 1966. Gaikaimen no yoshoku shisetsu (Culture Equipment in the Open Sea), *Suisan Doboku (Fisheries Engineering),* 3 (1), 17-20.

46. NAKAMURA, H., 1940. Ezo-awabi no tanki yoshoku (Short Term Culture of Ezo-awabi), *Suisan Kenkyushi (Journal of Fisheries Research),* 35 (4), 88-91.

47. NAKAJIMA, M., *et al.,* 1953. Report on the Survey of the Sea Surface for the Year 1953. Hokkaido Hogo Suimen Kanri Chosa Hokokusho.

48. NISHIKAWA, S., and K. KAWAMURA, 1958. Report on the Survey of the Sea Surface for the Year 1958. *Ibid.*

49. ODAKE, N., and H. TANIMOTO, 1965. Awabi-shubyo seisan ni tsuite (On the Production of Abalone Seedlings), *Tokushima ken Suishi jigyo Hokoku (Bulletin of the Tokushima Fisheries Experiment Station),* 1958-1961, 103-105.

50. OKAMURA, K., and H. SENO, 1918. Suisan koshusho takashima jikkenjo hanshoku hogo shiken seiseki dai-ichiji hokoku Suisan koshusho shiken hokoku (First Report on the Results of the Culture Experiments Carried out at Takashima), Experiment Culture for Fisheries Training Report, *Ibid.,* 14 (2), 1-6.

51. ——, ——, 1920. Second Report on the Results of the Culture Experiments Carried out at Takashima. Experiment Culture for Fisheries Training Report, *Ibid.,* 15 (4).

52. ONODERA, H. 1957. Miyagi-ken Hokubu kaiku ni okeru ezo-awabi no sanranki ni tsuite (On the Spawning Period of Ezo-awabi in the Northern Regions of the Miyagi Prefecture). Kesen-numa Branch of the Miyagi Fisheries Experiment Station.

53. SAITO, K., 1963. Ezo-awabi ni tsuite (On Ezo-awabi), Hokusuishi Geppo (Monthly Report of Hokkaido Fisheries Experiment Station), 20 (12), 425-439.

54. Saito, K., 1965. Okushiri-To no Awabi Ishoku Shubyo ni tsuite (On the Transplanted Seedlings of Abalones of Okushiri-To), *Suisan Zoshoku (Aquaculture)*, Special issue (5), 32-44.

55. ———, and K. Kawamura, 1959. Report on the Survey of the Sea Surface for the Production of Abalones for the Year 1959.

56. ———, and K. Tomita, 1965. Reibun-To senpaku no ezo-awabi ni tsuite (On the Ezo-awabi of Reibun-To), Hokusuishi Geppo (Monthly Report of the Hokkaido Fisheries Experiment Station), **22** (5), 243-259.

57. Sakai, S., 1959. Ezo-awabi to Wakame no seitai chosa (Study of the Ecology of Ezo-awabi and Wakame), Engan Gyogyo, Shuyaku Kei-ei Chosa Hokokucho (for the 2nd year). Miyagi Fisheries Experiment Station, 323-332.

58. ———, 1960. On the Formation of the Annual Ring on the Shell of the Abalone, *H. discus* var. *hannai* Iino, *Tohoku Jour. Agr. Res.* **11** (3), 239-244.

59. ———, 1962. Ezo-awabi no seitaigakuteki kenkyu: I. Shokusei ni kansuru jikkenteki kenkyu (Ecological Studies of Ezo-awabi: I. Experimental Studies on Feeding Habits), *Nissuikaishi (Bulletin of the Japanese Society of Scientific Fisheries)*, **28** (8), 766-779.

60. ———, 1962. Kaigara no shikisai to seicho oyobi shokusei to no sogokankei (Ecological Studies of Ezo-awabi: II. Correlation between Color of Shell, Growth, and Introduction), *Ibid.*, **28** (8), 780-783.

61. ———, 1962. Onagawa-Wan fukin ni okeru ezo-awabi no seisan kozo no kaiseki (Ecological Studies of Ezo-awabi: III. Analysis of the Production of Ezo-awabi in the Neighborhood of Onagawa Bay), *Ibid.*, **28** (9), 891-898.

62. ———, 1962. Seicho ni kansuru kenkyu (Ecological Studies of Ezo-awabi: IV. Study of Growth), *Ibid.*, **28** (9), 899-904.

63. Sano, T., and R. Batei, 1962. Ezo-awabi no seiiku ni oyobosu kenkyu joken ni tsuite (Environmental Conditions Affecting the Growth of Ezo-awabi), *Tohoku Suiken Kenkyu Hokoku (Bulletin of Tohoku Fisheries Experiment Station)*, (21), 79-86.

64. Shibui, T., T. Onodera, T. Chiba, and T. Imai, 1966. Ezo-awabi no jinko saibyo: II (Artificial Seedlings of Ezo-awabi, II), Nihon Suisan Gakkai Tohoku Shibu Taikai (Proceedings of the Conference of the Tohoku Branch of the Japanese Society of Scientific Fisheries).

65. Shiraishi, K., and T. Imai, 1964. Kaimen Suiso ni yoru awabi shubyo no ikusei Suisan zoyoshoku no shubyo seisan gijutsu ni kansuru kiso kenkyu hokokusho (Rearing of Abalone Seedlings in Water Tanks: Report on the Fundamental Research Related to Seedling Production Techniques for Fisheries Culture for the year 1963).

66. Tago, K., 1931. Nihonsan awabi-zoku no bumpu ni tsuite (Distribution of the Genus *Haliotis* of Japan), *Dozatshu (Zoological Magazine)*, **48** (508-510), 352-361.

67. Takahashi, K., and S. Aoyama, 1965. Aomori-ken Shiriya Tisaki no ezo-awabi ni tsuite (On Ezo-awabi of Shiriya Tisaki of the Aomori Prefecture), *Aomori Suishi (Bulletin of the Aomori Fisheries Experiment Station)*.

68. Takayanagi, T., 1966. Awabi no kizu to sono kagaku iryoho (Chemical Treatment of the Injuries of Abalones), *Yoshoku (Culture)*, **3** (3, 4).

69. TAKAYAMA, K., 1940. Mie-ken san awabi no seichodo ni tsuite (On the Growth of Abalones of the Mie Prefecture), *Suisan Kenkyushi (Journal of Fisheries Research)*, **35** (2), 99-100.

70. TAKAYANAGI, T., T. OBA, and T. NAKAMURA, 1966. Awabi no chikuyo ni kansuru kenkyu: I. Gyokaku ni yoru sonsho awabi no chikuyo to kizu ni kiinsuru kanosei shikkai no hasseiboshi ni tsuite (Study on the Rearing of Abalones: I. Storing of Injured Abalones and the Prevention of Infection Resulting from Such Injuries), Report from Chiba Fisheries Experiment Station for the Years 1963 and 1964, 168-179.

71. TAMURA, T., 1939. Kakushu kaisan kairui no kokyu ni oyobosu gaiihenka no eikyo (Influence of Changes in the Atmosphere on the Respiration of Various Marine Mollusks), *Suisan gaku zosshi (Journal of Fisheries)*, (43) 1-10; (44), 64-73.

72. TOKUHIMA, S., 1913. Awabi no gaiteki to shito no iwamushi (On Iwamushi, a Type of Abalone), *Suisan Kenkyushi (Journal of Fisheries)*, (9), 239.

73. Tokushima Fisheries Experiment Station, 1966. Report on the Index Surveys for the Year 1966.

74. UCHIDA, K., and K. YAMAMOTO, 1942. Chosen kinkai ni okeru awabi no bumpu (Distribution of Abalones in the Sea around Korea), *Venus*, **11** (4), 119-126.

75. UCHIMURA, K., 1881. Study on the Culture of Abalones. Sapporo Prefecture.

76. UEDA, S., and K. OKADA, 1939. Makigairui no tennenshiryo ni kansuru kenkyu: I. Awabi (Study on the Natural Effect of Mollusks: I. Abalones), *Nissui Kaishi (Bulletin of the Japanese Society of Scientific Fisheries)*, **8** (1), 1-5.

77. ——, ——, 1941. Study on the Natural Effect of Mollusks: II. Abalones, *Ibid.*, **10** (3), 139-142.

78. UNO, Y., 1967. Yogyogaku kakuron (Awabi) [Discussion on Fisheries Culture (Abalones)], *Suisan Gaku Zenshu* (Koseisha Koseikaku), 643-677.

79. Yamagata Fisheries Experiment Station, 1966. Report on the Index Surveys for the Year 1966.

80. YAMAMOTO, G., 1964. Mutsu-wan ni okeru hotategai zoshoku Suisan zoyoshoku gyosho (Culture of Scallops in Mutsu Bay: Report on the Fisheries Culture), *Nihon Suisan Shigen Nogi Kyokai*, **6**.

81. YAITA, K., 1939. Senkai zoshoku jigyo awabi zoshoku jigyo (Abalone Culture in Shallow-sea Water), Report on Culture from Hyogo Fisheries Experiment Station, 1939.

PART V
The Evolution of Prawn Culture

Chapter 1 of this discussion of the evolution of prawn culture is a compilation by Kuroda of all the data available in Japan on the subject of prawns; Chapter II presents a review of the progress Japan has made in prawn culture. Comparatively speaking, the research work done by Japan in the area of prawn culture is far more extensive than any other country's.

Some repetitions occur in Chapter II of the material given in Chapter I. These repetitions were deemed necessary for a clear-cut presentation of certain salient points.

That the "World Scientific Conference on the Biology and Culture of Shrimps and Prawns" was held under the auspices of the FAO on June 12-25, 1967 in Mexico City, with approximately 200 scientists representing 33 countries participating, is a very significant factor in pointing up the importance of prawns and shrimps in the field of world fisheries today. It is also interesting that the word "culture" figured in the title of the Conference.

(Hudinaga)

CHAPTER 1

Biological Research on Prawns

1. Introduction

The history of biological research on prawns begins with Fabricius who published his studies on *Penaeus* in 1798, and continues through more than 800 papers on various aspects of prawn biology, which were compiled and published by Chin and Allen in 1959. The research of the past 15 years has benefited greatly from these earlier studies.

Bate (1888) was the first to confirm that prawns belong to the family *Penaeidae*, and classification into subfamilies and genera was done by Burkenroad (1934a, b, 1936, 1939). This systematic work on prawns and shrimps was continued by Kubo (1949), Racek (1955), Dall (1957), Hall (1962), and Racek and Dall (1965), and consequently the classification *Penaeidae* is now very nearly complete. The problems relating to different species being classified together, or types of the same species being classified separately have been discussed in detail by Holthuis (1949, 1962a, b), Kubo (1954), Hall (1956), Gunter (1957, 1962), and Burkenroad (1963a, b). Kishigami (1900) subjected the prawns of Japan and its surrounding territory to detailed study and Kubo (1949) classified them. In fact, the classification of prawns in this chapter draws heavily from the work of Kubo published in 1956, and includes the prawns and shrimps of the Indian and Pacific Oceans.

As early as 1863, Müller discovered that the embryological development of prawns passes through *nauplius* and *zoea* (protozoea). Full details of the embryological development of prawns were reported only after 1930, and published mainly by Hudinaga (1935, 1942), Heldt (1938), and Pearson (1939). In Japan, Hudinaga (1942) carried out very extensive studies on the fundamental aspects of prawn culture, but a systematic study of the life history and ecology of prawns had already begun in 1930. Following World War II, the commercial importance of prawn resources blossomed, and the pace of research in this field accelerated, since the USA, India, Australia, and Japan are very rich in prawns. Biological studies of several varieties of prawns were made in Japan by Kubo (1956), followed by Maekawa (1961), Yasuda (1956, 1958), Ikematsu (1963), and others. Experiments and observations of the physiology and habitat of prawns were begun in the second half of 1950. Oxygen uptake was studied by Maekawa and Otsuka (1955), Egusa (1961), and Rao (1958).

Migratory behavior was studied by Onizuka (1914), Miura and Yamaguchi (1955), Egusa and Yamamoto (1961), and Fuss (1964). Osmotic pressure and ion adjustments were studied by Panikkar and Viswanathan (1948), Panikkar (1951), Williams (1960), and Dall (1964). The influence of temperature and salinity on growth and survival was studied by Zein-Eldin (1963), Zein-Eldin and Aldrich (1965), and Zein-Eldin and Griffith (1966). The endocrine glands were studied by Knowles (1953) and Dall (1965a). Dall (1965b, 1965c, 1965d) also studied the structure of the carapace as well as the hydrocarbon and calcium metabolism.

Prawn parasites have become the subject of study in recent times (Hutton, Sogand-ares-Bernal, Eldred, Ingle, and Woodburn, 1959; Kruse, 1959). Hiraiwa (1933, 1934, 1936) and Hiraiwa and Saito (1939) studied prawn parasites such as *Isopoda* in Japan. The study of prawn migration and growth under natural conditions by tagging, was later supplemented by staining, which has now become the more important method (Dawson, 1957; Costello, 1959, 1964; and Klima, 1965).

2. Species and distribution

2.1 Major world species and their distribution

The family *Penaeidae* belongs to the Order *Natantia* under the Class *Decapoda*. The development of *Penaeus* is said to be very typical of decapods (Calman, 1909; Gordon, 1955). A large number of species are found to be distributed in a wide area, from the warm belts and inner bays to deep waters. By 1960, 318 species divided into 4 sub-families were listed (including *Sergestidae*, Waterman and Chace, 1960). The subfamily *Aristaeinae* is easily distinguished from the other 3 subfamilies; the upper flagellum of the antennule is attached to the posterior edge of the third protopodite and is very much shorter than the lower flagellum. In the subfamily *Solenocerinae* the cephalic groove reaches the dorsal gland, a condition not seen in the other two subfamilies. *Sicyoninae* lack prosartema and differ from the subfamily *Penaeinae* by not having an endopodite on the abdominal appendage. The first two families are mostly deep water types, and the latter two normally dwell in coastal regions (Burkenroad, 1939). The two groups also differ in their development during the larval period (Gurney, 1942). In the seas around Japan, *Sicyoninae* are mostly found in waters 200 to 300 m deep (Kubo, 1949). The varieties that are used for commercial purposes are mostly coastal, and almost all of them belong to the subfamily *Penaeinae*.[1] For details of the characteristics of the genera and species of prawns, Kubo (1949), Racek (1955), Dall (1957), Racek and Dall (1965), and Jouvert (1966) are good references for species found in the Indian and Pacific Oceans; Anderson and Lindner (1943) and Voss (1955) list the species in the American waters; and Yoshida (1941) and Ikematsu (1963) present the species found in the seas around Japan.

Since *Penaeus* usually grow to large size and fetch a high market price, they are commercially important. The regions from which they are caught are shown in Figure 5.1, but the utilization of these resources varies according to fluctuating conditions; of the world resources, those in the seas around Japan show maximum utilization. The distribution range of *Penaeus* coincides more or less with the range where the mean surface water temperature in summer is 20°C or more (August in the northern hemisphere and January in the southern). In the northern hemisphere, the distribution range of *Penaeus* is very distinct compared to the other commercial type, *Pandalidae*, which is found in the northern regions with a lower temperature. It is

[1]Prawns in the deep waters of the Pacific Ocean around Shikoku and Kyushu are also caught for commercial purposes—*Metapenaeopsis latus, M. coniger, Parapenaeus lanceolatus*—from depths of 150 to 250 m (Yasuda, Sasaoka, Dohi Tonagai, 1962); *Hymenopenaeus robustus* living in depths of 360 to 500 m in Gulf of Mexico is also reported as an important commercial variety (Clifford, 1956).

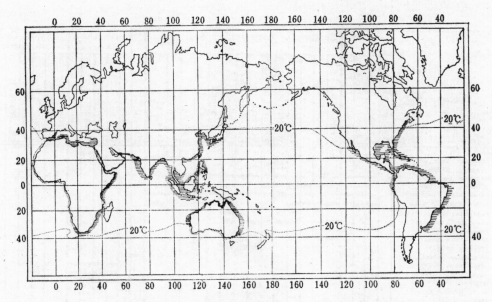

Figure 5.1. Commercial distribution of *Penaeus*.

Dark line—regions in which *Penaeus* are used commercially; *Dots*—regions in which *Penaeus* are used very little for commercial purposes, or where the distribution is limited; *Dotted line*—20°C isothermal line in the distribution of the mean surface water temperature in summer.

very interesting to note that the activities related to growth, spawning, feeding, etc., of *Penaeus* suddenly become accelerated when the water temperature reaches 20°C. On the other hand, the Atlantic and Pacific Oceans show a wider distribution in the western part than in the eastern; in the former the number of species is large and the regions are also very rich in *Penaeus*. According to the Annual Report of the FAO on fishing statistics, the average annual production of crustaceans throughout the world for the years 1961-1964 was 1,100,000 tons, of which *Penaeus* constituted a major part, i.e. 47%. The fishing rate for *Penaeus* is maximal in Mexico Bay (Atlantic Ocean off the USA and Mexico), India (particularly the western coast), and Japan. In these countries *Penaeus* comprised 60% of the total catch (Table 5.1).

It is very difficult to list accurately all the species that are caught for commercial purposes, but 52, belonging to 11 genera, are shown in Figure 5.2, of which *Penaeus* constitute a major portion. Judging from distribution, *Penaeus* can be divided broadly into 3 zones, namely, the Atlantic Ocean—the Middle Sea, the Pacific Ocean of America, and the Indian Ocean—the western Pacific Ocean (Dall, 1957). Except in some parts of the Panama Canal and the Suez Canal where the water currents mix, each zone has its own specific types. Distribution is firstly influenced by water temperature, and then within the optimum water temperature range, is slightly influenced by salinity and allied factors (Burkenroad, 1939). Gunter (1956, 1961a, b) and Gunter, Christmas, and Killebrew (1964) pointed up the particular importance of salinity. A detailed study of distribution reveals that characteristics of the sea floor

Species	Africa				Southern Asia				Australia					East Asia					Pacific Ocean of America						Atlantic Ocean of America						
	Western Africa	Southern Africa	Eastern Africa	Red Sea	Arabian Sea	Bay of Bengal	Indo-China	Indonesia	Western Australia	Southern Australia	Victoria	New South Wales	Queensland	Phillippines	Southern China	Taiwan	Yellow Sea	Japanese Islands	Hawaii	Mexico	Panama	Colombia	Equador	Northern Peru	Northern part of Argentina	Southern Brazil	Amazon-Olnok	Caribbean	Gulf of Mexico	Atlantic Ocean	Middle Sea

Penaeus
- *japonicus* Bate
- *orientalis* Kishinouye
- *monodon* Fabricius
- *indicus* H.M.-Edw.
- *latisulcatus* Kishinouye
- *semisulcatus* De Haan
- *esculentus* Haswell
- *merguiensis* De Man
- *canaliculatus* Olivier
- *penicillatus* Alcock
- *plebejus* Hess
- *marginatus* Randall
- *stylirostris* Stimpson
- *occidentaris* Streets
- *vannamei* Boone
- *brevirostris* Kingsley
- *californiensis* Holmes
- *setiferus* (Linnaeus)
- *duorarum* Burkenroad
- *aztecus* Ives
- *schmitti* Burkenroad
- *brasiliensis* Latreille
- *trisulcatus* Leach

Metapenacus
- *monoceros* (Fabricius)
- *ensis* (De Haan)
- *joyneri* (Miers)
- *burkenroadi* Kubo
- *dobsoni* (Miers)
- *affinis* (H.M.-Edw.)
- *brevicornis* (H.M.-Edw.)
- *macleayi*, (Haswell)
- *bennetae* Racek & Dall
- *endeavouri* (Schmitt)

Metapenaeopsis
- *acclivis* (Rathbun)
- *barbata* (De Haan)
- *novaeguineae* (Haswell)
- *palmensis* (Haswell)

Parapenaeus
- *longirostris* (Lucas)

Parapenaeopsis
- *tenella* (Bate)
- *stylifera* (H.M.-Edw.)
- *sculptilis* (Heller)
- *maxillipedo* (Alcock)

Trachypenaeus
- *curvirostris* (Stimpson)
- *anchoralis* (Bate)
- *constrictus* (Stimpson)
- *similis* (Smith)

Xiphopeneus
- *kroyeri* (Heller)
- *riveti* Bouvier

Artemesia
- *longinaris* Bate

Protorachypene
- *precipua* Burkenroad

Aristaeomorpha
- *foliacea* (Risso)

Hymenopenaeus
- *mulleri* (Bate)

Table 5.1. FAO Fisheries statistics annual report on marine prawn catch in the world for 1961-1964

Notes: 1) *Panulirus japonicus (Ise-ebi)*, the European *Crangon*, and *Pandalus* are not included.
2) Except for the USA (Pacific Ocean), the whole of Chile, and part of Japan and Korea, the data is mainly concerned with *Penaeus* (raw weight unit: 1,000 ton)

Name of country	1938	1948	1958	1961	1962	1963	1964	1961-1964, mean percentage
Grand Total[1]	[277.0]	[315.0]	420.0	450.0	500.0	533.0	[540.0]	100
Japan	17.5	33.4	55.6	73.5	79.3	86.8	78.6	15.5
Korea	38.5	33.0	16.4	22.9	20.4	14.2	18.1	3.7
Taiwan	—	—	4.0	5.5	6.4	9.0	9.8	1.5
India[2]	—	—	86.7	62.8	83.2	81.6	95.0	17.3
Thailand[3]	—	—	10.3	16.5	20.1	23.3	29.5	4.4
Pakistan	—	—	14.9	19.5	19.1	18.4	—	3.8
USA (Atlantic Ocean)	63.7	73.1	88.9	69.6	76.3	99.4	89.1	16.5
USA (Pacific Ocean)	1.2	1.7	8.1	9.5	10.4	9.6	5.0	1.7
Mexico	—	34.0	50.2	72.3	70.6	70.0	68.9	13.4
Brazil	—	—	17.4	25.0	35.2	—	—	6.9
Chile	—	—	12.9	10.5	12.3	12.8	16.6	2.6
Spain	—	11.4	13.7	15.0	14.1	13.0	12.7	2.7
United Arab	—	—	—	—	—	10.4	—	2.1

[1]Brackets include more than 10% of the estimation.
[2]Figures for 1958 include other marine crustaceans.
[3]Figures for 1958, 1961-1962 include marine crabs.

also exert an influence (Hildebrand, 1954, 1955; Springer and Bullis, 1954; Williams, 1958, 1965).

Of the 52 commercial varieties, 31 are found in the Indian Ocean and the western Pacific Ocean. The distribution range differs greatly for the various species of *Penaeus*, but this is not relevant to the commercial importance of the species. For example, *P. orientalis* of the Yellow Sea, *M. barbata* and *M. acclivis* of the Japanese coast, and *M. dobsoni* of India dominate the respective regions, but have little commercial value.

2.2 Major Japanese species and their distribution

Although there are 47 species of penaeids belonging to 16 genera in the Japanese

Figure 5.2. Distribution of important species of *Penaeus*. Thick lines indicate particularly thick population, and dotted lines indicate thin distribution.

Data obtained from Maki and Tsuchiya (1933), Burkenroad (1939), Yoshia (1941), Edmondson (1946), Kubo (1949), Kow (1955), Panikkar and Menon (1956), Qureshi (1956), Racek (1955, 1956), Djajadireja and Sachlan (1956), Dall (1957), Lindner (1957), U.S. Fish. and Wildl. Serv. (1958a, b), Cheung (1960), Shaikhmahmud and Tembe (1960), George (1961), Menon and Raman (1961), Tiews (1962), Kutkuhn (1962), Ikematsu (1963), F.A.O. (1964), Mistakidis and Neiva (1964), Racek and Dall (1965), Williams (1965), Bhimacher (1966), and Jouvert (1966).

waters (Kubo, 1949), only about 16 species belonging to 6 genera are commercially significant (Figure 5.3). Most of these species are found in the inner southern part of Tokyo Bay and its neighboring shallow waters. Although there are some species in the coastal regions of the Japan Sea, compared to the number of penaeids found in the

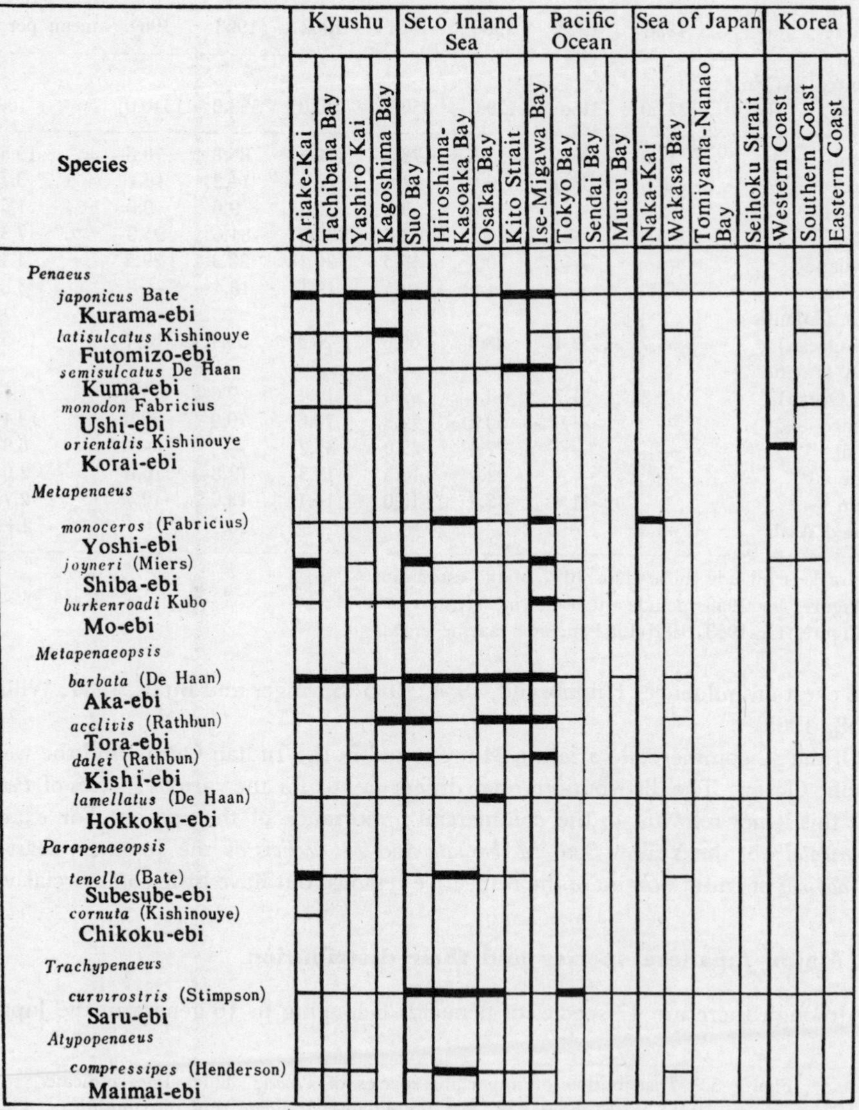

Species	Kyushu				Seto Inland Sea				Pacific Ocean				Sea of Japan				Korea		
	Ariake-Kai	Tachibana Bay	Yashiro Kai	Kagoshima Bay	Suo Bay	Hiroshima-Kasoaka Bay	Osaka Bay	Kito Strait	Ise-Migawa Bay	Tokyo Bay	Sendai Bay	Mutsu Bay	Naka-Kai	Wakasa Bay	Tomiyama-Nanao Bay	Seihoku Strait	Western Coast	Southern Coast	Eastern Coast
Penaeus																			
japonicus Bate **Kurama-ebi**																			
latisulcatus Kishinouye **Futomizo-ebi**																			
semisulcatus De Haan **Kuma-ebi**																			
monodon Fabricius **Ushi-ebi**																			
orientalis Kishinouye **Korai-ebi**																			
Metapenaeus																			
monoceros (Fabricius) **Yoshi-ebi**																			
joyneri (Miers) **Shiba-ebi**																			
burkenroadi Kubo **Mo-ebi**																			
Metapenaeopsis																			
barbata (De Haan) **Aka-ebi**																			
acclivis (Rathbun) **Tora-ebi**																			
dalei (Rathbun) **Kishi-ebi**																			
lamellatus (De Haan) **Hokkoku-ebi**																			
Parapenaeopsis																			
tenella (Bate) **Subesube-ebi**																			
cornuta (Kishinouye) **Chikoku-ebi**																			
Trachypenaeus																			
curvirostris (Stimpson) **Saru-ebi**																			
Atypopenaeus																			
compressipes (Henderson) **Maimai-ebi**																			

Figure 5.3. Distribution of some of the important Japanese species of Penaeidae. Thick lines indicate a very large number of Penaeidae. Data collected from Urita (1921, 1926), Yokoya (1930), Nagata, Tanizaki, and Nakazawa (1931), Nishimura (1939), Yoshida (1941), Aichi Fisheries Experiment Station (1942), Ota (1949), Kubo (1949), Kubo and Asada (1957), Yasuda (1956), Yasuda, Saraoka, Kobayashi (1957), Maekawa (1961), Ikematsu (1963), and Funada (1966).

Pacific Ocean, the number is quite small (Figure 5.4 and Table 5.2). The fishing season of penaeids differs according to the species and the place, but usually extends from late spring to autumn. In Japan penaeids are commonly caught either in late spring or at the beginning of summer, when they are mature, and in autumn when they are one year old. However, in Suo-nada (Maekawa, 1961) and in Ariake-Kai (Ikematsu, 1963), *Metapenaeus joyneri (Shiba-ebi)* is caught in the winter. In Ise and Migawa Bay, *Metapenaeus monoceros (Yoshi-ebi)* is caught from later autumn to winter (Aichi Fisheries Experiment Station, 1942). Recently, fishing has been done in winter even in Suo-nada (Maekawa, 1961).

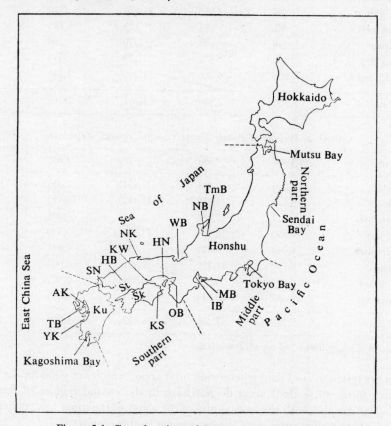

Figure 5.4. Coastal regions of Japan and penaeid habitats.

MB—Migawa Bay; IB—Ise Bay; OB—Osaka Bay; KS—Kit Strait; YK—Yashiro Kal; TB—Tachibana Bay; AK—Ariake-Kai; SN—Suo Nada; HB—Hiroshima Bay; KW—Kasaoka Bay; NK—Naka-Kai; HN—Harima Nada; WB—Wakasa Bay; NB—Nanao Bay; TmB—Tomiyama Bay; Ku—Kyushu; Sk—Shikoku; St—Setanaikai

As far as the distribution of penaeids in suitable places is concerned, it seems that the location of the growing places inside the bays is related to the availability of nutrition. According to Anderson, King, and Lindner (1949), the number of straits passing from the suitable growing places to the open seas, as well as the size of such growing places, greatly influence the yield of *Penaeus*. With regard to the latter, the suitable fishing

Table 5.2. Yield rate of *Penaeus* and other varieties of prawns in Japan
(From the Statistical Report of the Ministry of Agriculture and Forestry)

A. *Penaeus*

Unit ton

Sea regions	1959	1960	1961	1962	1963
Hokkaido	0	0	0	0	0
Pacific Ocean	980	1,149	1,297	1,081	763
Northern part	102	61	40	35	33
Middle part	796	1,005	1,126	910	599
Southern part	81	81	129	135	130
Sea of Japan	46	71	70	69	56
East China Sea	466	559	1,133	555	1,144
Seto Inland Sea	1,261	1,206	1,266	1,140	955
Total	2,756	2,986	3,767	2,846	2,920

B. *Other Varieties of Prawns [except Panulirus japonicus (Ise-ebi)]*

Sea regions	1959	1960	1961	1962	1963
Hokkaido*	1,924	2,214	1,924	2,255	2,462
Pacific Ocean	15,112	14,948	15,112	16,175	15,934
Northern part	2,717	3,492	2,717	4,204	2,201
Middle part	11,353	10,227	11,353	10,530	12,479
Southern part	1,041	1,227	1,041	1,441	1,253
Sea of Japan*	8,792	11,256	8,792	10,087	8,384
East China Sea	6,532	6,271	6,532	7,305	5,611
Seto Inland Sea	24,431	22,614	24,431	19,005	18,900
Total	56,792	57,306	56,792	54,829	51,292

*Most of the specimens belong to *Pandalidae*.

grounds are restricted to Okiai where wide tidal areas, which are good for the growth of *Penaeus*, form; these tidal areas do not form in the coastal regions of the Sea of Japan, a fact coupled with the low salinity of the water of the inner bay, which might account for the smaller number of *Penaeus* present in this region.

Most of the nurseries of *Penaeus* in the bays are confined to waters of low salinity. However, the actual role of salinity in the yield rate of *Penaeus* is not clear, nor whether low salinity is an essential condition of the growing region. Various views are held by researchers on these points. Furthermore, several reports claim collection of young *Penaeus* from high salinity waters (Gunter, 1950, 1961b; Lindner and Anderson, 1956; Hoese, 1960). Hoese (1960) concludes that salinity is not an important factor for the growth of young *Penaeus* since, if other conditions are suitable, they survive even in water of high salinity. On the other hand, Gunter (1961b) thinks that salinity is an important factor, and reports that although young *Penaeus* do not die in water of high salinity, they do not grow well for some unknown reason.

3. Life history

3.1 Development stages

Renfro (1963) divided the life history of *P. aztecus* into six stages on the basis of changes in morphology and ecology. *P. japonicus* can likewise be divided into six stages, namely, embryo, larvae, juvenile, young, immature, and mature. The embryonic stage begins soon after fertilization, the larval stage commences from hatching, and the juvenile stage from molting at the end of the larval period. One of the important morphological features of the juvenile stage is the fluctuating ratio between the various body parts (particularly the sixth abdominal segment). Hence, when the percentage of the body parts becomes stable, the specimen could be considered as having entered the young stage. According to the observations of Hudinaga (1942), the ratio of various body parts is more or less stable in the 15th and 16th instar, when the body length measures 23 to 26 mm. Maekawa (1961) collected *P. japonicus* from the mouth of Shogun Bay in August, when the specimens were supposed to be migrating to their growing places. The body length[2] of most of the specimens was 15 to 25 mm. Hence it is assumed that in *P. japonicus* the transition from the juvenile stage to the young stage occurs soon after the migration of the prawns to the growing places in the bays. The beginning of the immature stage is not very clear morphologically. However, ecologically it is said to be during the period when the prawns shift their growing places from the coastal regions to regions in the open seas. The gonads (particularly the ovaries) are either immature or have just started to mature, but spawning does not take place. According to Maekawa (1961), the size of the specimens when they migrate from the coastal growing regions to open seas is 110 to 120 mm for *P. japonicus*, about 70 mm for *Metapenaeus burkenroadi*, 90 mm for *Metapenaeus monoceros* and *Penaeus semisulcatus*, 50 mm for *Metapenaeus joyneri* and *Trachypenaeus curvirostris*, and 40 mm for *Metapenaeopsis barbata* and *Metapenaeopsis acclivis*. The mature stage begins when the ovaries have matured.

3.2 Reproduction

3.2.1 MATURATION

The gonads of *Penaeus* are found in the posterior half of the cephalothorax. They are ventral to the heart and dorsal to the liver. In females, a pair of leaf-like ovaries grow up to the abdomen (Figure 5.5). The wall of the ovaries consists of a thin outer epithelium, a relatively thick layer of connective tissues, and a germinal epithelium. There is no musculature. The germinal epithelium is localized and confined to the gonadal area and, in the elongated parts of the ovaries, consists of strips in the middle on the ventral side of the abdomen (King, 1948). The maturation of the ovaries is easily detected from their color and size. Usually, the ovaries can be divided into three stages, namely, immature (colorless, found soon after the liberation of ova), slightly mature (light yellow-green), and mature (bluish green) (Yasuda, 1956;

[2]Unless otherwise mentioned, prawn size will be expressed in terms of body length (length from posterior margin of the eyestalk to the tip of the telson).

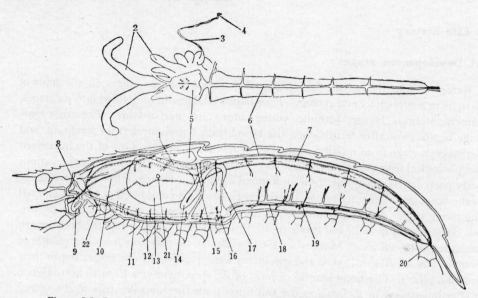

Figure 5.5. Position of various organs of *P. japonicus* (Hiroshima Bundridai-Takashi Seibutsu Gakkai, 1949; data partly modified).

1—Testes; 2—ovaries; 3—oviduct; 4—female genital aperture; 5—heart; 6—dorsal abdominal artery; 7—intestine; 8—brain; 9—green gland; 10—stomach; 11—anterior branch of thoracic artery; 12—thoracic ganglia; 13—liver (midgut gland); 14—vas deferens; 15—thoracic artery; 16—ejaculatory duct; 17—seminal vesicle; 18—ventral artery; 19—abdominal ganglia; 20—rectum; 21—opening of midgut gland; 22—circum-esophageal ring.

Ikematsu, 1963). When the ovaries have completely matured, they constitute 10% of the body weight (Table 5.3). King (1968) has divided the maturing condition of the ovaries of *P. setiferus* into the following five stages:

Table 5.3. Number of spawned ova and number of mature ova

Species	Mature ovary percent of body weight	Number of spawned ova (×1,000)	Body length of mother prawn (mm)	Reference
P. japonicus	10	700	200	Kajiyama (1935a), Hudinaga (1942)
Penaeus monodon	14.5	490	180	Ikematsu (1963)
Metapenaeus monoceros	6–15	125–494	100–130	,,
Metapenaeus joyneri	8–20	75–375	96–125	,,
Parapenaeopsis tenella (with long life history)	9–12	40	52–55	,,
Parapenaeopsis tenella (with short life history)	—	10–20	32–44	,,
Trachypenaeus curvirostris	—	162	47	,,
Metapenaeopsis barbatus	3–5.5	14–33	74–80	,,
Parapenaeopsis cornulus	6	78	86	,,
P. setiferus	—	860	172	Anderson, King and Lindner (1949)
P. trisulcatus	—	800	—	Heldt (1938)

Immature: Very small, transparent and difficult to distinguish;

Slightly mature: Black pigment scattered on the surface and slightly opaque;

Yellow stage: Yellow or yellow-orange in color; large amount of fatty yolk stored in ova;

Mature: Deep yellow-brown in color, filling almost the entire length of the body; rods form around the ova cytoplasm;

Liberation of ova: Ovaries change from ash green to an opaque milky white, and constrict. Externally, it is difficult to distinguish this stage from the mmature one; however, when the prawn is observed under a microscope, this final stage is characterized by a large number of nutritive cells.

The duration of various stages ranges from 1 month during the period of early maturation to 1 to 2 months in the yellow stage. According to Lindner and Anderson (1956), the ova are released within a month after the maturation of the ovaries. Histologically, the development of ova in *P. orientalis* can be divided into 8 stages (Oka and Shirahata, 1965). According to Hudinaga (1942), the jelly material in the mature ova of *P. japonicus* (which corresponds to the rods in Figure 5.6 as explained by King) is secreted soon after the liberation of the ova, and forms the membrane of the latter. Due to this jelly membrane, the ovum constricts and becomes pearl-like in appearance, measuring 0.24 mm in diameter. The eggs are not sticky. The eggs sink in calm water (heavier than sea water) but start to float with even the slightest water movement. Since there is no musculature in the ovary itself, the ova are laid by the constriction of the muscles in the cephalothorax and around the ovaries in the abdomen (King, 1948).

Maturity in males is not as apparent as in females. How-

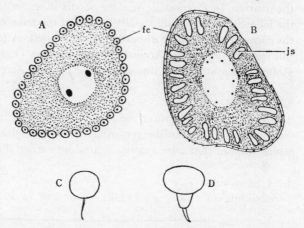

Figure 5.6. Reproductive cells in *P. japonicus* (Hudinaga. 1942; King, 1948).

A—young follicles; B—mature ova; C—spermatozoa of *P. japonicus;* D—spermatozoa of *P. setiferus;* fc—follicular cell; js—jelly substance.

ever, in a mature state, the ejaculatory duct at the end of the vas deferens swells and turns milky white in color. The genital aperture protrudes. Spermatophores form at the tip of the vas deferens in *P. setiferus* (King, 1948). Mature spermatozoa are found only in the vas deferens and are carried outside by the movement of the cilia present in its walls, but in *P. orientalis* they may be found even in the testes (Oka and Shirohata, 1964). In *P. setiferus* the spermatozoan consists of a round head, a middle section, and a tail (King, 1948). In *P. japonicus,* the middle section is absent. The diameter of the head is more or less equal to the length of the tail, i.e., 5.0 to 5.3 μ (Hudinaga, 1942).

3.2.2 MATING

Mating occurs before spawning and soon after molting when the carapace of the female is soft and the carapace of the male hard. The female receives the spermatozoa enclosed in spermatophores. By comparing the shape of the spermatophores in the seminal vesicle with those at the tip of the vas deferens, and supported by the fact that the spermatophores of the left seminal vesicle form at the tip of the left vas deferens, while those of the right form at the tip of the right vas deferens, King (1948) concluded that prawns mate while facing in opposite directions (the head of one faces the other's tail). Hudinaga (1942) reports that this is not so with *P. japonicus* however; instead the male and female lie tail to tail. The copulatory organ has been observed in many species but not in *Metapenaeus barbatus* (Ikematsu, 1963) or *P. orientalis* (Yoshida, 1949). No definitive data are available on the time lag between mating and spawning, but it does seem that this time lag differs from species to species. When the copulatory organ is absent in the female (e.g., *P. japonicus*), spermatozoa are stored in vesicles and seen throughout the year, regardless of whether the ovaries have matured. But in other prawns, spermatozoa are seen only during, or just before, the maturation of the ovaries. In *P. orientalis*, mating occurs in autumn and spawning in the spring of the following year (Yoshida, 1949). When the females molt, the spermatophores and the copulatory organ are discarded with the shell. After shedding, mating can usually, but not always, recur. According to Hudinaga (1948) the same female *P. japonicus* molted and mated three times, *viz.* on August 2, August 22, and September 15.

3.2.3 SEX RATIO

Up to sexual maturation the ratio of females in the catch usually fluctuates, but sometimes it is 1 : 1. In the spawning period the number of males suddenly decreases, which implies that after mating males die earlier than females (Ikematsu, 1963).

3.2.4 SPAWNING

According to Hudinaga (1942), spawning in *P. japonicus* occurs at night. The

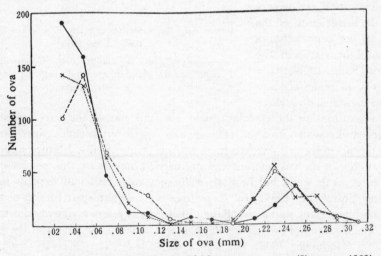

Figure 5.7. Distribution of ova by size of *Metapenaeus monoceros* (Ikematsu, 1963).

females swim about and release ova over a wide area. The spawning activity of *P. duorarum* is related to the lunar cycle, i.e. while spawning increases during full moon, it is almost nil during the new moon (Idyll and Munro, 1965). Unlike prawns belonging to *Caridea*, the ova of *P. japonicus* are directly released into the sea water. Each spawning takes 3 to 4 minutes, and the number of ova increases with the size of the specimen (Table 5.3). Data on the size of ova in mature ovaries (Figure 5.7) show that, in a number of species, a large number of small ova are present with mature ova. It has often been pointed out that the presence of these small ova indicates the possibility of a second spawning during the same spawning period (Yatsuyanagi and Maekawa, 1956a, b, 1957a; Ikematsu, 1955; King, 1948; Cummings, 1961; and Williams, 1965). However, this has not yet been confirmed. Another explanation holds that these small ova play an important role in the ejection of the mature ova at the time of spawning (Oka, 1967).

3.2.5 SPAWNING AREAS

Spawning generally takes place in open seas as most specimens do not mature in the warm water regions of inner bays. However, *M. stebbingi*, *M. monoceros* (Dakin, 1946), and *M. mastersii* (Morris and Bennett, 1952; Racek, 1956) spawn either in warm water basins of coastal regions or in inner bays. The spawning places of *M. affinis* are also located, for the most part, in the mouth of the bays approximate to the open seas (Aichi Fisheries Experiment Station, 1943; Oshima and Yasuda, 1932). Judging from the location of fishing grounds during the spawning period, the spawning places in the open seas differ from species to species, but most spawn in relatively shallow places. In *M. acclivis*, *M. dalai*, and *M. barbatus* (of Ariake-Kai) spawning occurs in the slightly deeper regions of 20 to 40 m (Maekawa, 1961; Ikematsu, 1963). Even in Gulf of Mexico, the spawning areas determined from the distribution of mature specimens and planktonic larvae, are located in 27 to 45 m water depth or slightly less (Heegaad, 1953; Temple, 1962; Temple, Harrington, and Fischer, 1963; Fischer, 1965). The small prawns of Indian waters usually spawn in water 8 to 20 m deep (Panikkar and Menon, 1956). However, in Australia *M. macleayi* and *M. inscipes* (=*M. ensis*) spawn in waters of 45 to 65 m depth, and *P. plebejus* in 90 to 100 m depth. Mature specimens of *P. esculentus* and *P. carinatus* (=*P. monodon*) were found in water 120 to 145 m deep (Racek, 1956).

3.2.6 SPAWNING PERIOD

The spawning period and its duration varies from species to species. Without exception prawns spawn in the summer in warm regions. *Caridea* spawn during the period between spring and autumn (Yasuda, 1958; Ikematsu, 1963). The spawning periods of *P. japonicus* of Shubo-nada and Ariake-Kai are shown in Figures 5.8 and 5.9, on the basis of seasonal changes in temperature and salinity. According to these figures, spawning occurs when the water temperature is about 20°C or more. Moreover, the water temperature in the final phase of spawning is considerably higher than in the initial phase. This trend is particularly striking in large specimens of *P. japonicus*, *M. monoceros*, *P. monodon*, and *M. joyneri*. The spawning period is more or less confined to the period when the water temperature shows a steady increase, and when the water temperature begins to decrease from September, spawning decreases correspondingly. But in small specimens, the spawning period is delayed, particularly in the case of *M. acclivis* of Shubo-nada for which spawning is at its maximum at the time the water temperature shows a decline. However, the same species spawns a little earlier in Kasaoka Bay and the spawning peak is in July and August (Yasuda, 1956, 1958). The spawning of *Parapenaeopsis tenella* and *Trachypenaeus curvirostris*

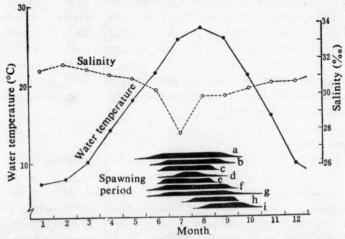

Figure 5.8. Seasonal changes in water temperature and salinity in the western part of the Seto Inland Sea and the spawning period of *P. japonicus* (Maekawa, 1961; and others).

a—*P. japonicus;* b—*M. monoceros;* c—*P. monodon;* d—*M. affinis;* e—*M. joyneri;* f—*M. barbatus;* g—*Trachypenaeus curvirostrus;* h—*M. dalei;* i—*M. acclivis.*

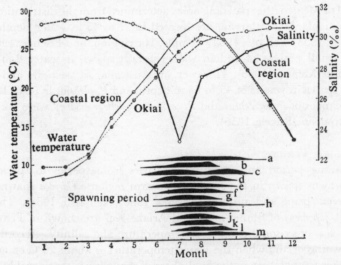

Figure 5.9. Seasonal changes in water temperature and salinity in Ariake-Kai and the spawning period of *P. japonicus* (Ikematsu, 1963).

a—Kuruma-ebi; b—Shiba-ebi; c—saru-ebi; d—*Parapenaeopsis tenella;* e—*Parapenaeopsis cornutus;* f—*P. monodon;* g—*M. monoceros;* h—*M. affinis;* i—*M. barbatus;* j—*M. dalei;* k—*M. lamellatus;* l—*M. acclivis;* m—*Atypopenaeus compressipes.*

occurs in late August or September, due to the short-term cycle of the one-year-old but mature individual. In many species, the commencement of spawning more or less coincides with the period when there is a sudden decrease in salinity. However, the relationship between the two is not yet clear.

Even in Gulf of Mexico, spawning of *P. duororum* is stimulated by a water temperature of 20°C and further spawnings are observed in the range of 19.6 to 30.6°C (Jones, Dimitriou, and Ewald, 1963). However, the spawning rate is high when the water temperature at the bottom layer is about 27°C or more, decreasing by 10% when the temperature falls below 24°C (Idyll and Munro, 1956). Hence, the changes in the temperature of the bottom water layer in the spawning places, influence both the spawning rate of the current year and the resources of the following years (Eldred, Ingle, Woodburn, Hutton, and Jones, 1961). The initiation and completion of spawning are closely related to water temperature, but the two phases are controlled more by the changes than by the absolute values. The sudden increase in the water temperature in spring coincides with the initiation of spawning at the bottom of the sea, and the sudden decrease in the water temperature in autumn usually brings about its completion, even when the water temperature is higher than it was when spawning commenced (Anderson, 1956). In Indian water basins in the tropical belt, spawning continues for a longer period than in the temperate regions. In these regions, juvenile *M. monoceros* and *M. dobsoni* are caught throughout the year, but the juvenile *P. indicus* is never seen during the monsoons. Spawning activity shows much variation but two peaks usually occur (November to December and April to May or July to August) (Panikkar and Menon, 1956; Menon and Raman, 1951; Rajyalakshmi, 1961). George (1962) opines that spawning activity is closely related to the changes in saline content.

3.2.7 FERTILIZATION

Fertilization takes place in water after the ova are liberated. The life history of the prawns begins here. The spermatozoan pierces the covering jelly membrane of the ovum; the number piercing an ovum may reach 10, and a corresponding number of fertilization cones develop. The substance of the egg cell does not flow simultaneously with the development of these cones, however. Initially, only one spermatozoan fuses with the cytoplasm of the egg cell and enters inside (Figure 5.10). At this time, the egg cell undergoes the first maturation division and produces the first polar body, which is followed by the formation of a fertilization membrane around the entire egg surface, and by a second maturation division. The time lag between the spawning and the first cleavage of the egg in *P. japonicus* is shown in Table 5.4.

3.3 Development

3.3.1 EMBRYO

The eggs of *P. japonicus* are mesolecithal and the cytoplasm is very thin on the surface. The nucleus is at a polar end surrounded by the cytoplasmic layer. According to Hudinaga (1942) and Kajishima (1951), the first cleavage divides the egg cell into two equal halves, and then continues at intervals of 15 to 20 minutes. Along with the progress of the second cleavage, the two blastomeres rotate in opposite directions. The

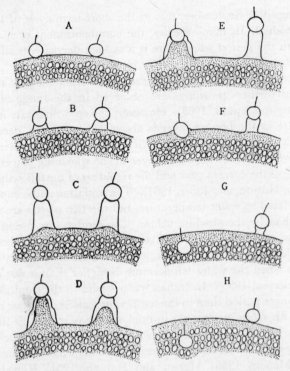

Figure 5.10. Course of spermatozoa entrance in *P. japonicus* (Hudinaga, 1942).

A—spermatozoa reach the egg surface; B—formation of fertilization cones; C—complete formation of fertilization cones; D—fusion of spermatozoa and flow of cytoplasm in cones; E—constriction of fertilization cones; F, G, and H—only one spermatozoan fuses with the egg cell; the rest remain on the egg surface.

Table 5.4. Changes in the fertilized egg of *P. japonicus* (Hudinaga, 1942)

Time of spawning: 0 hours, 15 minutes, July 21, 1940
Water temperature: 27.3°C; Salinity: 29.4°/oo

Time lag after spawning		Changes in the egg
Minutes	Seconds	
0	0	Spawning
0	45	Spermatozoa reach the egg surface
1	40	Egg becomes round in shape
2	20	Formation of fertilization cones
3	05	Fertilization cones at maximum size
4	20	Fertilization cones disappear
4	10	Liberation of first polar body
7	00	Granules of jelly layer disappear
11	20	Fertilization membrane begins to form
12	40	Liberation of second polar body
28	00	Jelly layer disappears
30	15	First cleavage of egg

nuclei in the four-celled stage remain at the center of the cytoplasm on the surface. When the blastula attains the 64-cell stage, it enters the gastrulation phase. Two of the endodermal cells invaginate perpendicular to the primary cleavage plane. The invagination of the mesodermal cells begins when the embryo has 124 cells. Thereafter the mesodermal layer forms below the ectoderm (Figure 5.11). Rudiments of appendages develop in the order of second antennae, mandible, and first antennae in 4 hours, 4 hours and 10 minutes, and 6 hours after spawning, respectively. Simple eyes are formed by 11 to 12 hours after spawning, and the larvae hatch in 13 to 14 hours (27 to 29°C).

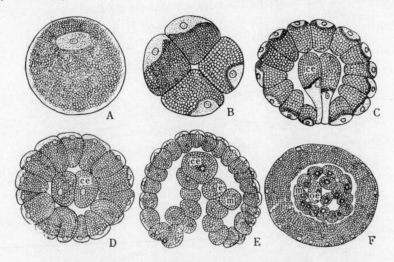

Figure 5.11. Early embryology of *P. japonicus* (Kajishima, 1951).

A—uncleaved egg; B—4-cell stage; C—64-cell stage; invagination of central endoderm; D—later phase of C (transverse section); E—invagination of mesodermal and endodermal cells; F—later phase of invagination (transverse section); ce—central endodermal cell; 3—endodermal cell; m—mesodermal cells

3.3.2 Larval Stages

P. japonicus undergoes metamorphosis 3 times, i.e. it passes through 3 larval stages, namely, *nauplius, zoea* (or *protozoea*) and *mysis* (or *zoea*).

1. *Nauplius* (Figure 5.12A, B) is a pear-shaped larva and a rudimentary, cephalothoracic carapace appears at a later stage. The caudal spines are symmetrical and straight. The setae develop either individually or in groups on the lateral margins of the appendages. No masticatory organ is present. The larvae swim by a pair of antennae and derive nutrition from the yolk in the body, feeding on no substance from outside. The changes during each molting are few, and restricted to the elongation of the body in the posterior direction, the development of a rudimentary second pair of maxillipeds, and the segmentation of the antennae. In the late nauplius stage, the rudimentary cephalothoracic carapace forms, with two groups of 7 spines each. In the case of *P. japonicus,* the nauplius stage lasts for 36 to 37 hours after hatching (27 to 29°C), but the number of larval stages varies from species to species, and the data published by researchers differs considerably (Table 5.5).

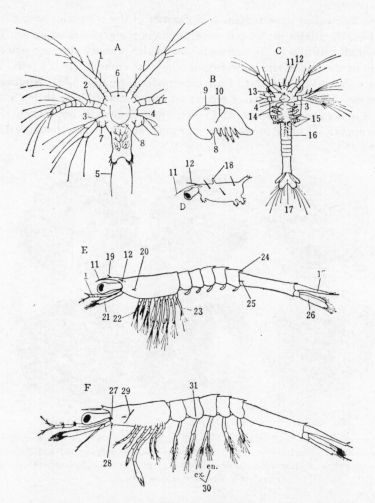

Figure 5.12. Larvae of *P. japonicus* (Hudinaga, 1942; Dobkin, 1961; and Cook, 1965).

A—ventral view of *nauplius*; B—side view of *nauplius*; C—ventral view of *zoea*;
D—side view of *zoea*; E—side view of *mysis*; F—side view of juvenile prawn.

1—first antenna; 2—second antenna; 3—mandible; 4—labium; 5—caudal spine; 6—simple eye; 7—basal part of mandible; 8—labium; 9—dorsal papilla; 10—rudiment of cephalothoracic carapace; 11—rostrum; 12—supra-orbital spine; 13—labral spine; 14—first and second maxillae; 15—1-3 maxillipeds; 16—rudimentary walking legs; 17—telson; 18—dorsal organ; 19—rostral spine; 20—hepatic spine; 21—branchiae on second antenna; 22—endopodite of thoracic appendage; 23—exopodite; 24—dorsal spine of abdominal segment; 25—lateral spine; 26—telson; 27—dorsal spine of antenna; 28—antero-lateral spine; 29—cephalic groove; 30—abdominal appendages; 31—abdominal pleuron.

2. *Zoea* (Figure 5.12C, D): This larval stage differs from *nauplius* in having compound eyes, a rudimentary cephalothoracic carapace, and functional appendages behind the second mandible. In the first stage, all the thoracic segments are articulate, but the abdominal segments are still fused; however, in *P. japonicus* the segments behind the second thoracic segment are free from the carapace. In the second stage

Table 5.5. Number of larval stages in *P. japonicus*

Species	Nauplius	Zoea	Mysis	Reference
Penaeus japonicus	6	3	3	Hudinaga (1942)
Penaeus trisulcatus	8	3	4	Heldt (1938)
Penaeus setiferus	5	3	2	Pearson (1939)
Penaeus setiferus	4–5	3	2	Johnson & Fielding (1956)
Penaeus duorarum	5	3	3	Dobkin (1961)
Metapenaeus monoceros	6	3	3	Hudinaga (1942)
Metapenaeus burkenroadi	6	3	3	Hudinaga (1942)
Metapenaeus dobsoni	3	3	3	Menon (1951)
Trachypenaeus constrictus	5	3	2	Pearson (1939)
Xyphopenaeus kroyeri	5	–	–	Renfro & Cook (1963)
Sicyonia brevirostris	5	3	4	Cook & Murphy (1965)
Sicyonia wheeleri	3	3	3	Gurney (1943)
Sicyonia stimpsoni	5	3	2	Pearson (1939)
Sicyonia carinata	8	3	4	Heldt (1938)

the compound eyes become stalked and separated from the carapace, and the rostrum and supra-orbital spine develop. The spines of the abdominal appendages develop in the third stage and, at the same time, the telson becomes differentiated. The first antenna is uniramous but the second is biramous and much larger than the first. The second antennae are the important organs for swimming. The mandibles undergo drastic changes and become masticatory organs. The two pairs of maxillipeds are more or less of the same shape. The exopodites serve as accessory swimming organs and the endopodites, together with the two pairs of maxillae, serve as feeding organs. In the third stage, rudiments of all the thoracic appendages are found, but those behind the third maxilliped are not functional. The labrum is very large.

It is now known that *zoea* molt three times: in the first stage the compound eyes are fused with the carapace, in the second stage they become separated from the carapace, and in the third the telson becomes differentiated. Since these morphological changes are similar to those of *Caridea* larvae, Gurney (1926) named this developmental phase *protozoea*.

3. *Mysis* (Figure 5.12E): The first and second antenna undergo drastic changes and lose the function of swimming; the latter activity is taken over by 5 pairs of walking legs and their exopodites. The mouth parts, including the first and second pair of maxillipeds, still retain the appearance of *zoea* in *Penaeinae*. The number of molts in this stage varies from species to species.

3.3.3 JUVENILE STAGE

When *mysis* molt for the final time, the first juvenile stage (Figure 5.12F) appears. The changes that are generally observed during this stage are rapid development of abdominal appendages and their functionality as organs of swimming. It is also during this stage that the exopodites of the walking appendages begin to degenerate. The shape of the body is more or less similar to that of the adult, but essential organs are still missing. Molting is repeated and the specimen approaches the adult stage.

The main morphological changes during this stage are related to the proportionate growth of various parts, modification of the telson, an increase in the rostral serrations, the development of gills, and the branching of endopodites in abdominal appendages.

3.3.4 IDENTIFICATION

The identification of various types of larvae and the early juvenile stages of *P. japonicus* from plankton collections is still very difficult as the data on Japanese varieties are very few and mostly confined to the morphology of the larval stages for *P. japonicus* (Hudinaga, 1942) and the morphology of *nauplius* for *Metapenaeus monoceros* and *M. affinis* (Hudinaga, 1941). Since the special characteristics of the juvenile prawn soon after metamorphosis are not yet clear, nor are some of the special morphological features in the juvenile and young stages (particularly those of *Atypopenaeus compressipes*; Yasuda, 1966), an identification chart is not always useful. One significant reason for the lack of data on the ecology of the planktonic life of prawns in nature has been an inadequate study of the classification of the larval stages. Now, however, study on the types of larvae of the species found in the Atlantic Ocean and in the Gulf of Mexico is in progress, and the morphological features of the larvae (Dobkin, 1961) and juvenile prawn (Williams, 1953, 1959) of all the important species have been assessed. Identification charts are available for the larval and juvenile stages of *Penaeus*, *Sicyonia*, *Parapenaeus*, *Solenocera*, *Trachypenaeus*, and *Xiphopenaeus* (Cook, 1965).

3.3.5 SURVIVAL OF LARVAE

There are several reports suggesting some relationship between the catch rate of *P. japonicus* and the amount of rainfall. A high degree or correlation between the amount of *P. setiferus* caught off the coast of Texas for a particular year, and the total rainfall in that year and the preceding two years was found by Hildebrand and Gunter (1953) and Gunter and Hildebrand (1954). Thomson (1956) also found the same correlation for the prawns of Australian waters. In this case the correlation was most significant between the amount of catch and the amount of rainfall for the 12 months, 18 to 6 months preceding the catch. The dependence of the catch rate on the amount of precipitation is usually found in the case of *P. japonicus*, which enter the water basins of the inner bays during the young stage. According to Subrahmaniam (1966), one of the reasons for the direct correlation between the catch rate of *P. monodon* in estuaries and the amount of river water flow of the previous year, is the difference in fecundity. A preliminary investigation established that in the years when the river inflow was high due to heavy rain, the number of juvenile prawns in the planktonic stage was large, while in the year when the inflow was low, the number of juvenile prawns and the amount of plant planktons were meager. Sometimes this type of correlation does not exist and, according to Ikeda (1961), the amount of rainfall is more closely related to the development rate, rather than to fecundity in the case of *P. orientalis*.

The catch rate of *P. japonicus* sometimes shows a high correlation with the amount of precipitation for the same year or, even, for the same month (Menon and Raman, 1961). In this case, the increase in the catch during heavy rain is, according to Racek (1959), due to the accumulation of different varieties of prawns. In other words, when rainfall is heavy the water from the rivers suddenly flows in at a high rate, and the prawns in the estuaries react initially by burrowing into the sea floor. When this is followed by a sudden decrease in salinity below a particular level, most of the prawn groups scuttle to the open seas.

3.4 Growth

3.4.1 PATTERN OF GROWTH

Several reports on the monthly changes in the length of the carapace of juvenile prawns in the bays exist, according to which the growth of penaeids can be divided into three categories. In the first category, *P. semisulcatus, Metapenaeus burkenroadi,* and *Metapenaeus joyneri* (Figure 5.13), the growth composition of the carapace of the same

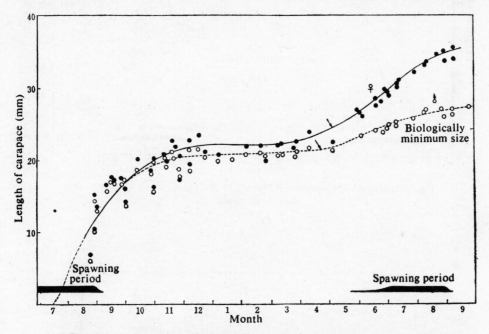

Figure 5.13. Growth of *Metapenaeus joyneri* (Category 1). Arrows indicate the biologically minimum size (Ikematsu, 1963).

age group (one-year-olds) shows a simple peak. In the second category, *P. japonicus, Metapenaeus monoceros, M. dalei, Trachypenaeus curvirostris, Metapenaeopsis barbata,* etc., (Figure 5.14), the growth of offsprings show two separate peaks, that is, one group develops early and the other later. However, even the group which develops early does not mature in the same year. The third category is more or less similar to the second, except that some individuals develop early, mature, and spawn in the same year, and constitute the so-called "short-term generation" (Yasuda, 1956; Ikematsu, 1963). *Parapenaeopsis tenella, Trachypenaeus curvirostris* (of Kasaoka Bay and Ariake-Kai), and *Metapenaeopsis barbata* (Kasaoka Bay) belong to this category (Figure 5.15).

3.4.2 GROWTH RATE

When the growth per day in natural conditions is calculated on the basis of the data

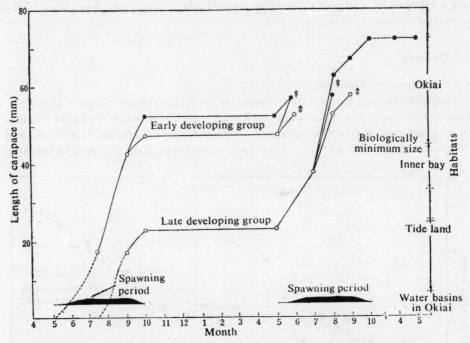

Fig. 5.14. Growth of *P. japonicus* (Category 2) (Yatsuyanagi and Maekawa, 1955; data partly modified).

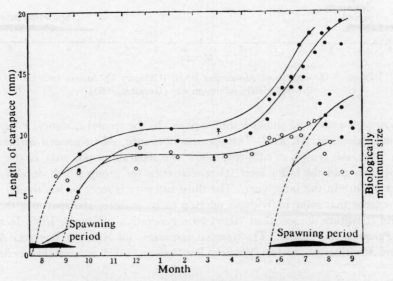

Figure 5.15. Growth of *Parapenaeopsis tenella* (Category 3) (Ikematsu, 1963; data partly modified).

published earlier, it is possible to obtain a rough value for the growth rate (Table 5.6). In the larger varieties, it is about 1 mm of carapace length per day, and in the smaller mostly 0.3 to 0.7 mm. However, the growth rate differs according to the development stage, the season, and, in the adult stage, sex. The growth rate of young and immature prawns is much higher than for adults, and for the latter, that of the females is higher than that of the males. In winter when the water temperature is low the growth rate is almost nil.

The growth rate of the juvenile prawns of the Gulf of Mexico is more or less the same as that explained in the preceding paragraph. For *P. setiferus* it is 1.2 mm, and for *P. aztecus* about 1.5 mm (Williams, 1955a). Loesch (1965) also obtained more or

Table 5.6. Apparent growth rate of penaeid shrimps in natural conditions (body length)

Species	Month	Stages of Growth	Growth rate (mm/day)	Reference
Penaeus japonicus	5–7	Young	0.7	Yatsuyanagi and Maekawa (1955)
	8–9	Young	1.0	Yatsuyanagi and Maekawa (1955)
	9–10	Immature-mature (♀)	0.8	Hiramatsu, Segawa, Tago, and Arima (1962)
	5–8	Immature-mature (♀)	0.9	Kubo (1955)
	5–8	Immature-mature (♂)	0.8	Kubo (1955)
P. orientalis	7–9	Young-immature	1.5	Kishigami (1970)
P. semisulcatus	8–10	Young	1.0	Hiramatsu, Segawa, Tago, and Arima (1962)
	8–10	Young	0.9	Yatsuyanagi and Maekawa (1956a)
Metapenaeus joyneri	9–10	Young	1.0	Yatsuyanagi and Maekawa (1954)
	8–11	Young-immature	1.3	Hiramatsu, Segawa, Tago, and Arima (1962)
	8–11	Immature-mature	0.4	Ikematsu (1963)
Metapenaeus monoceros	9–10	Young	1.4–1.7	Ikematsu (1963)
	8–10	Young-immature	1.7	Hiramatsu, Segawa, Tago, and Arima (1962)
	5–6	Mature	0.6	Ikematsu (1963)
	6–10	Immature-mature (♀)	0.4	Ota (1949)
	6–10	Immature-mature (♂)	0.3	Ota (1949)
M. burkenroadi	5–7	Young	0.6	Aichi Fisheries Experiment Station (1943)
	5–8	Immature-mature	0.2–0.3	Oshima and Yasuda (1932)
Metapenaropsis barbata	9–10	Young	0.5–0.7	Yasuda (1956)
	5–7	Immature-mature (♀)	0.5–0.7	Utsunomiya and Maekawa (1959)
M. acclivis	5–7	Young	0.7–1.0	Yasuda (1956)
Trachypenaeus curvirostris	9–12	Young-immature	1.3	Hiramatsu, Segawa, Tago, and Arima (1962)
	5–8	Immature-mature	0.3	Yasuda (1949)
Metapenaeopsis dalei	9–11	Young	0.3	Yatsuyanagi and Maekawa (1957a)
	6–8	Immature-mature (♀)	0.2	Yatsuyanagi and Maekawa (1957a)
Parapenaeopsis tenella	8–10	Young	0.6	Ikematsu (1963)
	5–8	Immature-mature	0.3–0.4	Yasuda (1956)

less the same values. In the early period of young prawns, the growth rate is about 1.7 mm (*P. aztecus*) or 2.2 mm (*P. fluviatilis = P. setiferus*). Lindner and Anderson (1956), after verifying the data by tagging the specimens, have reported that the growth rate of individuals of 100 mm length is 1.0 mm, and that of 170 mm individuals about 0.33 mm.

3.4.3 RATE OF GROWTH

The relationship between the length of carapace (1) and body length (L) after the young stage can be expressed by the following formula:.

$$L = a + bl.$$

Here *a* and *b* are constants (Yatsuyanagi and Matsuo, 1951; Maekawa, 1961; Ikematsu, 1963; and several others). In females, the value of these constants throughout the sexually mature condition varies greatly, and the rate of growth in the body length compared to that in the carapace is less than that in the young and male prawns. In other words, the body length of mature females is less than that of the males, even though the carapace length is the same.

The body weight (*W*) in relation to the growth of carapace length increases, and the relationship between the two can be expressed by the following formula (Ikematsu, 1963; Maekawa, 1961; Anderson and Lindner, 1958; and several others):

$$W = ml^n.$$

Here *m* and *n* are constants. The value of *n* differs from species to species, but is usually in the range of 2.4 to 3.1. Here, too, the value of the constants varies after sexual maturation in the female, and the value of *n* is smaller than that in males and young prawns. In other words, the body weight of mature females is less than that of males with the same carapace length.

The relationship between body volume (*V*) and carapace length can be expressed by the following formula (Dall, 1958):

$$V = p + q + rl^2.$$

Here *p*, *q*, and *r* are constants.

According to Kajiyama (1935a), the body weight of *P. japonicus* during the wintering period decreases by about 35%. Yatsuyanagi and Maekawa (1955) reported that corpulence (body weight[3]/carapace length × 1,000) increases after May and reaches its maximum in September. After attaining the maximum, corpulence decreases to a minimum in December to February. Hence there are seasonal variations in corpulence.

3.5 Longevity

Most prawns are said to live for one year (Table 5.7). However, some *Trachypenaeus curvirostris*, *Metapenaeopsis barbata*, and *Metapenaeopsis acclivis* live for 2 years, and some *P. japonicus* and *Metapenaeus monoceros* are thought to live the same length of time (Maekawa, 1961). The so-called "short-term" prawns live for only 2.5 to 4 months (Ikematsu, 1963). Longevity in all instances is determined by the availability/non-

Table 5.7. Longevity, biological minimum, and maximum body length (mm) of *Penaeus japonicus* **and other prawns**

Species	Longevity (months)	Biological minimum size		Maximum size		Reference
		♀	♂	♀	♂	
P. japonicus	12	120	100	180	160	Yatsuyanagi and Maekawa (1955), Yoshida (1960)
,,	24	135	—	230	—	Yatsuyanagi and Maekawa (1955), Maekawa (1961)
P. semisulcatus	12	122–141	—	220–240	170	Yasuda (1956), Ikematsu (1963)
Metapenaeus joyneri	12-15	106–110	84–91	133–152	120–140	Kishigami (1960), Ikematsu (1955, 1963), Maekawa (1961)
Metapenaeus monoceros	12	100–110	90	160–180	130–175	Yatsuyanagi and Maekawa (1956b), Ikematsu (1959, 1963)
,,	24	121	—	164	—	Maekawa (1961)
Metapenaeus burkenroadi	12	51	—	100	80	Oshima and Yasuda (1932)
,,	12	70–72	60	—	—	Yatsuyanagi and Maekawa (1957b), Ikematsu (1963)
Metapenaeus barbata	4	60	—	60	—	Yasuda (1956)
,,	12	64–70	—	90	80	Yasuda (1956), Ikematsu (1963)
Metapenaeus acclivis	12	63–68	—	100	—	Yasuda (1956), Ikematsu (1963)
Metapenaeopsis dalei	3-4	41–52	—	52	—	Yasuda (1956), Ikematsu (1963)
,,	12	58	—	90	—	Yasuda (1956), Ikematsu (1963)
Trachypenaeus curvirostris	12	—	—	57	43	Yatsuyanagi and Maekawa (1957a)
,,	20	47	—	63	58	Yatsuyanagi and Maekawa (1957a)
Parapenaeopsis tenella	4	28–29	—	52	—	Yasuda (1956), Ikematsu (1963)
,,	12	43–44	32	67–73	46	Yasuda (1956), Ikematsu (1963)
Atypopenaeus compressipes	12	38	—	56	—	Yasuda (1956)

availability of the animals in commercial fishing, as studies on the survival rate throughout all the stages of life are insufficient.

Idyll and Munro (1965) estimated that there were about 510,000,000,000 *P. duorarum zoea* in 1953 in the 1,000 square nautical miles designated for the survey of the catch in the Tortugas fishing grounds in Florida. The survival rate per day was 80.4%. Hence the survival rate during the 35-day transition period from larvae into juvenile prawns, which enter the coastal rearing grounds, is about 0.14% of the *zoea*.

3.6 Fluctuations in spawning

In spite of the fact that spawning is assumed to be a continuous process, it is not clear why the appearance of juvenile prawns on the rearing grounds shows two peak periods in most of the species. Kubo (1955) attributes this to the presence of peak periods in spawning *per se*. The other possible explanation is the presence or absence of predators. Since the habitats and the life history are more or less the same in all the species, when larger varieties of *Penaeus* exist on the same rearing grounds with prawns, they eat them. This was first pointed out with *P. duorarum* by Lindner and Anderson (1956) and Anderson (1956). With *P. japonicus,* the appearance of the second peak of one-year-olds on the rearing grounds in tidal areas, coincides with the migration of earlier groups to open seas (Yatsuyanagi and Maekawa, 1955). A number of investigators are working on this interesting aspect.

4. Ecology

4.1 Habits

4.1.1 VERTICAL DISTRIBUTION IN PLANKTONIC STAGES

Data on the life of prawns in the planktonic stage in natural conditions in the Japanese coastal areas are not available. According to Temple and Fisher (1965), the larvae of penaeid shrimps are found in large number at the 18 to 34 m layer, and in very small numbers on the surface layer of the northern part of the Gulf of Mexico. On the other hand, the juvenile prawns in their early stage are found in large numbers in a 2 m layer, and the number increases as the layer deepens. In both the larval and juvenile stages, there is a distinct daily vertical migration. The larvae and juvenile prawns float on the surface layer at night and sink to the bottom at mid-day. Compared to *zoea, mysis* and juvenile prawns float on the surface much earlier at night and remain there for a longer time in the morning. However, this type of vertical migration is not seen in November, and both the larvae and the juvenile prawns are found distributed more or less uniformly from the surface layer to the bottom layer (Figure 5.16).

4.1.2 BURROWING HABIT

One of the common habits of *Penaeus,* to burrow in the sea floor, appears gradually during the larval stage and becomes prominent with growth. This seems to be a self-protective step against enemies, as prawns which are in the process of burrowing show certain reactions to physical (external) stimuli, reactions allied to the escape behavior

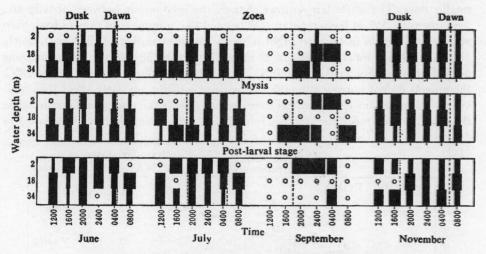

Figure 5.16. Vertical distribution of penaeid shrimps in planktonic stage in the northern part of Gulf of Mexico (Temple and Fisher, 1965).

when they are disturbed on emergence from their burrows. According to Dall (1958) and Egusa and Yamamoto (1961), the depth of the burrows differs according to the size of the prawn. Usually the dorsal surface of the prawn is within 3 cm of the surface of the bottom layer of sand. The respiratory current during the burrowing time passes through the respiratory tubes consisting of the right and left first antenna, the second antennary gills, and the accessory appendage of the mandible into the gill chamber. The chamber opens on the ventrolateral margins of the cephalothoracic carapace (Figure 5.17).

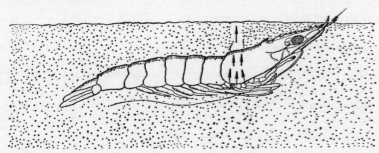

Figure 5.17. *Metapenaeus bennatae* (=*M. mastersii*) in burrow. Arrows show the direction of water flow (Dall, 1958).

The burrowing behavior of *Penaeus* is greatly controlled by light. Usually they come out of the sand after dusk and enter the burrows after dawn, a habit which is repeated daily (Onizuka, 1914; Miura and Yamaguchi, 1955; Fuss, 1964). According to the observations of Fuss and Ogren (1966), the critical light intensity to which *P. duorarum* reacts is about 1.01076/lumen/m². When light intensity exceeds this level, the animal burrows. The larger prawns react more rapidly to excessive light than the smaller ones. The water temperature changes the relationship between activity and light intensity. When the temperature is below 14°C, a number of prawns have been observed to remain in burrows, light intensity notwithstanding. This is particularly observed in large prawns of 140 mm or more. More than 50% of the specimens show normal activity between 14° and 16°C. Even when the temperature is about 28°C, the prawns do not burrow although the brightness might adversely affect their activities. The reactions to low temperature are not always the same. In autumn, when the water temperature decreases to about 16°C, activities drop sharply. When the low

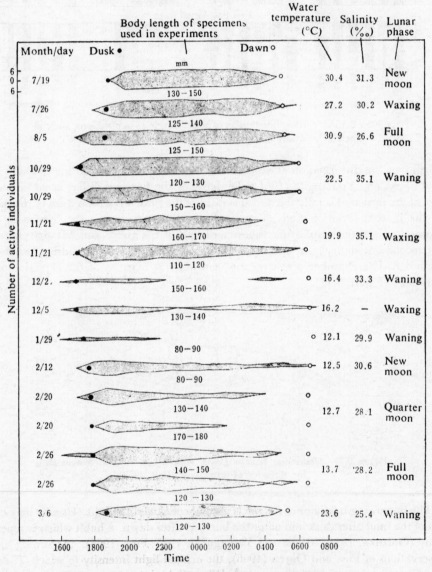

Figure 5.18. Seasonal variations in daily activities of *P. duorarum* (Fuss and Ogren, 1966; data partly modified).

temperature continues for 2 to 3 months, the proportion of active specimens even at
12 to 13°C is much higher than at 16°C. The length of the active period is controlled
by the length of the day and the number of active specimens is influenced by the
water temperature. Hence, in winter when the nights are long, a small number of
specimens become active slowly, and in summer when the nights are short, a larger
number of specimens become active much faster. The number of active individuals is
more or less the same from dusk to dawn in summer. In winter, the individuals become active just after dusk and simultaneously react to the changes in brightness.
However, the number of burrowing prawns increases owing to low temperature
(Figure 5.18).

Salinity, starving, fatigue, etc., do not influence the relationship between light
intensity and activity. Neither is the lunar phase related to activity, but the brightness on a clear night of a full moon does influence it. There is an inherent rhythm in
the daily activities of prawns which is not yet understood when darkness continues
for some time; however, when brightness continues, after about three days, the prawns
follow a rhythmic pattern of emerging from the sand at night and entering the burrows
during the day (Figure 5.19). Observations carried out by Dall (1958) indicate that
when darkness continued for some time, *M. mastersii* (=*M. bennetae*) resumed daily
rhythmic activity in about seven days.

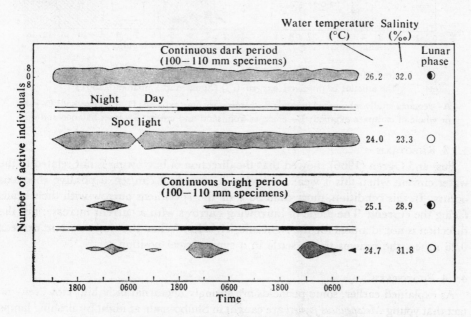

Figure 5.19. Influence of continuous dark and bright conditions on
the activities of *P. duorarum* (Fuss and Ogren, 1966).

According to experiments done by Egusa (1961), the burrowing habit of *Penaeus*
is greatly influenced by the amount of dissolved oxygen. When the amount of oxygen
is less than 1.0 cc/l, *Penaeus japonicus* emerge from the sand regardless of the light
intensity (Figure 5.20).

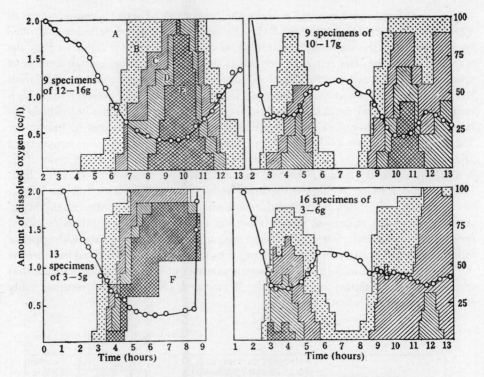

Figure 5.20. Changes in burrowing habits of *Penaeus,* due to the difference in the amount of dissolved oxygen (○) (Egusa and Yamamoto, 1961).

A—remains totally submerged in sand; B—respiratory pore exposed; C—eyes exposed; D—part or whole of carapace exposed; E—emerges from sand and creeps; F—loses balance and dies.

4.1.3 RHEOTAXIS

Fuss and Ogren (1966) showed that the direction of burrowing is not related to the water current when this is weak, but when it exceeds 0.2 m/sec, a positive rheotaxis occurs. In this condition, more than 90% of the specimens burrow with their heads facing the current. The angle of burrowing narrows with a current increase, but the direction is not identical for the right and left sides. When the water current exceeds 0.26 m/sec, the prawns try to settle in a symmetrical position.

4.1.4 PHOTOTAXIS

As explained earlier, some penaeids are strongly nocturnal in habit. However, the fact that young *Metapenaeus joyneri* are caught in Shubo-nada at night by flashing lamps during migration on the open sea, is somewhat contradictory to the aforementioned fact (Maekawa, 1961). Even in India young prawns are caught by flashing light in water fields (Menon, 1954). Hence, it can be said that young prawns tolerate light. if it is not too strong. This aspect needs further study.

4.1.5 SCHOOLING

According to Racek (1956), *Penaeus* of Australia can be divided into two distinct categories on the basis of schooling, those which tend to form groups in which prawns of the same age grow and form larger clusters, and those which consist of prawns that do not form a distinct group but remain scattered in their original places except at the time of migration to open seas during their young life. Except for *Alypopenaeus compressipes* (Yasuda, 1958), almost all the Japanese varieties of *Penaeus* belong to the latter category. In other words, the Japanese varieties do not display a schooling instinct. However, *P. orientalis*, *Metapenaeus joyneri* and, to a lesser extent, *Metapenaeus monoceros* show some gregariousness while wintering.

4.2 Habitat

Habitats of *Penaeus* can be divided broadly into three categories according to their development and growth: embryonic development, larvae, and early juvenile stage; late juvenile and young stage; and finally, habitats for the immature and adult stages. The first habitat is found in the offshore water basins. Life at this stage is planktonic so the horizontal distribution of prawns is greatly controlled by the water current of the sea. The vertical distribution shows variations according to the developmental phase. The planktonic stage is greatly controlled by water temperature. From the results of the rearing carried out by Hudinaga (1942) and Maekawa (1961), it was estimated that the planktonic stage takes about one month in summer (Table 5.8).

Table 5.8. Number of days required for the early development of *Penaeus*

Stages of development	Water temperature (°C)	
	19-25 (Maekawa, 1961)	26-29 (Hudinaga, 1942)
Embryonic period	0.7	0.7
Nauplius	2	1.6
Zoea	4-6	6
Mysis	7-8	6
Juvenile		
7→14 mm	11	—
5→ 9 mm	—	12
Total	25-28	26

The second habitat is found in the growing places inside the bays, in shallow places either near the tidal areas or the ebb tide line. Juvenile prawns begin floor-dwelling life in this habitat, which can be seen to widen up to places where the water is almost clear. The variations according to species are fairly clear and according to Maekawa (1961), the suitable habitats for young prawns in Shogun Bay are: (1) the warm water belt and the adjoining coasts (*P. japonicus*, *Metapenaeus burkenroadi*, *Metapenaeus joyneri*, and *Metapenaeus monoceros*); (2) the seaweed grounds and the surrounding muddy places (*P. semisulcatus*, *Metapenaeopsis acclivis*); and (3) the regions where the water of the open seas has strong influence (*Metapenaeopsis barbata*, *Trachypenaeus curvirostris, Parapenaeopsis tenella,* and *Alypopenaeus compressipes*). The larger the prawn, the nearer to the open seas is its habitat. However, prawns which live in tidal areas may be adversely affected by exposure. The distribution of young *P. japonicus* in tidal areas is not related to the floor qualities of the seas, but the distribution seems to vary according to the changes in tidal currents or waterways (Nagata, Tanizaki, and Nakazawa, 1931; Yasuda, Suzuki, and Morioka, 1957; Maekawa, 1961). Although prawns can live in tidal zones, the presence of predators in such areas is probably a discouraging factor, and hence their suitability as habitats for juvenile and young prawns is a separate problem. The living period in this habitat varies according to the development period; while it is usually 2 months in July and August for early developing individuals, it continues from September to October to April or May of the following year, or about 7 months for the late developing individual. Hence, prawns dwell in this habitat for much of the year, but during winter move toward the open seas, and during the optimum temperature period from spring to autumn, the distribution spreads toward the land. The change of habitats occurs twice in a year, once in July and again in late September or early October. Even in water basins outside Japan, the habitats of young prawns are found in warm water basins in the bays. This is true at least for the commercially grown *P. japonicus*. The Indian *P. stylifera* is an exception which grows throughout its life in open seas and never enters the bays even during the larval period (Panikkar and Menon, 1956).

The adult habitat is usually the open sea. The depth of habitats (Figure 5.21) differs. According to Utsunomiya (1959), Yasuda (1956), Maekawa (1961), and Ikematsu (1963), *Metapenaeus joyneri* and *Metapenaeopsis barbata* prefer a muddy bottom, while *Trachypenaeus curvirostris* and *Metapenaeopsis dalei* prefer a deep sandy bottom. However, *P. japonicus, Metapenaeus monoceros, Metapenaeus burkenroadi,* etc. are found mostly near the nursery grounds. Particularly in the case of *M. burkenroadi*, the life history is completed within the inner bay itself (Yatsuyanagi and Maekawa, 1957b).

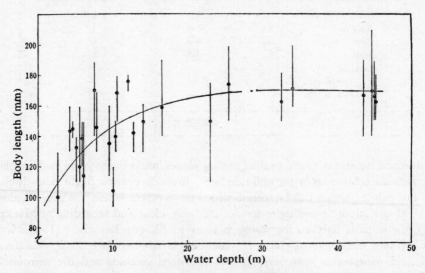

Figure 5.21. Changes in body length range (vertical lines) and mean (circles) of *P. japonicus* due to changes in water depth (Yoshida, 1960).

Trachypenaeus curvirostris, on the other hand, is found in various beds ranging from a muddy bottom (Suo-nada) to a sandy bottom mixed with gravel (Kasaoka Bay and Ariake-Kai). Within this range they occur at various depths. Hildebrand (1954, 1955) and Springer and Bullis (1954) report that the distribution of prawns in the Gulf of Mexico is closely related to the nature of the substratum, i.e. *P. setiferus* is found on a muddy bottom and *P. duorarum* on calcareous mud, a sandy bottom, or mud mixed with sand, while *P. aztecus* lives on a silt bottom. This type of selectivity in substratum is also found among specimens reared in a laboratory (Williams, 1958).

The distribution during winter spreads toward offshore regions, and during the optimum temperature period the distribution becomes restricted to shallow regions. At times the prawns re-enter the bays but the center of distribution is always in the open seas. Since in most fishing operations, a particular species is not sought, the distribution particularly in winter is not always accurately determined except in the case of *Metapenaeus joyneri* and *P. orientalis*. The wintering places of *M. joyneri* are found in the central part of Suo-nada (20 to 30 m), the Rinjima surroundings (30 to 40 m), and in Ariake-Kai (20 m). When the wintering place and the growing beds are widely separated (*viz. P. orientalis*), the adult habitat is widely spread, but in other species, the range is relatively narrow and restricted to waters of 50 m depth.

4.3 Migration

Migration on a fixed course is seen only when the prawns move from one habitat to another; otherwise it is irregular and believed to be related to floor feed availability (Anderson, 1956). Except when the growing place is far away from the wintering place, prawns are not found to move in groups even during the spawning period.

There is no definite explanation of the factors contributing to the movement of prawns from the first habitat to the nursery grounds inside bays. In the case of *P. japonicus* it is well known that the functions of the body become more and more active in the young stage according to the salinity of the sea water. However, this does not explain why juvenile prawns are often found in waters of low salinity. When the temperature, diet conditions, etc. are the same, the growth rate of juvenile prawns (Johnson and Fielding, 1956; Zein-Eldin, 1963), as well as the frequency of molting (Dall, 1956a), is not related to salinity. In warm water basins inside bays, although it is a fact that the areas are rich in nutrition, it is not possible to accept the explanation that the prawns migrate to these bays in search of food, as the animals have no means of knowing whether feed is present or not. In the northern parts of the Gulf of Mexico, the direction of movement of prawns is away from the coastal nursery grounds, and hence it would seem that the juvenile prawns come back to the coastal regions by instinct (Idyll and Jones, 1964). The return to the growing regions is found during the juvenile stage, and not during the larval stage, since the swimming capacity of larvae is poor. This shows that the movement is strongly controlled by some inborn factor. On this point, Aldrich (1963a) experimented and the results he obtained support this view. For example, the distribution layers of *P. duorarum* in the juvenile stage in water columns varying in salinity, change according to the water temperature; when it is high, the prawns shift toward waters of low salinity (Figure 5.22). The migration of prawns from the coastal regions to the inner bays and inland water basins

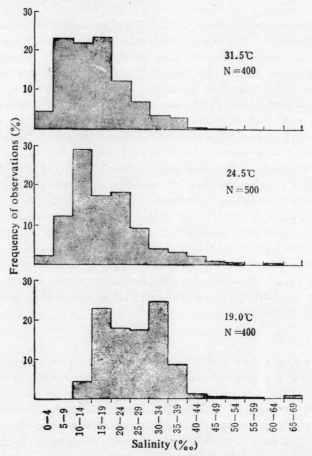

Figure 5.22. Influence of temperature on the selection of salinity
by juvenile prawns (Aldrich, 1963).

seems to indicate that the tides at night carry them, and this is also supported by the
plankton collection record from the mouth of the bay (Figure 5.23). According to
Williams (1955b), the distribution of young prawns in the coastal growing regions
of North Carolina is restricted to areas with weak and irregular tides.

Migration from the nursery grounds to the adult habitat occurs when the gonads
mature, and thus seems closely related to gonadal maturation; hence, it can also be
considered a spawning migration. However, the migration of prawns of the same age
in autumn seems to be related to the wintering period. This migration is more related
to the developmental stages or the size of the prawns, than to the water temperature.
In the case of *P. japonicus*, this migration is at peak at the end of spring or summer,
and once again in autumn. In the first instance, migration occurs when the water
temperature is still on the increase. The migration in autumn is accelerated when the
water temperature falls, but the migration toward the open sea is restricted to young
prawns and the post larvae do not leave their nurseries even during winter. When the
cold is very severe in winter, there is sometime great mortality among the post larvae
(Iverson and Idyll, 1960; Eldred, Ingle, Woodburn, and Jones, 1961).

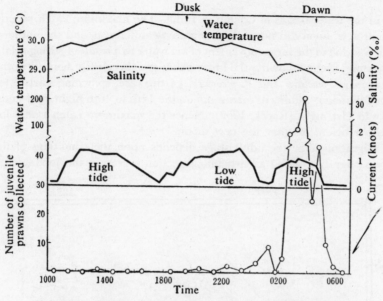

Figure 5.23. Number of juvenile *P. duorarum* collected in the straits of the southwestern Florida peninsula (Idyll and Jones, 1964).

The migration toward offshore regions for spawning is closely related to the lunar phase. According to Utsunomiya, Yatsuyanagi, Tomiyama, and Maekawa (1954), the catch rate of *P. japonicus* and other varieties is maximum on the day following the new moon or full moon. The maximum value is higher during full moon than during new moon (Figure 5.24). The same results were obtained by Menon and Raman

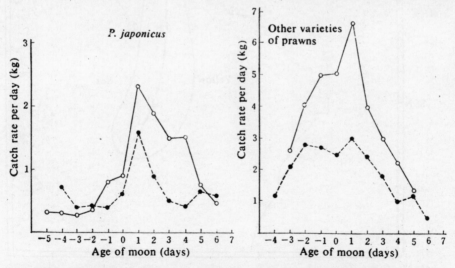

Figure 5.24. Relationship between the catch rate and lunar phase in prawns of Shogun Bay in the Yamaguchi Prefecture (Utsunomiya, Yatsuyanagi, Tomiyama, and Maekawa, 1954).

●=during full moon; ○=during new moon; horizontal lines indicate the number of days either before or after full or new moon (● or ○).

(1961) in the water basins of Cochin in India. The maximum catch rate during full moon and new moon did not show any striking difference, and the authors believed that this was due to the large movement of sea water as a result of strong tidal currents. The spawning cycles following the lunar phase are also clearly noticed in the Australian prawns (*P. macleayi* and *P. plebejus*). In this case, spawning starts 3 to 5 nights after the full moon, attains its maximum on the 12th to 16th night, and ends between the 19th to 21st night (Racek, 1959). Hence the maximum catch rate is found only once, either before or after the new moon.

Any migration after the adult stage depends upon the conditions of the region. No group migration along a particular direction is seen in the Japanese coast since

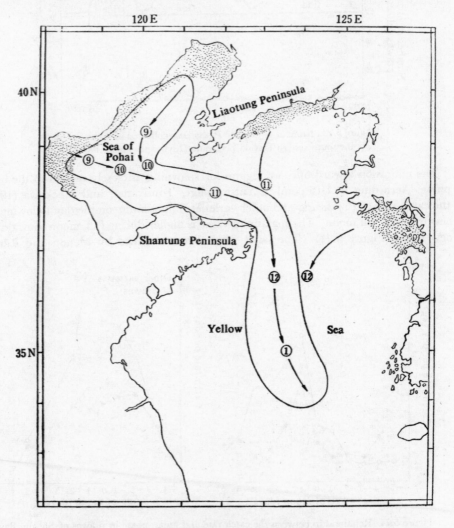

A. Migration southward.

Figure 5.25. Movement of *P. orientalis*. Number—Month; dotted areas—spawning and growing regions (Satouchi, 1937; Kasahara, 1948).

the wintering places are nearby. In the Yellow Sea, the migration for wintering is clearly seen in *P. orientalis*. According to the data published by the Eiko Fisheries Experiment Station (1938), Satouchi (1937), and Kasahara (1948), the movements are as shown in Figure 5.25A and B. In the various regions of Sea of Pohai and on the western coast of the Korean Peninsula, the prawns of the same age become the target of commercial fishing in September to October. However when the coastal water temperature drops to 20°C, the prawns migrate in groups southward, and by November to December, they descend to the Shantung Peninsula. In January, they are found scattered for a temporary period in the open seas west of Cheju-Do. This movement during the wintering period has been explained in detail by Ikeda (1962), but the fishing grounds are found restricted to regions within 123 to 123.5°E. In December to

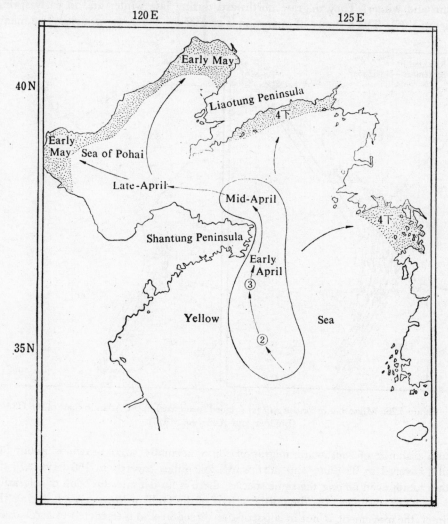

B. Migration northward.

Figure 5.25. Movement of *P. orientalis*.

February, the prawns move southward and are found at 35 to 37°N, and later reverse moving northward. The habitats during this period are covered by warm currents on one side (water temperature 9°C, salinity 18.00 ‰ or more) and coastal water basins (water temperature 6°C, salinity 7.50 ‰ or below) on the other side. The habitat centers are regions where the water basins are mixed. The movement of prawn groups northward continues up to April, and by early April they reach the Shantung Peninsula, and by late April, Sea of Pohai. This type of movement, northward and southward, is also found in *P. setiferus* of the Atlantic coast of the USA. According to Lindner and Anderson (1956), prawns of 133 m body length are found near the growing regions even during the winter, but when the body length is more than 130 mm, they migrate southward when the water temperature falls during autumn and winter. They migrate northward during late winter and in early spring (Figure 5.26). The distance of migration is greater for the larger prawns. The maxi-

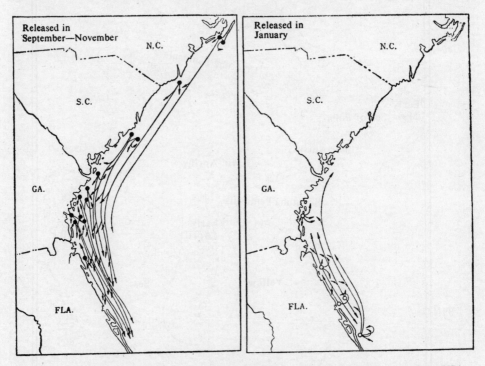

Figure 5.26. Migration of *Penaeus setiferus* (tagged specimens) in the Atlantic coast of the USA (Lindner and Anderson, 1956).

mum distance of southward migration for a normally sized prawn is about 360 miles covered in 95 days, and northward, 260 miles covered in 168 days. On the other hand, even among the same species, there is no definite direction of migration in spring or summer. In the northern part of the Gulf of Mexico, even during the winter, the movement is not in a particular direction and is restricted to a relatively narrow range between offshore and land (Figure 5.27). Hence, the movement does not show any characteristic features for a particular species, but seems to be controlled

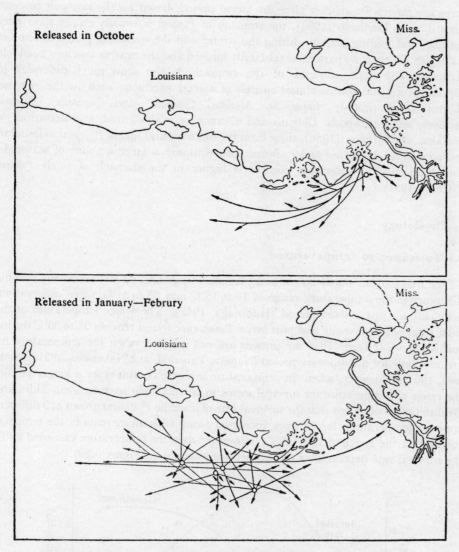

Figure 5.27. Migration of *Penaeus setiferus* of the northern part of Gulf of Mexico
(Lindner and Anderson, 1956).

by the temperature of the water of the habitats. Thus the migration of prawns during winter in any region is related to the temperature changes, and the temperature range during migration differs from species to species. It is also influenced by the reaction of the prawn at various stages of growth, to temperature as well as salinity at low temperature.

4.4 Feeding habits

There are no data on the natural feed of *Penaeus* during the planktonic period, but there are reports on its diet after the larval period, based on the stomach content. According to Ikematsu (1963), the stomach of *Penaeus* is usually empty soon after molting, and feeding activity during the winter and the spawning period in females is sluggish. There is no particular selectivity for feed and the prawns take any available organism. The main contents of the stomach do not show much difference in various species and consist almost entirely of animal products, such as the remains of crustaceans (mainly *Amphipoda, Mysidea, Copepoda,* and *Decapoda*), ciliates, bivalves, and gastropods. Detritus and diatoms were also found, and according to Panikkar and Menon (1956), these form the main natural diet of *P. japonicus* found in Indian waters. *M. dobsoni* and *P. indicus* also consume a large amount of seaweeds. The proportion of vegetable content is higher in the stomach of small *Penaeus* (Menon, 1951).

5. Physiology

5.1 Tolerance to temperature

The free movements of larvae and juvenile *P. japonicus* do not show any striking difference in the temperature range of 13 to 15°C and 33 to 34°C, when observations are made for just a short period (Hudinaga, 1942). The water temperature of the tidal areas where juvenile and post larval *Penaeus* are found reaches 34 to 38°C during mid-summer. It seems that the prawns are not affected when the temperature increase is only for a temporary period (Nagata, Tanizaki, and Nakazawa, 1931; Maekawa, 1961). However, when the temperature increase continues for a longer period, the range of temperature for survival seems to be relatively narrow. Zein-Eldin and Griffith (1966) observed that the survival rate of juvenile *P. aztecus* grown at 9 different constant temperatures for 28 days, increased along with an increase in the temperature within the range of 15 to 20°C. However when the temperature exceeded 25°C, the survival rate decreased as the temperature increased (Figure 5.28).

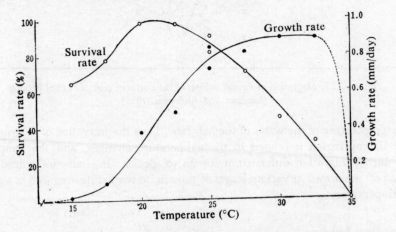

Figure 5.28. Changes in the survival rate for 1 month and 1 day of *Penaeus aztecus* at various temperatures (Zein-Eldin and Griffith, 1966).

5.2 Tolerance to salinity

The optimum salinity range of *Penaeus japonicus* during the embryonic stage is 27 to 39°/₀₀, but the range becomes wider with further growth and during the early juvenile stage is 23 to 47°/₀₀. During the period of *zoea*, the tolerance range narrows (Hudinaga, 1942). According to Yoshida (1960), *P. japonicus* of 53.1 to 101.9 mm die within 2 to 4 hours in fresh water, within 10 to 17 hours in water of 3.66°/₀₀ salinity, and within 18 to 27 hours in water of 6.26°/₀₀ salinity. At 11.53°/₀₀ and above, no mortality is seen within the first 24 hours. Zein-Eldin and Aldrich (1965) have shown through experiments that the tolerance to salinity among juvenile *P. aztecus* differs according to temperature, i.e. below 15°C, the tolerance is poor to salinity (Figure 5.29). The salinity range for survival for 28 days is narrower than that for 24 hours. In the case of *M. mastersii*, the tolerance to low salinity among young prawns in rivers is poorer among males than among females. During migration, the male prawns escape to sea water earlier than the females (Dall, 1958).

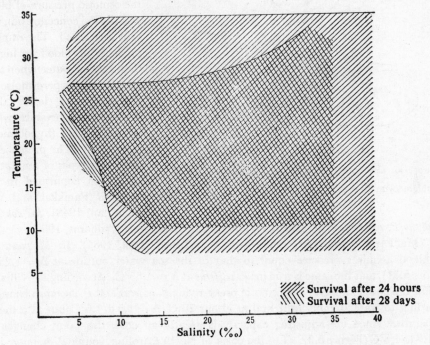

Figure 5.29. Combination of water temperature and salinity ensuring more than 80% survival of *Penaeus aztecus* (Zein-Eldin and Aldrich, 1965).

5.3 Osmotic pressure and ion control

The blood of crustaceans consists of plasma and corpuscles. The characteristics of these two differ according to the age of the specimen, nutritional condition, molting period, and sometimes sex also (Maynard, 1960). In *M. mastersii*, 3 types of blood corpuscles have been identified (Figure 5.30), which were formed in a pair of longitudinal tissue strands on the dorsal side of the stomach (Dall, 1964).

The regulating mechanism of blood ion concentration is well developed in crustaceans. The osmotic pressure among many marine animals is the same as the sea water

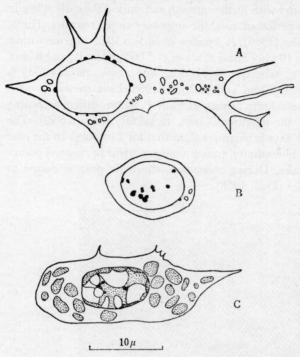

Figure 5.30. Cell entities in blood of *Metapenaeus bennatae*.*

A—thigmocyte; B—lymphocyte; C—amoebocyte with large granules (Dall, 1964).

outside. However, a capacity to control the osmotic pressure is well developed in freshwater species and in some marine animals (Robertson, 1960). The habitats of *Penaeus* vary considerably, ranging from a relatively high salinity of sea water in the open seas, to regions of fresh water. In all the specimens examined, the capacity to control both the osmotic pressure of blood and the ion concentration was well developed. The osmotic pressure of blood is higher than that of water when salinity is low (hyperosmotic regulation), and lower than the external pressure when salinity is high (hypoosmotic regulation). The presence of these two types of regulation have been confirmed in *M. monoceros* (Panikkar and Viswanathan, 1948), *M. dobsoni*,

P. indicus, P. carinatus (Panikkar, 1951), *P. aztecus, P. duorarum* (Williams, 1960), *P. setiferus* (MacFarland and Lee, 1963), and *M. mastersii* (Dall, 1964). In *M. mastersii*, the blood acquires pressure equal to that of the sea water outside at $\Delta = 1.25°C$ (Figure 5.31), and the same is true for *P. setiferus* at $\Delta = 1.55°C$. According to Williams (1960), the equalizing point of osmotic pressure for *P. aztecus* and *P. duorarum* changes according to water temperature and at about 28°C, $\Delta = 1.5$ to $1.6°C$, but at a lower temperature since the adjusting capacity is greatly affected, this point changes to $\Delta = 1.7$ to $1.8°C$ (Figure 5.32). On the coast of North Carolina, young *P. duorarum* are found even in winter but *P. aztecus* are found only in summer. This seems to indicate that the osmotic pressure control at low temperature is stronger in *P. duorarum* than in *P. aztecus*

According to Dall (1964), the concentration of K^+, Na^+ and Cl^- ions in the blood of *M. mastersii*, shows more or less the same changes according to the difference in the concentration of ions in the sea water. Ca^{++} ions, however, show a higher value than the surrounding sea water (Figure 5.33).

*Obviously an error in the original—Translator.

5.4 Oxygen consumption rate

The oxygen consumption rate of *Penaeus* differs according to the temperature, salinity, size of the specimen, and activity.

According to the results obtained by Egusa (1961), the oxygen consumption rate when there is a sufficient amount of dissolved oxygen in the water is as shown in Table 5.9. The oxygen consumption rate per unit weight of the specimen is higher for small specimens and highly active specimens. Oxygen consumption rate is also influenced by the amount of dissolved oxygen, and when the total dissolved O_2 is low, there is an increase in oxygen consumption rate. When the oxygen consumption is maximum, i.e. about 1 cc/l, the prawns emerge from the sand. When a further

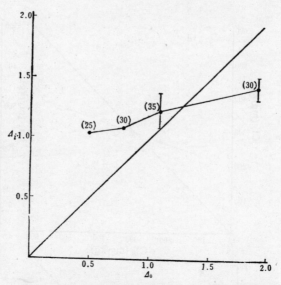

Figure 5.31. Relationship between osmotic pressure of blood (Δ_i) and osmotic pressure (Δ_o) of sea water observed at freezing point of *Metapenaeus bennatae*.

Vertical lines show the range of difference. Values for mixed blood taken from specimens (number of specimens used indicated in brackets) (Dall, 1964).

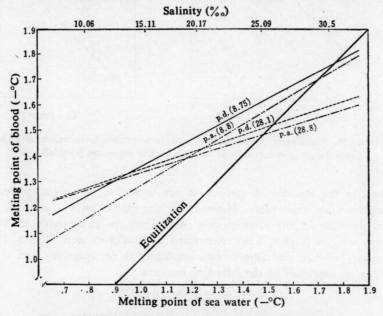

Figure 5.32. Relationship of the osmotic pressure of surrounding sea water to blood osmotic pressure at melting point.

p. a.—*Penaeus aztecus*, total length 42 to 110 mm; p. d.—*Penaeus duorarum* 35 to 110 mm; figures in brackets indicate the temperature (°C) (Williams, 1960).

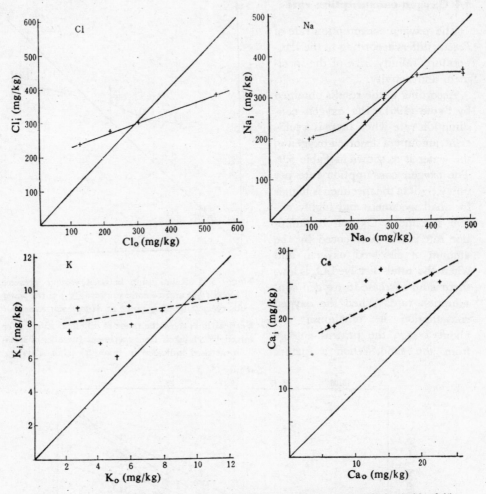

Figure 5.33. Relationship in *Metapenaeus bennatae* between the ion concentration in blood (*i*) and the ion concentration of the surrounding sea water (*o*) (Dall, 1964).

reduction in the amount of oxygen occurs thereafter, a sudden decrease in the consumption rate takes place. However, in the case of large specimens, there is a gradual decrease in the consumption rate of oxygen along with the decrease of dissolved oxygen (Figure 5.34). According to Maekawa and Otsuka (1955), the relationship between the oxygen consumption rate (*a*, cc/gr/hr) and temperature (*t* °C) can be expressed by the following formula:

$$a = 0.0029t - 0.0137$$

Experiments carried out by Rao (1958) showed that the oxygen consumption rate of *M. monoceros* was minimum if the salinity of the habitat was at the optimum level. Changes in salinity brought about variations in the osmotic pressure between the body fluid and sea water, and caused an increase in the oxygen consumption rate.

Table 5.9. Oxygen consumption rate of *Penaeus japonicus*
(Egusa, 1961)

Body weight (g) Mean	Range	Number of specimens examined	Mean temperature (°C)	Mean Salinity (‰)	Oxygen consumption rate (cc/kg/hr) Resting period	Active period	Approximate value of dissolved oxygen (cc/l)
3.1	2.4-3.7	13	23.0	15.8	135	—	4.3
			22.8	15.8	—	438	3.8
5.5	4.6-6.2	8	22.8	15.9	117	—	4.5
			22.7	15.7	—	423	3.8
11.8	11.4-12.6	5	23.0	15.7	106	—	4.2
			22.6	15.5	—	260	4.5
16.1	14.8-18.2	3	22.9	15.8	77	—	4.7
			22.6	15.6	—	312	4.2

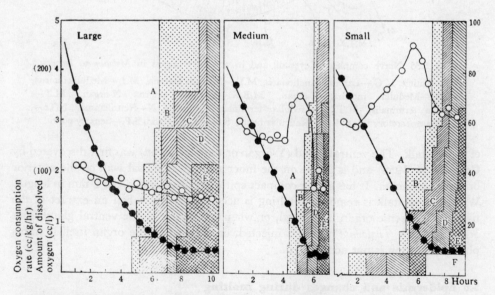

Figure 5.34. Relationship between the amount of dissolved oxygen (black spots), oxygen consumption rate during resting period (white spots), and burrowing conditions of prawns. The mean weight of the specimens examined was as follows: Large, 13.6 g; Medium, 5.4 g; Small, 3.5 g. Burrowing conditions are as shown in Figure 5.20 (Egusa, 1961).

5.5 Hormones

Studies have been conducted on the neuro-hormone tissues and hormone control of crustaceans (Knowles and Carlisle, 1956; Enami, 1957; Welsh, 1961).

The hormones of *Penaeus* have not yet been studied in detail but there are reports on the post-commissural organ of *P. brasiliensis* (Knowles, 1953), the eyestalk of *P. setiferus* (Young, 1956), and the endocrine organ and the central gland of the eyestalk of *M. mastersii* (Dall, 1954a). According to Dall (1956b), the position of the group of neuro-hormone cells in the eyestalk of *M. mastersii* is as shown in Figure 5.35. In living specimens a white sinus gland is clearly seen through the transparent epithelium

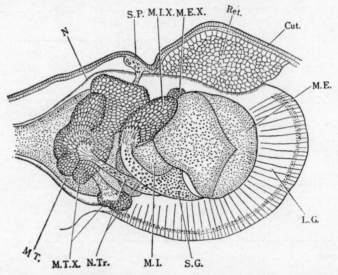

Figure 5.35. Nerve complex of eyestalk and its accessory organs in *Metapenaeus bennatae*.*

Cut.—Cuticle, L. G.—Lamina ganglionaris, M.E.—Medulla externa, M.I.—Medulla interna, M.I.X.—Medulla interna, X-organ, M.E.X.—Medulla externa X-organ, M.T.—Medulla terminalis, M.T.X.—Medulla terminalis X-Organ, N—Neurilemma, N.Tr.—Neurosecretory tract, Ret.—Reticular tissue, S.G.—Sinus gland, S.P.—Sensory pore.

of the eyestalk. The ventral gland (Y-organ or thoracic gland) was first discovered by Gabe (1953, 1956) and is found on the inner side of the dorsal wall of the anterior branchial chamber. It has a very compact epithelial tissue and is 1 to 2 mm in length. When the eyestalk is removed, molting is accelerated, but when an extract of the tissues of the sense organ is injected, or when an extract of the ventral gland from *Scylla serrata* or *Thalamita crenata* is injected, or even when the organ itself is transplanted, molting is not accelerated.

5.6 Epidermis and changes during molting

According to Dall (1965b), the cuticle of *M. mastersii* is made up of three layers,

* Obviously an error in the original—Translator.

but the pigmented layer is not distinct (Figure 5.36). The thickness of the cuticle in the soft parts of the walking legs is 5 μ and in the cephalothoracic part 75 μ. It contains 38.7% inorganic matter, of which 98.5% is calcium carbonate. During molting, chitin and protein of the old epidermis are absorbed (a total of about 39%) but the inorganic matter is not absorbed (Table 5.10). Just before molting (premolt: Passano, 1960), the cells of the epidermis form prominent striations. New epicuticle and exocuticle synthesize and separate from the old epidermis. Soon after molting, calcium is gradually deposited in the exocuticle and, in about 5 hours, the formation of a new

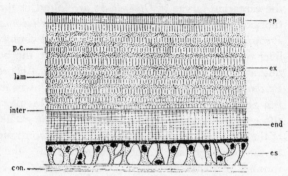

Figure 5.36. Cuticle during molting period (C-period) of *Metapenaeus bennatae* (Dall, 1965b).

ep—epicuticle; ex—exocuticle; end—endocuticle; es—epidermis; p.c.—pores; lam—lamina; inter—intermediate layer; con.—connective tissue.

Table 5.10. Composition of epidermis and molted cuticle in *M. mastersii* (mg per g of dry epicuticle) (Dall, 1965b)

Sections	Ca	Mg	CO³	P	Inorganic substance (%)	Chitin	Residue	Organic substance (%)	Chitin in organic substance (%)
Period of molting	190	1.35	191.5	4.95	38.7	210	402.7	61.3	34.3
Molted cuticle	247	2.20	263.0	5.35	51.7	159	323.5	48.3	33.0
77% value of molted cuticle*	190	1.70	202.0	4.10	—	122	239.0	—	—

*The weight of molted cuticle is 77% of the weight of the epidermis during molting.

cuticle is completed. This is more or less common to all crustaceans. However, no storage of glycogen in the cells of the epidermis is seen in the case of *Penaeus*. On the other hand, acidic mucopoly-saccharides form as the precursor of chitin, and are carried to the epidermal cells by the amoebocytes of the blood (Dall, 1965c). The exchange of calcium between body fluid and sea water is carried out through the gills at the rate of 90% intake and 70% ejection. The storage of calcium is continued throughout the molting period in the form of visible crystals in between the exocuticle and the endocuticle (Dall, 1965d).

6. Parasites

According to Kruse (1959), *sporozoa* such as *Cephalolobus, Nematopsis*, etc. and adult

Opecoeloides (trematodes) are found in the walls of the stomach and the intestine of *P. setiferus*, *P. aztecus*, and *P. duorarum* of the Gulf of Mexico. In all instances, the cercarian larvae of *Opecoeloides* are found in the connective tissue of the cephalothorax and around the intestine. *Thelohania (sporozoa)* are found in the muscles, and *Prochristianella* (cestode), and the larvae and young of *Contracaecum* (nematode) in the liver of these prawns. These parasites enter the body of the prawn soon after they migrate to the nursery grounds inside bays. According to Aldrich (1963b), parasitic *Prochristianella* were not seen in the young prawns first discovered in the bays in May, but about one month later, the larvae of this parasite were found in about 83% of those young prawns assumed to be from the earlier group. Each had more than 20 of the parasites. It seems that the parasites continue to enter the prawns as they grow in the bays and, according to Kruse (1956), the larger the prawn, the larger the number of parasites.

There are several parasites whose life history is not clearly understood. For *Prochristianella penaei*, prawns are an intermediate host and fishes *(elasmobranchia)* the final hosts. Recently, the parasitic cestode was discovered in an adult condition in *Dasyatis sabina* (a type of ray fish) (Aldrich, 1964). The rate of infection by these parasites differs according to the type of prawn, but it is found to be maximum in *P. duorarum*. In *P. setiferus,* the percentage of infection is irregular.

Kajiyama (1935a) reported that in *P. japonicus,* infection by the parasitic *Isopoda* in the branchial chamber, obstructs the development of gonads in the host. Hiraiwa (1933, 1934, 1936) and Hiraiwa and Sato (1939) studied infection by the parasitic *Epipenaeon japonicus (Bopyridae)* in *Penaeopsis akayebi* (= *M. barbata*) of Hiroshima Bay. The study revealed that both the male and female gonads, as well as the development of secondary sexual features, were affected by the infection. However, unlike infected crabs and hermit crabs, changes such as the development of ova in the testis and the appearance of secondary female features in the males were not evident. When the parasites leave the prawns, the gonads develop normally.

7. Tagging

Reports on the tagging of *Penaeus* are not yet complete, but practically no tagging occurs on the Japanese coasts. Lindner and Anderson (1956) tied celluloid discs of 3/8 inch diameter (5/16 inch for smaller prawns) to a nickel pin which was fixed on the first abdominal segment of 46,532 prawns. However, tagged prawns of less than 80 mm body length were not recovered; for prawns of 85 to 110 mm, recovery was less than 10%. McRae (1952) tagged 1,137 prawns on the coast of Texas and was able to recover 2.01% after 80 days. Recently, efforts have been made to tag small prawns with lighter materials (Figure 5.37), which is very helpful in distinguishing the specimens. However, since there is a high mortality among small prawns, this method of tagging has not yet been commonly adopted.

It was Menzel (1955) and Racek (1956) who first tried the staining method. Menzel injected Fast-green solution into *P. setiferus* and was able to differentiate the injected prawns from the uninjected even after 60 days. Racek found that the mortality rate was minimum when the prawns were dipped in Nile-blue sulfate solution and, furthermore, the color held longer. Dawson (1957) tried staining with 26 types of pigments in 99 shades on prawns of 30 to 175 mm body length. The results showed that some of the colors remained even after the lapse of 100 days and at least one molting. These were Fast-green FCF, Niagara-sky-blue 6B, Trypan-red, and Trypan-blue. The latter two dyes persisted for a very long period, i.e. above 200 days. When fish meat dyed with Trypan-red was fed to prawns, they absorbed the color which persisted for more than 230 days and through 3 molts. In the dipping method, the color did not persist for more than 4 days. According to Costello (1964), the mortality becomes high if air bubbles are not prevented from entering the body of the prawns during injection. It is necessary to use sterilized, redistilled water as a solvent, and filtered sea water or saline water should not be used since staining with these is never uniform. Unless the colored water is filtered before using, the effect will not be prominent. Since Trypan-blue and Trypan-red show a toxic effect in prawns when

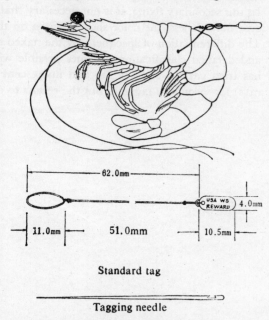

Standard tag

Tagging needle

Figure 5.37. Tagging method and Atkins type tag (Allen and Costello, 1962).

their solutions are kept for a long time at room temperature, these dyes should be used within three days of their preparation. The combination of optimum concentration of dye fluid and the injection dosage for *P. aztecus* is given in Table 5.11.

Table 5.11. Optimum concentration and dosage of 3 types of vital dyes for staining *P. aztecus* **(Costello, 1964)**

Total length (mm)	Trypan-blue		Trypan-red		Fast-green FCF	
	Concentration (%)	Dosage (cc)	Concentration (%)	Dosage (cc)	Concentration (%)	Dosage (cc)
75	0.125	0.020	0.250	0.020	0.500	0.030
95	0.250	0.025	0.500	0.040	0.500	0.070
115	0.250	0.035	0.500	0.060	0.500	0.100
135	0.250	0.070	0.500	0.090	0.500	0.180
155	0.500	0.050	0.500	0.140	0.500	0.250

The staining method has certain advantages in that no injuries are inflicted and it can be applied even to young specimens. However, only a few dyes can be used as of now. Hence, it is necessary to find other suitable ones and also discover a double-staining method in which, after staining as described above (wherein the color holds in the body), some design would be caused on the abdominal part (secondary tagging). In this secondary fixing, it is not necessary that the color be prominent to the naked eye; nevertheless, the color should persist on those parts where it has been applied. The differentiation of specimens by the naked eye is done with the primary tagging, and detailed classification becomes possible with the secondary one. Klima (1965) has tried various types of ink and fluorescent pigments (identifiable by ultraviolet rays) for secondary tagging, but the results to date have not been very encouraging.

(Kurata)

Bibliography

1. Aichi Fisheries Experiment Station, 1942. Ecological Survey of Important Varieties of Prawns in Aichi Prefecture, 1940, 1-37.
2. ——, 1943. Moebi seitai chosa (Ecological Study of *Metapenaeus burkenroadi*). Report for 1940, 1-20.
3. ALDRICH, D. V., 1963a. Behavior and Tolerances, *U. S. Fish. Wildl. Serv., Cir.*, **183**, 61-64.
4. ——, 1963b. Incidence and Potential Significance of *Prochristianella penaei*, a Cestode Parasite of Commercial Shrimp in Galveston Bay, *U. S. Fish. Wildl. Serv., Cir.*, **183**, 68-70.
5. ——, 1964. Shrimp Parasitology, *U. S. Fish. Wildl. Serv., Cir.*, **230**, 82-83.
6. ALLEN, D. M., and T. J. COSTELLO, 1962. The Use of Atkins-type Tags on Shrimp, *U. S. Fish. Wildl. Serv., Cir.*, **161**, 88-89.
7. ANDERSON, W. W., 1956. Observations upon the Biology, Ecology, and Life History of the Common Shrimp, *Penaeus setiferus* (Linnaeus), along the South Atlantic and Gulf Coasts of the United States, *Proc. I. P. F. C.*, **6** (III), 399-403.
8. ——, and M. J. LINDNER, 1943. A Provisional Key to the Shrimps of the Family Penaeidae with Special Reference to American Forms, *Trans. Amer. Fish. Soc.*, **73**, 284-319.
9. ——, and J. E. KING, 1949. Early Stages in the Life History of the Common Shrimp, *Penaeus setiferus* (Linnaeus), *Biol. Bull.*, **96**, 168-172.
10. ——, and M. J. LINDNER, 1958. Length-weight Relation in the Common or White Shrimp, *Penaeus setiferus, U. S. Fish. Wildl. Serv., Sp. Sci. Rept. Fish.*, **256**, 1-13.
11. BATE, S., 1888. Report on the Crustacea Macrura Collected by H. M. S. "Challenger" during the Years 1873-1876, *Rep. Sci. Res. H. M. S. "Challenger"*, 1873-1876.
12. BHIMACHER, B. S., 1966. Information on Prawns from Indian Waters: Synopsis of Biological Data, *Proc. I. P. F. C.* **10** (II), 124-133.
13. BURKENROAD, M. D., 1934a. The Penaeidae of Louisiana with a Discussion of Their World Relationships, *Amer. Mus. Nat. Hist., Bull.*, **68**, 61-143.
14. ——, 1934b. Littoral Penaeidae, Chiefly from the Bingham Oceanographic Collection, *Bull. Bing. Ocean. Coll.*, **4**, 1-109.
15. ——, 1936. The Aristaeinae, Solenocerinae, and Pelagic Penaeinae of the Bingham Oceanographic Collection: Materials for a Revision of the Oceanic Penaeinae, *Bull. Bing. Ocean. Coll.* **5**, 1-151.
16. ——, 1939. Further Observations on Penaeidae of the Northern Gulf of Mexico, *Bull. Bing. Ocean. Coll.*, **6**, 1-62.
17. ——, 1963a. Comments on the Petition Concerning Penaeid Names (Crustacea, Decapoda), *Bull. Zool. Nomencl.*, **20** (2), 169-174.

18. BURKENROAD, M. D., 1963b. Comments on the Petition Concerning Penaeid Names, *Bull. Zool. Nomencl.*, **20** (4), 247-248.

19. CALMAN, W. T., 1909. Appendiculata. Fasc. 3: Crustacea, Pt. 6 in *A Treatise on Zoology* edited by Ray Lankester. Adam. and C. B. Black: London.

20. CHEUNG, T. S., 1960. A Key to the Identification of Hong Kong Penaeid Prawns with Comment on Points of Systematic Interest, *Jour. Hong Kong Univ. Fish.*, **3**, 61-69.

21. CHIN, E., and D. M. ALLEN, 1959. A List of References on the Biology of Shrimp (Family Penaeidae), *U. S. Fish. Wildl. Serv., Sp. Sci. Rept., Fish.*, **276**, 1-143.

22. CLIFFORD, D. M., 1956. Marketing and Utilization of Shrimp in the United States, *Proc. I. P. F. C.*, **6** (III), 438-443.

23. COOK, H. L., 1965. A Generic Key to the Protozoan, Mysis, and Post-larval Stages of the Littoral Penaeidae of the Northwestern Gulf of Mexico, *U. S. Fish. Wildl. Serv., Fish. Bull.*, **62** (2), 437-447.

24. ——, and M. A. MURPHY, 1965. Early Developmental Stages of the Rock Shrimp, *Sicyonia brevirostris* Stimpson, Reared in the Laboratory, *Tulane St. Zool.*, **12**, 109-127.

25. COSTELLO, T. J., 1964. Field Techniques for Staining-recapture Experiments with Commercial Shrimp, *U. S. Fish. Wildl. Serv., Sp. Sci. Rept.*, **484**, 1-13.

26. CUMMINGS, W. C., 1961. Maturation and Spawning of the Pink Shrimp, *Penaeus duorarum* Burkenroad, *Trans. Amer. Fish. Soc.*, **90**, 462-268.

27. DAKIN, W. J., 1946. Life History of a Species of *Metapenaeus* in Australian Coastal Lakes, *Nature*, 158, 4003, 99.

28. DALL, W., 1957. A Revision of the Australian Species of Penaeinae (Crustacea, Decapoda, Penaeidae), *Aust. Jour. Mar. Freshw. Res.*, **8**, 135-231.

29. ——, 1958. Observation on the Biology of the Green Tail Prawn, *Metapenaeus mastersii* (Haswell) (Crustacea, Decapoda, Penaeidae), *Aust. Jour. Mar. Freshw. Res.*, **9**, 111-134.

30. ——, 1964. Studies on the Physiology of a Shrimp, *Metapenaeus mastersii* (Crustacea, Decapoda, Penaeidae): I. Blood Constituents, *Aust. Jour. Mar. Freshw. Res.*, **15**, 145-161.

31. ——, 1965a. Studies on the Physiology of a Shrimp, *Metapenaeus* sp. (Crustacea, Decapoda, Penaeidae): II. Endocrine and Control of Molting, *Aust. Jour. Mar. Freshw. Res.* **16**, 1-12.

32. ——, 1965b. Studies on the Physiology of a Shrimp, *Metapenaeus* sp. (Crustacea, Decapoda, Penaeidae): III. Composition and Structure of the Integument, *Aust. Jour. Mar. Fresh. Res.*, **16**, 13-23.

33. ——, 1965c. Studies on the Physiology of a Shrimp, *Metapenaeus* sp. (Crustacea, Decapoda, Penaeidae): IV. Carbohydrate Metabolism, *Aust. Jour. Mar. Freshw. Res.*, **16**, 163-180.

34. ——, 1965d. Studies on the Physiology of a Shrimp, *Metapenaeus* sp. (Crustacea, Decapoda, Penaeidae): V. Calcium Metabolism, *Aust. Jour. Mar. Freshw. Res.*, **16**, 181-203.

35. DAWSON, C. E., 1957. Studies on the Marking of Commercial Shrimp with Biological Stains, *U. S. Fish. Wildl. Serv., Sp. Sci. Rept., Fish.*, **231**, 1-24.

36. DJAJADIREDJA, R. R., and M. SACHLAN, 1956. Shrimp and Prawn Fisheries in

Indonesia with Special Reference to Kroya District, *Proc. I. P. F. C.*, **6** (III), 366-377.

37. DOBKIN, S., 1961. Early Developmental Stages of Pink Shrimp, *Penaeus duorarum*, from Florida Waters, *U. S. Fish Wildl. Serv., Fish. Bull.*, **61** (190), 321-349.

38. EDMONDSON, C. H., 1946. Crustacea in Reef and Shore Fauna of Hawaii, *Bernice P. Bishop Museum, Sp., Publ.*, **22**, 1-381.

39. EGUSA, S., 1961. Studies on the Respiration of the "Kuruma" Prawn, *Penaeus japonicus* Bate: II. Preliminary Experiments on Its Oxygen Consumption, *Bull. Jap. Soc. Sci. Fish.*, **27**, 650-659.

40. ———, and T. YAMAMOTO, 1961. Studies on the Respiration of the "Kuruma" Prawn, *Penaeus japonicus* Bate: I. Burrowing Behavior, with Special Reference to Its Relation to Environmental Oxygen Concentration, *Bull. Jap. Soc. Sci. Fish.*, **27**, 22-26.

41. ELDERD, B. R., M. INGLE, K. D. WOODBURN, R. F. HUTTON, and H. JONES, 1961. Biological Observations on the Commercial Shrimp, *Penaeus duorarum* Burkenroad, in Florida Waters, *Fla. St. Bd. Conserv., Prof. Ser.*, **3**, 1-139.

42. ENAMI, J., 1957. *On the Neura Secretion*. Kyodo Medical Book Publication: Tokyo. 208 p.

43. FABRICIUS, J. C., 1798. *Supplementum Entomologiae Systematicae (Hafniae)*. Copenhagen, Denmark.

44. FISHER, C. E., 1966. Distribution and Abundance of Shrimp Larvae, *U. S. Fish. Wildl. Ser., Cir.*, **246**, 10-12.

45. Food and Agriculture Organization, 1964. Report to the Governments of Brazil, Uruguay, and Argentina on Investigation and Assessment of Shrimp Resources, Based on the Work of M. N. Mistakidis, Rep. FAO, EPTA, 1934, 1-52.

46. FUNADA, K., 1966. Miyazu-Wan no taisei shiryo seibutsu bunpu chosa (Survey on the Distribution of Biological Organisms of Bottom-dwelling Types in Miyazu Bay), *Kyoto Suishi Gyoseki (Kyoto Fisheries Research Report)*, **27**, 81-108.

47. FUSS, C. M. JR., 1964. Observation on Burrowing Behavior of the Pink Shrimp, *Penaeus duorarum* Burkenroad, *Bull. Mar. Sci. Gulf Carib.*, **14**, 62-73.

48. ———, and L. H. OGREN, 1966. Factors Affecting Activity and Burrowing Habits of the Pink Shrimp, *Penaeus duorarum* Burkenroad, *Biol. Bull.*, **130**, 170-191.

49. GABE, M., 1953. Sur l'existence chez quelques Crustacés malacostracés d'un organ comparable a la glande de la mue des insectes, *C. R. Acad. Sci., Paris*, **237**, 1111-1113.

50. ———, 1956. Histologie comparée de la glande de mue (organ Y) des Crustacés malacostracés, *Ann. Sci. Nat.* (b) (11), **18**, 145-152.

51. GEORGE, M. J., 1961. Studies on the Prawn Fishery of Cochin and Alleppey Coast, *Indian Jour. Fish.*, **8**, 75-95.

52. ———, 1962. On the Breeding of Penaeids and the Recruitment of Their Post-larvae into the Backwaters of Cochin, *Indian Jour. Fish.*, **9**, 110-116.

53. GORDON, I., 1955. Importance of Larval Characters in Classification, *Nature*, **176**, 911-912.

54. GUNTER, 1950. [sic] Seasonal Population Changes and Distributions as Related to Salinity, of Certain Invertebrates of the Texas Coast, Including the

Commercial Shrimp, *Publ. Inst. Mar. Sci. Univ. Texas,* **1** (2), 7-51.

55. GUNTER, 1956. Some Relations of Faunal Distributions to Salinity in Estuarine Waters, *Ecology,* **37** (3), 616-619.

56. ——, 1957. Misuse of Generic Names of Shrimp (Family Penaeidae), *Syst. Zool.,* **6** (2), 98-100.

57. ——, 1961a. Some Relations of Estuarine Organisms to Salinity, *Limnol. Oceanogr.,* **6** (2), 182-190.

58. ——, 1961b. Habitat of Juvenile Shrimp (Family Penaeidae), *Ecology,* **42,** 598-600.

59. ——, 1962. Specific Names of the Atlantic American White Shrimp, *Gulf Res. Repts.,* **3,** 107-114, 118-121.

60. ——, and H. H. HILDEBRAND, 1954. The Relation of Total Rainfall in the State and Catch of Marine Shrimp (*Penaeus setiferus*) in Texas Waters, *Bull. Mar. Sci. Gulf Carib.,* **4,** 95-103.

61. ——, J. Y. CHRISTMAS, and R. KILLEBREW, 1964. Some Relations of Salinity to Population Distributions of Motile Estuarine Organisms, with Special Reference to Penaeid Shrimp, *Ecology,* **45,** 181-185.

62. Gurney, R., 1926. The Protozoal Stage in Decapod Development, *Ann. Mag. Nat. Hist.,* (9) **18,** 19-27.

63. ——, 1942. *Larvae of Decapod Crustacea.* Ray Society: London.

64. ——, 1943. The Larval Development of Two Penaeid Prawns from Bermuda of the Genera *Sicyonia* and *Penaeopsis, Proc. Zool. Soc. London,* Ser. B, **113,** 1-16.

65. HALL, D. N. F., 1956. The Validity of the Generic Names of Penaeid Prawns. *Proc. I. P. F. C.,* **6** (III), 450.

66. ——, 1962. Observations on the Taxonomy and Biology of Some Indo-West Pacific Penaeidae (Crustacea, Decapoda), *Fish. Publ. Colonial Off., London,* **17,** 1-229.

67. HEEGAAD, P. E., 1953. Observations on Spawning and Larval History of the Shrimp, *Penaeus setiferus* (L.), *Publ. Inst. Mar. Sci., Univ. Texas,* **3,** 73-105.

68. HELDT, J. H., 1938. La reproduction chez les Crustacés Décapodes de la famille des Pénéides, *Ann. Inst. Oceanogr. Monaco,* **18** (Facs. 2), 31-206.

69. HILDEBRAND, H. H., 1954. A Study of the Fauna of the Prawn Shrimp (*Penaeus aztecus* Ives) Grounds in the Western Gulf of Mexico, *Publ. Inst. Mar. Sci., Univ. Texas,* **3,** 233-366.

70. ——, 1955. A Study of the Fauna of the Pink Shrimp (*Penaeus duorarum* Burkenroad) Grounds in the Gulf of Campeche, *Publ. Inst. Mar. Sci., Univ. Texas,* **4,** 169-232.

71. ——, and G. GUNTER, 1953. Correlation of Rainfall with the Texas Catch of White Shrimp, *Penaeus setiferus* (Linnaeus), *Trans. Amer. Fish. Soc.,* **82,** 151-155.

72. HIRAIWA, Y. K., 1933. Studies on a Bopyrid, *Epipenaeon japonica* Thielemann: I. Morphological Studies in Both Sexes, *Jour. Sci. Hiroshima Univ.,* B, 1, **2,** 49-70.

73. ——, 1934. Studies on a Bopyrid, *Epipenaeon japonica* Thielemann: II. Reproductive and Excretory Organs, *Jour. Sci. Hiroshima, Univ.,* B, 1, **3,** 45-63.

74. ——, 1936. Studies on a Bopyrid, *Epipenaeon japonica* Thielemann: III. Development and Life Cycle, with Special Reference to the Sex Differentiation in

the Bopyrid, *Jour. Sci. Hiroshima Univ.*, B, 1, **4**, 101-141.

75. HIRAIWA, Y. K. and M. SATO, 1939. On the Effect of Parasitic Isopoda on a Prawn, *Penaeopsis akayebi* Rathbun, with a Consideration of the Effect of Parasitization on Higher Crustacea in General, *Jour. Sci. Hiroshima Univ.*, B (Zool.), **7**, 105-124.

76. HIRAMATSU, T., K. SEGAWA, N. TAGO, and I. ARIMA, 1962. Toyomaekai ni okeru ebi somono kaiikibetsu gyokaku tokusei to juyoshu no gyokaku keiko ni tsuite (On the Catch of Special Types and Important Varieties of Prawns in Togomaekai), *Fukuoka Toyomae Suisan Shiken Gyoho*, 1961, 1-48.

77. Hiroshima Bunridai and Takashi Seibutsu gakkaihen, 1949. Anatomical Illustration of Animals, Nihon Shuppansha, Osaka.

78. HOESE, H. H. D., 1960. Juvenile Penaeid Shrimp in the Shallow Gulf of Mexico, *Ecology*, **41**, 582-593.

79. HOLTHIUS, L. B., 1949. The Identity of *Penaeus monodon* Fabr., *Proc. K. Ned. Akad. Wetenschappen*, **52**, 1051-1057.

80. ——, 1962. On the Names of *Penaeus setiferus* (L.) and *Penaeus schmitti* Burkenroad, *Gulf Res. Repts.*, **3**, 115-118.

81. ——, 1962b. Penaeid Generic Names (Crustacea, Decapoda), *Bull. Zool. Nomencl.*, **19** (2), 103-114.

82. ——, 1963. Comments on the Petition Concerning Penaeid Names, *Bull. Zool. Nomencl.*, **20** (4), 245-247.

83. HUDINAGA, M., 1935. Kuruma-ebi no hassei (Development of *Penaeus japonicus*), *Sonai Suisan Kenkyuho (Sonai Fisheries Research Report)*, **1**, 1-15.

84. ——, 1941. Yoshi-ebi to moebi no no upuriusu *(Nauplius of Metapenaeus monoceros* and *M. burkenroadi)*, *Suikaiho*, **8**, 282-287.

85. HUDINAGA, M., 1942. Reproduction, Development, and Rearing of *Penaeus japonicus* Bate, *Jap. Jour. Zool.*, **10**, 305-393.

86. HUTTON, R. F., F. SOGANDARES-BERNAL, B. ELDRED, R. M. INGLE, and K. D. WOODBURN, 1959. Investigations on the Parasites and Diseases of Salt-water Shrimps (Penaeidae) of Sports and Commercial Importance to Florida, *Fla. St. Bd. Cons., Tech. Ser.*, **26**, 1-38.

87. IDYLL, C. P., and A. C. JONES, 1964. Abundance and Distribution of Pink Shrimp Larvae and Post-larvae in Southwestern Florida Waters, *U. S. Fish. Wildl. Serv., Cir.*, **230**, 25-27.

88. ——, and J. L. MUNRO, 1965. Abundance and Distribution of Pink Shrimp Larvae of the Tortugas Shelf of Florida, *U. S. Fish. Wildl. Serv., Cir.*, **246**, 19.

89. IKEDA, I., 1962. Kokai ni okeru koraiebi no gyokyo ni tsuite (On Fishing Conditions of *P. orientalis* in the Yellow Sea), *Saikai Suisan Kenpo*, **27**, 1-24.

90. IKEMATSU, W., 1955. Ariake-Kai san shibaebi no seikatsushi ni tsuite (On the Life History of *Metapenaeus joyneri* of Ariake-Kai), *Nissuikaishi (Bulletin of the Japanese Society of Scientific Fisheries)*, **20**, 969-978.

91. ——, 1959. Ariake-kaisan yoshiebi no seikatsushi ni tsuite (On the Life History of *Metapenaeus monoceros* of Ariake-Kai), *Saikai Fisheries Research Center, Ariake-Kai Kenpo*, **5**, 19-29.

92. ——, 1963. Ariake-Kai ni okeru ebi, Amirui no seikatsushi, Seitai ni kansuru kenkyu (On the Life History and Ecology of Prawns of Ariake-Kai), *Saikai*

Suikenpo, **30**, 1-124.

93. IVERSON, E. S., and C. P. IDYLL, 1960. Aspects of the Biology of the Tortugas Pink Shrimp, *Penaeus duorarum*, *Trans. Amer. Fish. Soc.*, **89**, 1-8.

94. JOHNSON, M. C., and J. R. FIELDING, 1956. Propagation of the White Shrimp, *Penaeus setiferus* (Linn.) in Captivity, *Tulane St. Zool.*, **4**, 175-190.

95. JONES, A. C., D. DIMITRIOU, and J. EWALD, 1963. Abundance and Distribution of Pink Shrimp Larvae on the Tortugas Shelf of Florida, *U. S. Fish. Wildl. Serv., Cir.*, **183**, 86-88.

96. JOUVERT, L. S., 1966. A Preliminary Report on the Penaeid Prawns of Durban Bay, *Oceanogr. Res. Inst. S. Afr., Invest. Rept.*, **11**, 1-32.

97. KAJISHIMA, K., 1951. Kuruma-ebi ran no bunrikakkyu no hassei (Embryology of *Penaeus japonicus*), *Dozatsu (Journal of Zoology)*, **60**, 258-262.

98. KAJIYAMA, E., 1935a. Kuruma-ebi no sanran narabini hatsuiku chosa (Spawning and Embryonic Development of *Penaeus japonicus*), *Hiroshima Suishiho (Bulletin of Hiroshima Fisheries Experiment Station)*, **12**, 134-159.

99. ——, 1935b. Kuruma-ebi no seicho ni tsuite (On the Growth of *Penaeus japonicus*), *Hiroshima Kensuisan Kaiho (Bulletin of the Hiroshima Prefectural Association of Fisheries)*, **13**, 20-24.

100. KASAHARA, N., 1948. Shina tokai kokai no sokobikiami gyogyo to sono shigen (Bottom Trawl Fishing and the Resources in the Yellow Sea), *Nissui Kenkyushoho*, **3**, 1-194.

101. KING, J. E., 1948. A Study of the Reproductive Organ of the Common Marine Shrimp, *Penaeus setiferus* (Linnaeus), *Biol. Bull.*, **94**, 244-262.

102. KISHIGAMI, K., 1900. Honposan kuruma-ebi-zoku chosa (Study on *Penaeus japonicus*), *Suichoho*, **8**, 1-34.

103. ——, 1917. Korai-ebi (On), *Suikaiho*, **2**, 79.

104. KLIMA, E. F., 1965. Evaluation of Biological Stains, Inks, and Fluorescent Pigments as Marks for Shrimp, *U. S. Fish. Wildl. Serv. Sp. Sci. Rept., Fish.*, **511**, 1-8.

105. KNOWLES, F. G. W., 1953. Endocrine Activity in the Crustacean Nervous System, *Proc. Roy. Soc. London*, B, **141**, 248-267.

106. ——, and D. B. CARLISLE, 1956. Endocrine Control in the Crustacea, *Biol. Rev.*, **31**, 398-473.

107. KOW, THAM AH, 1955. The Shrimp Industry of Singapore, *Proc. I. P. F. C.* 5 (II), 145-155.

108. KRUSE, D. N., 1959. Parasites of the Commercial Shrimp, *Penaeus aztecus* Ives., *P. duorarum* Burkenroad, and *P. setiferus* (Linnaeus), *Tulane St. Zool.*, **7**, 123-144.

109. KUBO, I., 1949. Studies on the Penaeids of Japanese and Adjacent Waters, *Jour. Tokyo Coll. Fish.*, **36**, 1-467.

110. ——, 1954. Systematic Studies on the Japanese Macrurous Decapod Crustacea: 2. On Two Penaeids, *Metapenaeus affinis* (H. Milne-Edwards) and *M. burkenroadi*, nom. nov., Erected on the Japanese Form Known as *M. affinis*, *Jour. Tokyo Univ. Fish.*, **41**, 89-93.

111. ——, 1956. A Review of the Biology and Systematics of Shrimps and Prawns of Japan, *Proc. I. P. F. C.*, 6 (III), 387-398.

112. KUBO, I., and E. ASADA, 1957. A Quantitative Study of Crustacean Bottom Epifauna of Tokyo Bay, *Jour. Tokyo Univ. Fish.*, **43**, 249-289.

113. KUTKUHN, J. H., 1962. Gulf of Mexico Commercial Shrimp Populations—Trends and Characteristics, 1956-1959, *U. S. Fish. Wildl. Serv., Bull.*, **62**, (212), 343-402.

114. LINDNER, M. J., 1957. Survey of Shrimp Fisheries of Central and South America, *U. S. Fish. Wildl. Serv., Sp. Sci. Rept.*, **235**, 1-166.

115. ——, and W. W. ANDERSON, 1956. Growth, Migration, Spawning, and Size Distribution of Shrimp, *Penaeus setiferus, U. S. Fish. Wildl. Serv. Fish. Bull.*, **56** (106), 555-645.

116. LOESCH, H., 1965. Distribution and Growth of Penaeid Shrimp in Mobile Bay, Alabama, *Publ. Inst. Mar. Sci. Univ. Texas*, **10**, 41-58.

117. MAEKAWA, K., 1961. Setonaikai, tokuni yamaguchi-ken enkai ni okeru gyogyo no chosei-kanri to shigenbaiyo ni kansuru kenkyu (On the Study of the Control of Fishing Operations and the Culture of Resources in the Coastal Regions of Yamaguchi Prefecture in Seto Inland Sea), *Yamaguchi-ken Naikai Suishi Choken, Gyoseki*, **11**, 1-483.

118. ——, and Y. OTSUKA, 1955. Kuruma ebi no sanso shohiryo ni kansuru kenkyu-I. Kuruma-ebi no sanso shohiryo o motomeru ichi, ni no kokoromi (On the Oxygen Consumption of *Penaeus japonicus* Bate: I. Some Experiments to Determine the Amount of Oxygen Consumption in *Penaeus japonicus*), *Yamaguchi-ken Naikai Suishi Choken, Gyoseki*, **7**, 33-38.

119. MAKI, S., and H. TSUCHIYA, 1933. Taiwansan jukyakurui zusetsu (On the Decapoda of Taiwan), *Taiwan Chuoken, Noho (Report from Taiwan Central Research of Agriculture)*, **3**, 1-215.

120. MAYNARD, D. M., 1960. Circulation and Heart Function in *The Physiology of Crustacea*, Vol. I, edited by T. H. Waterman, New York: Academic Press.

121. McFARLAND, W. N., and B. D. LEE, 1963. Osmotic and Ionic Concentrations of Penaeidean Shrimps of the Texas Coast, *Bull. Mar. Sci. Gulf and Carib.*, **13**, 391-417.

122. McRAE, E. D. JR., 1952. Progress Report on the Shrimp Investigation, Ann. Rept. Mar. Lab. Texas Game Fish. Comm., Sept. 1, 1950 to August 31, 1951.

123. MENON, M. K., 1951. The Life History and Bionomics of an Indian Penaeid Prawn, *Metapenaeus dobsoni, Proc. I. P. F. C.*, **3** (II), 80-93.

124. ——, 1954. On the Paddy Field Prawn Fishery of Travancore-Cochin and an Experiment in Prawn Culture, *Proc. I. P. F. C.*, **5** (II), 131-135.

125. ——, and K. RAMAN, 1961. Observations on the Prawn Fishery of the Cochin Backwaters with Special Reference to the Stake Net Catches, *Indian Jour. Fish.*, **8**, 1-23.

126. MENZEL, R. W., 1955. Marking of Shrimp, *Science*, **121**, 446.

127. MISTAKIDIS, M. N., and G. DE S. NEIVA, 1964. Occurrence of Two Penaeid Shrimps, *Artemesia ionginaris* Bate and *Hymenopenaeus mulleri* (Bate), and of Some Lesser Known Shrimps in Coastal Waters of South America, *Nature*, **202** (4391), 471-472.

128. MIURA, G., and M. YAMAGUCHI, 1955. Kuruma-ebi no sachu sennya kodo,

sono ta ni tsuite no 2, 3 no kansatsu (Sand Burrowing Habit of *Penaeus japoni-cus* Bate), *Suisan Zoshoku (Aquaculture)*, **2**, 20-26.

129. Eiko Fisheries Experiment Station, 1938. Survey on the Important Fisheries Groups, Report from Manshu Fisheries Experiment Station, **2**, 1-20.

130. MORRIS, M. C., and I. BENNETT, 1952. The Life History of a Penaeid Prawn *(Metapenaeus)* Breeding in a Coastal Lake (Tuggerah, New South Wales), *Proc. Linn. Soc. N. S. W.*, **76**, 164-182.

131. MÜLLER, F., 1863. Die Verwandlung der Garneelen, *Arch. Naturgesch.*, **29**, 8-23.

132. NAGATA, M., M. TANIZAKI, and K. NAKAZAWA, 1931. Kumamotoken san kuruma-ebi ni kansuru chosa kenkyu (Dai-ippo). [On the Study of *Penaeus japonicus* of Kumamoto Prefecture (Report No. 1)], *Kumamoto Suishi (Bulletin of Kumamoto Fisheries Experiment Station)*, 28.

133. NISHIMURA, S., 1939. Hokkaido oyobi Kitachishima kinkai san ebi kanirui (On Prawns and Crabs of Hokkaido and Kitachishima), *Suikenshi*, **34**, 382-385.

134. OKA, M., 1967. Korai-ebi no zoyoshoku ni kansuru kenkyu: II. Shubyo seisan ni tsuite (On the Culture of *P. orientalis*: II. Study on the Production of Seed-lings). *Suisan Zoshoku (Aquaculture)*, **15** (2), 7-32.

135. ——, and SHIRAHATA, S., 1964. Korai-ebi no kenkyu: I. Seishoku Kiko ni kansuru kenkyu (On the Study of *P. orientalis*: I. Study on the Reproduction of *P. orientalis*), *Nagasakidai Suikenpo (Bulletin of the Fisheries Department of Nagasaki University)*, **17**, 55-67.

136. ——, ——, 1965. Korai-ebi no kenkyu: II. ransoran no keitaiteki bunrui to ranso seijukudo ni tsuite (On *P. orientalis*: II. Study on the Morphological Aspect of Ova and Maturation of Ovaries), *Nagasakidai Suikenpo (Bulletin of Fisheries Department of Nagasaki University)*, **18**, 30-40.

137. OSHIMA, Y., and D. YASUDA, 1932. Moebi no seitai ni tsuite [On the Ecology of *Penaeopsis affinis* (Milne-Edwards)], *Nissuikaishi (Bulletin of the Japanese Society of Scientific Fisheries)*, **11**, 135-145.

138. OTA, S., 1949. Nakanoumi shinji-ko san yoshiebi seitai chosa (Ecological Study of *Metapenaeus monoceros* of Shinji-ko), *Suisancho, Chosa Shiryo*, **18**, 1-18.

139. ONIZUKA, M., 1914. Kuruma-ebi no shusei narabi ni chikuyoho (Culture of *Penaeus japonicus*), *Suikenshi*, 9, 509-540.

140. PANIKKAR, N. K., 1951. Physiological Aspects of Adaptation to Estuarine Condi-tions, *Proc. I. P. F. C.*, **2** (III), 168-175.

141. ——, and M. K. MENON, 1956. Prawn Fisheries in India, *Proc. I. P. F. C.*, **6** (III), 328-344.

142. ——, and R. VISWANATHAN, 1948. Active Regulation of Chloride in *Meta-penaeus monoceros* Fabricius, *Nature*, **161** (4082), 137-138.

143. PEARSON, J. C., 1939. The Early Life Histories of Some American Penaeidae, Chiefly the Commercial Shrimp, *Penaeus setiferus* (Linn.), *U. S. Bur. Fish. Wash., Bull.*, **49**, 1-73.

144. QURESHI, M. R., 1956. Shrimp Fisheries of Pakistan, *Proc. I. P. F. C.*, **6** (III), 359-362.

145. RACEK, A. A., 1955. Littoral Penaeinae from New South Wales and Adjacent Queensland Waters, *Aust. Jour. Mar. Freshw. Res.*, **6**, 209-240.

146. RACEK, A. A., 1956. Penaeid Prawn Fisheries of Australia with Special Reference to New South Wales, *Proc. I. P. F. C.*, **6** (III), 347-359.

147. ——, 1959. Prawn Investigations in Eastern Australia, *Res. Bull.* **6**, *State Fish., Chief Sec. Dept., N. S. W.*

148. ——, and DALL, W., 1965. Littoral Penaeinae (Crustacea Decapoda) from Northern Australia, New Guinea, and Adjacent Waters, *Verh. Akad. Wet. Amst.*, (b) **56**, 1-116.

149. RAJYALAKSHMI, T., 1961. Observations on the Biology and Fishery of *Metapenaeus brevicornis* (M.-Edw.) in the Hooghly Estuarine System, *Indian Jour. Fish.*, **8**, 383-402.

150. RAO, K. P., 1958. Oxygen Consumption as a Function of Size and Salinity in *Metapenaeus monoceros* Fab. from the Marine and Brackish Water Environments, *Jour. Exp. Biol.*, **35**, 307-313.

151. RENFRO, W. C., 1963. Life History Stages of Gulf of Mexico Prawn Shrimp, *U. S. Fish. Wildl. Serv., Cir.*, **183**, 94-98.

152. ——, and H. L. COOK, 1963. Early Larval Stages of the Seabob, *Xyphopenaeus kroyeri* (Heller), *U. S. Fish. Wildl. Serv., Fish. Bull.*, **63**, 165-177.

153. ROBERTSON, J. D., 1960. Osmotic and Ionic Regulation in *The Physiology of Crustacea*, Vol. I, edited by T. H. Waterman. New York: Academic Press.

154. SATOUCHI, S. 1937. Takai kokai teigyo gyojo no saininshiki (On Bottom Trawl Fishing Operations in the Yellow Sea) *Karatsu*, 142.

155. SHAIKHMAHMUD, F. S., and V. B. TEMBE, 1960. Study of Bombay Prawn, *Indian Jour. Fish.*, **7**, 69-81.

156. SPRINGER, S., and H. R. BULLIS, 1954. Exploratory Shrimp Fishing in the Gulf of Mexico: Summary Report for 1952-1953, *U. S. Fish. Wildl. Serv., Comm. Fish. Rev.*, **16** (10), 1-16.

157. SUBRAHMANIAM, M., 1966. Fluctuations in Prawn Landings in the Godavari Estuarine System, *Proc. I. P. F. C.*, **11** (II), 44-51.

158. TEMPLE, R. F., 1962. Larval Distribution and Abundance, *U. S. Fish. Wildl. Serv., Cir.*, **161**, 18-22.

159. ——, and C. C. FISCHER, 1965. Vertical Distributions of the Planktonic Stages of Penaeid Shrimps, *Publ. Inst. Mar. Sci., Univ. Texas*, **10**, 59-67.

160. ——, ——, and D. L. HARRINGTON, 1963. Larval Distribution and Abundance, *U. S. Fish. Wildl. Serv. Cir.*, **183**, 18-24.

161. THOMSON, J. M., 1956. Fluctuations in Australian Prawn Catch, *Proc. I. P. F. C.*, **6** (III), 444-447.

162. TIEWS, K. F. W., 1962. A Report to the Government of the Philippines on Marine Fishery Resources, *Phil. Jour. Fish.*, **6** (2), 107-216.

163. U. S. Fish. Wildl. Serv., 1958a. Foreign Shrimp Fisheries other than Central and South America, *Sp. Sci. Rept. Fish.*, **254**, 1-71.

164. ——, 1958b. Survey of the United States Shrimp Fishery, Vol. 1, *Sp. Sci. Rept., Fish.*, **254**, 1-311.

165. UTSUNOMIYA, T., 1959. Suo-nada no 4 gyojo ni okeru ebirui no sosei, bunpu nado no henia ni tsuite (On the Composition, Distribution, and Fluctuations of Prawns in Four Fishing Grounds in Suo-nada), *Yamaguchi Naikai Suishi Choken Gyoseki*, **10**, 39-46.

166. Utsunomiya, T., and Maekawa, K., 1959. Yamaguchi ken Seto nai kai ni okeru juyoseibutsu no seitaigakuteki kenkyu, dai-18, ho, akaebi [Ecological Study of the Important Organisms in Seto Inland Sea of the Yamaguchi Prefecture: Report 18 on *Metapenaeopsis barbates* (De Haan)], *Yamaguchi Naikai Suishi Choken Gyoseki*, **10**, 81-92.

167. ——, K. Yatsuyanagi, A. Tomiyama, and K. Maekawa, 1954. Naiwan kisui kuiki ni okeru shutsugen gyorui to sono shokusei ni tsuite (On the Fishes and Their Food Habits in Fishing Regions Inside a Bay), *Yamaguchi Naikai Suishi Choken Gyoseki*, **6**, 11-24.

168. Urita, T., 1921. Kagoshima-ken ni sansuru ebirui oyobi sono bunpu ni tsuite (On the Types and Distribution of Prawns in Kagoshima Prefecture), *Dozatsu (Journal of Zoology)*, (393), 214-220.

169. ——, 1926. Shinaaoshima san koraiebi ni tsuite (On *P. orientalis* of Aoshima), *Dozatsu (Journal of Zoology)*, **38** (457), 361-368.

170. Voss, G. L., 1955. Key to the Commercial and Potentially Commercial Shrimps of the Family Penaeidae of the Western North Atlantic and the Gulf of Mexico, *Fla. St. Bd. Cons., Tech. Ser.*, **14**, 1-23.

171. Waterman, T. H., and F. A. Chace Jr., 1960. General Crustacean Biology in *The Physiology of Crustacea*, Vol. I, edited by T. H. Waterman. New York: Academic Press.

172. Welsh, J. H., 1961. Neurohumors and Neurosecretion in *The Physiology of Crustacea*, Vol. II, edited by T. H. Waterman. New York: Academic Press.

173. Williams, A. B., 1953. Identification of Juvenile Shrimp (Penaeidae) in North Carolina, *Jour. Elisha Mitchell Sci. Soc.*, **69**, 156-160.

174. ——, 1955a. A Contribution to the Life Histories of Commercial Shrimp (Penaeidae) in North Carolina, *Bull. Mar. Sci. Gulf Carib.*, **5**, 116-146.

175. ——, 1955b. A Survey of North Carolina Shrimp Nursery Grounds, *Jour. Elisha Mitchell Sci. Soc.*, **71**, 200-207.

176. ——, 1958. Substrates as a Factor in Shrimp Distribution, *Limnol. Oceanogr.*, **3**, 283-290.

177. ——, 1959. Spotted and Brown Shrimp Post-larvae (Penaeus) in North Carolina, *Bull. Mar. Sci. Gulf Carib.*, **9**, 281-290.

178. ——, 1960. The Influence of Temperature on Osmotic Regulation in Two Species of Estuarine Shrimp (Penaeus), *Biol. Bull.*, **119**, 560-571.

179. ——, 1965. Marine Decapod Crustaceans of the Carolinas, *U. S. Fish. Wildl. Serv., Fish. Bull.*, **65** (1), 1-298.

180. Yatsuyanagi, K., and K. Maekawa, 1954. Yamaguchi ken seto naikai ni okeru juyoseibutsu no seitaigakuteki kenkyu, dai-7-ho, setonaikai san shibaebi no seitaigaku teki kenkyu (Ecological Aspects of Some Important Organisms in the Seto Inland Sea of the Yamaguchi Prefecture: Report No. 7. Ecological Aspect of *M. joyneri* of the Seto Inland Sea), *Yamaguchi Naikai Suishi Choken Gyoseki*, **6**, 1-9.

181. ——, ——, 1955. Yamaguchi ken seto naikai ni okeru juyoseibutsu no seitaigakuteki kenkyu, dai-8-ho, setonaikai san kuruma-ebi no seitai (Ecological Aspects of Some Important Organisms in the Seto Inland Sea of the Yamaguchi Prefecture: Report No. 8. Ecology of *P. japonicus* of the Seto Inland

Sea), *Yamaguchi Naikai Suishi Choken Gyoseki*, **7**, 1-15.

182. YATSUYANAGI, K., and K. MAEKAWA, 1956a. Yamaguchi ken seto naikai ni okeru juyoseibutsu no seitaigakuteki kenkyu, dai-10-ho, kumaebi no seitai (On the Ecological Aspects of Some Important Organisms in the Seto Inland Sea of the Yamaguchi Prefecture: Report No. 10. Ecology of *P. semisulcatus*), *Yamaguchi Naikai Suishi Choken Gyoseki*, **8**, 25-38.

183. ——, ——, 1956b. Yamaguchi ken seto naikai ni okeru juyoseibutsu no seitaigakuteki kenkyu, dai-11-ho, yoshiebi no seitai (On the Ecological Aspects of Some Important Organisms in the Seto Inland Sea of the Yamaguchi Prefecture: Report No. 11. Ecology of *M. monoceros*), *Yamaguchi Naikai Suishi Choken Gyoseki*, **8**, 39-51.

184. ——, ——, 1957a. Yamaguchi ken seto naikai ni okeru juyoseibutsu no seitaigakuteki kenkyu, dai-15-ho, kishiebi no seitai (On the Ecological Aspects of Some Important Organisms in the Seto Inland Sea of the Yamaguchi Prefecture: Report No. 15. Ecology of *M. dalei*), *Yamaguchi Naikai Suishi Choken Gyoseki*, **9**, 13-20.

185. ——, ——, 1957b. Yamaguchi ken seto naikai ni okeru juyoseibutsu no seitaigakuteki kenkyu, dai-16-ho, moebi no seitai (On the Ecological Aspects of Some Important Organisms in the Seto Inland Sea of the Yamaguchi Prefecture: Report No. 16. Ecology of *M. burkenroadi*), *Yamaguchi Naikai Suishi Choken Gyoseki*, **9,** 21-28.

186. ——, and K. MATSUKIYO, 1951. Shuki oyobi Shunki ni okeru akaebi, toraebi, saruebi no okisa, seihi oyobi tokyo kocho to taicho to no kankei (Relationships among the Size, Sex, and Length of the Carapace in Akaeki, Toro-ebi, and Saru-ebi in Autumn and Spring), *Nisuikaishi (Bulletin of the Japanese Society of Scientific Fisheries)*, **16**, 182-183.

187. YASUDA, J., 1949. Saruebi no seitai ni kansuru ni, san ni tsuite [Some Aspects of Ecology of *Trachypenaeus curvirostris* (Stimpson)], *Nissuikaishi (Bulletin of the Japanese Society of Scientific Fisheries, Zoology)*, **15**, 180-189.

188. ——, 1956. Shrimps of the Seto Inland Sea of Japan, *Proc. I. P. F. C.*, **6** (III), 378-386.

189. ——, 1956. Naiwan ni okeru ebirui no shigen seibut-sugakuteki kenkyu (II) Kakuron, kakushurui no seitai ni kansuru kenkyu [On the Various Theories of Biological Resources Pertaining to Prawns in Bays (II) with Special Reference to the Ecological Aspects of Various Types], *Naikaisuikenpo*, **9**, 1-81.

190. ——, 1958. Naiwan ni okeru ebirui no shigen seibitsugakuteki kenkyu (On the Biological Resources of Prawn in Bays), *Naikaisuikenpo hokoku*, **11**, 178-198.

191. ——, 1966. Maimai ebi no yotai ni tsuite [On the larvae of *Atypopenaeus compressipes* (Henderson)], *Ichiton Gakuen Tandai, Shizen Kakagaku Kenkyukai Kaiho*, 1, 1-6.

192. ——, H. SASAOKA, and U. KOBAYASHI, 1957. Setonaikai no ebi gyogyo no gorika ni kansuru kenkyu: II. Ebi no shurui, bunpu ido narabini sofei ni tsuite (On the Fishing Operations of Prawns in the Seto Inland Sea: II. Types, Distribution, and Migration of Various Prawns), *Naikaisuikenpo*, **10**, 28-36.

193. ——, ——, and M. SUZUKI, 1957. Seto naikai no ebi gyogyo no gorika ni kansuru kenkyu: I. Higata ni izon suru ebirui ni kansuru kenkyu, Tokuni

kurumaebi ni tsuite (On the Fishing Operations of Prawns in the Seto Inland Sea: I. Study of *Penaeus japonicus*), *Naikaisuikenpo*, **10**, 20-27.

194. YASUDA, J., H. SASAOKA, W. DOHI, and J. TONOGAI, 1962. Shikoku, kyusuu no taiheiyo engan ni okeru shin kai sokobiki-ami gyogyo no genkyo to shorai ni tsuite no hirotsu no kosatsu, Tokuni ebirui ni tsuite (On Deep-sea Trawling in Pacific Coasts along with Shikku and Kyushu, with Special Reference to Prawns), *Tokushima ken suishi hokoku (Bulletin of Tokoshima Fisheries Experiment Station)*, 1-17.

195. YOKOYA, Y., 1930. Report of the Biological Survey of Mutsu Bay: No. 16. Macrura of Mutsu Bay, *Sci. Rep. Tohoku Imp. Univ.*, **5** (3), 525-548.

196. YOSHIDA, H., 1941. Chosen kinkai san yuyo ebirui (Prawns of the Sea around Korea), *Chosen Suishi Hokoku (Bulletin of Fisheries Experiment Station in Korea)* **7**, 1-36.

197. ——, 1949. Koraiebi no seikatsushi ni tsuite (Life History of *P. orientalis*), *Nissuikaishi (Bulletin of the Japanese Society of Scientific Fisheries)*, **15**, 245-248.

198. ——, 1960. Ariake-Kai no ebi no seitai seikatsushi ni kansuru kenkyu (Ecology and Life History of Prawns of Ariake-Kai), 44.

199. YOUNG, J. H., 1956. Anatomy of the Eyestalk of the White Shrimp, *Penaeus setiferus* (Linn., 1758), *Tulane St. Zool.*, **3**, 171-190.

200. ——, 1959. Morphology of the White Shrimp, *Penaeus setiferus* (Linn., 1758), *U. S. Fish. Wildl. Serv., Fish. Bull.*, **59** (145), 1-168.

201. ZEIN-ELDIN, Z. P., 1963. Effect of Salinity on Growth of Post-larval Penaeid Shrimp, *Biol. Bull.*, **125**, 188-196.

202. ——, and D. V. ALDRICH, 1956. Growth and Survival of Post-larval *Penaeus aztecus* under Controlled Conditions of Temperature and Salinity, *Biol. Bull.*, **129**, 199-216.

203. ——, and G. W. GRIFFITH, 1966. The Effect of Temperature upon the Growth of Laboratory-held Post-larval *Penaeus aztecus, Biol., Bull.*, **131**, 186-196.

CHAPTER II

Technique of Prawn Culture

1. Seedling production

1.1 Introduction

Penaeus japonicus Bate is a most valuable marine species in Japan and its culture in some parts of the country was begun many years ago. However, actual experiments on prawn culture were started in relatively recent times. Most of the developments in the culture technique are based on the fundamental studies carried out by Hudinaga (1942), who was the first to experimentally induce spawning in Penaeus japonicus in 1934 in the Kumamoto Prefecture. He was also able to rear the larvae hatched from these eggs up to the mysis stage. Hudinaga later continued his research on these lines in the Yamaguchi Prefecture.

The larvae of P. japonicus pass through the stages of nauplius, zoea, mysis, and metamorphose to become adult prawns. The nutritional requirements during the nauplius stage are met by the yolk preserved in the body, but during the zoea stage, the larvae must feed from external sources. It is difficult to rear larvae on artificial feed during the zoea stage, but rearing is easier after the mysis stage, by which time the larvae have become fairly strong. Thus the most important problem in rearing larvae of P. japonicus is getting them through the zoea stage.

In 1938, Hudinaga used Matsue's culture of Skeletonema costatum (1954) as feed and was able to nurse a large number of larvae through the zoea to the mysis stage. In 1940, he succeeded in rearing the larvae hatched in June to the adult stage of 8 to 10 cm body length by October. Thus he was able to achieve a through culture of P. japonicus from eggs, but further experiments on this line had to be temporarily suspended because of World War II. The results of the Hudinaga experiments on the rearing of P. japonicus are given in Table 5.12.

Table 5.12. Survival rate of larvae (Hudinaga, 1942)

Stage of growth	Survival rate %
Eggs	70.6
Nauplius	92.7
Zoea	27.9
Mysis	86.5
Post-larval	85.4*

*Results after 12 days.

Hudinaga resumed his research work after World War II in the Chiba Prefecture. He introduced several new types of apparatus and equipment and made substantial progress, which enabled him to make a truly epoch-making report on culture techniques of Penaeid larvae (Patent No. 1959-27796).

Hudinaga first transformed the old method of using glass containers for larval culture into a modern one by using large water tanks (2 m × 1 m × 1 m) fitted with air-circulating equipment; this made it possible to rear larvae at a depth of more than 50 cm, resulting in the growth of 10×10^3 post-larvae per tank.

Larger amounts of feed became necessary due to the expansion in the rearing capacity, and hence *Skeletonema* were mass produced through the use of air-circulating facilities fitted to the tanks. In addition, *nauplius* of *Artemia* could be effectively used to feed the *mysis* and early post-larvae. The meat of short-necked clams could also be used to feed the larvae during their late post-larval period. Thus, it was possible to produce juvenile prawns on a commercial scale.

The Pacific Fishing Company Limited came into existence in 1959 on the basis of the technical achievements described above. The company was later renamed the *Penaeus* Culture Company. *Penaeus* culture became commercialized in Takamatsu of the Kagawa Prefecture after 1960, when the abandoned salt beds could be used for culturing. The production rate of juvenile prawns during the early period of commercialization is given in Table 5.13.

Table 5.13. Results of rearing larvae of *Penaeus*

(Data published by *Penaeus* Culture Company, 1960 to 1962)

	Year of observations	1960*	1961	1962
Mother prawns	Number of specimens obtained	1,774	3,604	9,364
	Number of days involved	94	93	107
Number of larvae ($\times 10^3$)	Number of eggs released	197,865	538,370	754,392
	Nauplius	—	282,827	273,663
	Zoea	—	137,912	77,371
	Mysis	—	62,630	32,948
	Post-larvae	—	30,359	17,290
	Late post-larvae	2,240**	8,400	8,311
	Juvenile	—	4,005	5,921
Yield rate of reared larvae (%)	Hatching rate	60	53	36
	Nauplius stage	62	49	28
	Zoea stage	44	45	42
	Mysis stage	61	48	52
	Early post-larval stage	29	28	48
	Late post-larval stage	—	48	71
Parent prawns	Number of juvenile prawns per mother prawn	1.26×10^3	1.11×10^3	0.63×10^3
Yield rate from egg to final stage (%)		1.13	1.56	1.10

*Yield rate for 1960 was taken from the mean values of culture tanks showing relatively better results.

**In 1960, the number of days for the rearing of late post-larvae was not constant. Late post-larvae treated as juvenile prawns.

In 1960, indoor rearing was carried out in 72 concrete rearing tanks ($3 m \times 1 m \times 1$ m) and 70 concrete culture tanks ($4 m \times 1$ m or $2 m \times 1$ m). But in 1961, the rearing tanks were made of wood($2 m \times 1 m \times 1$ m)and the number was increased to 140. In the same year, 104 new ponds were built for juvenile prawns (rearing ponds of $3 m \times 1.5 m \times 0.5$ m for late post-larvae).

With the increase in rearing facilities, the number of juvenile prawns increased from $2,000 \times 10^3$ in 1960 to $4,000 \times 10^3$ in 1961, to $6,000 \times 10^3$ in 1962. The number of juvenile prawns per mother prawn, however, decreased from 1,260 in 1960, to 1,110 in 1961, to 630 in 1962. Obviously, considerable technical difficulty with the rearing method still exists as indicated by the progressive decrease in the number of juvenile prawns.

Proper feed supply is also problematic and handicaps the rearing. This flaw is related to the difficulties involved in the culturing of *Skeletonema costatum* in all seasons. For a large-scale culture of *Skeletonema*, carefully processed sea water has to be used. If the sea water were better sterilized and the culture carried out at a regulated temperature, continuous culture might be possible. However, it involves massive efforts to sterilize a large amount of sea water and, even if this is possible, physiological changes in the cultured *Skeletonema*, depending upon the rearing conditions, may affect the water quality. Hence, an alternate feed item should be found

As shown in Table 5.13, the outcome of rearing the larvae by the method described above is not very encouraging, particularly during the early post-larval period, as the production of seedlings is uncertain. During this period, the most suitable diet for the larvae is *Artemia*, which is very costly since it has to be imported. Hence, the right type of feed should be carefully investigated, to establish a rearing method which would give better results.

To solve these problems, rearing experiments were carried out on a large-scale which increased every year to 1,000 and then to 3,000 tanks. These experiments revealed the facts mentioned below, and led to certain improvements in the rearing method.

When fertilized eggs or planktonic larvae of oysters (*C. gigas*) are given as feed to *zoea*, the results of rearing are the same as those reared on *Skeletonema*. Hence, it was found that *zoea* could be reared on zooplankton in addition to *Skeletonema*, and that the rearing result would be the same, even when fertilized eggs or planktonic larvae of other oysters are given as feed; however, *C. gigas* seems to be the ideal species since it is easily available during summer which is the peak spawning period of *Penaeus japonicus*. But, when the spawning of *Penaeus japonicus* becomes restricted to a particular period, the daily requirement of feed is very high, and thus it becomes very difficult to maintain the necessary amount of feed with ordinary methods. In order to overcome this problem, cold storage of mature oyster eggs was tried, but both the nutritional value and the floating quality of the eggs were poorer than in fresh eggs; therefore, the results of rearing by this method of feeding were not as good as those from rearing on fertilized eggs or planktonic larvae.

Efforts were simultaneously made to use ponds with a double-bottomed floor. In such ponds, various diatoms were cultured on the lower floor which was spread with gravel in the summer. During the day, diatoms were allowed to float to the surface of the water and form colonies or lumps, which were given as feed to the *zoea*.

The results were good. It was also found that these colonies of diatoms could be used as feed for larvae following the *mysis* stage. However, in order to allow plentiful growth of the diatoms, it was necessary to widen the area of the double-bottomed tank by more than ten times the total area of the rearing tanks. The diatoms that grew in the freshly gravelled beds included such varieties as *Nitzschia*, which are suitable for *zoea*, but those which grew in the old beds, were dominated by *Navicula*, which is not so suitable, because it does not float; hence the results of rearing were not successful.

Table 5.13 also presents the results of experiments in which the feed for *zoea* was *Skeletonema* up to June, 1960, artificially fertilized eggs of *C. gigas* after July, 1960, cold storage oyster eggs (*C. gigas*) up to June, 1961, cold storage oyster eggs (*C. gigas*) and diatoms after July, 1961, and mainly diatoms in 1962.

When feed is considered individually as plant and animal types, some difference in the economical production becomes evident. It is usually possible to culture the plant feed in a short time, and hence this type is suitable when large-scale rearing is done.

In the past, the larvae were reared indoors in order to avoid bright light. But since light suits diatoms, it appears that the room should be made bright for them; however, the degree of brightness must be simultaneously regulated so as not to affect the growth of larvae. This fact has been confirmed both by experiments and from indoor rearing examples under varying degrees of brightness.

Again, larvae used to be reared in rearing tanks inside a room up to the free-swimming stage, and then in open-air tanks known as juvenile prawn tanks when they entered the bottom-dwelling phase. This type of rearing might seem to be ideal as far as the ecology of the larvae is concerned, but since the tanks are used repeatedly, the quality of water became poor. Moreover, this method does not seem to be suitable when larvae at various stages of growth are to be released in the water tanks. Also, in this method, a decrease in the survival rate soon after the release and a temporary arrest of growth occurred. Even with bottom-dwelling larvae fed on the meat of bivalves, when the water temperature varies, it has now been found that rearing them in brighter conditions rather than in darkness is much better. In other words, it is ideal to rear the larvae in water tanks from the early larval stage to the juvenile stage in bright conditions suitable to the culture of diatoms.

As mentioned earlier, the demand for feed for the larvae of *Penaeus japonicus* has greatly increased; now it is possible to rear the larvae on a much larger scale than heretofore, because tanks have been expanded to a capacity of several tons.

Since 1962, the rearing of larvae has been shifted to open-air tanks with a much wider area than the indoor tanks; by using culture tanks or juvenile prawn tanks, rearing is carried out from the hatching stage. The outcome of this rearing technique is better than that with indoor rearing. In 1964, saline water from the salt beds in the Yamaguchi Prefecture ($10\,m \times 10\,m \times 2\,m$) was introduced, and this model experiment, which did not involve indoor rearing, resulted in $1,000 \times 10^3$ juvenile prawns per tank.

When large-scale seedling production was required in the past, large-scale equipment and complicated techniques were involved. When rearing is done according to the new method, since it is possible to feed the larvae on diatoms and various types of planktons grown after directly adding nutrients to the rearing water, the rearing becomes simpler and involves no complicated technique. Hence, since 1965, the small-scale industries in Japan and also the seedling production operations, follow the new program. *Penaeus* culture has thus become a commercially feasible venture.

1.2 Spawning

Mating in *Penaeus* occurs soon after molting. During this period the females receive spermatophores from males and store them in a seminal receptacle. The thelycum bends down during mating (Figure 5.38).

When the females mature, a dark green color spreads in the ovaries, which are on the dorsal part of the body, particularly between the carapace and the abdomen. This coloration can be seen with the naked eye.

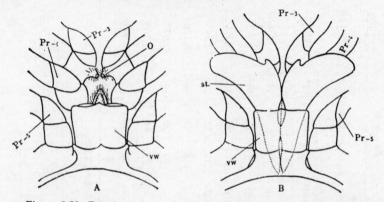

Figure 5.38. Female accessory reproductive organs (Hudinaga, 1942).
A—before mating; B—after mating; Pr-3, Pr-4, Pr-5—3rd, 4th, and 5th walking legs;
O—Gonopore; st.—style; vw—wall of the accessory reproductive organs.

Mature female prawns can be collected from mid-March to mid-October on the Japanese coast. Spawning is at its peak from mid-June to late August, and usually occurs at night, the prawns laying their eggs while swimming. As the eggs are laid, the spermatophores stored in the seminal vesicle are simultaneously released, and the eggs thus fertilized. Soon after release, the eggs become irregular in shape due to the gelatinous covering and float for a while. When they are fertilized they become covered by fertilization membrane and settle down to the sea floor. The size of the egg at this time (soon after the formation of the fertilization membrane) is 0.26 to 0.28 mm.

Nowadays, female prawns are selected from the commercial catch[1] and the rate of spawning studied on the basis of the development of the ovaries (see Table 5.14).

The period required for spawning of the mother prawns after their release in tanks is given in Table 5.15. In May, they spawn within 2-3 days of their stay in tanks but in July and August this period is reduced to a day. Since the water temperature of the rearing tank is maintained at 25 to 30°C, the maturation of female prawns caught during the cold season, is accelerated when introduced into the rearing tanks with a higher water temperature.

[1] Eggs were collected from *Penaeus* reared from the stage of hatching; however, it will be some time before eggs from prawns in natural conditions can be obtained.

Table 5.14. Rate of spawning and number of eggs released

(Data obtained from *Penaeus* Culture Company Limited for the year 1960-1962)

Month	Availability Number of days			Mother Prawn Number of prawns			Number of spawning individuals			Rate of spawning (%)			Number of eggs × 10³			Number of eggs per specimen × 10³		
	1960	1961	1962	1960	1961	1962	1960	1961	1962*	1960	1961	1962*	1960	1961	1962	1960	1961	1962*
Apr.	—	—	9	—	—	274	—	—	46	—	—	17	—	—	9,202	—	—	383
May	—	10	11	—	141	143	—	67	69	—	48	49	—	29,435	28,390	—	439	450
June	20	16	20	600	161	453	102	100	180	17	62	41	23,511	55,850	98,520	230	559	515
July	25	22	21	280	763	1,513	87	387	426	31	51	28	40,545	95,100	143,120	470	246	334
August	22	21	26	520	1,942	5,636	177	855	—	34	44	—	88,128	298,275	355,610	500	349	—
Sept.	20	23	20	344	592	1,345	110	228	—	31	39	—	42,973	59,710	119,450	390	262	—
Oct.	7	1	—	30	5	—	8	2	—	27	40	—	2,708	800	—	339	200	—
Total	94	94	107	1,774	3,604	9,364	484	1,639	—	33	46	—	197,865	539,170	754,392	409	329	—

*Number of spawning prawns is not confirmed for the period of August, 1962.

Table 5.15. Spawning rate of female prawns arranged according to the period of observation (number of days)

(Data obtained from *Penaeus* Culture Company, 1961)

Month	Number of days	Number of specimens observed	Number of spawning specimens	Rate of spawning, %
5	0	86	3	3.4
	1	83	1	1.2
	2	82	36	43.9
	3	46	18	39.1
	4	28	3	10.7
	5<	25	0	0
6	0	168	18	10.7
	1	150	17	11.3
	2	133	28	21.0
	3	105	13	12.3
	4	92	8	8.6
	5<	84	1	1.2
7	0	781	349	44.7
	1	432	35	8.1
	2	397	1	0.3
	3	396	8	2.0
	4	388	2	0.5
	5<	386	0	0
8	0	1955	793	40.5
	1	1162	23	1.9
	2	1139	3	0.2
	3	1136	0	0
9	0	587	172	29.3
	1	415	42	10.1
	2	373	10	2.7
	3	363	0	0

The rate of spawning is greatly influenced by the individual differences in the standard of selection of female prawns and the spawning rate during the peak period is about 40%, with 200×10^3 to 500×10^3 being the average number of eggs per specimen.

The spawning time of prawns changes according to season (see Table 5.16). A large part of the eggs are released between 18 to 22 hours in May, 20 to 0 hours in June, 0 to 4 hours in July, and 2 to 4 hours in August and September. It is very interesting to note that the spawning time is delayed with the advance of the season.

1.3 Hatching of larvae and metamorphosis

1.3.1 NAUPLIUS

When the water temperature is 27 to 29°C, the eggs of *Penaeus* hatch into *nauplius* within about 13 to 14 hours after spawning, which feed on the yolk stored in the body

Table 5.16. Spawning time

(Data obtained from *Penaeus* Culture Company, 1961)

Period of month	Time (hour)						
	18-20	20-22	22-0	0-2	2-4	4-6	Total
Late May	12	29	7	8	4		60
%	20.0	48.3	11.7	10.3	6.7		
Early June	5	22	4	1	1		23
Mid-June	5	23	17	6	10		61
Late June			2	1			3
Total	10	45	23	8	11		97
%	10.3	46.4	23.7	8.3	11.3		
Early July		2	6	8	1		17
Mid-July		1	5	63	65	20	154
Late July			2	59	38	5	104
Total		3	13	130	104	25	275
%		1.1	4.8	47.2	37.7	9.2	
Early August				21	95	66	182
Mid-August			2	28	140	88	258
Late August				29	111	41	181
Total			2	78	346	195	621
%			0.3	12.5	55.7	31.4	
Early September				8	70	50	128
Mid-September				10	34	17	61
Late September				6	7	6	19
Total				24	111	73	208
%				11.5	53.3	35.2	
Grand Total	22	77	45	248	576	293	1261
%	1.7	6.1	3.5	19.7	45.7	23.3	

and do not feed on an external diet. The *nauplius* has a strong phototaxis and moves about on 3 pairs of jointed appendages. At rest, it floats with its dorsal surface facing downward. *Nauplius* transform into *zoea* in 36 to 37 hours after hatching. During this naupliar period the individual molts six times, and the various parts of the body, as well as the joint appendages, change with each molting in the manner described in Table 5.17.

1.3.2 ZOEA

After the 6th molt, *nauplius* transform into *zoea*. The exo- and endo-podia of the mandibles fall off, and the swimming movement stops temporarily. However, the 4 pairs of jointed appendages which began to develop during the *nauplius* stage, assist in the swimming movement, but as the posterior half of the body becomes rapidly elongated, the swimming movement becomes sluggish. The larvae swim by flexing the jointed appendages. *Zoea* molt 3 times and then transform into *mysis*. Even when the water temperature is suitable for feeding, the duration of the *zoea* stage varies

Table 5.17. Morphological changes of *nauplius* **(Hudinaga, 1942)**

Stages	Time after hatching, hours	Number of stylets on second antennae	Number of caudal styles	Body length, mm
1st	0	5	1+1	0.32
2nd	3	6	1+1	0.34
3rd	6	7	3+3	0.37
4th	11	9	4+4	0.39
5th	16	10	6+6	0.44
6th	23	12	7+7	0.50

according to the nature and availability of the feed. When the water temperature is 27 to 29°C, the larvae can be reared on cultured *Skeletonema* for 4 days provided the quality of the water remains good. However, when fertilized eggs or planktonic larvae of *C. gigas* are given as feed, the larvae require only 3 days for rearing.

The morphological characteristics during the various stages of *zoea* are given below (Hudinaga, 1935).

Zoea, 1st stage: Simple eyes. Carapace covers head and extends up to half of the 3rd thoracic segment. The body beyond the 3rd thoracic segment very long, and segments up to 8th thoracic one can be easily counted. The body part between the 8th thoracic segment and the caudal segment bulges slightly, revealing 2 segments. The number of caudal styles is 7+7, the same as in the *nauplius* stage 6, i.e. after the 5th molt. Body length, 0.092 mm.

Zoea, 2nd stage: At the center of the anterior margin of the carapace a protrusion appears and another on each of the right and left sides. Compound eyes protrude. Five abdominal segments are prominent. Body length, 1.53 mm.

Zoea, 3rd stage: The 6th abdominal segment very long. Caudal segment develops and caudal appendage seen. Number of caudal styles increases to 8+8. Body length, 2.24 mm.

1.3.3 Mysis

When *zoea* molt for the 3rd time and change into *mysis*, the morphology resembles that of adult prawns. The activities also increase greatly. The head and thorax completely fuse with each other and become cephalothoracic. The carapace extends up to the 8th thoracic segment. The 3rd maxillae and the 5 pairs of walking legs, which were seen as vestiges in *zoea*, are now well developed. The body is bent at the juncture of the thorax and the abdomen, and the walking legs move rapidly up and down. The larvae swim backward.

When the water temperature is 27 to 29°C, if *Skeletonema* and *nauplius* of *Artemia* are given as feed, the larvae molt 3 times in 3 days and transform into the post-larval stage. Along with each molt, there are some changes in the body of the *mysis*. When the larvae enter the 2nd stage of *mysis*, the anterior 3 pairs of walking legs become sharp at the tips, a characteristic feature of *Penaeus*.

The body length is 2.83 mm in the 1st stage, 3.34 mm in the 2nd, and 4.34 mm in the 3rd.

Table 5.18. Number of larvae and survival rate arranged according to the month of spawning

(Data for 1960-1962 obtained from *Penaeus* Culture Company)

Stages

Month of spawning	Eggs			Nauplius			Zoea			Mysis			Post-larvae			Late post-larvae		
	1960	1961	1962	1960	1961	1962	1960	1961	1962	1960	1961	1962	1960	1961	1962	1960	1961	1962
4	—	—	9,202	—	—	3,330	—	—	1,974	—	—	493.8	—	—	189.2	—	—	45.0
5	—	29,435	28,390	—	11,577	10,567	—	6,709	5,725	—	3,287	2,329.4	—	1,962	696.2	—	762	88.8
6	—	55,850	98,520	—	21,774	26,310	—	15,570	17,855	—	6,783	5,443.0	—	2,904	1,766.8	—	1,130	877.7
7	—	95,100	143,120	—	56,259	43,081	—	33,571	14,632	—	15,136	6,939.0	—	9,407	2,599.0	—	3,096	1,520.4
8	—	298,275	355,610	—	162,187	133,449	—	70,718	31,073	—	33,185	15,227.8	—	14,498	9,611.0	—	3,230	5,000.3
9	—	59,710	119,450	—	31,030	37,570	—	11,344	6,122	—	4,239	2,515.0	—	1,588	1,350.0	—	182	778.4
10																		
Total	—	538,370	754,392	—	282,827	273,663	—	137,912	77,371	—	62,630	32,948.0	—	30,359	17,289.9	—	8,400	8,310.6

Number of larvae produced (×10³)

Stages

Month of spawning	Eggs			Nauplius			Zoea			Mysis			Post-larvae		
	1960	1961	1962	1960	1961	1962	1960	1961	1962	1960	1961	1962	1960	1961	1962
4	—	—	36	—	—	59	—	—	25	—	—	38	—	—	24
5	70	39	37	48	62	54	61	49	41	81	60	30	5	39	13
6	60	40	27	60	71	68	32	44	30	47	43	32	87	39	50
7	68	59	30	72	58	34	32	45	47	76	62	37	21	33	58
8	55	54	38	59	48	23	38	47	49	58	44	63	4	22	52
9	59	52	31	71	42	16	23	37	39	61	37	54	—	11	58
10	47			64			81			41					
Aggregate		53	36		49	28		45	42		48	52		28	48

Survival rate (%)

1.3.4 POST-LARVAE

When the *zoea* enter the post-larval period, the 5 pairs of abdominal swimming appendages (pleopods), which did not function in the *mysis* stage, become active; the walking legs are used only to collect food and to creep.

During the early stage, the post-larvae float like plankton, but after 3 or 4 molts, they are able to creep on the sides of vessels or on the floor. When the post-larvae molt 10 to 12 times, they are able to move on sand as the adult prawns do. The morphology of the body gradually changes with each molt of post-larvae and after 20 to 22 molts, the morphology of the body, as well as that of the appendages, resembles those of adults.

For the sake of convenience, particularly from the seedling point of view, post-larvae are classified into early post-larvae and late post-larvae. Early post-larvae are those which have molted only 3 or 4 times, and the late post-larvae are those after the 5th or 6th molting. Again, larvae which have molted about 20 times are called juvenile prawns.[2]

1.4 Rearing of larvae and survival rate

1.4.1 SEASONAL VARIATIONS IN THE SURVIVAL RATE OF REARED LARVAE

It is but natural that there would be some difference in the survival rate of reared larvae of *Penaeus* depending upon the rearing method, but there is also considerable difference in the survival rate of larvae reared by the same method and this difference is due to seasonal changes. In Tables 5.18 and 5.19, the data on the number of

Table 5.19. Number of juvenile prawns produced and their survival rate arranged according to the month of larval release

Month of larval release	Number of juvenile prawns produced ($\times 10^3$)						Survival rate (%)		
	Stages								
	Late post-larvae			Juvenile prawns			Late post-larvae		
	1960	1961	1962	1960	1961	1962	1960	1961	1962
4	—		3.9			2.6			70
5	—		34.4	—		25.4	—		74
6	—	214.1	312.5	—	135.4	243.9	—	63	78
7	—	1,990.9	1,687.0	—	1,056.6	1,035.4	—	53	61
8	—	2,995.0	2,638.1	—	1,303.2	2,179.4	—	44	82
9	—	1,667.8	3,707.0	—	923.0	2,434.4	—	55	64
10	—	1,532.1		—	586.8		—	38	
Total	—	8,400.0	8,382.9	—	4,005.0	5,921.1	—	48	70

[2]It is difficult to judge the age of post-larvae by the naked eye. Hence the larvae on the n day of metamorphosis, as post-larvae, are expressed as *pn* ($n = 1, 2, ...$).

The early post-larvae of 1961 were expressed as $P_1, P_2, ..., P_7$, and the late post-larvae as $P_7, P_8, ... P_2, P_6$ (number of rearing days, 25). The early post-larvae of 1962, 1963, were expressed as $P_1, P_2, ..., P_4$, and late post-larvae as $P_4, P_5, ..., P_{23}$ (number of rearing days, 22). Since 1964, the post-larval stage has not been divided into early post-larval and late post-larval stages, but expressed as $P_1, P_2, ..., P_{20}$ (number of rearing days, 19).

larvae and the survival rate arranged according to the years in Table 5.13, are re-arranged according to the months of spawning.

From Tables 5.18 and 5.19, it is concluded that the survival rates at various stages of larvae reared in model rearing tanks of size 2 m × 1 m × 1 m or larger, are as follows: at hatching stage—50%, *nauplius* stage—50%, *zoea* stage—40%, *mysis* stage—50%, early post-larval stage—50%, and late post-larval stage—70%. Hence the survival rate from the egg to the juvenile stage is $0.50 \times 0.50 \times 0.40 \times 0.50 \times 0.50 \times 0.70 = 0.0175 \ (=1.75\%)$. However, the survival rate improves greatly when the large-scale larval production method is employed, a technique which will be described later.

1.4.2 DIET AND SURVIVAL RATE

During the rearing of larvae of *Penaeus*, usually *Skeletonema* is given as feed for *zoea*, *nauplius* of *Artemia* as feed for *mysis* and floating post-larvae, and meat of bivalves as feed for bottom-dwelling post-larvae. Hudinaga and Kittaka (1966)

Table 5.20. Effect of various types of feed on the larvae of *Penaeus* (Hudinaga and Kittaka, 1966)

Classification of feed	Type of feed	Zoea	Mysis	Early post-larval	Late post-larval
Plant feed	Cultured *Skeletonema*	●	○		•
	Planktonic diatoms	○	△		
	Fixed diatoms	○	○	△	○
	Chlamydomonas	×			
	Pieces of *Ulva pertosa* and *Zostera marina*	×			
Animal feed	Eggs and larvae of *C. gigas*	○	○		
	Cold-storage eggs of *C. gigas*	○	△	×	
	Sperms of *C. gigas*	×			
	Rotifera	○	○	○	△
	Artemia		●	●	
	Daphnia			△	
	Copepoda			○	○
	Cold-storage *Copepoda*			△	
	Gammarus				△
	Cold-storage *Gammarus*				○
	Barnacles			○	○
	Nematoda			×	
	Annelida			△	○
	Meat of *C. gigas*				△
	Meat of short-necked clam			○	●
	Meat of prawn				○
Others	Compound diet				○
	Yeast	×			
	Fermented fluid	×			
	Milk and yolk of egg	×			

●—Standard feed; ○—Feed producing the same effect as standard feed; △—Results inferior to those of standard feed but still usable; ×—Feed ineffective.

carried out several experiments to find an alternate ꞏeed for rearing larvae. A part
of their results is given in Table 5.20. In other words, a number of alternate feeds,
which can be used in place of the standard feed, were identified.

From among the non-standard feeds, some were found suitable for large-scale
rearing and further experiments to improve them were done in water tanks used for
commercial rearing. By determining the survival rate of larvae reared on these feeds,
the possibility of adjusting the amount of feed was studied and the results are given
in Table 5.21.

**Table 5.21. Various types of feed and the survival rate of reared
larvae of *Penaeus* (Hudinaga and Kittaka, 1966)**

Feed	Stages of larvae		
	Zoea	*Mysis*	6 days of early post-larvae
Skeletonema	35.8-78.5 (52.2)		
Planktonic diatoms	50.0-86.6 (66.0)		
Adhesive diatoms	33.3-91.1 (56.0)	38.4-100 (72.1)	24.4-28.9 (26.7)
Eggs and larvae of			
C. gigas	33.3-100 (61.0)	21.7-98.5 (67.7)	
Cold-storage eggs of			
C. gigas	20.0-100 (56.8)	35.5	
Rotifera	0	25.0	16.1-33.3 (24.7)
Artemia			46.9-71.2 (59.0)
Meat of other prawns			21.1-23.3 (22.2)

Note: Figures in parentheses show the mean value.

For *zoea*, whose feed consists of planktonic diatoms, fixed diatoms, eggs and larvae
of *C. gigas*, or cold-storage eggs of *C. gigas*, the survival rate was the same as when the
feed consisted of *Skeletonema*. However, when the feed consisted of *Rotifera*, the results
(Table 5.20) were not good.

In the rearing experiments, for each larvae rearing tank, three tanks were used for
culture of *Rotifera*, but even then the necessary amount of feed could not be maintained. Hence, a more efficient method for *Rotifera* culture should be found so that
it can be commercially applied.

Regarding *mysis* larvae, the survival rate was good when the feed consisted of
adhesive diatoms and eggs, and larvae of *C. gigas*. Since adhesive diatoms grow in
double-bottomed culture ponds and form large colonies, these latter also included
other micro-organisms and were suitable as feed for *mysis* larvae, both from the viewpoint of size and nutritional value. Sometimes the adhesive diatoms firmly adhere
to the surface of the body of larvae and, as a result, swimming movement becomes
difficult. However, *mysis* larvae which move more actively than *zoea* are not so
adversely affected. Considered from the point of view of size, floating capacity, and
nutritional value, the eggs and larvae of oysters (*C. gigas*) are found to be the best
feed for *mysis* larvae.

The survival rate during the post-larval period is poor when they are fed on feed
other than *Artemia*. Even when *Artemia* is given as feed, the mean survival rate is

60%, which cannot be said to be good. In recent times, the results of rearing larvae in areas where *Artemia* could be grown easily and utilized as feed, have not been uniform (Kurata, 1967), particularly the rearing of larvae on California *Artemia*, considered the best variety. The survival rate did not differ much from that of the Utah *Artemia*, a variety considered inferior to California's. Hence it would seem in these instances that the survival rate is not influenced by the quality of *Artemia* when this is the only feed. That the post-larval stage is passed without mortalities when various types of zooplankton constitute the diet will be seen later when the large-scale seedling production method is presented.

1.4.3 FEED AND RESULTS OF REARING

Table 5.22 shows the results of rearing *zoea* on various feeds on a large scale in a number of tanks. In this case, the feed was classified as: (1) cold-storage eggs of *C. gigas*, (2) adhesive diatoms, (3) cold-storage eggs of *C. gigas* mixed with fixed diatoms, and (4) cultured *Skeletonema* or planktonic diatoms.

When the rearing experiments are carried out on a small scale, these feeds show more or less the same results as given in Table 5.21. But when the rearing is carried out on a large scale, the difference in the survival rate gradually widens. When adhesive diatoms are given as the only feed, the survival rate is extremely poor, due to the fact that one must stir the water in the tanks when such feed is supplied, a task which is difficult when several tanks are involved. When the feed is cold-storage eggs of *C. gigas*, the survival rate declines after August, because the eggs lose much of their nutritive value when preserved in refrigerators. When both diatoms and eggs of *C. gigas* constitute the feed, the survival rate becomes more or less stabilized, since the lack of quality in the one is compensated for by the other.

Table 5.23 shows the survival rate of larvae arranged according to the various periods of rearing. In this case, the diet of *mysis* larvae is classified as (1) cold-storage eggs of *C. gigas*, (2) adhesive diatoms, and (3) cold-storage eggs of *C. gigas* mixed with adhesive diatoms. While the results of rearing *mysis* are relatively better when only adhesive diatoms are used as feed, the results when only the cold-storage eggs of *C. gigas* are supplied are poor.

Table 5.24 is an analysis of the survival rate of larvae on the basis of the amount of *Artemia* given as feed for the early post-larvae. The amount of *Artemia* at various stages was as follows: (1) about 5 g per day per tank between P_1 and P_7, (2) half of (1) mixed with adhesive diatoms between P_1 and P_4 or P_7, (3) half of (1) mixed with adhesive diatoms or *Limnodrilus*, between P_1, P_2 and P_2 or P_3, and (4) *Artemia* excluded, adhesive diatoms and *Nematoda* used.

In cases (2) and (3), an effort to reduce the amount of *Artemia* was made. However, when the amount of *Artemia* is reduced to half, the survival rate of the larvae decreases by 36% of the rate in case (1). When *Artemia* is totally absent as in (4), the survival rate is extremely poor. However, when *Artemia* is used as in (1), the survival rate is not very high. It seems that factors such as environmental conditions, particularly lack of light in the rearing tanks, adversely affect the diatoms and lower the activity of the larvae. As will be explained later, the survival rate of post-larvae improves greatly by the large-scale seedling production method.

Table 5.22. Results of rearing *zoea* on various feeds

(Data for 1961 obtained from the *Penaeus* Culture Company)

	Feed													Total		
	Cold-storage eggs *C. gigas*			Cold-storage eggs *C. gigas* mixed with adhesive diatoms			Adhesive diatoms			*Skeletonema* or planktonic diatoms					
	Number of larvae × 10³															
Period of spawning	*Zoea*	*Mysis*	Survival rate %	*Zoea*	*Mysis*	Survival rate %	*Zoea*	*Mysis*	Survival rate %	*Zoea*	*Mysis*	Survival rate %	*Zoea*	*Mysis*	Survival rate %
1	2	3	4	5	6	7	8	9	10	11	12	13	14	15	16
Late May	6,040	2,940	48.6							669	347	51.9	6,709	3,287	49.0
Early June	6,255	2,300	36.7										6,255	2,300	36.7
Mid-June	7,995	3,631	45.4										7,995	3,631	45.4
Late June	1,320	852	64.1										1,320	852	64.1
Total	15,570	6,783	43.7										15,570	6,783	43.7
Early July	1,040	476	45.7	1,362	511	37.5	1,225	562	45.8				3,627	1,549	42.7
Mid-July				16,158	6,889	42.6	400	4	1.0				16,558	6,893	41.6
Late July				12,286	6,320	51.4	1,100	374	34.0				13,386	6,694	50.0
Total	1,040	476	45.7	29,806	13,720	46.0	2,725	940	34.5				33,571	15,136	45.1

Table 5.22—(Contd.)

1	2	3	4	5	6	7	8	9	10	11	12	13	14	15	16
Early August				30,928	15,272	49.3							30,928	15,272	49.3
Mid-August	8,850	2,800	31.6	21,410	11,258	52.5							30,260	14,058	46.3
Late August	7,680	3,255	42.4	1,850	600	32.4							9,530	3,855	40.4
Total	16,530	6,055	36.6	54,188	27,130	50.6							70,718	33,185	46.9
Early September	3,940	1.170	29.7	4,360	1,960	45.0				300	200	66.6	8,600	3,330	38.4
Mid-September	80	0	0	1,654	640	38.7							1,734	644	37.1
Late September				480	240	50.0	530	25	4.7				1,010	265	26.2
Total	4,020	1,170	29.1	6,494	2,840	43.8	530	25	4.7	300	200	66.6	11,344	4,239	37.3
GRAND TOTAL	43,200	17,424	40.3	90,488	43,690	48.2	3,255	965	29.6	969	547	56.4	137,912	62,630	45.4

Table 5.23. Results of rearing *mysis* on various feeds

(Data for 1961 obtained from *Penaeus* Culture Company)

Period of spawning	Feed											
	Cold-storage eggs *C. gigas*			Cold-storage eggs *C. gigas* mixed with adhesive diatoms			Adhesive diatoms			Total		
	Number of larvae ×10³											
	Mysis	Post-larvae	Survival rate %	*Mysis*	Post-larvae	Survival rate %	*Mysis*	Post-larvae	Survival rate %	*Mysis*	Post-larvae	Survival rate %
1	2	3	4	5	6	7	8	9	10	11	12	13
Late May	3,287	1,962	59.6							3,287	1,967	59.6
Early June	2,300	945	41.1							2,300	945	41.1
Mid-June	3,631	1,667	46.0							3,631	1,667	46.0
Late June	852	292	34.2							852	292	34.2
Total	6,783	2,904	42.8							6,783	2,904	42.8
Early July	864	476	55.1				685	526	76.8	1,549	1,002	64.6
Mid-July				6,893	3,756	54.4				6,893	3,756	54.4
Late July				6,344	4,299	67.7	350	350	100	6,694	4,649	69.4
Total	864	476	55.1	13,237	8,055	60.8	1,035	876	84.6	15,136	9,407	62.1

Table 5.23—(*Contd.*)

1	2	3	4	5	6	7	8	9	10	11	12	13
Early August	7,960			15,272	7,076	46.3				15,272	7,076	46.3
Mid-August		3,308	41.5	6,098	3,053	50.0				14,058	6,361	45.2
Late August	1,880	484	25.7	1,975	577	29.2				3,855	1,061	27.5
Total	9,840	3,792	38.5	23,345	10,706	45.9				33,185	14,498	43.6
Early September	1,080	350	32.4	2,250	929	41.3				3,330	1,279	38.4
Mid-September				62	48	77.4	582	206	35.4	644	254	39.4
Late September							265	55	20.7	265	55	20.7
Total	1,080	350	32.4	2,312	977	42.2	847	261	30.8	4,239	1,588	37.4
Grand Total	21,854	9,484	41.2	38,894	19,738	50.7	1,882	1,137	60.4	62,630	30,359	48.4

Table 5.24. Results of rearing post-larvae on various feeds

(Data for 1961 obtained from *Penaeus* Culture Company)

Period of spawning	Feed													Total		
	Artemia			Artemia and adhesive diatoms			Artemia and adhesive diatoms of Limnodrilus			Adhesive diatoms and Nematoda					
							Number of larvae $\times 10^3$								
	P_1	P_7	Survival rate %	P_1	P_7	Survival rate %	P_1	P_7	Survival rate %	P_1	P_7	Survival rate %	P_1	P_7	Survival rate %
1	2	3	4	5	6	7	8	9	10	11	12	13	14	15	16
Late May							1,962	762	38.8				1,962	762	38.8
Early June							945	265	28.0				945	265	28.0
Mid-June	1,667	762	45.7										1,667	762	45.7
Late June	292	103	35.3										292	103	35.3
Total	1,959	865	44.1				945	265	28.0				2,904	1,130	38.9
Early July	307	185	60.2	695	413	59.4							1,002	598	59.8
Mid-July				3,634	1,551	42.7	61	15	24.6	61	18	29.5	3,756	1,584	42.1
Late July	137	36	26.2	4,452	878	19.7				60	0	0	4,649	914	19.7
Total	444	221	49.9	8,781	2,842	32.3	61	15	24.6	121	18	14.8	9,407	3,096	32.9

Table 5.24—(*Contd.*)

1	2	3	4	5	6	7	8	9	10	11	12	13	14	15	16
Early August	2,369	695	29.3				2,928	705	24.0	1,779	98	5.5	7,076	1,498	21.1
Mid-August	1,584	491	30.9	3,918	969	24.7	858	225	26.2				6,361	1,685	26.5
Late August										1,061	47	4.4	1,061	47	4.4
Total	3,954	1,186	30.0	3,918	969	24.7	3,786	930	24.5	2,840	145	5.1	14,498	3,230	22.2
Early September										1,279	160	12.5	1,279	160	12.5
Mid-September										254	21	8.2	254	21	8.2
Late September										55	1	1.8	55	1	1.8
Total										1,588	182	11.4	1,588	182	11.4
Grand Total	6,359	2,272	35.7	12,699	3,811	30.0	6,754	1,972	29.2	4,549	345	7.6	30,359	8,400	27.6

1.5 Facilities for seedling production

The seedling production of *Penaeus* is usually carried out in separate land installations. Theoretically, one would assume that the eggs could be allowed to grow in ponds in which adult prawns are kept, but it has been confirmed that the rearing of larvae in a separate place is much more efficient.

The equipment for seedling production differs for rearing the hatched larvae up to the juvenile stage, for the culture of feed organisms required for the growing larvae, for supplying and draining the water, for aeration, for thermostatically maintaining the water temperature of the rearing tanks, and for filtering the water in the same.

1.5.1 TANKS FOR REARING HATCHED LARVAE

Water tanks used for rearing the hatched larvae of *Penaeus* are usually 2 m × 1 m × 1 m in size and kept indoors. Since the larvae of *Penaeus* are very sensitive to the quality of water, the quality must be carefully examined before rearing is started. The tanks are usually made of concrete or wood and, when of concrete, the inner surface is painted with waterproof mortar and any roughness smoothed. If the tanks are made of wood, the inner surface is repeatedly painted with vinyl paint (the product sold as Venylex is said to be good for this purpose).

The results of rearing the larvae of *Penaeus* in water tanks made of various materials are shown in Table 5.25.

While in concrete tanks, the larvae died soon after metamorphosing from *nauplius* to *zoea*; when *nauplius* were transferred to water tanks made of tiles, *zoea* grew well into the post-larval stage (Experiment Nos. 1 and 2). Results of rearing the larvae in acryl resin tanks were better than in concrete tanks (Experiment No. 3). A comparison of the results of rearing in wooden, acryl resin, and tile tanks, shows that the survival rate of larvae from egg to late post-larvae in these tanks were 0.24%, 0.7%, and 5.4% respectively. Thus tile tanks proved obviously the best type. This led to the conclusion, too, that the inner surface of the water tanks should be covered with tiles for rearing the larvae of *Penaeus*.

The foregoing results were obtained when the rearing was carried out in model tanks 2 m × 1 m × 1 m in late September, a period when the spawning of *Penaeus* is complete. This is also the period when rearing is possible indoors by using cultured *Skeletonema* as feed. The results would differ if the rearing method or the period differed. As will be explained later, the results of rearing are good even in concrete tanks if the large-scale seedling production method is applied.

When well-matured female *Penaeus* are kept in rearing tanks and the water temperature maintained around 25°C, spawning occurs in the night of the same day the prawns are placed in the tanks. The larvae hatch within 12 hours of spawning, and *zoea* emerge at 1.5 days after hatching. When a sufficient amount of feed is made available to the larvae, the larvae pass through the *zoea* phase in 4 days, and *mysis* in 3. Even after the transformation into post-larvae, it is better to prolong the rearing in tanks for at least 3 days during the period when they show active movement. In this way, the survival rate is improved.[3]

[3]According to the rearing method as of 1963.

Table 5.25. Results of rearing *Penaeus* larvae in water tanks made of various materials
(Data for 1962 obtained from *Penaeus* Culture Company)

Experiment number	Material	Capacity of water tank (tons)	Number of larvae $\times 10^3$						Survival rate from egg stage, %	Remarks
			Eggs	*Nauplius*	*Zoea*	*Mysis*	Post-larvae	Late post-larvae		
1	Tiles	0.2	—	—	5	5	2	3.5		Shifted as nauplius
	Concrete	8.0	1,400	640	0				0	Control group
2	Tiles	0.2			5	5	4	2.8		Shifted at mysis stage
	Tiles	0.2			5	5	3	2.4		Shifted at mysis stage
	Concrete	2.0	800	320	0				0	Control group
3	Acryl resin	2.0	1,000	500	350	200	105	25.8	2.6	Water not changed
	Concrete	2.0	500	0					0	
	Concrete	2.0	800	250	180	110	50	10.6	1.3	
4	Tiles	2.0	700	180	180	150	90	38.0	5.4	Water changed at mysis stage
	Acryl resin	2.0	1,000	470	350	160	10	7.0	0.7	
	Wood	2.0	800	350	230	70	40	7.3	0.9	
	Wood	2.0	600	110	20	0			0	
	Wood	2.0	500	90	30	1	1	0.1	0.02	
	Wood	2.0	300	40	10	1	0		0	
	Wood	2.0	150	60	40	20	6	0.4	0.3	

The ideal condition of rearing tanks for producing seedlings is to have 1 tank for spawning (expressed as E), 2 for the *nauplius* stage (N), 4 for *zoea* (Z), 3 for *mysis* (M), and 4 for post-larvae (P).[4] Thus it is possible to produce one tankful of post-larvae every day by allowing the prawns to spawn in one tank daily, and keeping $1+2+4+3+4=14$ tanks ready for rearing. If the average number of eggs released per tank is about 450×10^3, the number of late post-larvae produced per tank will be $45 \times 10^3 \times (0.50 \times 0.50 \times 0.40 \times 0.50 \times 0.50) \simeq 11 \times 10^3$. Hence, the number of juvenile prawns which can be expected under these conditions will be $11 \times 10^3 \times 0.7 \simeq 8 \times 10^3$.

[4]Three are enough, but after taking out the post-larvae, it requires one day to get the spawning tanks ready; hence $3+1=4$.

If, therefore, the number of rearing tanks is n and the number of days on which the mother prawns are available is t, the number of juvenile prawns, Y, producible in one year can be expressed as

$$Y = 8 \times 10^3 \times n/14 \times t.^5$$

For the years 1961 and 1962, as shown in Table 5.13, $n=140$, $t=100$. Hence the number of juvenile prawns which can be produced in one year is $8 \times 10^3 \times 140/14 \times 100 = 8,000 \times 10^3$. But the number of juvenile prawns actually produced in one year was, as shown in Table 5.13, about one-half to one-third the calculated yield. Hence, in order to produce seedlings according to plan, it is necessary to make arrangements for rearing an excess number of larvae in 2n to 3n rearing tanks.

1.5.2 REARING ROOMS FOR HATCHED LARVAE

Penaeus larvae are reared indoors to avoid direct sunlight and rain. During the early period of large-scale seedling production, it was assumed that *zoea*, which require bright conditions, collected at a particular place where there was sufficient light and, because of feces accumulation died. To prevent such mortality, darker conditions were considered best in the rearing rooms.

But experience has proven that optimum light intensity makes it possible to maintain the diatoms used as larval feed in good physiological condition for a longer period, and furthermore that the larvae are not adversely affected through dispersal when the air is circulated.

The survival rate of larvae reared in tanks under various conditions of brightness is given in Table 5.26.

When wooden tanks are used for rearing, the survival rate improves when the tanks are kept in bright rooms. Results differ with concrete tanks according to the depth of the water and the method of aeration, but nevertheless the survival rate was good when the tanks were placed in bright rooms. The results were even better when the tanks were placed outside in the month of August when the temperature is very high, and the light rays allowed to pass through the vinyl plate covers, outstripping the results of those reared in dark rooms.

1.5.3 FEED CULTURE TANKS

At present, the culture of feed organisms is carried out simultaneously in the larval rearing tanks. The need for separate culture tanks is thereby lessened, while the role of diatoms in the rearing of larvae acquires greater significance. It does not appear out-of-place to include a brief account here of the history of diatom culture.

Skeletonema costatum, the feed for *zoea,* was cultured in well-lighted rooms. Water tanks of more or less the same size as the rearing tanks were used, together with various methods of isolating the *Skeletonema* from natural sea water for culturing, but the following was more common.

One end of a glass tube of 5 mm inner diameter was blown into a fine pipette. *Skeletonema* taken from natural sea water were placed under a microscope and a long

---

[5]The number of mother prawns required for this purpose can be made available.

Table 5.26. Survival rate of larvae reared under various conditions of brightness

(Data for 1962 obtained from *Penaeus* Culture Company)

Number of eggs and larvae ×10³; figures in parentheses indicate survival rate (%)

Place of rearing tanks	Nature of rearing tanks	Period of spawning	Eggs	Nauplius	Zoea	Mysis	Post-larvae	Late post-larvae	Survival rate from egg stage, %	Tank size
Dark room	Wooden tank	Early August	77,670 (30)	23,460 (15)	3,472 (15)	530 (35)	183 (34)	61.9	0.8	2 m × 1 m × 1 m
		Mid-August	30,060 (12)	3,840 (2)	90 (44)	40 (75)	30 (33)	9.8	0	
		Late August	7,050 (14)	1,030 (71)	740 (27)	203 (90)	182 (55)	100.1	1.4	
		Early September	12,000 (14)	1,760 (45)	80 (37)	30 (66)	20 (50)	10.2	0.1	
		Mid-September	600 (23)	140 (0)	0	0	0	0	0	
		Late September	3,800 (33)	1,230 (45)	560 (38)	214 (34)	73 (29)	21.4	0.6	
Slightly bright room	Wooden tank	Early August	24,620 (28)	7,000 (23)	1,611 (34)	558 (41)	227 (47)	108.5	0.4	2 m × 1 m × 1 m
		Mid-August	31,250 (19)	5,919 (16)	940 (53)	500 (69)	345 (99)	342.0	1.1	
		Late August	10,600 (13)	1,390 (1)	20 (75)	15 (66)	10 (100)	11.9	0.1	
Open air	Wooden tank	Early August	1,800 (51)	1,090 (24)	260 (79)	206 (94)	194 (46)	89.7	5.0	2 m × 1 m × 1 m
		Mid-August	2,900 (67)	1,930 (41)	890 (48)	430 (72)	310 (59)	182.3	6.3	
		Late August	7,650 (31)	2,350 (23)	530 (60)	316 (65)	206 (40)	83.5	1.1	
		Early September	6,280 (25)	1,580 (43)	690 (4)	30 (100)	30 (57)	17.1	0.3	
		Mid-September	1,000 (22)	220 (0)	0	0	0	0	0	
Dark room	Concrete tank	Early August	5,760 (26)	1,500 (50)	750 (100)	880 (55)	482 (61)	292.6	5.1	2 m × 1 m × 1 m
		Mid-August	4,800 (44)	2,110 (2)	30 (40)	20 (30)	10 (100)	19.9	0.4	
		Late August	10,450 (19)	1,980 (0)	0	0	0	0	0	
		Early September	4,900 (17)	810 (7)	60 (0)	0	0	0	0	
		Mid-September	4,600 (17)	800 (33)	260 (62)	160 (63)	100 (39)	39.2	0.9	
		Late September	750 (7)	50 (4)	2 (100)	2 (50)	1 (0)	0	0	

Open air	Concrete tank	Mid-August	1,400 (20)	280 (29)	80 (100)	110 (50)	55 (76)	41.8	3.0	100 cm × 50 cm (cylindrical tank)
		Late August	2,000 (65)	130 (30)	40 (100)	40 (78)	78 (31)	10.4	0.5	
	Concrete tank	Mid-August	17,240 (47)	8,110 (10)	800 (24)	190 (100)	120 (55)	66.4	0.4	4 m × 1 m × 1 m
		Late August	7,300 (34)	2,480 (52)	130 (0)	0	0	0	0	
Indoors		Early August	3,800 (75)	2,840 (74)	2,100 (51)	1,070 (100)	2,046 (25)	504.4	13.3	4 m × 2 m × 1 m
		Mid-August	400 (10)	40 (0)	0	0	0	0	0	
	Concrete tank	Late August	16,200 (79)	12,720 (31)	3,980 (74)	2,950 (66)	1,950 (76)	1473.3	9.1	
		Early September	12,000 (55)	6,620 (12)	800 (50)	400 (80)	320 (80)	253.1	1.2	
		Mid-September	1,000 (60)	600 (17)	100 (0)	0	0	0	0	
Bright room	Concrete tank	Mid-August	4,900 (22)	1,110 (28)	310 (55)	170 (88)	150 (78)	117.6	2.4	4 m × 1 m × 1 m
		Late August	9,500 (87)	8,300 (43)	3,580 (68)	2,450 (50)	1,220 (59)	721.6	7.6	
		Early September	17,900 (50)	8,890 (87)	775 (55)	430 (41)	175 (40)	69.9	0.4	
		Mid-September	9,700 (16)	1,560 (33)	510 (69)	350 (69)	240 (54)	129.0	1.3	
	Concrete tank	Early August	3,500 (100)	4,400 (36)	1,600 (62)	985 (87)	855 (30)	256.9	7.2	3 m × 1.5 m × 0.5 m
		Mid-August	25,300 (70)	17,730 (34)	6,020 (49)	2,930 (56)	1,630 (21)	349.8	1.4	
		Late August	29,800 (58)	17,180 (18)	3,080 (20)	610 (39)	239 (39)	92.5	0.3	
		Early September	30,970 (32)	9,970 (19)	1,870 (20)	374 (46)	172 (62)	106.2	0.3	
Open air	Concrete tank	Late August	10,500 (63)	6,620 (0)	0	0	0	0	0	0 8 m × 2 m × 0.5 m
		Early September	8,800 (21)	1,880 (0)	0	0	0	0	0	

chain of their cells drawn out with the pipette. A culture fluid was then inoculated with these cells, and the flask stored in a room where sun rays could not fall directly on it. If the room were very dark, an electric bulb was used to provide enough light. When the *Skeletonema* growth was healthy, a drop of its culture fluid was drawn out and transplanted to another flask of culture fluid. This transplanted drop of *Skeletonema* was taken before the latter could grow beyond the desired limit. Depending upon the feed requirements the cultured *Skeletonema* in small flasks were transferred to larger flasks containing a freshly prepared culture fluid. An intermediate culture in glass water tanks of 10 to 12 liter capacity was grown before transferring to tanks of a two-ton capacity.

During flask culture, the culture fluid was sterilized by heating, and the flask shaken regularly to avoid the settling of diatoms. In tank culture, salt-nutrients were added to the culture fluid, which was filtered through a layer of sand, and diatoms kept from settling by air circulation.

According to the treatment method of Matsue (1954), salt-nutrients were added to the sea water during the flask culture stage as follows: 202 mg potassium nitrate (KNO_3), 20 mg potassium diphosphate (K_2HPO_4), 13 mg potassium silicate (K_2SiO_3), and 3 mg ferric chloride ($FeCl_3$) per liter of water. In large-scale culture, these amounts were reduced by one-half to one-fourth, and sometimes potassium silicate and ferric chloride were not added.

When the density of cultured *Skeletonema* became 50×10^3 to 100×10^3 cells per 1 cc, it could be used as feed for *Penaeus* larvae. The optimum feeding density of *Skeletonema* reared indoors was 10×10^3 cells per 1 cc. Although the *Skeletonema* were eaten by the larvae in the rearing tanks, a large portion were not and these settled to the bottom. Also, it is very difficult to maintain an optimum density of *Skeletonema* as they reproduce and grow depending upon the conditions of brightness in the rearing tanks.

The amount of *Skeletonema* provided for rearing larvae is 50 to 100 liters just before the *zoea* metamorphose, and afterward, 50 liters 3 to 4 times per day or an average of 200 liters daily.

When the culture tank and the rearing tank were $2 \, m \times 1 \, m \times 1 \, m$ in size, and the water depth 0.8 m, the number of rearing tanks in which larvae could be fed with *Skeletonema* was $2 \, m \times 1 \, m \times 0.8 \, m/0.2 \, m^3 = 8$ tanks. The condition at which *Skeletonema* attain a density in the culture tank suitable for feeding larvae depends upon the water temperature, intenstity of light, and density at the time of inoculation. This usually requires 3 to 4 days. Thus, if *Skeletonema* is to be used as feed on the third day following the inoculation, the number of culture tanks required per one rearing tank becomes $1/8 \times (1+3) = 1/2$ tank.

The condition discussed above applies to a dark rearing room, but when the rearing room is bright, *Skeletonema* grow healthily in tanks, and the number of culture tanks required for rearing becomes less. In rearing rooms where the roof is covered by semitransparent vinyl and brightness maintained even at night with electric lamps, only 1/10th of a tank of *Skeletonema* culture is required per larvae rearing tank.

1.5.4 EQUIPMENT FOR WATER SUPPLY AND DRAINAGE

The sea water used for growing larvae and culturing *Skeletonema* is generally drawn by pumps from the coastal areas in which the hatcheries are located. The sea water thus drawn is filtered, and only then run into the culturing or rearing tanks. Asbestos-cement tubes, hard vinyl tubes, polystyrene tubes, etc. which are harmless to the larvae, are used for water supply; tubes of cast iron, steel, lead, copper, or zinc cannot be utilized, or rather should not, but sometimes it is impossible to avoid using steel tubes or cast iron ones for plumbing, plugging, etc.

The sea water thus drawn is divided into (1) water for hatching, (2) water for growing *Skeletonema*, (3) water for late post-larvae, and (4) water for miscellaneous purposes. For (1) and (2), only filtered sea water is used[6] but for (3) and (4), even if the water contains other unwanted substances, it is usable.

In planned production, the maximum amount of sea water for one hour is determined from the amount of sea water used for one day. Again, the total amount of sea water run into the rearing tanks is determined from the plumbing plan, and on this basis the decision on the type of pump and its output made.

The types of pumps used for drawing water from the sea can be broadly classified as the eddy pump,[7] the axial-flow pump, and the slanting-flow pump. The difference between the three pumps lies in the shape of the runner. The direction of the flow of water from the runner is radial in the eddy pump, and axial in the axial pump. The curvature of water flow is gradual in the eddy pump; as the amount of water increases, so does the axial power. The curvature is sharp in the axial pump; as the amount of water increases, the axial power rapidly declines. The curvature in the slanting pump is medium; as the water increases, the axial power decreases but, compared to the axial pump, its decrease is less.

When the difference between the high and ebb tides is relatively large, if the water is to be drawn even during low tide, the eddy pump is commonly used.[8] However, when the difference between the high and ebb tides is relatively narrow, and the total level is low, the slanting pump is more economical.[9]

The volume of sea water in one rearing tank and one juvenile prawn pond is $v\mathrm{m}^3$ and $v'\mathrm{m}^3$ respectively. The necessary amount of sea water for one day when the number of rearing tanks and juvenile prawn ponds are n and n' respectively, is given below. Here the number of rearing tanks is multiplied by 14, this being considered the most suitable number (as explained under section 1.5.1).

Amount of sea water necessary for rearing tanks with hatched larvae...$1/14\ nv$ (1)

Amount of sea water necessary for fresh *Skeletonema* culture tank ...$1/4 \times 1/2\ nv$ (2)

Amount of sea water necessary for exchanging water in juvenile prawn tanks (water changed 4 times per day)...$4\ n'v'$ (3)

Amount of sea water used for temporary storage of mother prawns ...about $1/14\ nv$ (4)

[6] Filtering through sand has recently been established as satisfactory.

[7] Eddy pumps are of two types, the burette pump without runner and the turbine pump with runner; the latter is used in high-level areas, and the former for culture experiments.

[8] Since the automatic water drawing type does not have a covering lid in the strainer, no damage results and it is also convenient since it does not require vacuum pumps.

[9] This pump may be used at levels below 8 m.

Amount of sea water used for early post-larvae...about $1/14\ nv$ (5)

Amount of sea water used for juvenile prawns...about $1/20\ n'v'$ (5')

Amount of sea water used for washing the rearing tank or culture tank

   ...about $(1/14\ nv + 1/4 \times 1/2\ nv)$ (6)

Amount of sea water used for washing juvenile prawn ponds...about

   $1/20\ n'v'$ (6')

Amount of sea water used for washing the feed and implements...

   about $1/14\ nv$ (7)

   and about $1/20\ n'v'$ (7')

In the steps mentioned above, filtered sea water must be used for (1) and (2), but since it is better to have the same type of piping for (4), (5), (6), and (7), filtered sea water is used in these steps also. Hence, the necessary amount of filtered sea water per day is:

$$(1)+(2)+(4)+(5)+(6)+(7)=0.607\ nv \tag{8}$$

Since raw sea water can be used for (3), (5'), (6'), and (7'), the required amount per day is:

$$(3)+(5')+(6')+(7')=4.15\ n'v'=4.15 \times 20/14\ nv'=6\ nv' \left(\because n' = \frac{20}{14} \right) \tag{9}$$

Thus the necessary amount of sea water per day is:

$$(8)+(9)=0.6\ nv+6\ nv' \tag{10}$$

If $v=2\,\mathrm{m} \times 1\,\mathrm{m} \times 0.8\,\mathrm{m}$, $v'=3\,\mathrm{m} \times 1.5\,\mathrm{m} \times 0.4\,\mathrm{m}$,

   then $(10)=0.6 \times 1.6 + 6 \times 1.8\ n=12\ n$ (11)

While it is enough to pump in water for step (3) at intervals of 24 hours, for the other steps, water has to be pumped in at fixed hours each day. If it is assumed that the water is drawn at intervals of 6 hours, the necessary amount of sea water per unit time is:

$$[(1)+(2)+(4)+(5)+(6)+(7)+(5')+(6')+(7')] \div 6 + (3) \div 24$$

$$= \frac{1}{6} \left(0.6\ nv + \frac{3}{20}\ n'v' \right) + \frac{1}{24} \times 4\ n'v' = \frac{1}{6}\ (0.6\ nv + 1.15\ n'v')$$

$$= \frac{1}{6}\ (0.6\ nv + 1.64\ nv') = 0.65\ n \tag{12}$$

Given that $n=14, 28, 42,\ldots$ when 9 tons of sea water are required per hour, this amount must be increased by 9 tons for every 14 additional tanks.

1.5.5 AERATION EQUIPMENT

Aeration equipment is very essential for the production of *Penaeus* seedlings. By aeration the dissolution of oxygen is accelerated and the water stirred continuously, which improves the rearing environment.

Table 5.27 shows the rate of oxygen dissolution in sea water by aeration. When a turbo-blower with a low discharge force is used as the fan, the value of the coefficient of oxygen dissolution is much lower than when a compressor with a high discharge force is employed. If an air diffusion tube has been fitted to the aeration tube, the oxygen dissolution varies, i.e. the dissolution differs according to the size of the air bubbles. When the bubbles are very minute, the coefficient of oxygen dissolved is

Table 5.27. Rate of oxygen dissolved in sea water by different methods of aeration

Type of equipment	Water temperature (°C)	Water depth (m)	Amount of water m³	Amount of air blown in m³/hr.	Coefficient of oxygen dissolved/hr.	Remarks (Presence or absence of air strain)
	23.5	0.770	1.400	0.756	0.101	Absent
	29.0	0.440	0.734	0.786	0.387	Absent
Turbo-blower	24.0	0.385	0.700	0.756	0.384	Absent
	28.5	0.315	0.525	0.786	0.401	Absent
	25.0	0.440	0.742	0.503	0.598	Present
	26.0	0.800	1.427	1.078	0.834	Present
Compressor	27.0	0.613	1.094	1.078	1.090	Present
	29.0	0.440	0.769	1.078	1.701	Present
	25.0	0.600	1.049	1.397	0.510	Absent

high. When an air diffusion tube is not fitted, the coefficient of the oxygen dissolved is low, even though the amount of air blown in is greatly increased. Since oxygen dissolution differs according to the method of aeration, it is necessary to study the conditions under which rearing is planned.

Usually a compressor is used for circulating air in the rearing water. However, since the objective is to blow in air at a high discharge force, the amount of air discharge is low compared to the force used. On the other hand, with a turbo-blower, even though the amount of air discharged is high and the discharge force low, the efficiency is poor. The water depth of the rearing tank is 1 m or often even 2 m. Hence it is more effective to use an equipment with a discharge pressure of 0.2 to 0.3 kg/cm², and capable of blowing in a relatively large amount of air.

The turbo-blower seems to be suited to this purpose and the structure of it has certain advantages. For example, the rotating element is not welded and the air is blown from between the casing and the rotating element. There is no oil vapor in the circulating air and as a blower for the rearing water, it seems to be ideal.

The distributing pipes of the air circulator are made of gas tubes (zinc-steel tubing) or chloride (hard vinyl tubing) of 2 to 3 inches diameter. It extends to the rearing tanks and inside them passes through 2-way or 4-way cocks, connected by a chloride tube (soft vinyl) of about 4 to 5 mm inner diameter. The anterior end is fitted with the air diffusion tube[10] which is placed at the bottom of the water tank. The number of air diffusion tubes is calculated on the basis of 1 tube per 1 m² of bottom surface.[11] The amount of air which is required to be blown for one tank of 2 m × 1 m × 1 m, as calculated from Table 5.27 is $1.078/60 \times 2 = 0.036$ m³/min, if one circulating tube is used for 1 m² of area. Hence it is possible to determine the size of the blower.

1.5.6 TEMPERATURE MAINTENANCE

The optimum temperature suitable for spawning and rearing of larvae of *Penaeus*

---

[10] Commercial airstone can be used.

[11] In large-scale seedling production, the main tube is attached to a soft vinyl chloride tube of 9 mm diameter and fitted with a cylindrical, polished whetstone of 9 cm length × 3 cm diameter to the anterior edge. One tube is used per 5 to 10 m².

is 25 to 30°C. Hence, if larvae have to be reared in a season with low water temperature, it is necessary to increase the temperature of the rearing water. The increase as well as the maintenance of temperature can be either for the whole room in which the rearing is carried out, or only for the rearing tank.

Electric heaters are used as the source of heat for the rearing tanks when rearing is carried out on a small scale. For rearing tanks of $2 \text{ m} \times 1 \text{ m} \times 1 \text{ m}$, two types of heaters of 0.5 kw and 1 kw are used. The use of electric heaters for rearing is both convenient and clean, though not very economical. During large-scale rearing, some other source for heating and maintaining temperature should be found.

When a large amount of rearing water is to be temporarily heated, either steam from a boiler is directly introduced into the rearing water, or steam or hot water passed through coils in the water tank; in both instances, temperature is controlled. However, if the vinyl hose now commonly used in agricultural fields were utilized, it would prove more economical as solar heat could be obtained during the day and stored through the night.

The size of the steam boiler and the length of the tubes used for heating the water tanks can be determined in the following way:

The water temperature at the initial stage is taken as t_1 (°C), the temperature at which the water is to be maintained as t_2 (°C), and the amount of water used for rearing as Q (ton). The quantum of heat required for increasing the water temperature of Q (ton) water from t_1 (°C) to t_2 (°C) is,

$$Q \times 1,000 \times (t_2 - t_1) \text{ (kcal)}.$$

If the steam is directly introduced and the temperature increased in h (hours), since the quantum of heat for 1 atm of steam is 639 kcal/kg (at 100°C), the amount of steam required, W_0 (kg/h) is:

$$W_0 = Q \times 1,000 \times (t_1 - t_2)/(639 - t_2)h \text{ (kg/h)}.$$

However, heat loss from water surface, surroundings, and bottom occurs at the time of temperature increase. If the area of water surface is A (m^2), the heat escape rate from the water surface, due to direct cooling and vaporization, is k_1 (kcal/m², h, °C), and the temperature of the atmosphere ta (°C), then the heat loss from the water surface L_1 (kcal/h) is:

$$L_1 = k_1 \times A \times (t_2 - ta) \text{ (kcal/h)}.$$

When the major portion of the sides of the tanks is sunk into the ground, the heat loss L_2 (kcal/h) from the sides and bottom of the tank is as given below. Here, if the area of the sides and the bottom is take as B(m^2), the heat escape rate from the sides and the bottom as k_2 (kcal/m², h, °C), and the ground temperature as te (°C), then

$$L_2 = k_2 \times B \times (t_2 - te) \text{ kcal/h}.$$

Hence the total requirement of steam W_0' (kg/h) is:

$$W_0' = W_0 + \frac{L_1 + L_2}{639},$$

$$= Q \times 1,000 \times (t_2 - t_1)/(639 - t_2) \ h + [k_1 A(t_2 - ta) + k_2 B(t_2 - t_2)]/639 \text{ (kg/h)}$$

Now, if conditions for a large water tank of $10\,m \times 10\,m \times 2\,m$ are assumed as: $t_1 \simeq 10°C$, $t_2 \simeq 25°C$, $ta \simeq 20°C$, $te \simeq 14°C$, $k_1 \simeq 15$ kcal/m^2, h, °C, $k_2 \simeq 1.5$ kcal/m^2, h, °C, and $h = 24$ hours, while $Q = 160$ tons, $A = 100$ m^2 and $B = 180$ m^2, then

$$W_0' = \frac{160 \times 1,000\,(25-10)}{(639-25) \times 24} + \frac{15 \times 100\,(25-20) + 1.5 \times 180\,(25-14)}{639},$$

$\simeq 180$ kg/h.

The length of the heating tube when the heating area of the coil is F(m^2) and the heat escape rate of the heating tube k(kcal/m^2, h, °C), can be determined since

$$F = Q \times 1,000 \times (t_2 - t_1)/k(100 - t_2)\,h\ (m^2).$$

In the foregoing example, if a steel steam tube of 1 1/2 inch diameter is used, since $k = 400$ kcal/m^2, h, °C, the surface area of the tube 0.15268 m^2/m, and $F = 160 \times 1,000 \times (25-10)/400(100-25) \times 24 = 3.3$ m^2, the length of the heating tube is 3·3/0.15268 $\simeq 22$ m.

In some special cases, hot springs can be used for maintaining the temperature. At the Subtropical Fisheries Biological Research Station of Kagoshima, water from the hot springs, plentiful in Narikawa-hama in Yamakawa Machi, is passed through copper tubes in the water tanks. The results of rearing juvenile prawns in these "hot spring" tanks were said to be very good.

1.5.7 FILTRATION TANKS

Formerly the sea water used for rearing the larvae is cleaned thoroughly by filtering, but now raw sea water is widely used.[12] Nevertheless, depending upon the region and the season, it is sometimes necessary to clean the water.

Sea water is generally filtered with sand or pebbles, but sometimes sedimentation, flocculating, or biological processes are used to remove microorganisms, minute mud particles, organic matter, etc. Of these methods, flocculation is important as it holds the microorganisms in the filter membrane.

The filter is rectangular and a few square meters large, with a water collecting tube at the bottom. The filter layer is about 2 m thick with pebbles on the lower part, and fine sand in the upper parts.[13] The water is allowed to flow through grooves on the sides of the tanks and the filtered water collected in the collecting tube at the bottom. By using the control valve, the filtered water is drawn out.

The amount of filtered water, Q (m^3/day) is determined on the basis of the area of the filter bed A (m^2) and the rate of filtration V (m/day).

$$Q = A \times V\,(m^3/day).$$

Since V = 100 to 150 m/day, the rate of filtration per hour V' (m/h) is V' $\simeq 5$ m. The amount of filtered sea water Q' (m^3/h) required per hour for a rearing tank is

$$Q' = 1/6 \times 0.607\ nv \simeq 0.1\ nv.$$

[12]Raw sea water drawn through a pump is filtered in sieves of 60 to 100 mesh whereby large objects and large animals, which feed on *Penaeus* larvae, are removed. The animal and plant planktons which serve as food for *Penaeus* larvae are retained.

[13] In some filtering tanks, sand particles of the same size are used in all the filter layers.

Here, n is the number of rearing tanks and v is the amount of water in the rearing tanks. If $n=14$ and $v=1.6$ ton, then

$$A=Q'/V'=0.1\times14\times1.6/5\cong0.45\ m^2.$$

Thus the size of filter tank suitable for a rearing tank can be determined.

Since the efficiency of filtering becomes poor due to the settling of the sand layer used in the filter tanks, it is necessary to pass compressed air and water through the bottom of the tanks to stir the pebbles and sand.

1.6 Mass culture method of seedlings

The rearing of larvae of marine animals is difficult particularly in their early stages. In fact, at present *Penaeus* is the only marine species for which seedlings are produced on a commercial scale. However, the fundamental principles for rearing most marine larvae are the same: (1) the water used is carefully filtered sea water, (2) direct light or luminous conditions are avoided, and (3) the rearing of early larvae is carried out by providing standard feed (for *Penaeus*, cultured *Skeletonema*).

As far as the rearing of larvae under strict conditions is concerned, the size of the rearing tank is usually restricted to $2\ m\times1\ m\times1\ m$ or a few times larger. When it is necessary to increase the number of juvenile prawns, the equipment for feed production is also larger and so it is difficult to lower the cost of seedling production.

Hudinaga and Kittaka (1966) carefully studied the demand and supply position of feed for the early larvae of *Penaeus* and the rearing environment. The results showed that (1) raw sea water can be used for rearing the larvae, (2) when *Skeletonema* are used as feed, bright conditions in the rearing tanks are better and the larvae can be reared in the open air, and (3) the standard feed of the larvae used in the past need not be strictly adhered to, as various alternates can be given to the larvae at various stages.

The relationship between the larval stage and its feed is shown in Table 5.20, and it is clear that both the *zoea* and *mysis* of *Penaeus* may be given plant and animal feed, provided these float and are rich in nutrition. During the early post-larval period when the larvae are actively floating, animal feed capable of floating can be effectively used, and during the late post-larval period when the larvae are bottom dwelling, animal feeds which settle to the bottom (containing organic residual matter) are effective.

Hence, various methods of rearing can be devised by selecting feed suitable to the varying larval stages, and making a combination of these feeds for the whole series. Hudinaga and Kittaka (1967) succeeded in rearing the larvae of *Penaeus* in the same tank in which the feed organisms were also cultured. They succeeded by using various types of plankton found in raw sea water and effectively using them as a fertilizer in the rearing water. As compared to the old method of rearing, this one is simpler and more efficient. Hence this method has gained in popularity and at present is widely used in most of the commercial seedling producing areas of Japan. Although there are some fine differences in the methods adopted in different areas or in different seasons, broadly speaking the principles are identical. Details of the Hudinaga and Kittaka experiments of 1964 follow.

Open tanks of $10\,m \times 10\,m \times 2\,m$ filled with raw sea water filtered through a synthetic sieve of 60 to 100 mesh, were used for rearing outdoors. Mature specimens of *Penaeus* were introduced into the tanks and allowed to spawn. The water was not changed until a few days after the larvae metamorphosed into post-larvae. Later, 1/5th of the total water was changed daily. Throughout the rearing period, air was passed through the tank. During rearing, feed organisms were not separately cultured, but directly introduced into the rearing water and nutrient-salts added. The nutrient-salts consisted of potassium nitrate as a nitrogen source and potassium diphosphate as a phosphorous source. After the metamorphosis into post-larvae, depending upon the density of larvae and the rate of growth, *Artemia*, short-necked clam, etc. were provided as feed. The results are given in Table 5.28.

Of the 22 tanks used in the experiments, rearing in 8 tanks was stopped midway because of bad results. The results in 3 tanks were particularly poor. In all the others, 200×10^3 larvae attained the post-larval stage. Of 14 rearing tanks in which the larvae were reared up to the juvenile prawn stage, 4 tanks had less than 800×10^3 juvenile prawns. Rearing was first begun in 3 of these 4 tanks. One tank required more than 100 days for the larvae to attain the post-larval stage. In other tanks, 800×10^3 to $1,250 \times 10^3$ juvenile prawns, averaging $1,000 \times 10^3$ per tank, were produced.

The relationship between the total amount of feed and the number of juvenile prawns produced in various tanks is shown in Figures 5.39 and 5.40. Even when the minimum amount of *Artemia* (1.3 kg) was provided, the production of juvenile

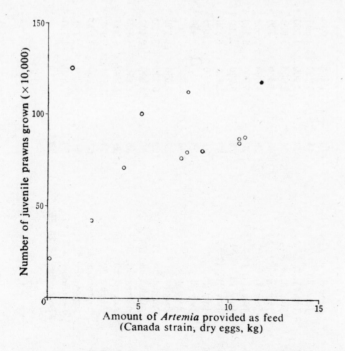

Figure 5.39. Relationship between the amount of *Artemia* used as feed and the number of juvenile prawns grown (Hudinaga and Kittaka, 1962).

Table 5.28. Experiments on seedling production in a large water tank (Hudinaga and Kittaka, 1967)

Experiment number	Date of mother prawns introduced (1964)	Number of mother prawns introduced	Number of larvae ($\times 10^3$)					Aggregate survival rate, %	Stages juvenile prawn	Remarks
			Nauplius	Zoea	Mysis	Post-larvae	Juvenile prawn			
1	June 30	18	1,500	230	300	150	209	14	P_{27}	
2	June 30	44	2,900	1,800	1,430	770	420	11	P_{30}	
3	July 1, 2	39	3,900	1,560	750	1,010	710	18	P_{28}	
4	July 3, 4	36	3,600	1,800	2,030	1,730	1,001	28	P_{30}	
5	July 5, 6	43	1,350	700						300×10^3 at P_{11}
6	July 7, 8, 10	55	3,000	2,550						Removed at Z_3
7	July 11, 12	70	3,900	3,210	2,100	900	807	21	P_{40}	
8	July 13, 14, 15	28	5,220	980	690	600				500×10^3 at P_6
9	July 16, 17	89	11,400	6,150	1,500	4,110	1,137	10	P_{34}	
10	July 26	31	3,600	2,100	1,430	990	1,251	35	P_{33}	200×10^3 at P_{11}
11	July 29	53		450	390	150				10×10^3 at P_4
12	August 6	198								
13	August 8, 9	117	7,350	6,990	3,300	1,500	892	12	P_{74}	
14	August 10, 11	109	6,300	3,150	3,600	830	770	12	P_{106}	
15	August 14, 15	58	3,900	2,400	2,400	1,200	884	23	P_{103}	
16	August 19, 20	36	1,350	750						300×10^3 at P_{11}
17	August 20	80	2,100	1,500	680	900	1,100	52	P_{63}	
18	August 30	74	1,130	900		150	859	76	P_{64}	
19	September 1, 2	67	2,250	1,800	1,800	990	809	36	P_{50}	
20	September 6	41	1,200	230						200×10^3 at P_7
21	September 9	30	1,650	1,730			1,202	73	P_{44}	
22	September 24, 25, 26, 29	38								Did not metamorphose into mysis

prawns was at its maximum, i.e. $1,250 \times 10^3$. However, by increasing the feed to 17.7 kg, the results were not appreciably different. When 18 kg of short-necked clams were provided up to P_{21} stage, $1,000 \times 10^3$ juvenile prawns were produced. The number of juvenile prawns did not vary appreciably even when the amount of clams was increased to 100 kg. From the figure, it is clear that 5 kg of *Artemia*[14] or 80 kg of clam meat is required to produce $1,000 \times 10^3$ juvenile prawns. Compared to the older methods of rearing, this amount is very small.[15]

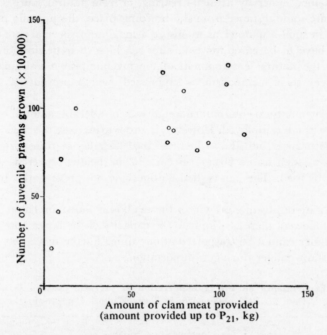

Figure 5.40. Relationship between the short-necked clam meat used as feed and the number of juvenile prawns grown (Hudinaga and Kittaka, 1967).

According to this method of rearing, even when there is an excess of feed organisms in the rearing tanks, either they are directly used by other larger-sized organisms or, when this is not the case, they rapidly decompose without harming the juvenile prawns. It is not necessary to be concerned about the deterioration of water quality due to the residual feed organisms, as the latter are ingested by *Skeletonema* and other organisms. This shows that when nutrient-salts are added for the culturing of *Skeletonema*, the organic nutritional substances can also be used directly in the early larvae just as are the inorganic salts mentioned earlier. Moreover, the standard feed of zoea→mysis, and floating post-larvae→bottom-dwelling post-larvae, i.e. *Skeletonema*→ *Artemia*→clam meat, were studied in view of the results shown in Table 5.20, and a wider range of rearing techniques subsequently devised.

[14] *Artemia* of Canada: hatching rate, 40%. Hence when converted to *Artemia* of America, the amount comes to 2.5 kg.

[15] In the older methods of rearing, 17.5 kg of *Artemia* (hatching rate, 40%) and 180 kg of clam meat were required for the production of $1,000 \times 10^3$ juvenile prawns (Hudinaga and Kittaka, 1967).

It was assumed in the past that special techniques and experience were required to rear the larvae of *Penaeus*, but according to the foregoing method, these larvae can be reared to the juvenile stage with little effort.

1.7 Transportation of juvenile prawns

In places where facilities for the production of juvenile prawns do not exist, these have to be obtained either by artificial rearing, or from natural sources. When the culture ponds are some distance from the breeding place, the juvenile prawns must be transported in such a manner as to ensure a maximum survival rate and, unlike adult prawns, have to be cared for even after reaching their destination. For these reasons, either the natural environment of the juvenile prawn is simulated during transportation, or their metabolism is suppressed, which prevents a lowering of their vitality.

Since a simulated natural environment requires a considerable amount of sea water, rafts or boats are usually employed. However, in order to increase the capacity for holding the juvenile prawns and also to increase the efficiency of transportation, a net box made of a wooden frame 60 cm × 40 cm × 10 cm (height) covered with a synthetic fiber net, is used. The net is tied to one side to prevent the prawns from escaping.

The survival rate of juvenile prawns in the net box is shown in Table 5.29. When this method is followed, there is a limit to the capacity of the boxes and thus a large amount of seedlings cannot be trasported at one time. Furthermore, loss due to predators is a constant danger during transportation.

Table 5.29. Experiments on the transportation of juvenile prawns in net boxes
(Data for 1962 obtained from *Penaeus* Culture Company)

Experiment number	Temperature of water (°C)	Duration of experiment (h)	Average weight of prawns used in experiment (g)	Number of prawns used in the experiment	Number of prawns when the boxes were opened	Loss (%)	Remarks
	22.5	24	0.02	5,800	4,997	13.8	4,371 prawns survived (87%) 6 days after release
1	22.5	24	0.02	3,000	2,936	2.1	2,879 (98%) prawns survived 6 days after release
	22.5	24	0.02	2,000	1,919	4.1	1,820 (94%) prawns survived 6 days after release
2	22.3	24	0.02	5,000	3,986	20.3	
3	22.0	24	0.03	5,000	4,368	12.7	
	22.0	24	0.03	3,000	2,917	2.8	
	22.0	24	0.03	859	750	12.7	

Table 5.30. Transportation experiments on juvenile prawns in bags in which oxygen is introduced
(Data for 1963 obtained from *Penaeus* Culture Company)

Experiment number	Temperature of water (°C) Outset	Minimum value	Amount of oxygen used in experiment (cc/l) Outset	At the end	Duration of experiment (h)	Average weight of prawns used in experiment (g)	Number of prawns used in the experiment	Number of prawns when bags were opened	Loss (%)	Remarks
1	17.8	14.0	5.31	2.60	18	0.04	4,466	4,447	0.4	4,368 prawns survived 3 days after release
2	17.9	14.0	5.33	2.17	18	0.04	4,538	4,511	0.6	4,249 prawns survived 3 days after release
3	18.7	12.1	5.28	15.10	17.5	0.03	6,655	6,438	3.2	
4	18.7	12.1	5.36	9.03	19.5	0.03	8,186	7,870	3.8	

When juvenile prawns are transported by the second method, either vinyl bags or polyethylene bags (50 cm × 40 cm × 10 cm) are used, containing 6 to 8 liters of sea water. The prawns are placed inside and, after forcing the air from the bag, about 15 liters of oxygen are introduced to maintain normal pressure in the bag. The trucks used for transporting these bags have been equipped to maintain the water temperature at 14 to 15°C. The survival rate of juvenile prawns transported by this method is given in Table 5.30. The results of transporting juvenile prawns in a cold-storage truck are given in Table 5.31.

Table 5.31. Transportation experiments on juvenile prawns in cold-storage truck (Miyamura, 1963)

Bag number	Temperature of water (°C)		Amount of oxygen used in experiment (cc/l)		Duration of experiment (h)	Average weight of prawns used in experiment (g)	Number of prawns used in the experiment	Number of prawns when bags were opened	Loss (%)
	Outset	At the end	Outset	At the end					
1	16.7	12.1	6.90	18.82	12	0.70	300	282	6.9
2	17.5	12.1	7.25	16.40	12	0.03	5,000	5,180	2.4
3	17.0	12.1	6.93	16.63	12	0.03	5,000	5,343	0.8
4	17.0	12.1	6.45	17.36	12	0.01	7,898	7,535	4.5

Since a large number of juvenile prawns can be transported efficiently, with a relatively low percentage of loss, by the second method, Miyamura (1963) has suggested that all transportation be handled in this manner.

2. Culture[16]

2.1 Growth

2.1.1 WATER TEMPERATURE AND GROWTH

Since *Penaeus* live in warm sea water, the rate of growth increases with an increase in water temperature. The results of rearing *Penaeus* with the water temperature fixed at 29°C, 26°C, and 23°C with sufficient feed provided, are shown in Table 5.32. The daily increase in body weight shows a gradual decline when the water temperature decreases.

When the water temperature decreases below a certain point, the feeding activity of *Penaeus* stops. They creep into the sand and hibernate. It is assumed that the water temperature at which *Penaeus* hibernate differs according to the size of the specimen, but when the water temperature gradually decreases due to seasonal variation, the resistance to low temperature becomes stronger. However, it is believed that the activity of prawns stops when the water temperature is around 5°C.

---

[16] Much of the information on culture was gathered from Professor S. Uno.

Table 5.32. Water temperature and growth

Experimental water temperature (°C)	29	26	23
Number of rearing days	0→62	0→62	0→62
Number of specimens	100→55	100→92	100→62
Total weight (g)	970→1,060	1,050→1,780	1,040→1,010
Average weight (g)	9.70→19.63	10.50→19.34	10.40→16.29
Daily increase in body weight (%)	1.09	0.95	0.71

Comments: Water tank was 2 m × 1 m × 0.7 m (water depth) and double-bottomed; feed was short-necked clam.

2.1.2 NUMBER OF PRAWNS RELEASED FOR REARING AND GROWTH

When the experiments were carried out with 300, 150, and 75 prawns released per 1 m² area (the number of specimens were changed to 200, 100, and 50 after one month) for which sufficient feed was provided in a water temperature of 26 to 29°C, the results were as shown in Table 5.33.

There is a clear relationship between density of specimens and growth; growth deteriorates as density increases. Regardless of the density, however, the value of the food conversion rate is relatively constant for various periods, i.e. 9.4 to 10.2 for the first 29 days, 7.68 to 8.04 for the next 22 days, and 9.94 to 10.28 for the last 21 days. In this investigation, the feed was carefully measured and care was taken to see that there was no residue left in the water tank. Hence, it is clear from the above that

Table 5.33. Number of prawns released and growth

Experiment number	1	2	3
Number of rearing days	0→29→51→73	20→29→51→73	0→29→51→73
Number of specimens	600→492 400→359→353	300→221 200→185→181	150→129 100→92→89
Total weight (g)	96→770 630→1,325→1,980	48→550 500→1,070→1,620	24→322 260→630→960
Average weight (g)	0.16→1.56 1.58→3.74→5.60	0.16→2.49 2.50→5.97→8.95	0.16→2.50 2.60→6.84→10.78

Water tank number	1			2			3		
Rearing period	0→29	29→51	51→73	0→29	29→51	51→73	0→29	29→51	51→73
Increase in group weight (g)	674	714	636	502	606	515	298	370	330
Total amount of feed provided (g)	34,350	27,400	31,650	24,050	23,600	26,450	13,990	14,900	16,800
Food conversion rate	50.9	38.4	49.7	47.9	39.0	51.4	47.0	40.2	50.9

Comments: Tanks were 2 m × 1 m × 0.7 m (water depth) and double-bottomed; feed was Blue Sea mussel; water temperature was 26 to 29°C.

while density does not affect the rate of conversion of food consumed, the amount of food consumed is controlled by the density. Therefore, when the density is high, the food intake is less and consequently the growth is poor.

2.1.3 TYPE OF FEED AND GROWTH

Under natural conditions, *Penaeus* feed mainly on various types of bottom-dwelling animals. However, when culture is carried out in ponds with a greater density of prawns than in natural conditions, artificial feed must be provided. Although *Penaeus* eat most animal organisms, they prefer bivalves such as the short-necked clam or mussel. In large-scale culturing, feed is conveniently supplemented with bivalves (henceforth called raw feed), frozen feed, or artificial feed. Though the feed given should be studied from a nutritional point of view first, this is presently difficult to do as the nutritional requirements of *Penaeus* are still not clearly understood. But on the basis of rearing experiments done in water tanks of $2 m \times 1 m \times 1 m$, the effect of various types of feed is discussed below.

1. *Raw feed and frozen feed:* Mussel meat, crushed just before feeding, was used as raw feed and frozen prawns, as frozen feed in Experiment Nos. 1 to 5, and the results of rearing prawns on these two feeds are given in Table 5.34.

Table 5.34. Comparison of the effect of raw and frozen feed on the growth of *Penaeus*

Experi-ment number	Diet Number rearing days	Raw feed Number speci-mens	Total weight (g)	Average weight (g)	Frozen feed Number speci-mens	Total weight (g)	Average weight (g)	Remarks
1	2	3	4	5	6	7	8	9
	0	200	960	4.80	200	930	4.65	Size of tank, 2 m × 1 m × 0.5 m (water depth)
	32	118	1,210	10.25	86	690	8.00	Water temperature, 30°C
1	60	41	580	14.14	12	140	11.66	Amount of dissolved oxygen 3-4 cc/l
	86	41	740	18.05				Water exchange rate , 1/4
	105	32	620	19.40				
	0	200	940	4.70	200	950	4.75	Double-bottomed tank, 2 m × 1 m × 0.5 m (water depth)
	32	159	2,030	12.76	38	265	7.00	Water temperature, 30°C
2	60	83	1,160	13.97	4	35	8.75	Amount of dissolved oxygen, 3-4 cc/l
	86	74	1,390	18.78				Exchange rate, 1/48
	105	71	1,585	22.32				
	0	200	910	4.55	200	880	4.40	Double-bottomed tank 2 m × 1 m × 0.5 m (water depth)
3	32	176	1,470	8.35	166	1,200	7.23	Water temperature, 25-20°C
	60	170	1,680	9.89	150	1,190	7.93	Amount of dissolved oxygen, 3-4 cc/l
								Exchange rate, 1/48

TECHNIQUE OF PRAWN CULTURE

Table 5.34—(*Continued*)

1	2	3	4	5	6	7	8	9
	0	200	1,140	5.70	200	1,140	5.70	Double-bottomed tank
	31	109	1,110	10.18	45	390	8.67	Water temperature, 30°C
4	50	72	970	13.47	10	85	8.50	Amount of dissolved oxygen, 2-3 cc/l
	75	69	1,120	16.23				Exchange rate, 1/48
	94	61	1,150	18.85				
	0				200	1,545	7.73	Double-bottomed tank
	32				90	1,050	11.67	Water temperature, 30°C
5	39				29	345	11.89	Amount of dissolved oxygen more than 4 cc/l
								Exchange rate,1/48
	0	600	96	0.16	600	96	0.16	Double-bottomed tank
6	29	{492 / 400	{770 / 630	{1.56 / 1.58	159	100	0.63	Water temperature, 26-29°C
	51	359	1,325	3.74				Exchange rate, 1/48
	73	353	1,980	5.60				
	0	300	48	0.16	300	48	0.16	Double-bottomed tank
7	29	{221 / 200	{550 / 500	{2.49 / 2.50	{121 / 200	{100 / 150	{0.83 / 0.75	Water temperature, 26-29°C
	51	185	1,070	5.97	114	205	1.80	Exchange rate, 1/48
	73	181	1,620	8.95	42	120	2.85	
	0	150	24	0.16	150	24	0.16	Double-bottomed tank
8	29	{129 / 100	{322 / 260	{2.50 / 2.60	90	80	0.89	Water temperature, 26-29°C
	51	92	630	6.84	74	220	2.99	Exchange rate, 1/48
	73	89	960	10.78	18	60	3.33	

Remarks: (1) Experiment Nos. 1 to 5, raw feed=mussel meat; frozen feed=frozen prawns. Experiment Nos. 6 to 8, raw feed=mussel; frozen feed=frozen mysids.

(2) The results of rearing prawns on raw feed in Experiment Nos. 6 to 8 are the same as the results of Experiment Nos. 1 to 3 given in Table 5.33. However, they are repeated in this table.

When the conditions of rearing were the same, except for feed, the growth and survival rates were better when raw feed was provided than when frozen feed was used. When the water temperature was low, the difference in the effect of these two feeds was not striking, but when the water temperature was high, i.e. 30°C or more, the loss and deterioration in growth were considerable in prawns reared on frozen feed within a month of commencing rearing. Hence, rearing prawns on frozen feed only is not recommended, particularly when the temperature is high.

Crushed mussel meat was used as raw feed and frozen mysids as frozen feed for the juvenile prawns, and the results of the effect of these two feeds (Experiment Nos. 6

Table 5.35. Effect of combined raw and frozen feed on the growth of *Penaeus*

Experiment number	Proportion of feed (%)		Number of rearing days	Number of prawns	Total weight (g)	Average weight (g)	Remarks
	Raw feed	Frozen feed					
	100	0	0	100	870	8.70	Double-bottomed tank, 2 m × 1 m × 0.7 m (water depth)
			16	89	1,130	12.70	Water temperature 26-29°C
			32	70	1,000	14.28	Raw feed, mussel
			65	37	670	18.11	Frozen feed, mysids
	75	25	0	100	930	9.30	
			16	89	1,130	13.29	
			32	57	780	13.68	
			65	19	330	17.37	
1	50	50	0	100	920	9.20	
			16	92	1,200	13.19	
			32	79	1,150	14.74	
			65	39	745	19.10	
	25	75	0	100	890	8.90	
			16	90	1,100	12.36	
			32	80	1,130	14.30	
			65	61	1,100	18.03	
	0	100	0	100	880	8.80	
			16	100	1,080	11.02	
			32	22	220	10.00	
			65	2	25	12.50	
	20	80	0	100	570	5.70	Double-bottomed water tank, 2 m × 1 m × 0.7 m (water depth)
			26	95	780	8.22	Water temperature 29°C
2			49	89	900	10.11	Raw feed, short-necked clams
	10	90	0	100	550	5.50	Frozen prawn
			26	96	740	7.71	
			49	72	620	8.61	
	75	25	0	80	910	11.38	Double-bottomed tank, 2 m × 1 m × 0.7 m (water depth)
			29	73	1,140	15.20	Water temperature 29°C
			51	72	1,255	17.43	Raw feed, short-necked clams
3	50	50	0	70	835	11.93	Frozen prawn
			29	57	970	17.01	
			51	57	1,090	19.12	
	25	75	0	81	960	11.85	
			29	76	1,150	15.13	
			51	72	1,290	17.91	

to 8) in a water temperature of 26 to 29°C are show n in Table 5.34. In prawns reared on raw feed, the body weight increased from 0.16 g to 5.6 to 10.8 g in 2.5 months, but in prawns reared on frozen feed, the body weight increased only 2.9 to 3.3 g during the same period.

2. *Combination of raw and frozen feed:* Probably due to the fact that the nutritive value of other prawns and mysids is insufficient, or that their quality deteriorates during storage, the results of rearing *Penaeus* on frozen feed are not good. However, when frozen feed is mixed with a raw feed, this insufficiency in the frozen feed is compensated and the results are better.

The results of rearing *Penaeus* on mixed raw and frozen feed are given in Table 5.35. The growth of *Penaeus* when the diet contains 75% of less of frozen feed and 25% or more of raw feed does not differ from that of prawns grown on 100% raw feed. The growth deteriorates when the frozen feed is 90% and the raw feed 10%. Hence, it seems that the insufficiency in the nutritive value of frozen feed can be compensated by making a combination diet containing 75% or less frozen feed and 25% or more raw feed.

3. *Crushed bivalves:* When bivalves are used as feed, the whole animal is crushed. The shells gradually sink to the bottom and, if left, prevent the prawns from burrowing. Moreover, the shell pieces spoil the bottom material and hence must be removed every now and then. However if the shells are crushed to the texture of sand or smaller, it is not necessary to remove them from the bottom.

The effect of the degree to which clam shells are crushed on the rearing of *Penaeus* is shown in Table 5.36.

Even when the amount of feed is the same, the increase in the weight of prawns reared on pounded whole clams is inferior to that of prawns reared on clams whose shells are crushed (the viscera is slightly crushed). Prawns reared on pounded clams

Table 5.36A. Experiments in rearing *Penaeus* **with crushed short-necked clams**

Experiment number	Number of rearing days	Crushed clams			Pounded clams			Remarks
		Number of specimens	Total weight (g)	Average weight (g)	Number of specimens	Total weight (g)	Average weight (g)	
1	0	100	121	1.21	100	128	1.28	Double-bottomed tank, 2 m × 1 m × 0.7 m (water depth)
	22	98	310	3.16	92*	240*	2.61*	Water temperature 27°C
	38	91	640	7.03	87*	340*	3.91*	*4″ α—starch added to pounded clams
2	0	100	99	0.99	100	122	1.22	Ordinary water tank, 2 m × 1 m × 0.7 m
	22	83	245	2.95	79	180	2.28	Water temperature 27°C
	38	83	550	6.63	81	405	4.05	

Table 5.36B. Experiments in rearing *Penaeus* **with crushed short-necked clams**

Feed	Short-necked clams (coarsely crushed)			Short-necked clams (finely crushed)		
Number of rearing days	15	14	20	15	14	20
Number of specimens	157→153	153→151	151→149	145→145	145→145	145→145
Total weight (g)	100→290	290→570	570→1,280	96→280	280→520	520→1,140
Average weight (g)	0.64→1.89	1.89→3.80	3.80→8.59	0.66→1.93	1.93→3.64	3.64→7.86
Amount of feed (g)	3,600	5,102	10,659	3,550	4,924	9,724
Increase in weight (g)	190	280	710	184	240	620
Food conversion rate	19.0	18.2	15.0	19.3	20.5	15.7

Remarks: Double-bottomed water tank, 2 m × 1 m × 0.7 m (water depth).

were grown in double-bottomed water tanks and ordinary tanks. When the rearing in these two tanks is compared, the growth in the double-bottomed tank is better. This shows that the pounded clams are more easily dispersed in the first tank, making it more difficult for the prawns to reach them. To prevent this dispersal, α-starch was added to the pounded clams, which not only did not prevent the dispersal but adversely affected prawn growth.

Clam crushing is of two types: when the feed contains about 30% shell pieces of less than 3 mm, growth is better than when the feed contains 60% shell pieces. The food conversion rate in the first instance was 16.3% and in the latter, 17.4%. It is conjectured that when shells are finely crushed, the meat part is likewise finely crushed and hence a major part of it is carried away by the water movement before reaching the prawns. Thus it is more economical not to crush bivalves too finely when these are used as feed. The floor material of the tanks can be maintained by other measures.

4. *Cold-storage bivalves:* Bivalves can be stored in a raw condition until used as feed. If the nutritive value did not deteriorate in cold storage, bivalves would be more advantageous to use than other feeds. Hence the effect of storing crushed clams for a long or short period before use was studied and the results are given in Table 5.37. Compared to clams crushed just before feeding, those which were crushed and preserved at 0°C for two days showed no ill effect.

Table 5.37. Feeding experiments with cold-storage short-necked clams

Feed	Clams crushed just before feeding		Clams crushed and kept for half a day in cold storage		Clams crushed and kept for two days in cold storage	
Number of rearing days	15	14	15	15	15	15
Number of specimens	157→153	153→151	155→149	149→149	156→150	150→147
Total weight (g)	100→290	290→570	115→320	320→630	115→310	310→580
Mean weight (g)	0.64→1.89	1.89→3.80	0.74→2.16	2.16→4.23	0.74→2.07	2.07→3.94

5. *Combined feed:* Although bivalves have certain very desirable qualities as feed for rearing *Penaeus,* due to their limited availability, this feed is not economical on a commercial basis. Hence a substitute feed, particularly an artificial one, is badly needed.

As shown in Table 5.38, the growth of *Penaeus* on a combined feed[17] in which fish meal is the main source of protein, is not good. Nor does the growth rate improve

Table 5.38. Combined feed experiments with fish meal as the main protein component

Feed	Combined feed	Combined feed	Combined feed and crushed clam	Crushed clam
Number of rearing days	18	18	18	18
Number of specimens	145→145	149→148	150→149	145→141
Total weight (g)	1,140→1,360	1,280→1,530	1,180→1,660	1,130→1,370
Average weight (g)	7.86→9.45	8.59→10.34	7.86→11.14	7.79→9.72
Amount of feed (g)	1,568	1,762	Combined 1,211 Crushed clam 7,526	12,103
Increase in weight (g)	220	250	480	240
Food conversion rate	7.12	7.05		50.4

Remarks: Double-bottomed water tank, 2 m × 1 m × 0.7 m (water depth); feed dipped in clam meat extract just before feeding; 75% combined feed and 25% clam meat; feed given on alternate days; water temperature 25°C.

when the combined feed is made more attractive by dipping it first in an extract of clam meat. But growth does improve when the combined feed and clam are given together (75% combined feed and 25% clam meat). However, the components of the feed have to be considerably modified since the value of the food conversion rate is abnormally high compared to the food conversion rate in some fish cultivations in which combined feed of this type is used.

The growth of *Penaeus* is not good when they are reared on feed containing fish meal. Details of the food assimilation rate[18] are not yet clear for *Penaeus,* but it seems that combined feed containing protein fluid from the viscera of marine animals is more easily assimilated than fish meal.

Yamashita (unpublished) prepared a feed composed of protein fluid from the viscera of cuttlefish, a particular type of viscous agent, some nutritional improving agents, etc. and solidified it with gelatin.[19] The results of rearing *Penaeus* on this diet are given in Table 5.39 and, as experimental results, are very good. Unfortunately, since this feed contains gelatin, it swells and decomposes within a short time in water of high temperature, and is therefore unsuitable for rearing ponds.

Unlike fish, *Penaeus* use a number of appendages for catching food and hence the chances of their feed disintegrating are more. It is particularly important therefore

[17]Nichihai product: pellet diet used for salmon.

[18]Measurements taken by Mr. Takasei are available.

[19]Publication of Mr. Yamashita's experimental work is expected in the immediate future.

Table 5.39. Experiments with artificial diets
(Yamashita, unpublished)

Experiment number	Components of diet	Artificial, 100% clam, 0%	Artificial, 75% clam, 25%	Artificial, 50% clam, 50%	Artificial, 0% clam, 100%	Remarks
1	Number of rearing days	24	24	24	24	Double-bottomed
	Number of specimens	102→102	100→98	100→98	100→96	water tank,
	Total weight (g)	160→340	150→390	155→430	155→350	2 m × 1 m × 0.7 m
						(water depth)
	Average weight (g)	1.60→3.33	1.50→3.98	1.55→4.38	1.55→3.64	Water tempera-
	Amount of feed (g)	1,040	1,040	1,040	1,040	ture, 27°C
	Increase in body weight (g)	180	240	275	195	
	Food conversion rate	5.78	4.34	3.75	5.34	
2	Number of rearing days	34	34	34	34	Same as above
	Number of specimens	102→95	98→95	98→96	96→95	
	Total weight (g)	340→670	390→760	430→910	350→565	
	Average weight (g)	3.33→7.05	3.98→8.00	4.38→9.50	3.64→5.95	
	Amount of feed (g)	1,938	2,281	2,515	2,047	
	Increase in body weight (g)	330	370	480	215	
	Food conversion rate	5.87	6.16	5.24	9.50	
3	Number of rearing days	19	19	19	19	The number of
	Number of specimens	95→98	95→97	96→97	95→93	specimens be-
	Total weight (g)	670→800	760→855	910→1,185	565→750	ing larger at
	Average weight (g)	7.05→8.16	8.00→9.73	9.50→12.20	5.95→8.06	the end of some
	Amount of feed (g)	1,291	1,482	1,634	1,330	experiments than
	Increase in body weight (g)	130	95	275	185	the number ac-
	Food conversion rate	9.95	15.60	5.95	7.20	tually used, is due to some error in calculation.

to select a suitable material as a viscous agent, i.e. one which will hold together, for combined feed.[20]

6. *Natural diet:* The aforementioned experiments with feed were carried out indoors in double-bottomed water tanks of 2 m × 1m × 1m. The water quality and the quality of the bottom materials in these tanks were assumed to differ from the larger rearing ponds. The natural growth of various animals and plants, both in type and amount, likewise differs in tanks. In natural sea water conditions, *Penaeus* live on natural feed obtained from various types of plants and animals. This natural feed in rearing ponds differs according to the topographical conditions and water depth, but in water tanks this natural feed does not differ much from the original. Hence, from the point of view of feed, water tanks can be regarded as good as a natural environment.

A combination of raw and frozen feed or raw and artificial feed has yielded better results than simple feed containing either frozen or artificial components. From this

[20]Sodium alginate is a better solidifier. but this substance obstructs food absorption. Materials belonging to a synthetic resin group are in the experimental stage.

it is clear that *Penaeus* reared in cultivation ponds with both artificial and natural feed are in better conditions, nutritionally speaking, than those grown in experimental tanks.

To establish the value of natural feed, the latter was collected in experimental ponds used for rearing *Penaeus*. As can be seen from Table 5.40, *Penaeus* grown on

Table 5.40. Experiments with natural feed
(Maeda, unpublished)

Experiment number	Feed	Natural feed		Frozen feed	
1	Number of rearing days	24	24	24	24
	Number of specimens	160→151	160→121	160→154	160→129
	Total weight (g)	340→650	340→440	350→755	335→470
	Average weight (g)	2.12→4.30	2.12→3.63	2.19→4.90	2.09→3.64
2	Number of rearing days	40	40	40	40
	Number of specimens	150→142	119→92	149→130	127→110
	Total weight (g)	650→1,040	435→490	730→1,070	465→630
	Average weight (g)	4.30→7.32	3.64→5.32	4.90→8.23	3.64→5.72
	Remarks:	Double-bottomed water tank, 3 m × 1.5 m × 0.4 m	Ordinary water tank, 3 m × 1.5 m × 0.4 m	Double-bottomed water tank, 3 m × 1.5 m × 0.4 m	Ordinary water tank, 3 m × 1.5 m × 0.4 m

Natural feed: *Caprella, Gammarus*. Frozen feed: *Penaeus* (Tai-ebi).

frozen feed showed a higher rate of weight increase than those grown on natural feed. It is conjectured that this is due to the insufficient nutritive value and an insufficient amount of the latter, due to its limited collection from the drainage water path of the cultivation ponds. The feed contains a limited amount of crustaceans such as *Caprella, Gammarus,* etc. However, it is interesting that even when only natural feed with limited types of organisms is used, the weight increase of *Penaeus* is more or less the same as when artificial feed is given.

2.1.4 AMOUNT OF FEED AND GROWTH

Table 5.41 shows the relation between the amount of feed per day and body weight when crushed, short-necked clams are utilized. The relation between the amount of feed and the increase in body weight is directly proportional when the amount is high, i.e. when it is more than 25% for 1 g of mean weight of *Penaeus*, and more than 20% for 10 g of mean weight. However a direct relation is not seen when the amount of feed is small or when no feed is given. Figure 5.41 gives the results obtained when the daily feed amount for mean body weight of *Penaeus* during the experimental period is calculated and, on this basis, the relation between amount of feed and increasing rate of body weight per day determined. The relation is more or less linear.

When the planned amount exceeds 20%, the daily dosage of feed is determined from the amount and condition of the feed on the previous day. The calculation is

Table 5.41. Relation between amount of feed and growth of *Penaeus*

Experiment number	Amount of feed given (%)	Beginning of experiment			End of experiment			Total amount of feed provided (g)	Actual feed utilized (%)	Increased body weight per day (%)	Remarks
		Number of specimens	Weight (g)	Average weight (g)	Number of specimens	Weight (g)	Average weight (g)				
1	5	200	220	1.10	195	245	1.26	130	4.31	1.04	Double-bottomed tank,
	10	200	190	0.95	195	255	1.30	260	9.18	2.39	2 m×1 m×1 m
	15	200	200	1.00	197	305	1.54	390	12.41	3.28	Water temperature 28°C
	20	200	190	0.95	197	330	1.67	520	16.58	4.23	Number of rearing days,
	25	200	200	1.00	198	370	1.86	650	19.25	4.62	13
	30	200	220	1.10	200	405	2.02	780	21.04	4.53	
2	5	50	590	11.80	46	560	12.17	465	4.72	0.18	Double-bottomed tank,
	10	50	650	13.00	47	710	15.10	1,023	8.82	0.89	2 m×1 m×1 m
	15	50	570	11.40	47	680	14.46	1,333	12.54	1.39	Water temperature 28°C
	20	50	630	12.60	50	820	16.40	1,911	15.50	1.54	Number of rearing days, 17
3	0	50	580	11.60	50	590	11.80	0	0	0.10	Double-bottomed tank,
	5	50	615	12.30	46	620	13.47	465	4.42	0.53	2 m×2 m×1 m
	10	50	615	12.30	44	640	14.54	961	9.08	0.98	Water temperature 25°C
	15	50	610	12.20	48	770	16.04	1,426	12.16	1.59	Number of rearing days, 17
	20	50	605	12.10	49	790	16.12	1,891	15.96	1.09	
4	0	44	640	14.52	43	550	12.79	0	0	−0.91	Double-bottomed tank.
	5	48	770	16.04	46	820	17.82	539	4.84	0.75	2 m×1 m×1 m
	10	49	790	16.12	48	900	18.75	1,035	8.76	1.07	Water temperature 25°C
	15	50	820	16.40	44	870	19.77	1,599	13.16	1.33	Number of rearing days,
	20	47	710	15.10	45	850	18.88	1,988	18.20	1.59	14

Remarks: Feed=crushed, short-necked clams; amount given calculated on the basis of the amount of meat determined from the weight of crushed clams × 0.20.

made in such a way that no excess feed will be left, but when feed is mechanically distributed, the line in Figure 5.41 might bend downward in the vicinity of a dosage exceeding 20%.

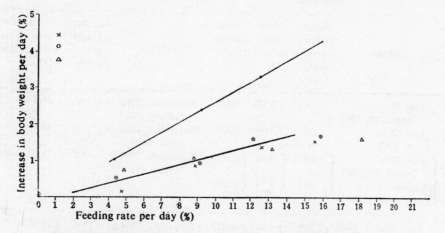

Figure 5.41. Relation between amount of feed and increased body weight.
• = body weight 1-2 g, water temperature 28°C; × = body weight 11-16 g, water temperature 28°C; ⊙ = body weight 11-16 g, water temperature 25°C; △ = body weight 16-20 g, water temperature 25°C.

In any case, when the amount of feed is less than the amount consumed by the *Penaeus*, a direct correlation is seen between the amount of feed and the increased rate of body weight. Hence the dosage of feed must be fixed for the cultivation of *Penaeus* by studying the environmental conditions, such as water temperature, etc. and the organism's rate of growth; the dosage should not allow for residue.

2.1.5 Growth of *Penaeus* in Cultivation Ponds

Figure 5.42 shows the growth of *Penaeus* in cultivation ponds on the coast of the Seto Inland Sea where the water temperature is 23°C in June, 28°C in July, 30°C in August, 27°C in September, 23°C in October, and 16°C in November.

When post-larvae weighing less than 0.1 g are used as seedlings, the sizes of the specimens at the time of catch can be determined on the basis of the time of release of the seedlings. Hence the mean weight of the specimens at the time of catch is about 12 to 15 g when the seedlings are released in June of the same year, about 12 to 13 g when the seedlings are released in July of the same year, and about 11 g when the seedlings are released in August of the same year. The mean weight of the specimens is about 21 g when late-hatched seedlings of the previous year are released in June.

As explained earlier, the growth of *Penaeus* differs considerably according to the conditions of rearing. Hence the growth rate given in Figure 5.42 is not always the standard growth trend in rearing ponds. It is quite difficult to rear the seedlings of the same year to a weight of 20 g under the conditions which prevail in the coastal regions of the Seto Inland Sea.

The growth of seedlings of the same size released at the same time, differs accord-

Table 5.42. Results of
(Data obtained from the Seto Inland

Pond No.	Cultivation				Catch			
	Period	Number of specimens (10³)	Mean body weight (g)	Total weight (tons)	Period	Number of specimens (×10³)	Mean body weight (g)	Total weight (tons)
1	June 19 to August 19, 1963	519	3.18	1.653	September, 1963 to January, 1964	320	21.3	6.81
2	July 12 to 26, 1963	1,247	0.06	0.071	December, 1963 to April, 1964	529	12.6	6.66
3	July 30 to August 21, 1963	1,236	0.06	0.080	November, 1963 to March, 1964	757	11.0	8.33
	September 22 to 27, 1963	144	2.70	0.389				
1	June 5 to 15, 1964	2,308	0.01	0.023	December, 1964 to April, 1965	440	11.96	5.14
	December, 1964 to January, 1965	10	10.0	0.102	August, 1964 to March, 1965*	903	4-10	4.59
	June 7 to 19, 1964	416	8.0	4.325	July, 1964 to August, 1964	261	14.0	3.65
2	August 9 to 19, 1964	533	4.0	2.131	December, 1964 to April, 1965	799	12.73	10.17
	August 26 to 29, 1964	434	6.0	2.602	December, 1964 to January, 1965	10	10.0	0.10
3	June 25 to July 17, 1964	4,102	0.01	0.041	January to April, 1965	718	12.3	8.82
					August 26 to 29, 1964*	434	6.0	2.60
					January to March, 1965	143	10.0	1.43
4	September 11 to 12, 1964	311	6.0	1.868	December, 1964 to April, 1965	297	15.56	4.61
	November, 1964	43	10.0	0.430				

ing to the density of the larvae or the quality and amount of feed.

The results of rearing *Penaeus* in the coastal regions of the Seto Inland Sea are shown in Table 5.42. When the density of *Penaeus* at the time of catch was 12 to 25 specimens/m², no difference was observed in their growth rate. Moreover, since the feed contained more than 70% clam meat, no ill effect was possible due to the quality and quantity of feed, but the duration of use of the rearing pond might have had some influence on rate of growth. For example, pond No. 1 of 1964 was used for nearly 180 days, from June to the end of November: in September, *Penaeus* from this pond were transferred

rearing *Penaeus* **in ponds**

Fisheries Development Company)

Feed (tons)					Density of cultivation (specimens/ m²)	Density at time of catch (specimens/ m²)	Survival rate (%)	Food conversion (tons)	Food conversion rate	Area of pond (m²)	Remarks
Short-necked clams	*Metapenaeopsis barbata*	Fish	Miscellaneous diet	Total**							
598.8	0.3			96.11	17.3	10.2	61.7	5.16	18.6	30,000	**Amount of clam meat calculated after multiplying by 0.16
537.1				85.94	41.6	17.6	42.4	6.59	13.0	30,000	
478.8				76.62	41.2	25.2	54.8	7.86	9.7	30,000	
875.0	27.66	5.35	8.80	181.81	65.9	12.6	57.9	8.94	20.3	35,000	*Transferred to other ponds after catch
					11.9	7.5	62.8				
627.9	37.52	4.94	0.90	143.81	32.2	26.6	83.8	5.54	25.9	30,000	Cultivated after collecting from pond Nos. 1 and 3
875.4	53.92	5.85	0.90	200.73	91.1	16.0	31.5	12.80	15.6	45,000	Transferred to other ponds after catch
349.9				55.98	15.6	14.8	83.7	2.31	24.2	20,000	After collecting from pond No. 1

to pond No. 4 which had been in use for about 80 days. The size of the *Penaeus* at the time of transfer was about 12 g for pond No. 1 while those of pond No. 4 were 15.5 g. Again, pond No. 2 contained prawns transferred from pond Nos. 1 and 3, but this pond had been used for a shorter period, and hence its specimens showed a better growth rate than those from pond Nos. 1 and 3—in spite of the fact that the density of the prawns at the time of catch was high, 26/m².

The foregoing facts prove that a pond loses its quality when used for a long period, resulting in a poor growth rate for the prawns reared in it.

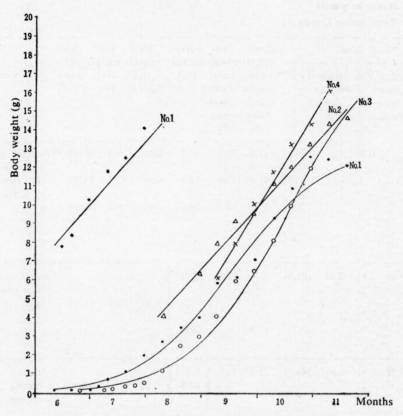

Figure 5.42. Growth of *Penaeus* in cultivation ponds (1964).

2.2 Survival rate

2.2.1 INFLUENCE OF VARIOUS FACTORS ON SURVIVAL RATE

1. *Dissolved oxygen:* One of the most common environmental factors which cause death among cultured *Penaeus,* is a decrease in the amount of dissolved oxygen. As shown in Figure 5.43, the amount of oxygen at night in the running water in which *Penaeus* are cultured, is influenced by water temperature, exchange rate of water, and the number of *Penaeus.* The influence of these factors on *Penaeus* is shown in Table 5.43. Whether oxygen tension results in the death of *Penaeus* depends upon the water temperature. An oxygen tension of 1.67 cc/l in a water temperature of 25°C is safe, while an oxygen tension of 1.60 to 2.00 cc/l at 27°C sometimes results in mortalities; likewise, a 30% mortality occurs when the oxygen tension is 1.84 cc/l if the water temperature is 30°C.

2. *Water temperature:* Since *Penaeus* tolerate very high temperatures, it cannot be assumed that a high water temperature in a natural environment is a mortality factor. However, a high water temperature could increase the oxygen consumption of *Penaeus* and thereby cause a decrease in the amount of dissolved oxygen, which factor could result in death. Again, although not significantly, a high water temper-

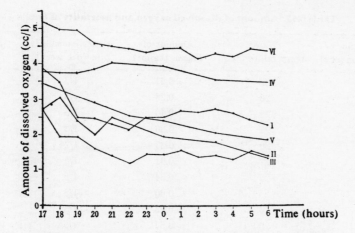

Experiment number	Water temperature (°C)	Water exchange rate	Amount of *Penaeus* (kg/m²)	Mortality (%)
I	19	1/3.12*	1.0	0
II	19	1/6.24	1.0	1.3
III	28	1/2.8	0.5	30.0
IV	25	1/5.4	0.25	0
V	23-29	1/23.6	0.22	7.0
VI	Flooding of the tank			

*The fraction represents the water replenishment in terms of the tank capacity in the given time (e.g. one-third tank capacity replenished in 12 hours)—General Editor.

Figure 5.43. Changes in the amount of dissolved oxygen at night in experimental tanks with running water.

ature weakens the physiological condition of *Penaeus,* which sometimes leads to mortality. As Table 5.43 reveals, the survival rates of *Penaeus* reared for about two months at different water temperatures—30°C and 20 to 25°C—were 42% and 85% respectively, when the diet consisted of raw feed. The survival rates changed to 2% and 75% respectively when the diet consisted of frozen feed. But when the water temperature increased, the survival rate decreased significantly.

Penaeus have little resistance to low water temperature; when the temperature drops to 10°C or below, they lie on their sides and sometimes even die. However, this is primarily the situation when the temperature falls abruptly; when the water temperature gradually decreases due to seasonal variations, only a few specimens are found dead at about 5°C. Resistance to low water temperature also varies according to the feed: As shown in Table 5.44, resistance to cold conditions is stronger in *Penaeus* fed on combined feed than in those fed on artificial feed (*vide* Table 5.39) or clam meat only.

3. *Density of cultivated Penaeus:* The survival rates of *Penaeus* reared in various densities—200, 100, or 50 per m²—are given in Table 5.33. In this study the mean

Table 5.43. Amount of dissolved oxygen and mortality of *Penaeus*

Minimum value of oxygen tension (cc/l)	Water temperature (°C)	Amount of *Penaeus* (kg/m²)	Water exchange rate (per week)	Mortality* (%)
1.37	32	0.51	1/3	30
1.60	28	0.23	1/5	0
1.62	29	0.46	1/5	1
1.67	25	0.80	1/3	0
1.84	30	0.23	1/24	7
1.97	27	0.91	1/5	10
2.01	25	0.34	1/5	0
2.11	25	0.80	1/3	0
2.12	24	0.23	1/3	0
2.16	28	0.57	1/5	0
2.25	25	0.57	1/5	0

*At night (about 12 hours).

Table 5.44. Cold tolerance of *Penaeus**
(Yamashita, unpublished)

Experiment number	Feed (%)		Mean weight of *Penaeus* (g)	Number of *Penaeus*	Number of dead specimens	Mortality rate (%)	Remarks
	Artificial feed	Clam meat					
1	100	2	13.17	48	17	35.4	Before experiment: density of *Penaeus* in double-bot- tomed water tanks high; 75 specimens/m²
	75	25	15.03	122	29	23.8	
	50	50	14.33	128	27	21.1	
	0	100	13.43	90	29	32.2	
2	100	0	14.59	48	6	12.5	Before experiment: *Penaeus* reared in ordinary water tanks; 25 speci- mens/m²
	75	25	16.22	49	3	6.1	
	50	50	16.22	49	6	12.1	
	0	100	14.00	45	5	11.1	

*Water temperature at which these experiments were conducted has not been given in the original—Translator.

weight of the *Penaeus* was below 10 g, but after 2.5 months of rearing, the growth rate showed considerable difference due to density; nevertheless, the survival rate remained constant at approximately 70%. Thus the survival rate does not deteriorate even when the density is high, if the environmental conditions are suitable. Unfortunately, high density usually pollutes the environmental conditions, and this results in a deterioration of the survival rate. The results of growing *Penaeus* with an average weight of less than 10 g at densities of about 90, 45, and 25 per square meter are given in Table 5.45. The Table clearly illustrates that the survival rate gradually decreases as density increases.

Table 5.45. Influence of density during rearing of *Penaeus*

Density of *Penaeus* during rearing (specimens/m²)	26.5	44.5	90.5
Structure of water tank	Ordinary tank	Double-bottomed tank	Double-bottomed tank
Water temperature (°C)	27	29	29
Feed	Crushed clam	Crushed mussel	Crushed mussel
Number of rearing days	0→26→50→70	0→27→56	0→27→56
Number of specimens	53→48→50→50	89→78→71	181→132→104
Total weight (g)	300→450→740→910	960→1,170→1,350	1,620→1,710→1,800
Mean weight (g)	5.66→9.37→14.80→18.20	10.78→15.00→19.00	8.95→12.95→17.31
Survival rate (%)	94.3	80.0	57.4

2.2.2 Survival Rate in Cultivation Ponds

Table 5.42 showed the survival rate of *Penaeus* in rearing ponds: the rate varies considerably according to the period of seedling release and seedling size. With post-larvae of 0.01 to 0.06 g, the survival rate is 30 to 50%; when specimens weighing more than 3 g are transferred to other ponds, the survival rate is 60 to 80%.

2.3 Food conversion rate

Crushed clams were given as feed to *Penaeus* of various sizes in water temperatures of 25 to 28°C, and the prawns reared for 14 to 30 days. The food conversion rate in this experiment is shown in Figure 5.44 and in Table 5.46.

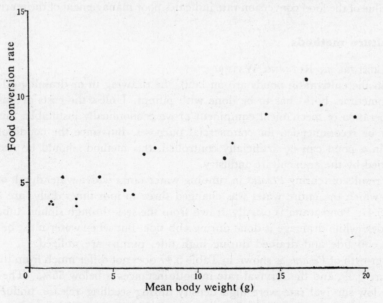

Figure 5.44. Relation between mean body weight and food conversion rate (calculations made after rearing two weeks).

Table 5.46. Food conversion rate in *Penaeus*

Number of experiments	Range of mean body weight (g)	Range of food conversion rate	Mean food conversion rate
7	1.27- 4.21	17.80-26.50	21.00
4	5.75- 7.52	21.70-35.00	28.05
3	11.63-14.12	32.20-44.55	37.75
3	16.50-19.40	44.25-54.50	48.60

Table 5.46 shows the food conversion rate for every 5 g of increase in mean body weight. When this table is used and rearing carried out up to 20 g, the food conversion rate can be calculated as follows:

$$\frac{(5 \times 21.00 + 5 \times 28.05 + 5 \times 37.75 + 5 \times 48.60)}{20} = 33.85 .$$

If the yield rate on clam meat is taken as 20%, this value can be converted into meat value:

$$33.85 \times 0.2 = 6.77.$$

When the food conversion rate in a cultivation pond is based on Table 5.42, the mean value is 18.2 (since the range is 9.7 to 25.9) and, compared to the aforementioned value, is very large; a larger figure shows that the feed is ineffective.

While a part of the feed in the cultivation ponds is consumed by other animals, the conversion rate of food into meat, and the inflow of natural diet, are better in rearing ponds than in narrow experimental water tanks. Hence the aforementioned high value of the food conversion rate indicates poor management of the rearing tanks.

2.4 Culture methods

2.4.1 CULTURE IN RUNNING WATER

When the cultivation ponds are on land, the drawing in or draining off of water (and sometimes both) has to be done with pumps. Unless the costs of installation and operation of mechanical equipment prove economically justifiable, this method cannot be recommended for commercial purposes. But since the conditions of the water in a pond can be artificially controlled, this method should be experimentally tried by the aquaculture industry.

The results of rearing *Penaeus* in running water tanks with a water depth of 40 to 50 cm, in which the entire water was changed three to four times daily, are shown in Table 5.47. Pond water is usually drawn from the seas through siphon tubes during high tides, while drainage is done during ebb tide. But when water must be supplied during ebb tide and drained during high tide, pumps are utilized.

The growth of *Penaeus* as shown in Table 5.47 does not differ much from that shown in Figure 5.39, and the survival rate in both instances is below 35%. The reasons for the low survival rate were high density during seedling release, undue proportion of frozen clam meat in the feed, etc. It would seem that the rearing techniques in these experiments were not satisfactory. Nevertheless, the results show that rear-

Table 5.47. Results of rearing *Penaeus* **in running water ponds**

(Data for 1962 obtained from the *Penaeus* Culture Company)

	Ordinary pond						Control pond: double-bottomed
Area of pond (m²)	1,100	765	365	2,550	1,250	1,250	552
Rearing period	May 10 to August 28	May 28 to September 3	May 16 to September 12	June 15 to August 27	June 22 to September 10	June 26 to September 7	May 26 to August 25
Number of rearing days	110	98	119	73	80	73	91
Number of seedlings released (×10³)	30.7	73.3	33.0	68.9	112.4	63.2	50.4
Number of specimens removed (×10³)	5.8	6.6	4.0	24.7	16.2	13.8	18.2
Density of seedlings at time of release (specimens/m²)	28	96	4.3	27	90	51	91
Density of seedlings at time of removal (specimens/m²)	5	9	5	10	13	11	33
Survival rate (%)	19.2	9.1	12.1	35.8	14.4	21.8	36.1
Body weight at time of release (g)	6.3	7.3	3.9	1.9	3.7	3.4	2.1
Body weight at time of removal (g)	14.6	12.8	14.3	9.1	8.8	8.5	10.2

ing in shallow water ponds of 40 to 50 cm depth, though adequate for experimental purposes, is not suited to large-scale production.

The special advantage of rearing in running water is an adequate supply of oxygen for the *Penaeus*. In addition, water exchange and movement are satisfactory and eliminate problems arising from stagnant water. Some experiments are presently underway to make the water of a pond flow in a particular direction, as it does in rivers. Moreover, in places like the coastal regions of the Sea of Japan where the difference between low and high tides is very little or almost nil, the drawing in and draining off of sea water has to depend upon mechanical devices. Hence an improvement could also be made along this line based upon the experience gained.

2.4.2 REARING IN DOUBLE-BOTTOMED WATER TANKS

The floor of a rearing pond deteriorates due to the settling and accumulation of planktons and algae, and the accumulation of residual feed and waste matter. As a result of this accumulation, hydrogen sulfide forms which hinders the growth of *Penaeus* and may even cause mortalities.

To prevent such deterioration in floor conditions, a number of measures have been suggested; of these the double-bottomed pond seems the most effective. The structure of such a pond is shown in Figure 5.45. In this instance, a supporting plate has been placed slightly above the floor, which has a number of minute perforations that do not permit sand leakage. Sand is spread above this plate to form a bed on which the prawns can crawl. Water is circulated from top to bottom and through the sand layer by an airlift mechanism attached to tubes of various sizes.

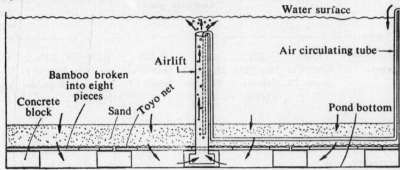

Figure 5.45. Structure of a double-bottomed pond (Hudinaga).

The results of rearing *Penaeus* in double-bottomed water tanks are shown in Table 5.48. At the beginning of air circulation, ammonia nitrogen greatly increases and, if the ammonia oxidation process is activated in the sand layer, along with a decrease in the ammonia nitrogen, nitrogen in the form of nitrate increases and causes an acidity in the sea water whereby the pH value decreases. However, *Penaeus* do not show any abnormal growth consequently. On the contrary due to the consumption of oxygen by the bacteria in the sand layer, though the amount of dissolved oxygen in the sea water decreases, mortality due to oxygen deficiency is avoided by increasing air circulation. In this way it is possible to rear prawns for about a month at a density of 2 kg per m^2. However, a gradual decrease in the number of *Penaeus* occurs after this period and the final results are very poor. When the sand layer of double-bottomed water tanks of this type becomes the bed for burrowing *Penaeus*, a decomposition of organic matter occurs. Hence at 20°C, this sand layer becomes a suitable environment for the growth of aquatic bacteria. The latter even enter the gills of *Penaeus*[21] and turn them black, a symptom of "gill-rot." Due to this disease, the mortality rate increases.

The formation of aquatic bacteria is closely related to water temperature. Bacteria do not form at high temperatures, so the results of rearing *Penaeus* during the summer in double-bottomed water tanks should be better than those shown in Table 5.48. But as Table 5.49 shows, the results of rearing *Penaeus* in double-bottomed ponds with an area of about 90 m^2 and a water depth of approximately 0.5 m, are not strikingly better than those obtained with ordinary ponds (Table 5.47).

2.4.3 REARING OF *Penaeus* UTILIZING DIFFERENCE BETWEEN HIGH AND LOW TIDES

Running water and double-bottomed water tanks are methods meant for intensive rearing of *Penaeus*, but neither the growth nor the survival rate by these methods is good. Hence it seems that a rearing method which, leaving aside the technical aspect

[21]According to Mr. Egusa.

Table 5.48. Water quality and results of rearing *Penaeus* **in double-bottomed water tanks**

		1	2
	Experiment number	1	2
	Number of rearing days	31	28
	Water temperature, °C	20-23	20-23
At time of release	Number of specimens	300	411
	Weight (g)	837	3,938
	Mean weight (g)	2.79	9.58
	Density (kg/m²)	0.42	1.97
At time of removal	Number of specimens	292	372
	Weight (g)	1,574	4,197
	Mean weight (g)	5.30	11.28
	Density (kg/m²)	0.79	2.10
	Survival rate (%)	97.3	90.5
At time of final collection	Total number of rearing days	140	
	Number of specimens	136	
	Weight (g)	1,550	
	Mean weight (g)	11.40	
	Density (kg/m²)	0.78	
	Ammonia concentration (γ/l)	~2,550	~4,200
	pH	7.9-7.4	7.7-7.4
	Amount of dissolved oxygen (cc/l)	1.7-3.5	1.6-3.8

Table 5.49. Results of rearing *Penaeus* **in sectioned corfs (Ozaki, 1965)**

Area (m²)	Time of release	Density at time of release (specimens/m²)	Mean weight at time of release (g)	Time of removal	Density at time of removal (specimens/m²)	Mean weight at time of removal (g)	Weight per unit area at time of removal (g/m²)	Survival rate (%)
10	August 25	240	0.6	December 20	200	18	3,600	83
90	July	255	16	December 20	155	30	4,650	61
50	July	116	37	December 20	80	50	4,000	69

for the time being, does not involve expensive installation of equipment, nor high maintenance costs, i.e. a cruder method, might be more profitable.

Some rearing ponds utilize the tidal difference for drawing and draining sea water, and in the past places with a considerable difference in high and low tides were used for storing and rearing *Penaeus*. These places do not require pumps or other implements for drawing water from the sea, and in some there is room for expansion. The environmental conditions of the rearing ponds in which the area has been greatly expanded, differ in many ways from the running water type of rearing pond. The various problems related to the selection of rearing places, the structure of rearing ponds, and the environmental conditions, will be discussed later.

2.4.4 CULTIVATION IN CORFS

The foregoing method, which requires a large area and large-scale equipment, cannot be utilized for small-scale production. Hence experiments are being done with rearing *Penaeus* in sectioned corfs. Figure 5.46 shows the structure of such corfs. Frames of $5.4\,\text{m} \times 5.4\,\text{m} \times 1.5\,\text{m}$ (height) are welded together and covered with a vinyl

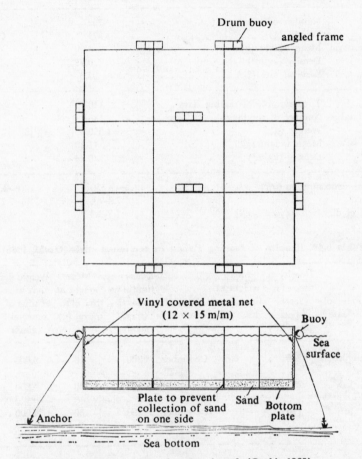

Figure 5.46. Structure of sectioned corfs (Ozaki, 1965).

metal net; a layer of sand is spread on a bottom plate. To avoid loss of sand due to currents, the plate is placed slightly above, and the sand layer around, the lower margin of the frame. Then, to avoid a collection of sand on one side of the frame due to raft movement, the bottom surface is divided into several sections by plates of the same height as that of the sand layer. Drum buoys are attached to rafts in regions with good tidal currents, and clam meat and horse mackerel used as feed. Table 5.49 shows the results of this experimental method of rearing *Penaeus*, revealing that the capacity of the corf per square meter is much more than that of earlier rearing methods.

The corfs used for the cultivation of *Penaeus* are different from the large rafts used for other marine animals such as yellowtails, etc. Since the mesh size of the net used in corfs cannot be increased, the mesh becomes choked with adhering substances and blocks the passage of the water current, a problem which occurs especially in the summer when the water temperature is high. Replacing the nets is difficult work. Furthermore, since *Penaeus* are not as active as fish, the water in the corfs cannot circulate much as a result of the animal's movements. In addition, *Penaeus* require sand and hence the structure of the corf must be strong; for this reason the installation of the raft becomes expensive. Hence sectioned corfs are advisable only in regions where facilities for land rearing are not available, and even then improvement in the structure and material of corfs is still needed.

2.4.5 TANK CULTURE

According to the large-scale seeding method already described, the environment for the rearing of *Penaeus* even after the post-larval stage seems to be favorable since the feed consists of bottom-dwelling organisms and planktonic elements, which grow easily and abundantly in water tanks. Again, the production rate of large post-larvae is 425 to 570 g/m^2 (Hudinaga and Kittaka, 1967). Large *Penaeus* can also be reared at a higher density by circulating air and changing the water. Ishida (1967) carried out an investigation in which the tank rearing method of larvae was utilized with adult prawns subsequent to the post-larval stage; his results are given in Table 5.50.

Table 5.50. Results of rearing *Penaeus* **in tanks (Ishida, 1966)**

Material investigated	Period of rearing (months)				
	Up to 1	1	2	3	4
Number of specimens	10,000	9,600	8,640	8,175	8,050
Mean weight (g)	0.15	3.0	8.4	13.9	16.7
Total weight (kg)	1.5	28.8	72.6	112.8	144.4
Amount of feed per month (kg)		93.9	225	354	227

Remarks: Period of rearing—July 25 to November 16, 1966; composition of feed—clam 70.3% *Metapenaeopsis barbata*, 26.6%, fish 3.1% (percentage refers to meat).

In Ishida's experiment, the ground allowed for the burrowing of the *Penaeus* comprised one-tenth of the total double-bottomed area. The sand did not turn black as the *Penaeus* kept it stirred with their burrowing and creeping. Thus it would seem that the results obtained would be unchanged in a single-bottomed tank, a fact confirmed in experiments carried out later (Ishida, unpublished). However, like the raft method, this method involves considerable expense and should be used only when post-larvae are not available.

2.5 Selection of culture grounds

In selecting a suitable area for culture ponds, various conditions such as the topography of the surroundings, the quality of the soil, the water temperature, quality, and supply and drainage, and availability of feed and transportation should be considered. In other words, the same factors which influence fish culture, affect pond culturing which utilizes the difference in water levels during ebb and high tides.

2.5.1 WATER TEMPERATURE

When the spawning of *Penaeus* is delayed, the offspring pass the winter either in a juvenile or slightly more advanced stage. Since the animal has a poor tolerance for low temperature, regions such as the Seto Inland Sea, once popular for *Penaeus* culture, are not always suitable. The coastal regions of the Sea of Japan, however, are influenced by warm currents and thus the water temperature is relatively high even during the winter. Here then the juvenile prawns can pass the winter without much difficulty, but as the temperature rise during the summer is very small, body weight does not increase easily. In regions influenced by Black Current, both the water temperature and the atmospheric temperature are high; good *Penaeus* growth can be expected throughout the year.

2.5.2 TIDES

Changing water in the ponds requires considerable effort in regions where the difference in tides is small. But when difference between low and high tides is great, changing the water is easily done by taking advantage of the difference and utilizing siphon tubes.

Figure 5.47 gives the difference in the tidal levels of various regions of the Japanese coasts. The difference is particularly large in the coastal regions of Ariake-kai and in the central part of the Seto Inland Sea. In these places the difference between high and low tides is more than 3 m. Compared to these regions, the difference in the coastal regions of the Pacific Ocean is small, and still smaller on the coastal regions of the Sea of Japan.

2.5.3 FLOOR QUALITY

Since *Penaeus* have the habit of burrowing in the sand, the bottom of ponds intended for culturing should have a clean, fine layer of sand. As *Penaeus* tend to collect in the sand layer in large numbers when the sectional raft culture method or tank culture method is employed, the quality of the sand should be good, the layer even, and the floor free of rocks or pebbles.

2.6 Structure of culture ponds

The structure of a culture pond differs from method to method. With intensive culture, the pond requires a special structure, but with extensive culture, such conditions as frequent water changes and maintenance of sand quality must be considered.

2.6.1 AREA AND SHAPE

The qualities of the ponds in which post-larvae of 0.01 to 0.02 g are reared are quite

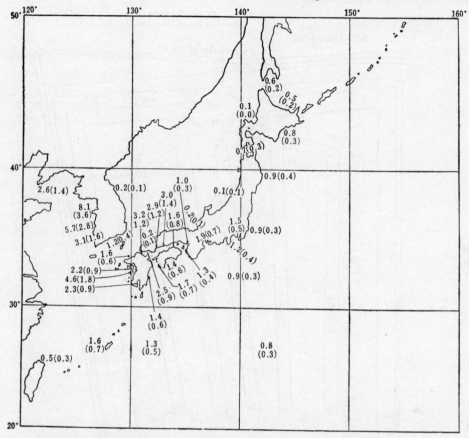

Figure 5.47. Difference in the tides (m) on the Japanese coasts: figures in parentheses indicate the level of low tides, and figures outside parentheses indicate the level of high tides.

distinct from the qualities required for rearing larger larvae up to the size of 1 to 2 g. Rearing marketable larvae requires a long time and rearing the two sizes of larvae just mentioned in the same pond spoils the bottom quality quickly. Furthermore, there is much uncertainty in the yield rate of the adult specimens when these larvae are grown in the same pond, since the exact amount of feed required is not easily determined.

There is no fixed standard for the area or shape of a culture pond. In culture places where 10 to 50 tons (700×10^3 to $3,300 \times 10^3$) of adult prawns are expected, it is necessary to have a culture pond of $20 \times 10^3 \, \text{m}^2$ area for the release of post-larvae, and more than one pond of an area 30×10^3 to $50 \times 10^3 \, \text{m}^2$ for rearing. Pond location should be such that the prawns can be easily fed and the specimens readily collected for marketing. At present, an area of 30×10^3 to $50 \times 10^3 \, \text{m}^2$ is considered sufficient.

Figure 5.48 illustrates a culture ground in which the difference between high and low tides is utilized. When the culture ground is of this size, the culture pond is either square or rectangular in shape. When the area is larger and the pond narrow and long, the water stagnates at the corners, and the water quality and floor quality gradually deteriorate. To overcome this deterioration, a drainage path is located at the corners of the pond to facilitate water changes.

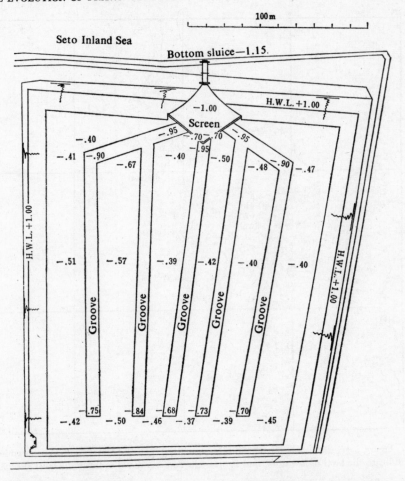

Figure 5.48. Structure of culture pond (front view) (Akiho Culture Center).

2.6.2. WATER DEPTH

The water depth in running water and double-bottomed types of rearing ponds is around 50 cm. Hence water temperature is easily influenced by atmospheric temperature, from high in the summer to low in the winter. As a result, *Penaeus* grown in these ponds are adversely affected and losses occur.

Even in ponds which utilize the difference between high and low tides, the water of the pond, except during the transition from low to high tide, is usually at a standstill. Not only is the water temperature influenced by external factors; the amount of dissolved oxygen is greatly reduced at night. To solve this problem, at least to some extent, the depth of the water and its level should be maintained at 2 m.

Figure 5.49 shows the relation between the water level of culture ponds in the western part of the Seto Inland Sea, where there is a considerable difference between high and low tides and the tidal level of the open sea. The bottom of the pond is higher than the level of the water entrance and the water entrance of the pond is usually kept at a higher level than the mean ebb-tide level. Even when the bottom of the

pond is lower than the mean ebb-tide level, there is no adverse effect on the growth of
the *Penaeus*, but since the water in
ponds of this type can only be drained completely by pumps, it is better
to avoid such levels.

2.6.3 POND FLOOR AND POND WALLS

A pond is dug with bulldozers to
the planned depth, the sand and
mud being piled on the sides to
form the walls; the latter are slanted
at about 8° to simulate the slope of
the sea. Such is the construction of
a pond which utilizes the difference
between high and low tides.

This type of construction is less
expensive and has the following
advantages during rearing: water
can be drained from part of the
pond twice daily and made to flow
along the sides; hence the bottom
quality of the pond does not deteriorate, remaining suitable for burrowing. When the water temperature
drops, *Penaeus* avoid the floor of the
pond and collect at the sides to
burrow in the sand, which makes
removal of them easy durin gwinter.
However, unless proper precautions
are taken, mortalities will arise if
the prawns are exposed to the cold.

Since the pond must be drained
completely after the *Penaeus* are
removed, the grooves at the bottom
of the pond are very important.
However, in ponds which utilize
the difference between high and
low tides for changing water, mud

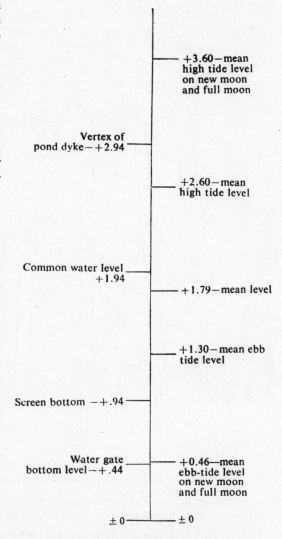

Figure 5.49. Relation between the level of water in a
pond and the tidal level in the sea (Akiho Culture
Center).

collects readily in these grooves. In Figure 5.50, the number of prawns seen at various
places of the pond is shown, and it is very clear that prawns seldom collect in the
grooves. Therefore production might increase if the mud were removed from the
grooves with pumps because, since the grooves presumably improve the floor
conditions in this type of pond, they must for now be considered essential.

2.6.4 SLUICE AND SCREEN

In culture ponds in which the difference between low and high tides is utilized,

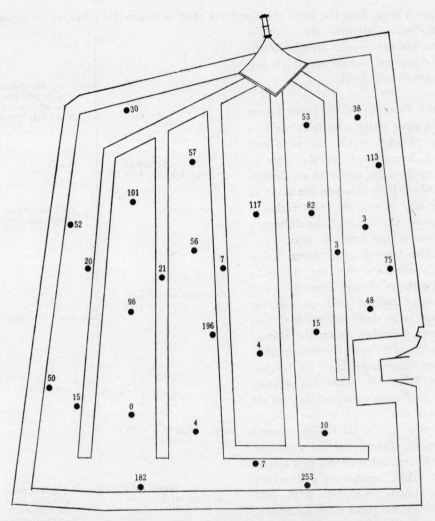

Figure 5.50. Number of specimens at various places in a rearing pond (data obtained from Akiho Culture Center). Figures indicate the number of specimens per 1.4 m².

Pond floor, 20 points, 45 specimens per m²; grooves 3 points, 7 specimens per m²; sides of pond 7 points, 77 specimens per m².

the water of the pond is changed through a sluice. The number and size of sluices differ according to the overall size of the pond. In culture ponds of 30×10^3 to 50×10^3 m² along the coast of the Seto Inland Sea, one sluice with a sliding door of 2 m $\times$ 1.5 m is constructed. This sliding door is opened when the level of the water in the sea is higher than that in the pond, and is closed when the level is lower. Although an automatic sliding door could be installed, as yet doors are manually operated. The structure of a sluice is shown in Figure 5.51.

When water from the open sea is drawn in through the sluice, waste matter of large dimensions, seaweed, predators of prawn larvae, and other unwanted organisms are

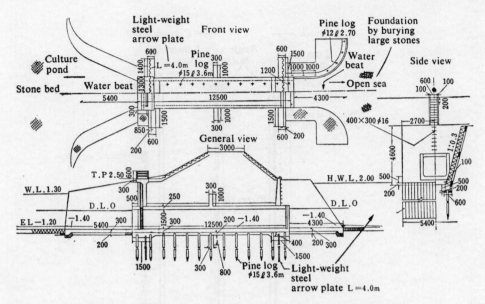

Figure 5.51. Structure of a sluice (used in Akiho Culture Center).

also drawn in. Conversely, when the water is drained off, some *Penaeus* escape. To avoid these situations, weirs which prevent the entrance of waste, etc. are placed outside the sluice, and net screens which prevent the escape of *Penaeus* and the entrance of predators are placed on the inner side. Since a large amount of water is changed in a short period in this type of culture tank, the screen has to be very strong to withstand the high water pressure.

As shown in Figure 5.45, in a culture pond of about 30 to $50 \times 10^3\,m^2$, the water drawing area covers about 600 m² at the center of the sluice. This area is shaped somewhat like the sliding door in the pond.

The screen is a metal net fitted to angled rods which are fixed in a concrete foundation. As shown in Figure 5.52, the angled rods are arranged in two rows at intervals of about 50 cm. Metal nets of 2 cm mesh are used on the open sea and metal nets of 1 cm mesh are used on the pond side. A synthetic pest control net is spread over the pond side of the metal net screens when post-larvae are released for rearing. When the post-larvae have grown, the net is gradually removed.

One of the major difficulties in using this type of screen is the accumulation of adhering organisms which choke the meshes. Examples of the adhering organisms to be found on the screen are listed in Table 5.51. When the nets become blocked with unwanted organisms, not only is the changing of water in the pond affected, but a difference in the water level between the inner side and the outer side of the screen occurs; sometimes the pressure is so great on the screen that it breaks or gives way.

Another disadvantage with this type of screen is that it retards the flow of water, and hence water does not reach the corners of the pond. Water paths of a few meters in length are made in some ponds, which connect the sluice to the corners of the pond, facilitating water exchange even in the corners.

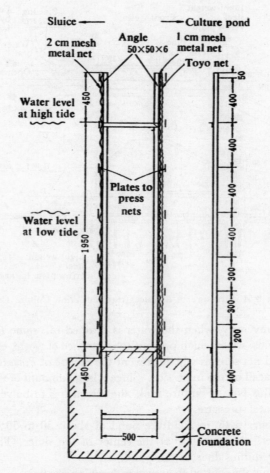

Figure 5.52. Side view of a screen.

Table 5.51. Organisms adhering to screen: number of organisms/m²
(Shimono and Ishida, unpublished)

Types	Places of collection		
	Outside sluice	At the center	On the side of culture pond
Mussels	187,800	272,200	930
Gammarus	133	90	
Caprella	20	16	
Cypridina	3		

Remarks: Date of investigation, August 31, 1964; screen positioned on August 25.

2.7 Management of farming process

2.7.1 Preparations Before Culture

The period of rearing *Penaeus* in ponds differs according to the size of the larvae and the time of their release. The culturing period usually extends from 6 to 18 months. During that time, the floor of the pond becomes polluted with a heavy accumulation of plankton, seaweed, residual feed, and *Penaeus* excreta. A heavy collection of mud further adds to the deterioration of the pond quality. Thus upon culture completion the water must be thoroughly drained from the pond after the *Penaeus* have been removed, and the mud which has collected in the grooves is removed with pumps. The floor of the pond should then be exposed to sunlight and fresh air and subsequently turned over with bulldozers.

When bivalves are used as feed, a large amount of shell pieces collect at the bottom of the pond and the *Penaeus* find it difficult to burrow. Bulldozing the floor of the pond mixes these shell pieces thoroughly with the sand, making the floor soft enough for the prawns to burrow into.

The predatory fish which sometimes enter the shallow grooves on the pond floor during culture can be killed with poisons such as gelan.[22]

The screen should be carefully examined every now and then, and a new net used for every culture.

2.7.2 Seed Release

After fertilization, the larvae attain an average weight of 0.02 g in one month, at which time they are able to burrow and strong enough to tolerate environmental changes. Hence these larvae can be transferred directly from the hatching tank to the rearing pond. However care must be exercised to see that the water temperature of the rearing pond corresponds to that of the hatching tank before the larvae are transferred. If the seedlings are packed in cold vinyl bags for transportation from the hatching tank to the culturing pond, then the water temperature inside the vinyl bags must be gradually increased to that of the rearing pond before the seedlings are released.

Healthy post-larvae begin feeding soon after their release in the culturing ponds. Finely chopped clam meat is used as feed initially. However, in rearing ponds in which the water has been changed frequently, the adhering diatoms will have grown at the bottom and will serve as a natural diet; in that case clam feed need be given only once or twice a day.

If the predators of *Penaeus* have been eliminated, and the escape of the post-larvae prevented, the survival rate of these seedlings in the rearing pond will presumably be closely related to the density of the crop at the time of release. According to past results obtained in rearing *Penaeus* in double-bottomed culture ponds, the survival rate after 1 month is 26.6% when the density of the post-larvae at the time of release is 777 specimens/m^2, and 51.0 to 61.4% when the density is 111 to 555 specimens/m^2. However the influence of density on survival rate need not be considered in culture

[22]Gelan is made of powdered Derris roots. It is more effective when it contains rotenon (5% or more).

ponds which utilize the difference between high and low tides, since the density does not generally exceed 40 specimens/m².

On the other hand, this density seems too low for rearing. If the number of larvae released per m² is 50, even if the feeding rate is 100% per day, it amounts to only $0.02 g \times 50 = 1.0 g/m²$ daily. Distributing such a small amount of feed uniformly over a wide area is very difficult. Hence the net used for the prevention of larval escape is divided into several sections and the released larvae reared in these sections for a few weeks.

When post-larvae are reared to the adult stage in the same pond, the floor of the pond becomes very soiled. Moreover, since the counting of specimens[23] in such cases is not so accurate, the difference between the planned production and that obtained can be quite large.

When *Penaeus* have attained a size of 1 to 2 g, counting them by the weighing method is easy.[24] Since the survival rate up to the adult stage is more or less constant, about 80%, the count of *Penaeus* at this stage is considered reliable. Hence, when 100 post-larvae are released per m² and reared for 1 to 2 months or a size of 1 to 2 g, an accurately counted number of seedlings can be released into the culture ponds.

2.7.3 COUNTING

When *Penaeus* are cultured in the same pond from the post-larval period to the adult stage, the specimens in the pond must be periodically counted. To do so, the pond is usually divided into sections and a few sections randomly selected for sampling. Before counting, a net is spread on a frame to prevent the *Penaeus* from escaping. This frame is kept at a particular place in the pond, and the *Penaeus* that collect in it removed by hand or net and counted. Figure 5.50 shows the results of counting *Penaeus* by various methods; according to this data, the number of *Penaeus* is maximum at the peripheral region of the pond, and a little less for the flat bottom surface, while the number in the grooves is very small. Hence to count the number of *Penaeus* accurately, the pond should be divided into three sections: periphery, grooves, and flat floor (ungrooved surface).

In very large ponds, it is impossible to count the number of prawns by exposing the pond floor. Hence the prawns have to be caught at a considerable water depth and then counted, a procedure which takes much time and involves much labor. Furthermore, *Penaeus* are not uniformly distributed over the pond floor and hence, to make the count as accurate as possible, the number of points assigned this area should be increased. However, even if the accuracy of the counting is improved through this suggestion, the profitability of the culture will still be assessed on the basis of the errors in the counting! Thus in actual practice the number of specimens is estimated on the basis of the counting done when the larvae weigh 1 to 2 g. Thereafter the number of specimens is estimated on the basis of past experience, the amount of feed provided, and the growth rate.

[23]The counting of post-larvae is usually carried out by the beaker method (Hudinaga and Kittaka, 1967). The deviation in the counting of specimens per person is less than 10%.

[24]After weighing a few thousand specimens, the mean weight of the *Penaeus* is calculated on the basis of the actual number. The value of the total weight is divided by the mean weight and the number of specimens thus determined.

2.7.4 HARVEST

When *Penaeus* attain a weight of about 20 g, they are caught and marketed. When the rearing pond is small, the water is completely drained off and the prawns harvested by digging them out of the mud or passing an electric current through the floor. When the pond is large, harvesting takes a fairly long time. The water in the pond is maintained at the same level as during the rearing period and the prawns in the sand burrows collected during the day by pump-fitted nets; night collection is done with hand nets as prawns usually swim after dark. The latter can only be done however if the water temperature is high as prawns do not swim freely when the water temperature is low; collection in low water temperatures is done only during the day with pump-fitted nets.

The pump-fitted nets are located at the back of the pump to which a number of nozzles have been connected. When water is sprayed on the floor of the pond, the prawns jump out of the sand and are caught in the nets. Since this method of collection stirs the sand in the pond, the quality of the floor is improved both by the stirring and by the removal of algae such as sea lettuce, etc.

In ponds which are 30×10^3 to 50×10^3 m², one to three sets of fykes are laid in the water. *Penaeus* come out of the sand at sunset in search of food, returning by dawn to the sand burrows. When the water temperature is high, activity is vigorous and from sunset to dawn collection by fykes is easy. But when the water temperature is low, the feeding activity is restricted to a short period after sunset and collection can only be done then. Since the prawns are very active and their respiratory rate increases at night, some die when caught due to the high density in the net, or to bad handling at the time of collection. Such is not the case when the water temperature is suitable

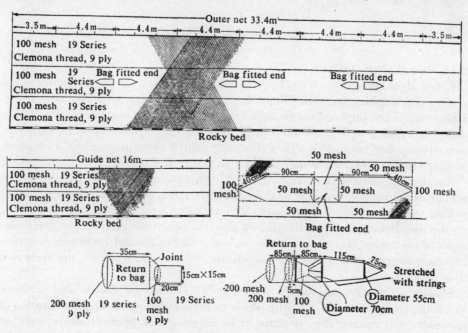

Figure 5.53. Structure of a fyke (Omura).

for collection. The structure of the fyke is shown in Figures 5.53 and 5.54.

In running water ponds the habit of prawns to move along the water current is utilized by placing nets at the mouth of the drain after removing the screen. In ponds in which the difference between low and high tides is utilized for rearing, prawns are collected by holding the net in the direction of the water current caused by the water wheels.

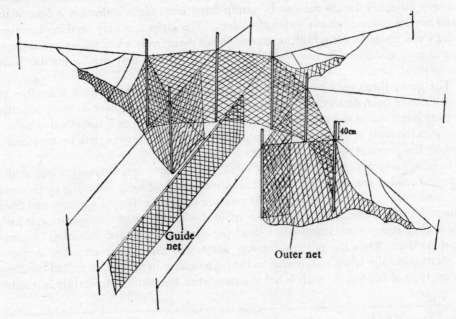

Figure 5.54. Placement of fyke.

2.7.5 REMOVAL OF ALGAE

Green algae such as sea lettuce, green string lettuce, etc. grow rampantly in the rearing ponds during the summer and, when they decompose, spoil the floor quality. Hence one of the important operations in culturing *Penaeus* is algae control. This is done by dragging nets across the floor of the pond once a day. This manually performed operation, however, does not effectively control algae in large and wide ponds in which the difference between low and high tides is utilized.

An algicide (commercial name "Delrad") containing dehydro (ab)ethyl-amine acetate as the main component was recently developed. Arasaki and Nozawa (1958) studied the effect of this drug on algae. "Delrad" is strongly adhesive and, by reacting on the metabolism of the algae, hinders algal nutrition and destroys the green pigment of the leaves. The drug rapidly loses its effect in natural conditions however, and moreover, not only does it have an adverse effect on fish, it seems quite likely that it is poisonous.

Shimono (unpublished) studied the effect of "Delrad" on *Penaeus* and seaweed. While the concentration of the drug which causes white rot and the disintegration of seaweed varies considerably, it seems to be effective at 2 ppm. However, even at 4 ppm *Penaeus* feed, molt, and grow normally. By utilizing this difference in the resistive strength of *Penaeus* and seaweed, Shimono estimated the effective removal of algae with this drug. In experiments with "Delrad" in rearing ponds which utilize the difference between low and high tides, it was found that its concentration could not be maintained because it does not readily dissolve in sea water. Furthermore, the concentration increases in some regions due to wind direction and water flow, which causes prawn mortalities; hence this drug cannot safely be employed on a commercial scale just now.

Brightness and transparency of pond water generally influence seaweed growth and experience has shown that growth is more rampant if the pond water is very clear. Where diatom growth is profuse, dulling the water's transparency, seaweed growth is controlled. It would appear therefore that maintaining a culture of diatoms is an effective method of controlling seaweed growth.

2.7.6 FISH FOUND IN CULTURE PONDS

It has long been known that even a relatively low concentration of Derris roots causes mortalities among fish. Hence before releasing the post-larvae in rearing ponds, this substance can be placed at various spots on the pond floor and the resultant dead fish easily collected.

The influence of powdered Derris roots on fish and *Penaeus* is shown in Table 5.52 (Shimono, unpublished). While the drug causes death among gobies at a concentration of 0.5 to 1.0 ppm, *Penaeus* survive even at a concentration of 5 ppm.

Table 5.52. Effect of powdered Derris root on gobies and prawns

Concentration of Derris powder (ppm)	Gobies		Prawns	
	Body length (cm)	Time taken for dying (minutes)	Body length (cm)	Time taken for dying (minutes)
10	2	18	2	Survived
	3	18	3	8 died in 60 min
5	2	75	2	Survived
	4	75	2	Survived
1	2	120	2	Survived
	4	120	4	Survived
0.5	2	200	2	Survived
	3	Survived	4	Survived

When the pond is wide, the removal of all *Penaeus* predators with Derris roots is difficult. According to an experiment in which post-larvae and gobies (*Tridentiger obscurus*) were grown together in water tanks ($2 \, m \times 1 \, m \times 1 \, m$), one goby ate about 1.7 post-larvae of *Penaeus* per day (Hudinaga and Kittaka, 1964). However, the relative density of post-larvae and gobies in a rearing pond is usually much lower than in this experiment, and hence the number of post-larvae generally eaten by gobies would be fewer.

Even after releasing post-larvae, eggs and fingerlings of predators enter the rearing ponds through the screen, but since there is not much difference in the rate of growth between the fish and the prawns, the chances of the latter being eaten are slim. But the entrance of fish such as the gizzard shad, mullet, etc. should be prevented as prawns are definitely not safe with these.

2.7.7 MAINTENANCE OF THE POND FLOOR

When the floor quality of the pond deteriorates from long usage, the entire activity of the *Penaeus* is adversely affected; this results in delayed growth, increased mortality rate, and a reduction in market value. Hence it is very important to prevent floor deterioration when prawns are cultured, and various measures have been considered such as water drainage (Yamamoto *et al.*, 1960), double-bottomed ponds, sectioned corfs, and tank culture.

When the area is small any of these methods apply, but with large ponds they involve considerable expense. So far no standard method for maintaining the quality of a large pond floor has been established, but some methods are said to be partially effective:

1. *Mechanical method:* Equipment which sprays water into the pond stirs the sand at the bottom and thereby secondarily oxidizes the floor materials. Unfortunately the effect of this method is relatively short-lived. Furthermore, stirring the floor materials in a pond which has greatly deteriorated can cause a severe shortage of oxygen temporarily, and hence the drag net operation must be resorted to daily.

2. *Chemical method:* Organic substances which collect at the bottom of the pond rapidly decompose when the water temperature increases during the summer and cause pollution. Hydrogen sulfide in particular, which is produced from black rot mud, uses up the dissolved oxygen in the water and simultaneously causes damage to the *Penaeus*.

Since hydrogen sulfide combined with free iron oxide changes into a harmless iron sulfide, the deterioration of a pond bed can be prevented by introducing iron oxide into the pond. Various mineral residues, used as silicate fertilizers in agriculture, are effective as iron oxide, particularly furnace waste (20 to 30% iron).

The amount of iron oxide required depends upon the degree of pollution, but usually 1 kg or more is required for an area of 1 m² (Yamashita, unpublished). The number of applications differ according to the length of the rearing period, but in summer when the temperature is high, iron oxide should be applied at least twice daily.

2.8 Conditions of rearing ponds

2.8.1 WATER TEMPERATURE

The mean water temperature of rearing ponds on the Seto Inland Sea coast, observed for 1500 hours during the day and 0500 hours in the early morning, is shown in Figure 5.55. The mean water temperature in the open sea and the atmospheric temperature during the same time period are included as a frame of reference.

The water temperature in the pond is greatly influenced by that of the open sea, which differs according to seasons, and is somewhat affected by the atmospheric temperature. The mean temperature at 1500 hours during the day is 2 to 3°C higher than the mean at 0500 morning hours. The water temperature is maximum in August and minimum in January and February. The maximum and minimum mean water

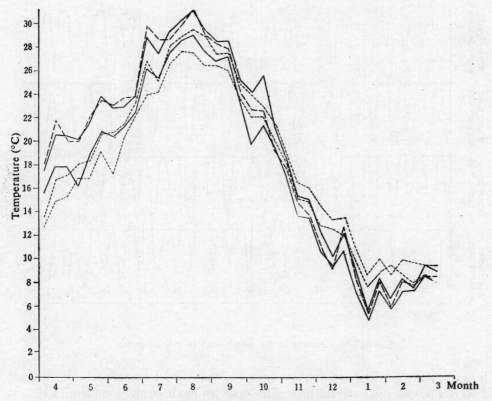

Figure 5.55. Mean water temperature of rearing ponds.

———water temperature of pond (upper line shows readings at 1500 hours, lower line shows at 0500 hrs; - - - - water temperature of open sea (upper line at 1500 hours, lower at 0500 hrs. — — — — atmospheric temperature (at 1500 hours).

temperatures during the day are 31.1°C in mid-August and 5.7°C in mid-January. The corresponding temperatures in early morning are 21.1°C in mid-August and 5.8°C in early February. The maximum and minimum water temperature measured for the entire year are 33.0°C for the 15 day hours on August 18, and 2.5°C for the 5 early morning hours on January 13. Thus the influence of the low water temperature during the winter on *Penaeus* will be significant.

2.8.2 Interchange of Sea Water

In a running water pond, the rearing of *Penaeus* is greatly controlled by the amount of sea water exchange. In rearing ponds utilizing the difference in low and high tides, the production of *Penaeus* is not directly controlled by the amount of sea water interchange. However, various environmental factors influence production and among these are some which affect the amount of dissolved oxygen, and thereby indirectly control production.

Some of the changes in the water level of rearing ponds of about 30×10^3 m² (given in Figure 5.48) are shown in Figure 5.56, in which the water is changed with the

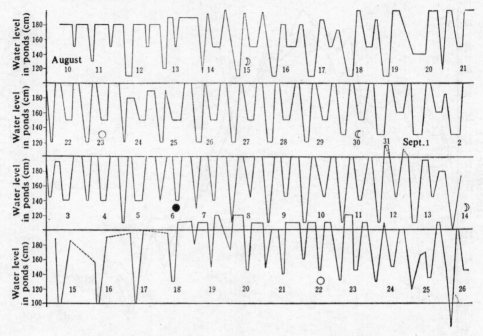

Figure 5.56. Changes in the water level in rearing ponds.

help of a sluice gate of 2 m width and 1.5 m height. The relation between the water level of a rearing pond and the tides of the open sea has been given in Figure 5.49.

The water level in these rearing ponds is at its maximum at high tide and at its minimum at low tide. *Penaeus* are active during the night and lie dormant in the floor sand during the day. Hence the water of the pond should be changed during the day, and a high water level maintained during the night. The amount of water which can be exchanged in the pond depends upon the difference between the levels in the pond and in the open sea. As the onset of the tides is delayed on each successive day by about 50 minutes, the exchange of pond water is restricted during the period when the high tides occur at night.

In summer when the temperature rises, or in winter when the temperature falls, if the water level decreases too much during the day, prawn rearing is adversely affected. Hence the water level should not be allowed to decrease below a particular level. Usually the water depth in the rearing ponds, even during ebb tide, should be maintained at about 1 m in the summer and about 2 m in the winter. The conditions shown in Figure 5.56 are those when the water of the pond is changed by an efficient use of a sluice gate in relation to the tides in the open sea. The variations in the water level shown in this figure are at a time when such can be controlled, so as not to adversely affect the prawns during the course of rearing.

The following is a detailed account of pond water exchange, based on the variations in the water level just described:

If the low water level on a particular day is expressed as h_1 (cm), the next rise in the water level as h_2 (cm), and the next lower and higher levels as h_1' (cm) and h_2' (cm) respectively, the increasing rate of water level, Δh (cm) [5], per day is

$$\Delta h = (h_2 - h_1) + (h_2' - h_1') .$$

The amount of sea water inflow per day, and the amount of sea water outflow, are approximately expressed by $\dfrac{\Delta h}{100}$ (m) $\times$ area of pond (m²). The daily variations of Δh are given in Figure 5.57. The value is at its maximum, 130 to 170 cm, during

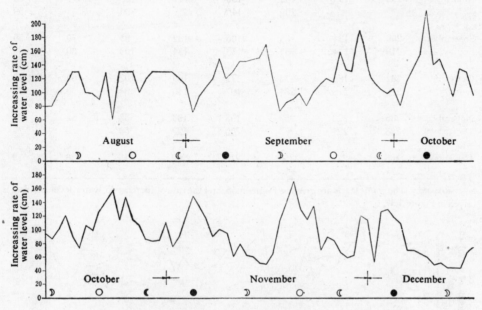

Figure 5.57. Increasing rate of water level (Δh) in rearing ponds
and its daily variations.

high tide, and minimum, 60 to 110 cm, during low tide. By determining the mean water level h (cm) from the maximum and minimum water levels on a particular day, and by determining $\dfrac{\Delta h}{h}$ of the increasing rate of water level, it is possible to obtain the approximate ratio of water exchange. The values thus determined are given in Table 5.53 with 0.8 during high tide, and 0.6 to 0.65 during low tide in August and September when the water level is low, and 0.5 to 0.7 during high tide, and 0.25 to 0.45 during low tide in October and November when the water level is high.

2.8.3 AMOUNT OF DISSOLVED OXYGEN

The nightly variations in the amount of dissolved oxygen are shown in Figure 5.58. As already explained, the amount of dissolved oxygen at 10:00 p.m. is often more than 5 cc in rearing ponds in which the water is exchanged, and rarely 4 cc or less. The amount of dissolved oxygen at 4:00 a.m. is less than at 10:00 p.m. the

[25] More accurately, the increasing rate of water level per one sunny day.

Table 5.53. Exchange of water in a pond during high and low tides

Month	Water level during high tide (cm)		Water level during low tide (cm)		Increasing rate of water level (cm) per day		Exchange rate (%)	
	Maximum	Minimum	Maximum	Minimum	High tide	Low tide	High tide	Low tide
August	193	135	185	130	117	109	71	69
			200	140		100		59
September	200	134	183	100	132	83	79	59
	210	144	185	133	134	103	80	65
October	282	198	272	172	169	102	69	46
	286	216	281	197	140	84	56	35
November	286	220	252	193	133	58	52	26
	288	224	263	200	127	63	50	27
December	286	227	253	210	99	45	38	19

Remarks: The water levels are given in Figure 5.56, and the rate of increase in water level per day is given in Figure 5.57.

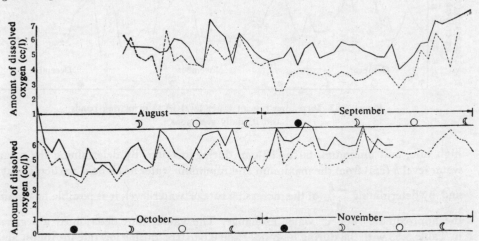

Figure 5.58. Changes in the amount of dissolved oxygen at night.

previous night. The amount of oxygen decreases to about 3 cc/l during the period from early September to mid-September, and the minimum value is 2.5 cc/l. This type of variation in the amount of dissolved oxygen is found not only in rearing ponds for *Penaeus* but in most rearing ponds for other varieties of fish as well.

Since, as a rule, pond water is not changed at night, the difference in the amount of oxygen at 10:00 p.m. and 4:00 a.m. indicates the amount of oxygen consumed. The mean decrease in dissolved oxygen for three nights during both low and high tides, is shown in Table 5.54. The decrease in the amount of dissolved oxygen at

Table 5.54. Decrease in the amount of dissolved oxygen at night during high and low tides

Month	High tide			Low tide		
	10:00 p.m.	4:00 a.m.	Amount of decrease	10:00 p.m.	4:00 a.m.	Amount of decrease
August				5.43	4.74	1.10
	5.47	4.68	0.58	5.22	4.65	0.57
September	5.10	3.80	1.06	5.20	3.29	1.91
	5.35	4.37	0.98	7.80	6.28	1.55
October	4.48	3.75	0.73	5.60	4.43	1.17
	5.16	4.61	0.34	5.07	4.56	0.51
November	6.28	5.33	0.92	5.83	5.04	0.79

Remarks : Measured values of dissolved oxygen are given in Figure 5.58.

night is not so drastic in July and August; it becomes maximum in September and is particularly appreciable during low tide in September and October.

Regarding the problem of maintaining the oxygen level in the rearing ponds, valuable studies have been done by Matsue and Sano (1958) on rearing ponds with a constant water depth. Inoue (1965) has studied the same problem in saline water rearing ponds in which the water depth varies according to tidal conditions.

In rearing ponds which utilize the difference between low and high tide, except when the water is changed in the pond, no variation in water depth occurs. But with reference to the maintenance of the oxygen level in such ponds, no reliable data on which calculations could be made are available as yet. Hence such a study should be made by taking into account the report on the measured values of dissolved oxygen (Figure 5.58) and the control of the water level in ponds (Figure 5.56).

The number of specimens and their weight during high and low tides in September as observed in a particular rearing pond (pond No. 2 of Table 5.42), are shown in Table 5.55. The total amount of oxygen consumed per unit-time in unit-area of the

Table 5.55. Oxygen uptake in rearing ponds

Date	Tides	Estimated weight (g)	Estimated number of specimens	Estimated weight (kg)	Estimated weight per unit area (g/m²)	Mean water level (cm)	Decreasing rate of oxygen content* (cc/l)	Consumption of oxygen per unit-time in unit-area (cc/m²/h)	Amount of oxygen consumed by *Penaeus* per unit-time in unit-area** (cc/m²/h)	Ratio of oxygen consumed by *Penaeus* (%)
September 6	High	7.0	927 × 10³	5,670	189	200	1.06	354	56.7	16
September 14	Low	8.0	810 × 10³	6,480	216	180	1.91	572	64.8	11
September 22	High	8.8	810 × 10³	7,128	237	183	0.98	298	71.1	24
September 29	Low	9.6	810 × 10³	7,776	259	185	1.55	477	77.7	17

*The rate of decrease in six hours, between 10:00 p.m. and 4:00 a.m.
**The amount of oxygen consumed by *Penaeus* is taken as 300 cc/kg/h.

water basin, and the amount of oxygen supposedly consumed by *Penaeus*, calculated on the basis of the difference in the amount of dissolved oxygen at 10:00 p.m. and 4:00 a.m. (Table 5.54), are shown in Table 5.55.

The ratio of oxygen uptake by *Penaeus* to the total oxygen depletion in the pond, does not exceed 25%. Compared to high tide conditions, the exchange of water in the pond during low tide is poor. Moreover, in addition to the increased demand for oxygen in sea water, there is also an increase in the consumption of oxygen by the abundant diatoms in the pond. Hence, the oxygen available to *Penaeus* decreases during low tide. The concentration of dissolved oxygen does not go below 2.5 cc/l and hence, as far as the oxygen level is concerned, rearing is still possible. This is one of the salient features of rearing ponds which utilize the difference between high and low tides.

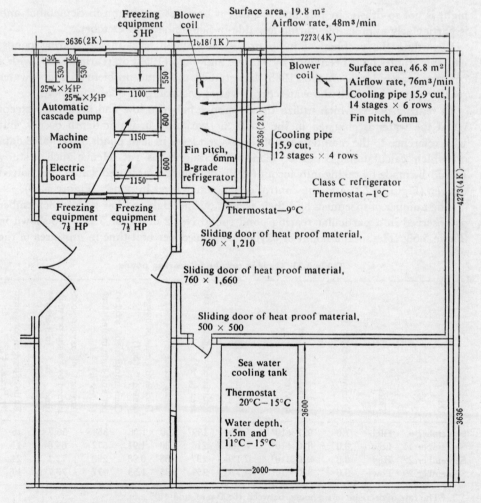

Figure 5.59. Model of packing equipment placement (Akiho Culture Center).

2.9 Packing

2.9.1 METHOD OF PACKING

The price of *Penaeus* varies greatly according to the animal's freshness. Hence considerable effort has been given to the improvement of packing since the inception of *Penaeus* culturing. At present, wooden or corrugated cardboard boxes of suitable size are used, in which the *Penaeus* are packed in sawdust. The *Penaeus* thus packed live for a few days when the atmospheric temperature is low, but since some die during transportation when the temperature is high, the temperature inside the box must be controlled as this controls the metabolism of the animal.

Penaeus should be kept at a low temperature before packing so as to accustom them to it and thus hold down loss during transportation. In winter there is no harm in packing *Penaeus* directly, but in summer they must first be transferred to a cold-water pond at a temperature of 5 to 10°C less than the temperature of the rearing pond, which should gradually be lowered to about 15°C prior to the transfer. Packing is then done in a room at 5 to 10°C.

Hence in a model culturing center, facilities such as a cold-water tank, a packing room, and storage chambers are essential. The equipment required for packing a few hundred kg of *Penaeus* per day is shown in Figure 5.59.

The method of packing differs according to the consignment season and the conditions before packing. Corrugated cardboard boxes of 22.0 cm × 21.5 cm × 29.5 cm (height) in which 1 cm thick styrol sheet is spread at the bottom are used. A few centimeters of sawdust, already dried and kept inside a room at a low temperature, are spread on the styrol sheet and the prawns arranged in this and covered with more sawdust. Corrugated cardboard boxes of the foregoing dimensions can hold 70 to 150 *Penaeus* weighing about 2 kg; hence 3 to 4 layers of prawns and sawdust are placed in them, with a final styrol sheet spread over the uppermost layer before the lid is closed. Adhesive tape of about 5 cm width is used to seal the box. In summer, a vinyl bag containing about 200 to 300 g of crushed ice is wrapped in paper and placed just under the lid to maintain the low [box] temperature.

2.9.2 TRANSPORTATION

The survival rate of *Penaeus* packed in sawdust at the time the box is opened, is given in Table 5.56. These survivors live for about 1.5 days, but 50% die after 3 days. Studies to determine the living rate of packed prawns were done in late September, and it was learned that prawns grown in this season at a relatively high temperature, lived for 1.5 days during transportation, even when the temperature during transportation was maintained at 10 to 14°C.

At high temperatures, the number of prawns molting is large, and prawns which have a soft carapace cannot tolerate transportation even when packed in sawdust. This is one reason why packing in the summer does not yield good results.

2.9.3 EXPERIMENTS IN PACKING

1. *Oxygen gas:* The survival of *Penaeus* which are packed in sawdust or other materials and kept in airtight boxes filled with oxygen or carbon dioxide, is shown in Table 5.57 (according to Akamine, unpublished). Oxygen or air is, of course, essen-

Table 5.56. Experiment with the transportation of *Penaeus*
by refrigerated truck (Miyamura, 1963)

Experiment number	Number of specimens	Weight (kg)	Mean weight (g)	Time lag (h)	Temperature (°C)	Number of specimens living	Number of specimens dead	Survival rate (%)
	10			34.5		9	1	90
	10					4	6	
	10			68.0		7	3	53
	10					5	5	
1	10			92.0		5	5	
	10					6	4	
	10					7	3	38
	10					1	9	
	10					0	10	
	224	4	17.8	40.3	12.6	213	11	95
2	211	4	18.9	40.6	12.8	198	13	94
	291	4	13.8	41.0	12.6	261	30	90
	228	4	17.5	41.3	11.9	216	12	95

Remarks: Date of Experiment No. 1, September 23-27, 1963; sawdust packing.

Table 5.57. Experiments with the transportation of *Penaeus* **in**
oxygen filled containers (Akamine, unpublished)

Experiment number	Method of packing	Experiment time (h)	Temperature of experiment (°C)	Number of specimens used	Number of living specimens	Number of dead specimens	Survival rate (%)	Remarks
1	Sawdust pack	17	11.0-15.2	30	21	9	70	
	Sawdust pack, carbon dioxide	17	11.0-15.2	30	0	30	0	
2	Sawdust pack	15	11-15	20	20	0	100	
	Sawdust pack, oxygen	15	11-15	20	20	0	100	
	Sawdust pack	20	20-25	30	27	3	90	
	Sawdust pack, oxygen	20	21.2-28.3	30	29	1	97	
3	Styrol pack	20	19.0-27.8	20	9	11	45	
	Styrol pack, oxygen	20	19.0-27.8	30	25	5	83	
	Cotton pack, oxygen	24	20.0-28.0	30	27	3	90	
	Sea water, 4 *l*, oxygen gas, 26 *l*	18	17.5-13.8	56	54	2	97	10-12 g of prawns kept in
4	Sea water, 6 *l*, oxygen gas, 16 *l*	18	18.4-14.6	100	86	14	86	a polyethylene bag

tial while packing *Penaeus*. Akamine carried out further experiments with *Penaeus* by packing them in vinyl bags containing some sea water, filling the bags with oxygen, and then closing the bags tightly. The living rate for this method of packing is shown in the lower line of Table 5.57.

The results of transporting *Penaeus* in polyethylene bags filled with sea water and oxygen, together with those from experiment No. 2 given in Table 5.56, are given in Table 5.58. During the experiment, oxygen gas was introduced into the sea water; when the weight of the bag increased, a decrease in the survival rate occurred. As against the survival rate of 93% in the sawdust packs (experiment No. 2 in Table 5.56), the survival rate in sea water and oxygen is 75%.

Table 5.58. Experiment with the transportation of *Penaeus* **in refrigerated truck (Miyamura, 1963)**

Number of specimens	Weight* (kg)	Mean body weight* (g)	Time (h)	Temperature of container (°C)	Number of living specimens	Number of dead specimens	Survival rate (%)	Amount of oxygen at start (cc/e)	Amount of oxygen at end (cc/e)
30					30	0	100	6.17	18.09
50					50	0	100	6.17	15.00
50			12	16.2-12.1	50	0	100	6.17	15.45
60					60	0	100	6.17	15.79
30					30	0	100	18.09	18.26
50					41	9	82	15.00	18.42
50			39	12.1-13.5	47	3	94	15.45	16.12
60					24	36	40	15.79	14.58

Remarks: Date of experiment, September 23-25, 1963; method of packing, polyethylene bag filled with 8 liters of sea water and 12 liters of oxygen.

* Data not given in the original text—Translator.

If 4 kg of *Penaeus* are packed in a corrugated cardboard box filled with sawdust with a 30-liter capacity (=22 cm ×45 cm ×30 cm), the packing area and the packing weight per unit weight of *Penaeus* are

$$30 \ l/4 \ kg = 7.5 \ l/kg, \ (0.23^{26} \times 30 + 4) \ kg/4 \ kg = 2.7 \ kg/kg.$$

On the other hand, when sea water and oxygen are used, the values are

$$(8 \ l + 12 \ l)/approx. \ 1 \ kg \simeq 20 \ l \ kg, \ (1 \times 8 + 0 \times 12) \ kg/approx. \ 1 \ kg \simeq 8 \ kg/kg.$$

Thus the values in this instance are higher, which means that sawdust packing for transportation is more economical.

2. *Filling materials other than sawdust:* When the weight of packed *Penaeus* is increased, the amount of filling to be used also increases. Along with this increase, such problems as drying, etc. arise, and hence the possibilities of using filling materials other than sawdust should be investigated. One of the characteristic features of sawdust is that it is a poor conductor of heat, and since it contains minute air spaces, respiration is not obstructed. Moreover, the material is also lightweight. A commer-

[26] Weight of one liter of sawdust taken as 0.23 kg.

cial product called "Parite,"[27] which contains these same qualities, is also available now.

Corrugated cardboard boxes are filled with "Parite" and, after cooling to about 5°C, kept at 24°C room temperature and 51% humidity. By studying the temperature changes in the box for some time, efficiency of maintaining a low temperature for "Parite" was compared with that of sawdust packing; the results are given in Table 5.59.

Table 5.59. Comparison of sawdust and "Parite" packing
(Mitsui Kinzoku Kogyo, K. K.)

Sample	Water		Weight of specimens used in the experiment, kg/l	Volume ratio (to sawdust)	Duration of keeping (h)	Cooling efficiency ratio (for sawdust)
	Volume, %	Weight, %				
Sawdust		12.6	0.228	1.00	12.5	1.0
"Parite"						
	0	0	0.075	0.33	9.5	0.8
	1.70	24.0	0.093	0.41	9.5	0.8
	3.60	48.0	0.111	0.49	9.5	0.8
	7.16	95.0	0.147	0.64	12.5	1.0
	15.42	209.0	0.228	1.00	15.0	1.2

When "Parite" is added with water, the material remains cold for a long period. When 95% water is added to weight percentage, "Parite" is 36% lighter than sawdust. When more water is added and the volume equal to that of sawdust, the material remains cold for a longer period of time. The survival rate of *Penaeus* in "Parite" is better than in sawdust. However, since "Parite" is costlier than sawdust, it is not yet used on a commercial basis.

3. Commercial culturing of prawns

The history of prawn cultivation is not so old and culturing has not yet been established either technically or commercially. Much improvement is needed as several difficult problems exist in high density rearing and, at present, only low density rearing is done in a rough manner. Certain improvements in rearing methodology are currently under investigation.

Feed accounts for a major portion of the expense in rearing prawns. From the results of a number of experiments carried out on various feeds, the most effective feed and method of feeding are likely to be determined in the near future, and feed expense thereby reduced.

[27] Pearl rock is finely powdered and rapidly heated; the particles swell and minute air bubbles form. The material is white in color and very lightweight.

The most effective feed for *Penaeus* is clam; abou 10 tons or more of clam are required for the production of 1 ton of *Penaeus*.[28]

There are several problems which have to be solved by further investigations, that relate to the cultivation of prawns. Mussels and another mollusk, *Brachidontes senhousia (hototogisu)*, are being considered as substitutes for clam. In some cultivation grounds the rearing of these mollusks has been introduced and thus dependence on clam as feed is gradually decreasing.

Considering the heavy feed requirements, it might be thought that prawn seedlings would be released in natural water basins and allowed to grow on natural feed, the adults being caught at a suitable time. However, large-scale production of seedlings has now been established and, since seedling production is also stabilized, the culture of *Penaeus* has spread to the coastal regions of Japan with the Seto Inland Sea as the center.

The regions suited for the cultivation of *Penaeus* can be described as those which have sandy beds in which young bivalves and polychaetes are plentiful. From the viewpoint of prawn production and efficiency of transportation, it is better to keep the size of the seedlings small at the time of release, but on the other hand, the yield rate is better if the size of the seedlings is larger. Thus there is some contradiction as to the suitable size of seedlings, but in order to have the size which will satisfy both yield rate and transport efficiency, it seems that the size which allows the post-larvae to escape from predators, i.e. the size at which they are able to burrow inside the sand (0.02 g mean weight), ought to be adhered to.

The effect of tagging is mainly ascertained on the basis of harvest statistics. At times the effect of tagging can also be found on the basis of the body length and body weight at various times before release. However, due to insufficient experience and poor scientific data, the effect of tagging cannot be accurately determined. It is presently known that the growth of *Penaeus* is rapid in water basins suited to their rearing. It seems that the production rate is more or less fixed in the summer regions and, in order to increase efficiency, some measures which would support the growth of natural feed in these regions needs to be found.

(Kittaka)

---

[28] The food conversion rate is 8 when the prawns have grown to 20 g. Since the meat content of clams is 15%, the food conversion rate for clam with shell is $8 \times \dfrac{100}{15} = 53$. Hence, when only clam is given as feed to *Penaeus*, about 50 tons are required for the production of 1 ton of *Penaeus*. But, if the clam meat content in the feed is 25%, the growth and yield rate of the prawns do not differ from the growth and yield rate when the feed consists of 100% clam meat. The minimum requirement of clam for the production of 1 ton of *Penaeus* is $53 \times 0.25 = 13$ tons.

Bibliography

1. ARASAKI, S., and K. NOZAWA, 1958. Delrad ni tsuite (On the Effect of Delrad), *Nisuikaishi (Bulletin of the Japanese Society of Scientific Fisheries)*, **23** (10), 599-603.
2. HUDINAGA, M., 1935. Kuruma-ebi-zoku *(Penaeus)* no kenkyu, dai-ippo: Kuruma-ebi *(Penaeus japonicus* Bate) no hassei (1) (Study of Penaeus: Report No. 1. Development of *Penaeus japonicus* Bate), *Wakkanai Suisan Kenhokoku (Bulletin of Hayatomo Fisheries Experiment Station)*, **1**, 1-51, pls. 1-26.
3. ——, and J. KITTAKA, 1964. Kuruma-ebi no yoshoku (Cultivation of *Penaeus japonicus*); *Food Science*, **3** (10), 77-81.
4. ——, ——, 1966. Kuruma-ebi yosei no hetai to shiryo (Metamorphosis of the Larvae of *Penaeus* and Their Diet), *Nihon Plankton Kenkyu Renkakukaiho*, **13**, 83-94.
5. ——, and M. MIYAMURA, 1962. Kuruma-ebi no yoshoku (Cultivation of *Penaeus*: Collection of Papers Covering 20 Years of Studies by Japan Oceanographic Association), 694-706.
6. HUDINAGA, M., 1942. Reproduction, Development, and Rearing of *Penaeus japonicus* Bate, *Jap. J. Zool.*, **10**, 305-393, pls. 16-46.
7. ——, and J. KITTAKA, 1967. The Large-scale Production of Young Kuruma Prawns, *Penaeus japonicus* Bate, *Inform. Bull. Plank. Japan*. Commemorative No. for Dr. Matsue's 60th Birthday, 35-46.
8. INOUE, H., 1965. Senkai yogyo shisetsu ni okeru kaisui-koryu to hoyomitsuda ni tsuite (On the Flow of Sea Water and Density of Cultivation in Shallow Water Culture Zones), *Suisan zoshoku (Aquaculture)* (Special Issue), 461-477.
9. ISHIDA, D., 1966. Sunaba o setchi shita konkurito Ike ni yoru kuruma-ebi shiiku (Rearing of *Penaeus* in Concrete Ponds with Sand Beds), *Suisan zoshoku (Aquaculture)*, **14** (3), 127-133.
10. KURATA, H., 1967. Burain Shurinpu ran ni kansuru shiryo (Data on Shrimp Eggs), *Suisan zoshoku (Aquaculture)*, **14** (4), 205-219.
11. LEONARD, J. W., 1939. Notes on the Use of Derris as a Fish Poison, *Trans. Amer. Fish. Soc.*, **68**, 269-280.
12. MAEKAWA, K., 1960. Kuruma-ebi no chikuyo (On the Cultivation of *Penaeus*), *Suisan zoshoku (Aquaculture)*, **7** (4).
13. MATSUE, Y., 1954. Kaisanfuyu keiso sukeretoonema no baiyo [Culture of *Skeletonema costatum* (Grev.) Cleve], *Suisangaku no Gaikan (Outlines of Marine Biology)*. Japan Association for the Promotion of Science.
14. MIYAMURA, M., 1963. Toyo samokontororu kabushiki gaisha: Reito kontenatorakku ni yoru tane-ebi maki yuso shiken (Toyo Thermo Control Company, Ltd.: Transportation Experiments on Seed Prawns by Frozen Container Trucks).
15. ——, 1967. Kuruma-ebi (Study on *Penaeus*), Yogyogaku Kakuron, 727-752. Tokyo: Koseikaku Publications.

16. OZAKI, Y., 1965. Kuruma-ebi no kowari yoshoku ni tsuite (Small-scale Cultivation of *Penaeus*), Technique and Management of Fisheries, October, 1965, 45-50.

17. SANO, K., and Y. MATSUE, 1958a. Yomanchi no sanso taisha kiko sanso taisha ni motozuku yogyo kijun to sono riron (Rearing Standards in Ponds Judged on the Basis of Oxygen Metabolism), *Suisan zoshoku (Aquaculture)*, **6** (1), 43-49.

18. ——, ——, 1958. *Ibid.*, II (Application of Rearing Theories in Pond Culture), *Ibid.*, **6** (1), 50-55.

19. YAMAMOTO, T., T. ICHIMURA, N. TACHINO, and Y. ISHIKAWA, 1960. Kuruma-ebi no yoshokuchi ni okeru angkyohaisui no mokuteki o seko oyobi sono koka ni tsuite (Reasons for Draining the Water from *Penaeus* Rearing Ponds and Its Effect), *Suisan zoshoku (Aquaculture)*, **8** (2), 133-137.

PART VI
Artificial Culture of Shellfish

The technique of shellfish culture has been progressing since 1940 at more or less the same pace in Britain, America, and Japan. Studies initially concentrated on the life history and ecology of shellfish, then branched into types of feed and their nutritional value and rearing conditions; with these studies, a rapid development of the techniques of large-scale culturing was possible, from the collection of eggs and fertilization through rearing to a young stage. Progress in culturing has now attained a proficiency wherein present investigations are mainly directed toward further improvement in the quality of the product, based on the various aspects of the fundamental biology of the shellfish. On the commercial side, there is much hope that an efficient technique for the production of seedlings will soon be pinpointed.

Japan began laboratory experiments in 1960 to devise a technique for artificial seedling production, which the Oyster Research Center put into effect with abalones on a commercial scale. Soon a country-wide effort to produce seedlings of other important varieties of shellfish took hold. Each type of shellfish experimented with yielded good results in the production of seedlings. Projects are currently underway for the production of abalone seedlings at various Fisheries Research Stations, and studies on the effect of transplantation on the same are progressing. Experiments modeled on the work of these Fisheries are being simultaneously carried out on bivalves such as scallops, *Pecten,* etc.

Various mechanical engineering techniques are presently being applied to artificial seedling culture and, along with studies on mass production methods, the question of technical progress directed toward improving the quality of the product, is also under vigorous investigation.

In this section, the data compiled from various incorporated Oyster Research Centers are related to seedling production techniques, but these should be considered in the context of the fundamental principles of feed and rearing.

(Imai)

CHAPTER I

The Culture of Microorganisms Used as Feed

1. Types and food value

The earliest report on the artificial culturing of the larvae of shellfish with food organisms comes from Prytherch (1923), who reared the larvae of *Crassostrea virginica* on the microorganisms in running water. Hori and Kusakabe (1926, 1927) reared the larvae of *Crassostrea gigas* and *Ostrea lurida* on *Chlorella pacifica*. Numerous other reports on the rearing of larvae of various shellfish have since been made and, in recent years, the technique for a pure culture of microorganisms has greatly advanced, resulting in the use of microalgae as feed for shellfish. Table 6.1 shows some of the important food organisms used for the rearing of shellfish larvae.

According to the type of life led, shellfish larvae can be classified as planktonic (oysters, etc.) and sessile (abalones, etc.). Planktonic larvae usually prefer flagellates of a few micron in size and planktonic diatoms, while sessile larvae prefer sessile, colored flagellates and diatoms.

It is an established fact that a mixed diet containing more than two types of food organisms is much better than a diet consisting of a single food organism. This is true not only for shellfish, but also for crustaceans. A single type of organism does not provide the nutrition required by shellfish, but more than two types assure a balanced nutrition through a mutual adjustment among the organisms. According to Davis and Guillard (1958), the growth of the larvae of *Crassostrea virginica* and *Venus mercenaria* is much better when a mixture of four types of organisms, namely, *Isochrysis*, *Monochrysis*, *Platymonas*, and *Dunaliella* is given, than when only a single type of organism is provided. Even in the crustacean *Tigriopus japonicus*, rearing can be sustained for a longer period by providing a mixed diet of *Rhodomonas* and *Isochrysis*, rather than giving each separately (Provasoli, Shiraishi, and Lance, 1959).

2. Culture of microorganisms

2.1 Nutritional requirements of microalgae and the composition of culture media

Several facts concerning the vitamin requirements, etc. of algae have become clear (Provasoli, 1964). Many of the useful, single-celled algae on which shellfish feed, require the vitamins shown in Table 6.2, particularly B_{12}, thiamin, and biotin. In many cases these vitamins can be effectively added to the culture media. When only a single type of alga is grown, only the vitamins required by that type are added, but when several types of algae are cultured together, it is convenient to prepare

Table 6.1. Food organisms of shellfish larvae

Food organisms	Species of shellfish	Reference
Carteria sp.	*Mercenaria mercenaria*	Loosanoff and Davis, 1963
Chaetoceros simplex	*Haliotis discus*	Sagara, Iino and Ai, 1961
Chlamydomonas sp. (D)	*Mercenaria mercenaria*	Davis and Guillard, 1958
Chlamydomonas sp.	*Mytilus edulis*	Hirano and Oshima, 1962
Chlorella sp. (580)	*Mercenaria mercenaria*	Davis and Guillard, 1958
Chlorella sp. (UHMC)	*Crassostrea virginica*	Davis and Guillard, 1958
Chlorococcum sp.	*Mercenaria mercenaria* *Crassostrea virginica*	Davis and Guillard, 1958
Chromulina pleiades	*Crassostrea virginica*	Walne, 1956
Cryptomonas sp.	*Crassostrea virginica*	Loosanoff and Davis, 1963
Cyclotella sp. (O-3A)	*Mercenaria mercenaria* *Crassostrea virginica*	Loosanoff and Davis, 1963
Dicrateria inornata	*Crassostrea virginica*	Loosanoff and Davis, 1963
Dicrateria sp. (BII)	*Mercenaria mercenaria*	Loosanoff and Davis, 1963
Dunaliella euchlora	*Mercenaria mercenaria* *Crassostrea virginica*	Davis and Guillard, 1958
Dunaliella sp.	*Mercenaria mercenaria* *Crassostrea virginica*	Davis and Guillard, 1958
Hemiselmis refescens	*Crassostrea virginica*	Loosanoff and Davis, 1963
Isochrysis galbana	*Mercenaria mercenaria* *Crassostrea virginica* *Ostrea edulis*	Davis and Guillard, 1958 Walne, 1956
Monas sp.	*Crassostrea gigas* *Pinctada martensii* *Haliotis gigantea* *Mactra sachalinensis* *Ostrea edulis* *Ostrea lurida* *Pteria penguin*	Imai and Hatanaka, 1949 Kobayashi and Yuki, 1952 Iino, 1952 Imai, Hatanaka, Sato, and Sakai, 1953 Imai, Sakai, and Okada, 1953 Imai, Sakai, Okada, and Yoshida, 1954 Kagoshima Fisheries Experiment Station, 1959
Monochrysis lutheri	*Mercenaria mercenaria* *Crassostrea virginica* *Ostrea edulis* *Ostrea lurida* *Mactra sachalinensis* *Pecten yessoensis*	Davis and Guillard, 1958 Imai, and others.
Olisthodiscus sp.	*Mercenaria mercenaria*	Loosanoff and Davis, 1963
Platymonas sp. (No. 5) (= *Tetraselmis maculata*)	*Haliotis discus hannai*	Imai, and others.
Platymonas sp. (No. 1)	*Mercenaria mercenaria* *Crassostrea virginica*	Davis and Guillard, 1958
Phaeodactylum tricornutum	*Mercenaria mercenaria* *Crassostrea virginica*	Davis and Guillard, 1958
Pyramimonas grossi	*Crassostrea virginica*	Loosanoff and Davis, 1963
Rhodomonas sp.	*Mercenaria mercenaria*	Loosanoff and Davis, 1963
Skeletonema costatum	*Mercenaria mercenaria*	Loosanoff and Davis, 1963
Stichococcus sp. (O-18)	*Mercenaria mercenaria*	Loosanoff and Davis, 1963

Table 6.2. Vitamin requirements o algae

Type of algae	Number of species used in the study	Number of species requiring vitamins	Number of species requiring only B$_{12}$	Number of species requiring only thiamin	Number of species requiring B$_{12}$+biotin	Number of species requiring B$_{12}$+thiamin	Number of species requiring biotin+thiamin	Number of species requiring B$_{12}$+biotin+thiamin
Chlorophyta	70	46	13	13		20		
Euglenophyta	11	11	2	1		7		1
Cryptophyta	11	11	2	2		7		
Pyrrophyta	19	18	13				1	4
Chrysophyta	23	22	1	7	2	9	1	2
Bacillariophyta	45	22	16	3		3		
Phaeophyta	3							
Rhodophyta	7	6	6					
Cyanophyta	16	7	7					
Total	205	143	60	26	2	46	2	7

(Provasoli, 1964)

vitamin mixture III (see Table 6.3). About 0.1 to 0.3 ml of this mixture are added to 100 ml of culture fluid.

Algae also require minerals in minute quantities for growth. Like the vitamin mixture, minerals can be prepared in the form of a mixture convenient for use (Tables 6.4 and 6.5).

The most commonly used culture fluid for growing marine microalgae is the modified type of Miquel media in which salt nutrients and elements are added to sea water. This type of fluid can be economically used when a large quantity is required. However, since the composition of sea water is usually not constant, it is necessary to collect it from such places and depths where the composition remains constant. Even though the same type of nutritive additives are used, the growth of cultured algae will not be the same because of the differences in the sea-water composition. When the variations are too many, a large quantity of sea water is collected and kept in a dark place for quite some time, until the organic substances break up and the water attains a stable condition. Suspended particles and bacteria should also be removed. A millipore filter can be used if the amount of sea water required is small, but when the culture is carried out on a large scale, a continuous centrifuge is used to remove the particles and germs.

If an artificial synthetic culture fluid is prepared with various chemicals instead of the culture fluid with natural sea water, it is possible to maintain the same composi-

Table 6.3. Composition of culture fluid

I) MSW media (modification of Miquel sea water) (Shiraishi)

Sea water	100 ml
Miquel A fluid[1]	0.2 ml
Miquel B fluid[2]	0.1 ml
Mineral mixture P6[3]	0.2 ml
Vitamin mixture III[4]	0.1 ml
Trisaminomethane	50 mg
pH = 7.8	

[1]Miquel A fluid

KNO_3	20.2	g
Water	100	ml

[2]Miquel B fluid

$Na_2HPO_4 \cdot 12H_2O$	4	g
$CaCl_2 \cdot 6H_2O$	4	g
HCl	2	ml
Water	80	ml

[3]Mineral mixture P 6 (to be explained later)

[4]Vitamin mixture III (to be explained later)

Culture fluid prepared by adding minerals, vitamins, and trisaminomethane to Miquel sea water. This fluid is very easy to make and is suited for growing several types of algae.

II) ES solution (Provasoli, personal communication)

Water	100 ml
$NaNO_3$	350 mg
Sodium glycerophosphate	50 mg
Fe (as EDTA: 1:1)[1]	2.5 mg
Mineral mixture PII[2]	25 ml
Vitamin B_{12}	10 μg
Thiamin	0.5 mg
Biotin	5 μg
Trisaminomethane	50 mg
pH = 7.8	

This solution is sterilized and added to sterilized sea water in the ratio of 2 ml solution to 100 ml water.

[1]$Fe(NH_4)_2(SO_4)_2 \cdot 6H_2O$	351 mg
Na_2 EDTA	330 mg
Water	500 ml

1 ml of the foregoing solution is equivalent to 0.1 mg Fe.

[2]Mineral mixture PII(to be explained later).

Solution is kept ready and added to the sea water when required. It can be used for several types of algae, since it is a common culture fluid.

III) ASW III media (Provasoli, McLaughlin, and Droop, 1957)

Sea water	100 ml
KNO_3	20 mg
K_2HPO_4	2 mg
Mn (as Cl)[1]	0.04 mg
Fe (as Cl)	0.01 mg
Vitamin mixture No. 8[2]	0.1 ml
Liver extract[3]	1.0 mg
Soil extract	4.0 ml
Sodium glutamate	50 mg
Glycine	50 mg
Trisaminomethane	100 mg
pH = 7.5	

[1]This is expressed in terms of the atomic weight of Mn. Hence 1 mg of Mn (as Cl) is equal to 3.6 mg of $MnCl_2 \cdot 4H_2O \cdot 3 \cdot$ Fe (as Cl) is also expressed in the same way.

[2]Vitamin mixture No. 8 (to be explained later).

[3]Liver Extract Concentrate 1:20, prepared by the Nutritional Biochemical Corporation, is commonly used in culture media.

IV) DVM media (Modification of DC media as published by Provasoli, McLaughlin, and Droop, 1957) (Shiraishi)

Water	100 ml
NaCl	1.8 g
KCl	60 mg
$NaNO_3$	50 mg
$MgSO_4 \cdot 7H_2O$	0.5 g
Ca (as Cl)	10 mg
K_2HPO_4	3 mg
$Na_2SiO_3 \cdot 9H_2O$	20 mg
Mineral mixture PII[1]	3 ml
Fe (as Cl)	0.01 mg
Vitamin mixture III[2]	0.1 ml
Vitamin B_{12}	0.2 μg
Trisaminomethane	100 mg
pH = 7.5	

[1]Mineral mixture PII (to be explained later).

[2]Vitamin mixture III[2] (to be explained later).

This is an artificial synthetic media prepared without sea water, which was used in the past to grow diatoms, but can also be used to grow several types of algae.

Table 6.3—(*Continued*)

V) ASP 12	[Provasoli (from Iwasaki, 1961)]	
Water	100	ml
NaCl	2.8	g
$MgSO_4 \cdot 7H_2O$	0.7	g
$MgCl_2 \cdot 6H_2O$	0.4	g
KCl	0.07	g
Ca (as Cl)	40	mg
$NaNO_3$	10	mg
K_3PO_4	1	mg
Sodium glycerophosphate	1	mg
$Na_2SiO_3 \cdot 9H_2O$	15	mg
Vitamin B_{12}	0.02	μ g
Biotin	0.10	μ g
Thiamin	10.00	μ g
Mineral mixture PII[1]	1	ml
Mineral mixture SII[2]	1	ml
Trisaminomethane	0.1	g
	pH=7.8-8.0	

[1]Mineral mixture PII (to be explained later).

[2]Mineral mixture SII (to be explained later).

Salinity higher than DVM.

VI) STP media	(Provasoli, McLaughlin, and Droop, 1957)	
Sea water	80	ml
Water	15	ml
Soil extract	5	ml
Sodium glutamate	50	mg
KNO_3	20	mg
K_2HPO_4	1	mg
Glycine	10	mg
DL alanine	10	mg
Vitamin mixture No. 8[1]	0.1	ml
Trypticase[2]	20	mg
Yeast autolysate[3]	20	mg
Sucrose	100	mg
	pH=7.5	

[1]Vitamin mixture No. 8 (to be explained later).

[2]Trypsin-broken casein, Baltimore Biol. Lab. product.

[3]Yeast autolysate, Albimi Lab. product.

Rich in organic substances and can be used for aseptic media. If germs grow, it gives rise to a white suspension, and hence can be used for testing aseptic conditions.

Table 6.4. Vitamin mixtures

I) Vitamin mixture No. 8	(Provasoli, and others)	
Water	100	ml
Thiamin	20	mg
Nicotinic acid	10	mg
Putrescine	4	mg
Calcium pantothenate	10	mg
Riboflavin	0.5	mg
Pyridoxine	4	mg
Pyridoxamine	2	mg
Para-aminobenzoic acid	1	mg
Biotin	0.05	mg
Coline	50	mg
Inositol	100	mg
Thimine	800	mg
Orotic acid	26	mg
Vitamin B_{12}	0.005	mg
Folic acid	0.25	mg
Folinic acid	0.02	mg

II) Vitamin mixture III	(Shiraishi)	
Water	100	ml
Vitamin B_{12}	0.1	mg
Thiamin	10	mg
Biotin	0.1	mg

Table 6.5. Mineral mixtures

I) Mineral mixture PII	(Provasoli, and others)	
Water	100	ml
$Na_2 \cdot$ EDTA	100	mg
Fe (as Cl) 1 mg=$FeCl_3 \cdot 6H_2O$	4.9	mg

B(as H_3BO_3) 20 mg=H_3BO_3	114 mg
Mn(as SO_4) 4 mg=$MnSO_4 \cdot 4H_2O$	16.4 mg
Zn(as SO_4) 0.5 mg=$ZnSO_4 \cdot 7H_2O$	2.2 mg
Co(as SO_4) 0.1 mg=$CoSO_4 \cdot 7H_2O$	0.48 mg
pH=7.5	

Table 6.5—(*Concld.*)

II) Mineral mixture P6	(Provasoli, and others)	III) Mineral mixture SII	(Provasoli, and others)
Water	100 ml	Water	100 ml
$Na_2 \cdot EDTA$	300 mg	Br(as Na)	100 mg
Fe(as Cl)	8 mg	Sr(as Cl)	20 mg
Mn(as Cl)	12 mg	Rb(as Cl)	2 mg
Zn(as Cl)	1.5 mg	Li(as Cl)	2 mg
Co(as Cl)	0.3 mg	I(as K)	0.1 mg
Cu(as SO_4)	0.12 mg	Mo(as Na)	5 mg
B(as H_3BO_3)	60 mg		
	H=7.5		

tion of the media. Hence in culture experiments requiring considerable repetition, a synthetic culture media should be prepared. Likewise, in experiments concerned with the nutritional requirements of algae, only an artificial synthetic culture fluid is usable. The chemicals used in the preparation of an artificial synthetic culture fluid sometimes bring about unexpected results due to the presence of impurities. Hence the chemicals must be of a very high grade, which makes the cost of preparing an artificial media high.

The culture fluid commonly used in experiments is given in Table 6.3.

2.2 Culture equipment

2.2.1 Test Tubes

Test tubes are used when a relatively small quantity of media for algae is required, and cotton-plugged test tubes made of silica glass are used generally. As will be explained later, when inoculation is done with a Pasteur pipette, the test tubes commonly available in the market are too long and hence have to be cut to about 14 cm. Since algae take a longer time to grow than bacteria, it is necessary to prevent the evaporation of water during this period, by placing a cotton plug, held in place with a rubber band, in the tube. Test tubes fitted with plastic screw caps can also be used, and are particularly useful when the media have to be transported.

2.2.2 Culture Vessels

Since a large area of the culture fluid should be exposed to air for an effective exchange of gases, conical flasks are commonly used for a still culture. The fluid layer should not be very thick and should occupy about one-third of the volume. When a large quantity of the media is required, round-bottomed flasks of 3-liter capacity are useful. When the quantity of culture fluid is even larger, hard glass bottles with a 10-liter capacity can be useful. In special cases, "Rubin" flasks, branch-fitted flasks, and flasks of other shapes can be considered.

2.2.3 Pipettes

If the pipettes used for aseptic treatment have very narrow outlets, it takes a long time to drive out the fluid. Since bacterial pollution can occur during this period, the mouth of the pipette is widened. Measuring pipettes are commonly used, since the quantity drawn in and expelled can be measured. Drawing pipettes, "Pasteur" (capillary) pipettes, etc. are used when small quantities of a fluid are to be taken for transplantation, but to maintain the aseptic conditions, all pipettes should have cotton plugs, and should be wrapped in paper and sterilized by dry heating. When several are used together, the pipettes can be bundled and placed in aseptic boxes for sterilization. Various pipettes and other equipment used in culture fluids are shown in Figure 6.1.

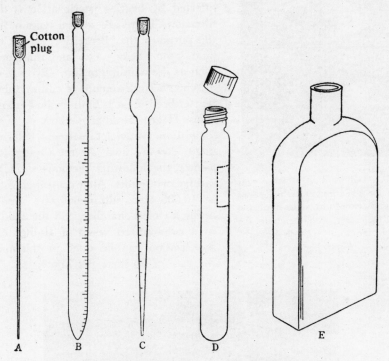

Figure 6.1. Instruments used for cultures.

A—"Pasteur" pipette; B—measuring pipette; C—whole pipette;
D—screw-lid test tube; E—bottle.

The glass equipment is sterilized either by dry heat or by high pressure before use. Plastic equipment is also used for the culturing of algae, but since most of it cannot be subjected to heat treatment, sterilization is done with chemicals such as 0.1% aqueous solution of $MgCl_2$, 70% methanol, etc. after which the equipment is washed with aseptic water.

2.3 Sterilization of culture fluid

The culture fluid can be sterilized in an autoclave at 2 atm and 120°C for 20 minutes. Sometimes the culture fluid changes its quality and sometimes condensation occurs; in these instances, the fluid is heated repeatedly for 30 minutes at a temperature just below the boiling point and then cooled.

Another method of sterilization is to divide the fluid into parts, sterilize each, then mix the parts in an aseptic manner. If condensation occurs, the pH of the culture

Figure 6.2. Culture fluid, which is centrifuged for a continuous period, passes into a tightly closed ultraviolet sterilizing vessel in the center, and then into the vessel on the right (Oyster Research Center).

fluid also changes. Hence it is necessary to take out a small sample of fluid and check its pH before and after sterilization. Miquel sea water can be treated with 0.05 to 0.1% tris (hydroxymethyl) aminomethane to prevent precipitation.

When the contents of the fluid are easily affected by heating, sterilization is done by filtration. There are several types of filters for this purpose: the "Berkefeld," the "Chamberland," the "Seitz," etc. and millipore filters such as the diaphragm, the "Pyrex," etc.

When a large amount of culture fluid has to be sterilized, heat is both expensive and laborious. Hence the filter method, or centrifugal separation, ultraviolet treatment, chlorine treatment, etc. are used. At the Oyster Research Center, the following methods seem to have satisfactory results. After centrifuging the fluid at 10,000 rpm, ultraviolet rays are applied inside a closed chamber, and the fluid placed in a presterilized vessel of 10-liter capacity. Sterilization can be done in this manner at the rate of 20 liters per hour.

Figure 6.3. Ventilated culture in large vessels (10-liter glass bottles) (Oyster Research Center).

2.4 Conditions of culture

To provide the best conditions for the growth of algae, temperature, luminosity, and other environmental conditions must be controlled. The temperature can be readily maintained at a constant level in a room by some simple device, but for small-scale culture, either temperature-controlled tanks or culture vessels kept in tanks through which water with a suitable temperature passes, are more economical. While the temperature suited to algal growth differs according to species, the general range is between 10 and 40°C, but in specific instances only between 15 and 25°C.

Table 6.6. Influence of temperature on the growth of *Monas*
(Imai and Hatanaka, 1950)

Temperature (°C)	Maximum density of *Monas* (number of cells/ml)	Number of culturing days
30 (29-32)	50,000	6
25 (24-26)	860,000	4
20 (19-21)	1,080,000	7
15 (14-17)	1,230,000	6
10 (8-13)	127,000	7
5 (2- 6)	20,000	11

Luminosity can usually be controlled if it is derived from artificial light sources. The luminosity required for the photosynthesis of algae is 400 to 700 mμ, which is sufficient for viewing. However, the effective wavelength differs according to the color of the algae. A light source with a suitable wavelength such as white fluorescent lamps, a tungsten bulb, etc. should be used. Since fluorescent lamps have a wavelength distribution suited to photosynthesis in green plants, these are suitable for culturing green algae. Regarding the degree of luminosity, the rate of growth increases with the strength of light, but within a restricted range of luminosity. However, the rate of growth shows no change beyond the saturation point of luminosity. The strength of light at the point of saturation differs according to the type of algae and temperature. For *Chlorella ellipsoidea*, it is 10,000 lux at 25°C, and for *Monochrysis lutheri*, 4,500 lux at 20°C. If the balance between the temperature and the strength of light is not maintained, the plants bleach and photosynthesis does not occur, a condition which often happens at low and high brightness.

Table 6.7. Conditions of culture for various algae
(Yanase and Imai, 1968)

Types	Optimum luminosity	Optimum temperature
Monochrysis lutheri	4,500 - 8,000 lux	20 - 25°C
Platymonas sp.		
(= *Tetraselmis tetrathela*)	4,500 - 8,000 lux	23 - 25°C
Nitzschia closterium	4,500 - 12,000 lux	23 - 29°C
Chaetoceros calcitrans	4,500 - 12,500 lux	23°C

To supplement the supply of O_2 and CO_2 (required for the growth of algae, and to prevent the settling of algae to the bottom, as well as to maintain the uniform degree of brightness required for photosynthesis, the culture is shaken, stirred, or aerated. Of these three processes, shaking is the easiest because it causes the minimum disturbance to the sterilized condition and, particularly when organic media are used for a culture, gives little opportunity for bacterial pollution. But large vessels are used in large-scale culturing and hence shaking becomes difficult. Hence stirring or aeration is done. While the aeration method, like the stirring method, has the advantage of assisting the exchange of gas, it is difficult to completely sterilize the air. A cotton-fitted filter tube is generally used and, in large-scale cultures, filters in which glass-wool is tightly fitted to the metal. Before passing the gas into the culture vessels, passing it through a solution of copper sulfate or mercuric chloride is recommended, as this not only prevents pollution, but also minimizes evaporation.

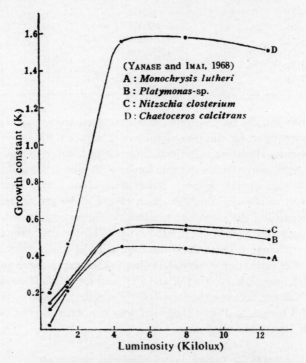

Figure 6.4. Relationship between the coefficient of growth and luminosity in various algae.

2.5 Large-scale culture methods

2.5.1 COLORLESS FLAGELLATES (*Monas*)

Colorless flagellates such as *Monas* are cultured along with bacteria, in media enriched with an organic substance such as glucose (Imai and Hatanaka, 1950). Oxygen, however, must be supplemented either by shaking the culture or by passing air through it. It is also possible to culture colorless flagellates via a suspension of dry bacteria separately cultured. This method is suitable for large-scale culturing since it is simple; only the bacteria which serve as feed for the colorless flagellates are cultured in the organic media and collected by centrifuging, dried, and preserved. When required, the dry bacteria are suspended and sterilized by heating. This sterilized suspension is added to sterilized sea water, and the colorless flagellates inoculated. After the inoculation, aerated culturing is done as it is more convenient in large-scale production.

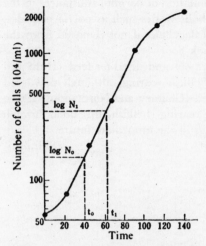

Figure 6.5. Growth curve of *Monochrysis lutheri* under continuous brightness (4,500 lux, 20°±1°C). Growth constant $K = \log N_1 - \log N_2/t_1 - t$. (Yanase and Imai, 1968).

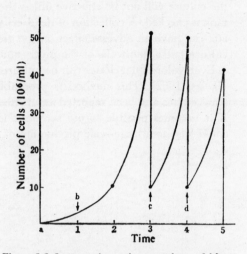

Figure 6.6. Large-scale continuous culture of *Monas* in 10-liter bottles with openings at the bottom.

a—2 g of dry bacteria suspended in 3 l of 30% sea water, the *Monas* inoculated, and aerated culture carried out; b—3 g of dry bacteria added to 30% sea water, and 5 l culture made; c—4 l taken out of the lower opening and used as feed for shellfish; 4 l of 30% sea water and 4 g of dry bacteria added to culture; d—step "c" repeated.

Table 6.8. Culture conditions of *Monas*
(Imai and Hatanaka, 1950)

Chlorine	5-10‰, about 30-50% sea water
Glucose	about 100 mg/liter
KNO_3	40-400 mg/liter
$Na_2HPO_4 \cdot 12H_2O$	5-50 mg/liter
$CaCl_2 \cdot 6H_2O$	5-50 mg/liter

Remarks: (1) Temperature of culture, 15-20°C; (2) When temperature of culture fluctuates, the concentration of the additives is increased by 10 times.

2.5.2 COLORED FLAGELLATES

Colored flagellates can also be cultured on a large scale in large vessels by the aerated culture method (Figure 6.3). Here, instead of only air, a mixture of air and 1 to 5% CO_2 is more effective. If an excess of CO_2 is used, the pH of the culture fluid decreases; hence the concentration of CO_2 suited for the purpose should be determined priorly. When large-scale culturing is done, the process has to be simplified as far as possible, for the more complicated the process the greater the chances of its becoming polluted by bacteria and other organisms. To simplify the process therefore, large tanks are used, a part of the culture already grown removed, and freshly prepared culture fluid added. This procedure is repeated so that the culture can be continued without interruption (Wisely and Purday, 1961). However, even in this method, the culture will not be effective unless the accumulation of harmful substances in the tank is checked. A collection of dead cells, metabolic wastes such as polysaccharides, etc. can have an adverse effect on the growth of flagellates if not removed from the culture tanks, the walls of the pipes, and the tank floor.

Even colored flagellates can be cultured separately and dried for feed (Hidu and Ukeles, 1962). This method is profitable in shellfish rearing, although powdered seaweed has also been reported as an effective feed (Chanley and Normandin, 1967). If it becomes possible to use synthetic feed for rearing shellfish, this breakthrough could lead to a large-scale production of seedlings in the immediate future.

(Shiraishi)

Bibliography

1. CHANLEY, P., and R. F. NORMANDIN, 1967. Use of Artificial Foods for the Larvae of the Hard Clam, *Mercenaria mercenaria* (L.), *Proc. Nat. Shellfish Ass.*, **57**, 31-37.
2. DAVIS, H. C., and R. R. GUILLARD, 1958. Relative Value of Ten Genera of Microorganisms as Foods for Oyster and Clam Larvae, *Fish. Bull.* 136 from *Fish. Bull.: Fish and Wildlife Service*, **58**, 293-304.
3. ——, and R. UKELES, 1961. Mass Culture of Phytoplankton as Foods for Metazoans, *Science*, **134**, 562-564.
4. HIDU, H., and R. UKELES, 1962. Dried Unicellular Algae as Food for Larvae of the Hard Shell Clam, *Mercenaria mercenaria*, *Proc. Nat. Shellfish Ass.*, **53**, 85-101.
5. HIRANO, R., and Y. OSHIMA, 1963. Kaisan dobutsu yosei no shiiku tosono shiryo ni tsuite (On the Rearing and Feed of the Larvae of Marine Animals), *Nisuikaishi (Bulletin of the Japanese Society of Scientific Fisheries)*, **29** (3), 282-297.
6. HUNTER, S. H., and L. PROVASOLI, 1964. Nutrition of Algae, *Ann. Rev. Plant Physiol.*, **15**, 37-56.
7. IMAI, T., and M. HATANAKA, 1950. Studies on Marine Non-colored Flagellates, *Monas* sp., the Favorite Feed of Larvae of Various Marine Animals: I. Preliminary Research on Culture Requirements, *Sci. Rep. Tohoku Univ.*, 4th Ser. **18** (3), 304-315.
8. IWASAKI, H., 1961. The Life Cycle of *Porphyra tenera* in vitro, *Biol. Bull.*, **121** (1), 173-187.
9. LOOSANOFF, V. L., and H. C. DAVIS, 1963. Rearing of Bivalve Mollusks, *Advance in Marine Biol.*, **1**, 1-136.
10. MYERS, J., and J. R. GRAHAM, 1959. On the Mass Culture of Algae: II. Yield as a Function of Cell Concentration under Continuous Sunlight Irradiance, *Plant Physiol.*, **34** (3), 345-352.
11. ——, ——, 1961. On the Mass Culture of Algae: III. Light Diffusers: High vs. Low Temperature *Chlorellae*, *Plant Physiol.* **36** (3), 342-346.
12. OFF, F. D., 1965. Synthetic Media and Techniques for the Xenic Cultivation of Marine Algae and Flagellates, *Virginia J. Sci.*, **16** (3), 205-218.
13. PROVASOLI, L., 1958. Growth Factors in Unicellular Marine Algae in *Perspectives in Marine Biology* edited by A. A. Buzzati-Traverso, 385-403.
14. ——, 1958. Nutrition and Ecology of Protozoa and Algae, *Ann. Rev. Microbiol.*, **12**, 279-308.
15. ——, 1963. Organic Regulation of Phytoplankton Fertility in *The Sea*, Vol. 2, edited by M. N. Hill, 165-219.
16. ——, 1964. On the Productivity of Sea Water, Resources Conservation Association. Japanese translation by Shiraishi.
17. ——, J. J. A. McLAUGHLIN, and M. R. DROOP, 1957. The Development of Artificial Media for Marine Algae, *Arch. für Mikrobiol.*, **25**, 392-428.
18. ——, K. SHIRAISHI, and J. R. LANCE, 1959. Nutritional Idiosyncrasies of *Artemia*

and *Tigriopus* in Monoxenic Culture, *Ann. New York Acad. Sci.*, **77**, 250-261.

19. WISELY, B., and C. PURDAY, 1961. An Algal Mass-culture Unit for Feeding Marine Invertebrate Larvae, Div. Fish. Ocean. Tech. Paper No. 12. Commonw. Sci. Ind. Res. Org., Australia.

20. YANASE, R., and T. IMAI, 1968. The Effect of Light Intensity and Temperature on the Growth of Several Marine Algae Useful for Rearing Molluscan Larvae, *Tohoku J. Agr. Res.*, **19** (1), 75-82.

CHAPTER II

The Rearing of Larvae and Seedlings of Bivalves

The author proposes to discuss in this chapter the technical aspects of the production of seedlings of shellfish as obtained in the Oyster Research Centers of Japan.

1. Preparation for artificial seedlings

The Oyster Research Center in Maine Bay has both the characteristics of the open sea and of the coastal regions. The bay is at the base of Karakuwa Hanto in the Miyagi Prefecture.

The seasonal variations in the water temperature range from minimum (5 to 6°C) in March and maximum (23 to 25°C) in late August. The salinity is 16.7 to 17.0 °/$_{oo}$ at a 1 m depth during the rainy season; the annual mean salinity at a 5 m depth is 18.0 to 18.5 °/$_{oo}$. The water in the bay is transparent up to an 8 to 9 m depth in May to June, the growing months of plankton, and about 11 m the rest of the year as the bay is very calm.

Sea water used for rearing larvae of bivalves is drawn from a depth of 5 m by a 3-HP chemical pump, and stored in water tanks of about 1.5-ton capacity. After filtering through sand, this water is used with larvae reared on rafts floated on the surface of the sea. The sea water is drawn through drawing pipes attached to the rafts, and hence the depth of the drawer can be adjusted above and below 5 m. The sea water supplied to the rafts is filtered through sand and a cartridge filter before its introduction into the rearing tanks. Filtration is necessary to remove small plankton, eggs, larvae, etc.

The sea-water drawing pipes (Chloridized vinyl) are removed and cleaned of adhering organisms once a year.

The rearing rafts, placed on rearing tanks, are fitted with steel tubes (Figure 6.7). Each raft covers 15 water tanks, and hence 12 rafts yield 180 tanks. The walls of the rearing tanks are made of polyethylene and each tank has about a one-ton capacity (2 m × 1 m × 0.8 m); the outer side of the tanks is covered with black vinyl sheets, by which the brightness inside the tank can be regulated at the same time that added support is given.

The polyethylene tanks, which look like "water buses," are supported by wooden frames suspended from the rearing rafts. When filtered sea water is poured into them, they elongate due to the outside water pressure. They are also referred to as "sea-surface water tanks," because their special feature is that they can be easily filled and detached. Moreover, their internal temperature is maintained by the natural sea environment, and sudden changes of the atmospheric temperature do not directly influence that inside the tanks.

Figure 6.7. Floating larvae-rearing rafts.

All water tanks are fitted with air pipes through which air is passed during larval rearing to stir the water. Water pipes are also fitted to the water tanks so that setting spat can be reared in the running water. The rafts protect the tanks from direct sunlight and dilution by rain water.

Culture rafts, used for the intermediate rearing of setting spat, are connected to the rearing rafts, of which 15 are kept inside the bay and 3 in the open sea. Hence a total of 30 rafts are utilized in Maine Bay (12 rearing, 18 culturing).

The types of bivalves for which seedlings are prepared by using the equipment just described are:

Ostrea edulis (European oyster)
Patinopecten yessoensis
Anadara broughtonii
Crassostrea virginica (American Virginia oyster)
Crassostrea angulata (Portuguese oyster)
Ostrea lurida (Olympia oyster).

2. Maintenance of spawning oysters and their treatment

As explained earlier, one of the characteristic features of the "sea-surface water tank" is that it maintains a constant water temperature inside regardless of changes in atmospheric temperature, due to the natural sea environment. But when sea conditions become abnormal, basins of cold and stagnant water form and, consequently, the rearing of larvae is adversely affected. To illustrate: the natural spawning of *Patinopecten yessoensis* in Maine Bay occurs in late March when the water temperature rises to 8.0 to 8.5°C, but sometimes the water temperature after the spawning period drops to 5 to 6°C, which has a detrimental effect on the larvae.

The larvae of European oysters are released in late June when the water temperature rises to 18°C. However, the rainy season begins around this time causing considerable change in the water temperature (Figure 6.8). When artificial rearing is to

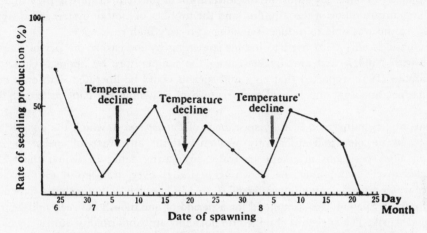

Figure 6.8. Influence of low temperature on production of artificial seedlings of European oysters (Oyster Research Center, Maine Bay, 1962, unpublished).

be carried out under such adverse conditions, spawning is purposely delayed by keeping the mother oysters in water of a lower temperature than spawning requires, and then returning them to water of a spawning temperature when the conditions for larval growth are optimal. To clarify: the mother oysters of *Patinopecten* are kept in sea water at 5°C during the spawning period, and the eggs collected when the water temperature increases to 12°C, which is suited to larval tank rearing. The layer-wise differences in water temperature in bays are utilized with European oysters, and the hanging depth of the mother oyster regulated. By doing this, maturation and spawning are induced at a time which allows the larvae to be released when the water temperature increases to 20°C in mid-July. On the other hand, spawning is delayed in the American and Portuguese oysters in Maine Bay, due to an insufficiently integrated temperature. These oysters are kept in water of a temperature 2 to 3°C higher than the natural sea water, and the maturation of the mother oysters thus accelerated.

3. Inducement of spawning

Although there are some bivalves in which it is very easy to cause artificial fertilization by cutting the ovary and releasing the eggs in sea water, in genera such as *Patinopecten*, *Anadara*, etc. it is very difficult to cause artificial fertilization. While it is possible to obtain larvae through artificial fertilization, maintaining them on a large scale at one time is not easy.

Galtsoff (1930) induced the liberation of eggs and sperms in *Crassostrea virginica* by raising the water temperature, and reported that the eggs were more effectively released thereby if a suitable amount of spermatic fluid were added to the medium. The inducement of spawning by heat stimulation is a practical method for obtaining a large amount of mature eggs within a fixed period, from which seedlings can be produced. Kanno (1962) used *Chlamys* for experimentation and found that the induced rate of spawning is higher when heat stimulation is repeated several times than when it is applied just once. By adjusting the lower limit of the temperature of the stimulus, the temperature range of the stimulus, and the number of mother oysters in his water tanks, Kanno was able to induce spawning at a very high rate.

Kinoshita *et al.* (1943) tried to induce spawning by increasing the pH of the sea water with NaOH, and also by increasing the temperature by about 5°C. Later, Yamamoto (1951) reported that eggs and sperms could be liberated with only heat stimulation, and that the critical temperature of spawning for *Patinopecten* is 8.0 to 8.5°C.

Spawning is induced in healthy specimens of *Patinopecten* in which the gonads are bright red in color and sufficiently swollen without any trace of sperm or egg release. The specimens are kept in water-circulating tanks in which the water is cooled to 5°C, and after the spawning period is over, transferred to the "sea-surface water tanks" containing water of 10 to 12°C, which action induces spawning (Figure 6.9). Spawning is stimulated for a period of one month between mid-March and mid-April. The eggs thus obtained do not show any abnormality during development. It seems that spawning can be better induced, however, if a low temperature treatment is applied beforehand. Yamamoto (1965) also reported that the spawning rate is better when the specimens are kept in water at 4 to 9°C for half a day prior to induced spawning.

The problem with *Anadara* is that the condition of the gonads cannot be observed externally. Hence it is difficult to select the mature clams in which spawning can be induced.

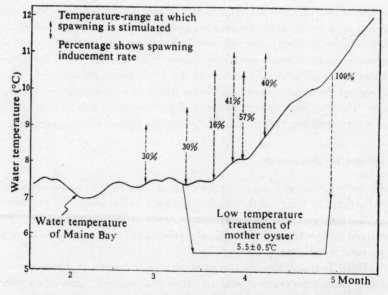

Figure 6.9. Inducement of spawning in mother oysters in natural conditions after a low temperature treatment (Oyster Research Center, 1966, unpublished).

Kanno and his colleagues (1965) studied the relation between the results of induced spawning in *Anadara* of Aomori for four years, and the integrated water temperature of the surface layer of the sea. They found that it was possible to induce spawning if the integrated water temperature was above 20°C. By studying this condition, the approximate degree of maturation of the mother oysters could be judged from the temperature conditions of the water in which they were found. The mature clams which had not yet matured were subjected to heat treatment for rapid maturation.

Kanno (1965) was able to induce spawning in *Anadara* by repeated heat stimulation. After rearing, mature clams were kept for 7 days in water-circulating tanks at 14 to 20°C; they were stimulated with heat in which the lower limit was held at 21°C, but the temperature varied by 6°C. Three to six mature clams began spawning after stimulation was repeated 3 times within 180 minutes.

At the Oyster Research Center, spawning was induced by subjecting the mature clams to heat treatment first, then placing them in cool sea water of 16.5°C for about 20 hours, and finally stimulating them at 21°C. The temperature of stimulation varied up to 7°C, and stimulation was given to 20 specimens. After 95 minutes, 3 males released spermatozoa, and after 3 hours and 30 minutes, 2 females began to spawn (Figure 6.10). It was interesting to note in this regard that no spawning occurred in any specimen with which heat stimulation had been tried.

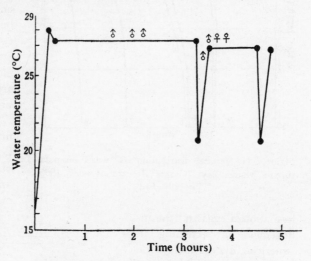

Figure 6.10. Inducement of spawning in *Anadara* (Oyster Research Center, 1967, unpublished).

Since European mother oysters release larvae in the form of veliger, it is necessary to know the time of release much in advance. The water temperature for the planktonic larvae and for gonadal maturation must also be regulated. Walne (1965) suggested the following equation to express the relation between the rate of larvae development and the water temperature for European oysters:

$$X \cdot Y = K,$$

where X = number of days required for the development of the embryo up to a shell
          length of 160 μ inside the mother's body;
        Y = temperature of sea water, $-10°C$ (the biological freezing point of the
          European oyster is taken as 10°C);
        K = constant 66.6.

Since some of the larvae growing inside the mother oyster can be released by gently shaking the oyster, the development period and the water temperature of the medium, as well as Walne's equation, assist in the prediction of the time of larval release. Hence by classifying the mother oyster according to the stage of larval development, and also by adjusting the hanging depth, larvae can be obtained at any time (Figures 6.11 and 6.12).

If mother oysters with larvae just about to be released are obtained, the larvae are

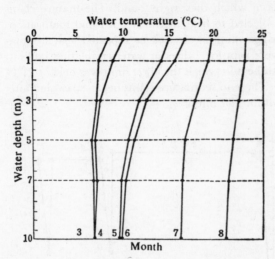

Figure 6.11. Vertical distribution of water temperature in Maine Bay (Oyster Research Center, 1965, unpublished).

removed by opening the shells with knives, and reared in water tanks. The larvae just before release show various coloration: those black in color are known as "black sick;" those not completely grown or gray in color, as "gray sick;" and those which are grown but still immature and white in color, as "white sick." When the larvae are removed from the mother oysters, only those which are more developed than "gray sick" can be reared in tanks.

Just as it is possible to obtain mature eggs by artificial fertilization from *C. gigas,* so one can obtain them from the American and Portuguese oysters which are oviparous. Detailed reports on the

formation of eggs and the inducement of spawning in American oysters have been published by Loosanoff (1945), and Galtsoff (1930). According to Loosanoff (1945), it is possible to obtain mature eggs of oysters by regulating the temperature throughout the year. In other words, by keeping the oysters in sea water in which the temperature is artificially regulated, the integrated temperature required for their maturation can be gradually supplied. Mother oysters are thereby obtained at a particular and requisite stage of maturation. To illustrate: mother oysters capable of

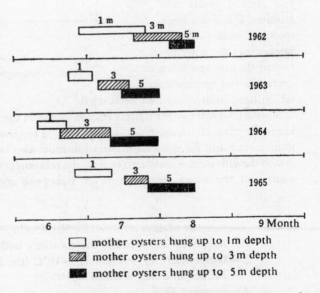

mother oysters hung up to 1m depth
mother oysters hung up to 3 m depth
mother oysters hung up to 5 m depth

Figure 6.12. Hanging depth of European mother oyster and the period of larval release (Oyster Research Centre, 1962, unpublished).

spawning can be obtained by keeping them for 60 days at 15°C, or 15 days at 20°C, or 9 days at 25°C. Loosanoff makes it clear that there are physiological strains in the American oyster, which show wide differences in the temperature required for maturation and spawning. American oysters reared at the Oyster Research Center, which had been obtained from the Sheepslot River in the state of Maine, were found to spawn at 21°C. Galtsoff (1930) showed that spawning could be easily induced in American oysters and *C. gigas* by rapidly increasing the temperature. According to him, when the water temperature was suddenly increased from 20°C to 33 to 34°C, 40% of the oysters spawned and, if a suitable quantity of spermatic fluid of the same species was also added, spawning as a result of heat stimulation was much more pronounced.

4. Rearing larvae in tanks

To remove residual sperms, immature eggs, body fluid, etc. the supernatant water is gently drawn off after the fertilized eggs have settled. Since healthy fertilized eggs settle rapidly, it is better to remove those which take too much time in doing so. The removal of such eggs is done by changing the water. If mucus and excreta are to be removed with gauze, etc. cleaning must be done within 2 to 3 hours of fertilization. Water exchange is repeated continuously until such time as the eggs are able to float. The density of the eggs up to the time of hatching should be below 30,000/liter. If the density is too high, the normal development of the embryo is affected. Since the larvae hatching from the eggs start moving actively near the water surface, hatching time is carefully noted and the larvae transferred with much care to other water tanks. This process makes it possible to separate the eggs which hatch late, unfertilized eggs, and other organic residue.

The early-D-stage larvae are released in "sea buses" of about one-ton capacity at the rate of 300 to 500 larvae per liter of sea water. A mixture of *Monochrysis lutheri* and *Chaetoceros calcitrans*, obtained from a pure culture, is given as feed at the rate of 500 cells per day per larva during the early stage. Feeding is done twice a day. As growth occurs, the feed is gradually increased to 10,000 to 20,000 cells per larva per day. Air is circulated in the water tanks through "Roots" blowers and the water gently stirred during rearing. By using "Muller" gauze selected according to the size of the larvae, the water in the rearing tank is completely changed at intervals of 3 to 4 days, and the tank itself replaced with a fresh one. Water tanks are changed to prevent any adverse effect from diatoms and bacteria which might adhere to the tank walls. To avoid the growth of such adhering organisms and plankton, the brightness inside the tank can be maintained at a lower level. Kanno (1965) reported that the survival rate and growth rate of larvae were better when a part of the water was changed during the course of rearing *Mactra* sp. This emphasizes the importance of changing the rearing water. At the Oyster Research Center, it was found that the growth of larvae of European oysters could be improved by merely changing the water tanks (Figure 6.13). This only proves the magnitude of the influence of the metabolic products of the organisms growing on the walls of the rearing tanks.

Walne (1958) reared the larvae of European oysters in water to which antibiotics were added to prevent the growth of bacteria. He added 50,000 I.U. of penicillin and 50 mg of streptomycin to 1 liter of water, and discovered that the larval growth rate accelerated in this medium. Larval growth is also favorable with antibiotics such as chloromycetin, streptomycin, oleomycin, etc. says Walne. But since anti-biotics are costly, using them in large-scale oyster rearing is not very economical.

The growth and survival rates of larvae are observed by collecting 100 ml of the rearing water every 3 days, counting the number of larvae, and measuring the size of the largest and smallest. The general condition of the larvae is checked also by observing their growth and luster and the condition inside the alimentary canal. On the basis of this examination, larvae are classified as healthy (H), medium healthy (MH), medium (M), medium weak (MW), and weak (W). Larvae categorized as weak are removed from the water tanks and fresh groups of larvae substituted. The mortality of larvae is highest when the vertex protrudes and the growth stops for a temporary period. However, growth after the protrusion of the vertex is rapid, and the growth curve assumes a sigmoid shape. Planktonic larvae display both free swimming activity and creeping motility until they ultimately reach the setting stage. At this point, the foot can be recognized. At the base of the foot, a black spot rich in pigment, i.e. the so-called eye spot, can be seen. The appearance of this spot signifies the inception of larvae setting.

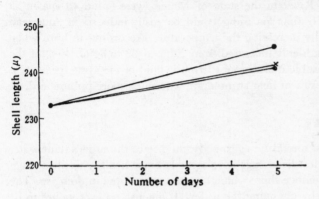

Figure 6.13. Rearing method and growth of European oyster (Oyster Research Center, 1965, unpublished).

●—daily exchange of rearing tanks; ○—daily exchange of half the rearing water; ×—complete exchange of water every two days.

Larval growth is mainly affected by the type and amount of food organisms and the conditions of the rearing temperature during the planktonic period.

The conditions suited for feeding, as suggested by Davis and Guillard (1958) are that the plankton should be able to carry out photosynthesis, the cell wall must be very thin or absent, and the metabolic products should not be toxic. *Monochrysis lutheri* and *Isochrysis galbana* satisfy these conditions. Loosanoff and Davis (1963) have suggested the following organisms, which can be used as feed for the larvae of the American oyster.

Suitable	Less suitable
Monochrysis lutheri	*Dunaliella euchlora*
Isochrysis galbana	*Platymonas* sp.
Chromulina pleiades	*Cyclotella* sp.
Dicrateria inornata	*Dunaliella* sp.
Pyramimonas grossi	*Chlorococcum* sp.
Hemiselmis refescens	*Chlorella* sp.
	Phaeodactylum tricornatum
	Cryptomonas sp.

Davis and Guillard (1958) studied the food value of 10 types of microscopic plankton by feeding them to the larvae of the American oyster, *Venus mercenaria*. When a single type of organism was given, *Isochrysis* and *Monochrysis* were found the best feed. But a mixture of *Isochrysis*, *Monochrysis*, *Dunaliella*, and *Platymonas* yields better results than a diet of *Isochrysis* or *Monochrysis*.

In preparing a diet composed of a mixture of organisms, the nutritional requirements of the larvae must be studied in detail at various stages of growth. Minute, colored flagellates such as *Monochrysis* and *Isochrysis*, which lack cell walls, are very essential during the early stage of larval development as the animals have a very weak digestive capacity. However, when the larvae begin to grow, they are capable of digesting *Chaetoceros calcitrans*, which has cell walls, and small *Chlorella*. Sometimes these prove more effective feed than the flagellates. *Phaeodactylum tricornatum* are diatoms with a cell wall measuring 3.5×27 μ. According to Davis and Guillard (1958), these diatoms can be used as feed for the larvae of the American oyster after it attains a shell length of 200 μ. The survival rate of larvae when grown on these diatoms is reportedly good. On the other hand, Takeda (1966) found that these diatoms could not be used as feed for the larvae of *Anadara* when their shell length was 90 μ.

Thus the feed of the larvae of bivalves has to be studied in greater detail, particularly the size and shape of the organisms and the nutritive elements in them, to determine the most suitable feed.

The next important step in larval rearing is to determine the right feed dosage. Needless to say, the optimum feed dosage increases with the larval growth. Dosage differs greatly according to the rearing water temperature, population density, type of feed, and the mode of feeding. For example, when the population density is low, considering the swimming capacity of the larvae, the feed dosage should be higher than what is actually required. When the population density is high, the mode of feeding must be considered, or the concentration of feed may become very high at the time of input. Under these conditions, the larvae lose their capacity to feed and excrete mucus. This ultimately leads to the settling of feed as waste.

When *Isochrysis galbana* was given as feed to the larvae of European oysters, Walne (1965) found that the growth rate was best when the density of larvae was 5 to 10/ml and the density of *Isochrysis*, 50,000 to 100,000 cells/ml. The effect of feeding did not improve by further increasing the density of the feed organisms, but the consumption of *Isochrysis* per larva per day was 60,000 cells when the shell length of the larva reached 260 μ, or 3 times the feed consumption when the shell length was 180 μ. Davis and Guillard (1958) have reported that when the larvae of the American oyster were reared at a density of 10 to 15 larvae/ml, the optimum concentration of diet was 400,000 cells of *Isochrysis*, and 250,000 cells of *Monochrysis*, per milliliter of water. On the other hand, when the concentration was increased to 500,000 cells/ml the growth of larvae was restricted. Loosanoff *et al.* (1953) used two types of *Chlorella* sp. differing in size (8 μ and 3 μ) for the larvae of *Mercenaria mercenaria* and found that the survival and growth rates were best when the concentration of the larger variety of *Chlorella* was 50,000 cells/ml, or when the concentration of the smaller variety was 400,000 cells/ml. When the concentration of the larger variety of *Chlorella* was increased to 300,000 cells/ml, 90% of the larvae died within a short period.

In feeding experiments with the larvae of the European oyster (mean shell length,

204.5 μ) at the Oyster Research Center, the animals were kept at a density of one larva per milliliter and fed 5,000 to 15,000 cells/ml of *M. lutheri* and *Ch. calcitrans*. It was found that a larva could consume about 15,000 cells within 10 hours. The growth of the larvae was best when the concentration of *Monochrysis* and *Ch. calcitrans* was 30,000 cells/ml (Table 6.9). When the larvae of *Patinopecten* (mean shell length

Table 6.9. Feeding experiments with the larvae of the European oyster

Density of food organism	*Monochrysis lutheri* shell length (μ)			*Chaetoceros calcitrans* shell length (μ)		
	Maximum	Minimum	Mean	Maximum	Minimum	Mean
5,000 cells/ml	239.5	199.0	222.0	239.5	196.0	219.0
10,000 cells/ml	251.0	210.5	225.0	247.0	207.5	226.0
20,000 cells/ml	247.0	210.5	229.0	272.0	218.0	241.0
30,000 cells/ml	254.0	215.0	233.5	275.0	232.0	244.0

Remarks: Shell length at initiation of experiment: maximum 218 μ, minimum 189 μ, mean 204.5μ; density of larvae 1 larva/ml; duration of rearing 72 hours; Oyster Research Center, 1963 (unpublished).

129 μ) were fed on *M. lutheri*, it was found that 1,500 to 6,000 cells were consumed by one larva in 10 hours. When the larvae were reared at a density of one larva per milliliter, the optimum food dosage per larva per day was found to be 5,000 to 6,000 cells (Figure 6.14).

The following is an account of the influence of the rearing water temperature on the growth of larvae. The growth of larvae usually improves with an increase in water temperature within the optimum range. The improvement is due to the higher feeding activity and higher rate of food assimilation, which a higher temperature induces (Walne, 1965). The growth of larvae is adversely affected when a sudden change in temperature occurs, particularly a sudden decrease; the latter affects not only the growth but the survival rate. The effect is more severe when the sudden change in temperature occurs at the time of vertex protrusion or at setting. Loosanoff and Davis (1963) reared larvae of the American oyster for 12 hours at various water temperatures, *viz.* 10, 15, 17.5, and 20°C. They found that the larvae did not feed at 10 to 15°C. Moreover, even after trans-

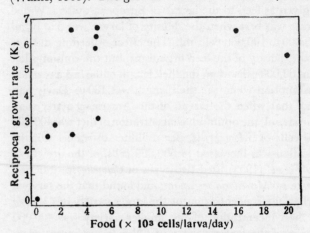

Figure 6.14. Reciprocal growth rate of larvae of *Patinopecten* reared on various concentrations of *Monochrysis lutheri* (Oyster Research Center, 1965, unpublished).

ferring them to sea water of 24°C, the larvae which had been kept at 10°C did not recover; those which had been kept at 15°C showed some recovery and began to feed. The larvae which were kept at 17.5°C took a little food and, when transferred to sea water of 24°C, completely recovered in the sense that they fed actively and showed rapid growth. In view of this fact, it can be said that a temperature lower than 15°C is lethal for the larvae of the American oyster.

The size of the larvae at the time of setting differs according to the maximum and minimum water temperature. For example, the size of the larvae of *Patinopecten* at the time of setting is 240 to 260 μ when the temperature is 12 to 15°C, but the size increases to 300 to 350 μ when the temperature is 8 to 10°C (Figure 6.15). If the water

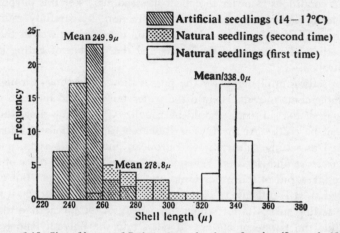

Figure 6.15. Size of larvae of *Patinopecten* at the time of setting (Ito *et al.*, 1965).

temperature can be maintained at a relatively higher level, growth is rapid and setting earlier. Hence the seedlings can be collected more efficiently. At the time of rearing the larvae though, it is better to increase the water temperature gradually within a definite range.

5. Setting of larvae and seedlings

A few days before the free-swimming larvae enter their sedentary life, a pair of circular, black-colored growths can be seen at the dorsal part, particularly at the base of the gills. This is called the "eye spot" because it purportedly functions as a visual organ. It is characteristic of larvae about to enter a sessile existence. Hence, by observing this eye spot, the time for introducing the substratum to which the larvae adhere can be determined.

Larval setting is controlled by various environmental conditions such as water temperature, salinity, etc. The salinity of sea water is an important controlling factor for the development of the pedal gland required for setting, and the physical characteristics of its secretion. Imai *et al.* (1954) used the shells of *Patinopecten* for preparing seedlings of Olympia oysters, and found that the number of oysters which set on the outer surface of the shells exceed those which set on the inner surface. It was concluded that the rough, ridged, outer surface is preferred to the smooth inner surface of shells.

It is not clear whether planktonic larvae select a particular place and material for setting. Yamamoto (1964) studied the setting of *Patinopecten* larvae and reported that physical conditions such as currents and eddies are the main ecological factors controlling larval setting. The fact that these physical conditions of the sea water influence larval setting, i.e. the larvae are guided by the flow of sea water toward the setting substrata, is very interesting. Eddies eventually form in water basins where a mild water current flows through the obstructions caused by the setting implements, and subsequently the chances of the larvae setting in these places are far greater.

When artificial seedlings of shellfish are prepared on a large scale, it is necessary to utilize this type of larval setting, and yet it is simultaneously imperative to control the environmental conditions in order to collect the seedlings. For this purpose, the circulation of rearing water must be made uniform and, by carefully estimating the number of larvae ready for setting on the substrata provided, as well as the sea available for the setting, the density of the larvae and the number of setting implements carefully controlled.

When using setting implements for the preparation of seedlings through artificial means, the implements introduced into the water tanks should have no toxic effect and the surfaces of the substrata should be rough. At the same time, the materials used should not be costly. At the Oyster Research Center, shells of *Patinopecten* are commonly used as setting substrata for European and other oysters, and polished planks for *Patinopecten* and *Anadara*. When 20 oysters settle on one piece of *Patinopecten* shell, they are taken out of the water tanks and hung in sea water. But with *Patinopecten* and *Anadara*, it is difficult to observe the setting spat and hence it is better to introduce the setting substrata for which the density of the larvae had already been estimated. While the larvae of oysters do not set in the polyethylene water tanks, *Patinopecten*, *Anadara*, etc. seem to settle down on anything available because of the presence of byssus. The development of more efficient setting implements for collecting seedlings is an important field for experimentation.

6. Development and metamorphosis of larvae

6.1 Larvae of Patinopecten

The eggs of *Patinopecten* are completely opaque while the cytoplasm is light yellow in color. The size of the egg is 55 μ. The head of the spermatozoan is triangular in shape and measures 5 μ. The tail of the spermatozoan is 50 to 60 μ in length (Yamamoto and Nishioka, 1943). After fertilization, the egg releases the first polar body in 2 to 3 hours. The second polar body is released 30 minutes after the first, and the polar lobe forms. At this point, the first cleavage occurs. The embryo is transformed into a blastula 20 hours after fertilization. In the next 24 hours, it becomes a trochophore and, 90 hours later, flattens if the water temperature is 11 to 17°C. Several cilia have formed at the vertex by this stage, which disappear in the early stage of veliger (Table 6.10). Ito *et al.* (1965) studied the relationship between the time involved and the water temperature required for the development of the embryo. In the range of 9 to 16°C, development becomes more rapid as the water temperature increases. The optimum temperature range for the development of *Patinopecten* is 6 to 20°C, but the

Table 6.10. Course of embryonic development of *Patinopecten* (Yamamoto, 1964)

Course of development	Time after Fertilization	Water temperature, °C
	Hours	
Formation of 1st polar body	5	8.5- 9.5
Formation of 2nd polar body	6	8.5- 8.7
1st cleavage (2 cells)	8	7.8- 8.0
2nd cleavage (4 cells)	10	7.0- 8.0
3rd cleavage (8 cells)	16	7.0- 8.0
4th cleavage (16 cells)	20	7.8- 8.7
Blastula	40	7.5- 8.3
	Days	
Gastrula	2	8.0- 8.9
Trochophore	4	7.8- 9.2
Veliger D-type larvae, 72×98 μ	5- 7	7.3- 9.5
Veliger D-type larvae, 104×87 μ	8-10	7.5- 8.0
Veliger D-type larvae (2nd stage of larval shell), formation of vertex, 118×104 μ	15-17	7.8- 9.5
Veliger D-type larvae, 180×168 μ	22-28	9.0-13.0
Veliger larvae, 200×180 μ	30-35	11.5-13.7
Veliger larvae (fully grown), 313×325 μ	40	12.2-14.2

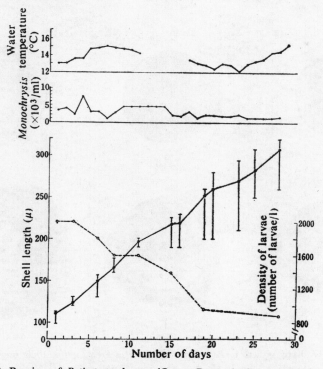

Figure 6.16. Rearing of *Patinopecten* larvae (Oyster Research Center, 1965, unpublished).

range between 10°C and 15°C is better suited, and 12°C seems to be the optimum. The best salinity is between 30.0 and 40.0°/₀₀ with 37.0°/₀₀ the optimum. Within 2 weeks of embryonic development, the shell length attains 120 μ and the right and left valves become asymmetrical. When the shells enter the umbo stage and attain a size of 150 μ, the rate of mortality is very high. Once the embryo passes this stage, the growth becomes rapid (Figure 6.16). When the embryo attains the size of 230 to 240 μ, the foot develops and the embryo enters the stage of setting. The larvae which are fully grown and capable of setting differ in size according to the temperature of the rearing water. When the water temperature is around 10°C, larvae with a shell length of 300 to 350 μ enter a sessile life, and at 13 to 15°C, larvae of 260 to 290 μ shell length enter a sessile existence. According to Ito *et al.* (1965), larvae with a mean shell length of 249.9 μ enter a sessile life at 13 to 17°C water temperature. When the larvae commence their sedentary life, a very thin and transparent conchiolin film forms around the shell, except at the vertex. This is what is known as the spat shell. The formation of a characteristic auricular-shaped adult shell is seen when the shell length of the larvae is 1 mm. The radiating ribs of the shells are formed one by one when the shell length becomes 3 mm. The sessile life ends when the size of the larvae increases to 6 to 10 mm, at which point the animals enter the next stage, i.e. become bottom

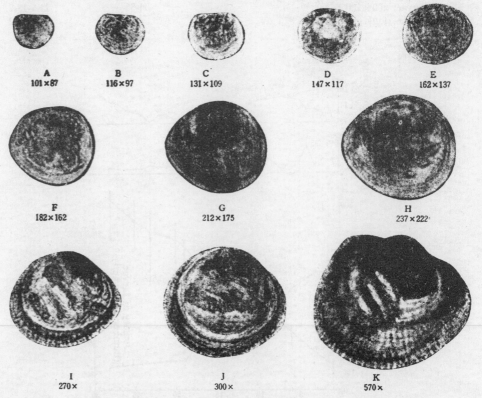

Figure 6.17. Early growth of *Patinopecten* (Oyster Research Center, 1965, unpublished).
A to H—planktonic larvae; I to K—sessile spat. (Unit, μ).

dwellers. But when the larvae begin their bottom-dwelling life differs according to the environmental conditions. If the environmental conditions are undisturbed, larvae as large as 2 to 3 cm have been found in a sessile state (Figure 6.17).

6.2 Larvae of Anadara

The eggs of *Anadara* are circular in shape, light pink in color, and vary in size from 56 to 65 μ, with an average of 61 μ. The head of the spermatozoan is oval in shape and measures 5 μ in length. About 45 minutes after fertilization at 22 to 25°C water temperature, the first polar body is released. The polar lobe forms after the second polar body is released, and the first cleavage occurs 1 hour 30 minutes after fertilization.

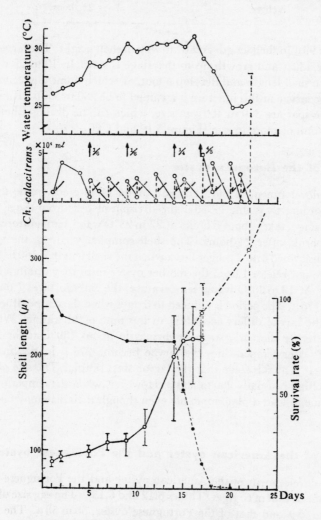

Figure 6.18. Artificial seeding of *Anadara* (Kanno, unpublished).

↑— amount of water exchange;   o—o—planktonic larvae;   o- - -o—sessile spat.

Gastrulation begins 7 hours after fertilization, the trochophore forms 15 hours later, and the early veliger form 22 hours thereafter (Table 6.11). The D-type larvae

Table 6.11. Early development of *Anadara*
(water temperature 22-25°C, salinity 17‰)

Course of development	Time after fertilization
Formation of 1st polar body	about 45 minutes
1st cleavage	about 1.5 hours
2nd cleavage	about 2 hours
Formation of gastrula	about 7 hours
Trochophore	about 15 hours
Veliger	about 22 hours

measure 83 to 99 μ in shell length and 67 to 74 μ in shell height. The larvae at the umbo stage measure 130 μ and growth up to this stage is slow. In 19 to 23 days the larvae attain 190 μ in shell length and develop a foot, at which point they enter a sessile life. The size of the larvae at this juncture is assumed to be 230 to 300 μ. Radiating ribs are seen when the spat are 360 to 400 μ in size, which can be distinguished clearly when they attain 1 mm or more in size (Figure 6.19 and 6.20).

6.3 Larvae of the European oyster

The early development of the European oyster takes place inside the body of the mother. According to Walne (1965), the division of cells from the 4-celled stage to the 16-celled stage, takes about 6 hours at 22 to 23°C water temperature. The embryo becomes a morula after 24 hours. The shell completely covers the velum and the embryo is transformed into a veliger larva when the shell length is 160 to 170 μ. Usually the larvae are not released from the mother oyster until they attain a shell length of 170 to 180 μ. At 19 to 20°C water temperature, the embryo takes 7 days to attain a size of 160 to 170 μ. The period is reduced to 6 days when the temperature is 22 to 23°C. The size of the larvae differs according to the time of liberation. While the size is 200 to 210 μ during the early stage, it shrinks to about 180 μ at the late stage. At 18 to 20°C of water temperature, the larvae become 270 μ in size in 15 to 16 days after release. It is at this stage that the larvae start setting. The size of the larvae at the time of setting is usually 280 to 300 μ. However, when the temperature is low, the larvae continue to lead a planktonic life even though their size may be 310 μ (Figure 6.21).

6.4 Larvae of the American oyster and the Portuguese oyster

The early development of the American oyster and the Portuguese oyster is more or less the same as that of *C. gigas* (Tables 6.12 and 6.13). The egg size of the American oyster is 50 to 55 μ and that of the Portuguese oyster, 50 to 58 μ. The embryo grows into a D-type larvae 24 hours after fertilization, with a shell length of 68 to 75 μ in the American oyster, and 75 μ in the Portuguese. The vertex develops when the shell

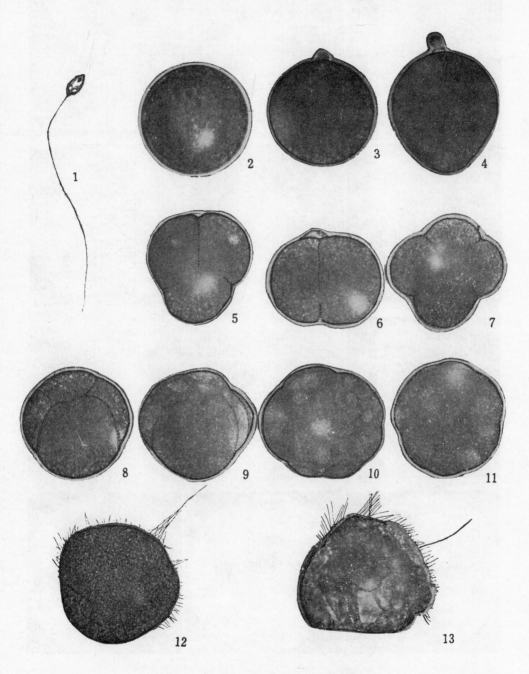

Figure 6.19. Early development of *Anadara*.

1—spermatozoan; 2—mature ovum; 3—release of 1st polar body; 4—formation of polar lobe;
5 and 6—1st cleavage; 7 and 8—2nd cleavage; 9 to 11—multicellular stage; 12—trochophore;
13—veliger.

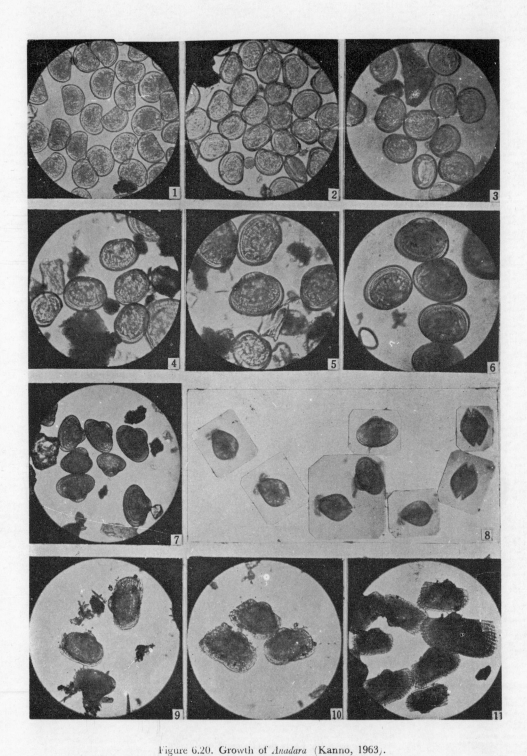

Figure 6.20. Growth of *Anadara* (Kanno, 1963).

1—D-type veliger larvae, shell length 83-99 μ, shell height 67-74 μ; 2 to 4—D-type veliger larvae; 5 and 6—larvae in umbo stage; 7—larvae just before setting; 8—larvae on the setting substratum; 9 to 11—spat.

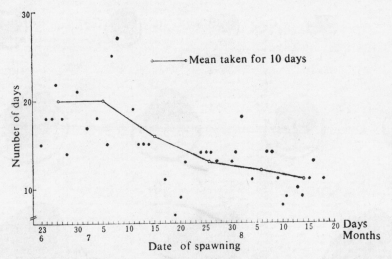

Figure 6.21. Relation between the date of spawning and the number of days taken by the larvae to settle in the European oyster (Oyster Research Center, 1962, unpublished).

Table 6.12. Early development of the American oyster (water temperature, 23-25°C) (Galtsoff, 1964)

Course of development	Time after fertilization
Formation of fertilization membrane	5 minutes
Release of 1st polar body	40 minutes
Release of 2nd polar body	1 hour, 10 minutes
1st cleavage	1 hour, 12 minutes
2nd cleavage	...
3rd cleavage	2 hours, 10 minutes
Morula	...
Gastrula	4 hours
Trochophore	5 hours

Table 6.13. Early development of the Portuguese oyster (water temperature, 23-24°C) (Oyster Research Center, 1965, unpublished)

Course of development	Time after fertilization
Release of 1st polar body	40 minutes
1st cleavage	1 hour, 20 minutes
2nd cleavage	1 hour, 40 minutes
Morula	3 hours
Gastrula	5 hours, 20 minutes
Trochophore	6 hours, 20 minutes
Veliger	20 hours

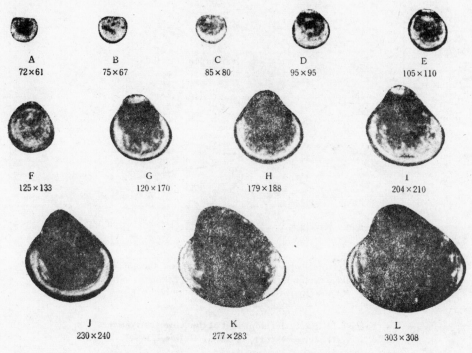

Figure 6.22. Early development of the American oyster
(Loosanoff and Davis, 1963).

length increases to 95 to 100 μ and, at this stage, the heightwise growth of the shell is much more rapid than its lengthwise growth. When the shell length reaches 270 μ, the eye spot can be seen, at which stage the larva begins to set. The shell length of setting larvae is 275 to 315 μ in the American oyster and 304±9.2 μ in the Portuguese (Figure 6.22).

(Nishikawa)

Bibliography

1. DAVIS, H. C. and GUILLARD, R. R., 1958. Relative Value of Ten Genera of Micro-organisms as Foods for Oyster and Clam Larvae. *Fish. Bull.*, **136** (58), 293-304.
2. GALTSOFF, P. S., 1930. The role of chemical stimulation in spawning of *O. virginica* and *C. gigas*. *Proc. Nat. Acad. Sci.* **16**, 555-559.
3. ——, 1933. Physiology of Reproduction of *O. virginica*. I. Spawning Reactions of the Female and Male. *Biol. Bull.* **74** (3), 461-489.
4. IMAI, T., S. SAKAI, and T. OKADA, 1954. Breeding of the Olympia Oyster in Tanks and Culture Experiments in Japanese Water. *Tohoku J. Agr. Res.*, **5**(1), 13-25.
5. ——, M. HATANAKA, R. SATO, S. SAKAI, and R. YUKI, 1950. Artificial Breeding of Oyster in Tanks. *Ibid.*, **1** (1), 69-86.
6. ——, ——, ——, ——, ——, 1953. Tank Breeding of the Japanese Surf Clam, *Mactra sachatinensis* Schrenk. *Sci. Rep. Res. Inst. Tohoku Univ.* D-**4**, 121-131.
7. ITO, S., S. KANNO and S. AKABOSHI, 1965. Hotategai no jinkosaibyo shiken (Experiment on Artificial Seedlings of *Patinopecten*), Mutsu-Wan *Suizoken. Shiryo* S. 40-No. 1.
8. KANNO, H., 1962. Ondo no hanpuku shigeki ni yoru kairui no sanran yuhatsu (Spawning of Shellfish Induced by Repeated Heat Stimuli), *Tohoku Suiken Kenkyu Hokoku (Bulletin of Tohoku Regional Fisheries Research Laboratory)*, (20), 114-120.
9. ——, 1963. Akagai no suiso saibyo (Preparation of the Seedlings of *Anadara bronghtoni* (Schrenk) in Water Tanks). *Ibid.*, (23), 108-116.
10. ——, 1965. Hotsukigai no suiso saibyo ni kansuru nisan no jikken (Experiments on the Preparation of Seedlings *Spisula sachalinensis* (Schrenk). *Ibid.*, (25), 131-141.
11. ——, 1965. Nimaigai no shubyo seisan ni kansuru kenkyu. Suisan seibutsu no zoyoshoku gijutsu no kakuritsu ni kansuru shiken kenkyu (On the Preparation of Seedlings of Bivalves. Investigations on the Culture Technique of Marine Organisms). Report from Fisheries Department, Govt. of Japan. Print, 69-79.
12. ——, S. AKAHOSHI, T. SASAKI, S. NAKAO, and T. TAMURA, 1965. Akagai no zoshoku ni kansuru kenkyu (On the Cultivation of *Anadara*), Research Report of Mutsu Bay Aquaculture Research Station, Aomori Prefecture, (10), 75-148.
13. KINOSHITA, T., S. SHIBUTANI, and J. SHIMIZU, 1943. Hotategai no sanran yuhatsu ni kansuru shiken (Yoho) [Experiments on the inducement of spawning in *Pecten yessoensis* Jay (Preliminary Report)]. *Nissuikaishi (Bulletin of the Japanese Society of the Scientific Fisheries)*, **11** (6-5), 168-170.
14. KORRINGA, P., 1952. Recent Advances in Oyster Biology. *Quart. Rev. Biol.* **27**, 266-308.
15. LOOSANOFF, V. L., 1945. Precocious gonad development in oysters induced in midwinter by high temperature. *Science*, **102**, 124-125.

16. Loosanoff, V. L., and H. C. Davis, 1952. Temperature Requirements for Maturation of Gonads of Northern Oysters. *Biol. Bull.*, **103** (1), 80-96.

17. ——, ——, 1963. Rearing of Bivalve Mollusks. Advances in Marine Biology **1**. 1-136. Academic Press, London, New York.

18. ——, ——, and P. E. Chanley, 1953. Behavior of Clam Larvae in Different Concentrations of Food Organisms. *Anat. Rec.* **117**, 586-587.

19. Takeda, K., 1966. Akagai fuyu yosei ni taisuru shiryo shiken (Feeding Experiments on the Planktonic Larvae of *Anadara*). Bulletin of Mutsu Bay Aquaculture Research Station, Aomori Prefecture, Print.

20. Walne, P. R., 1965. Observation on the Influence of Food Supply and Temperature on the Feeding and Growth of Larvae of *Ostrea edulis*. (L.). *Fish Invest. Lond.*, Ser, II, **24** (1), 1-45.

21. ——, 1966. Experiments in the Large-scale culture of the Larvae of *Ostrea edulis* (L.) *Ibid.*, **25** (4), 1-53.

22. ——, 1958. The Importance of Bacteria in Laboratory Experiments on Rearing the Larvae of *Ostrea edulis* (L.). *J. mar. biol. Ass.* U.K. **37**, 415-425.

23. Yamamoto, G., and U. Nishioka (1943). Jinko jusei ni yoru hotategai no hassei ni tsuite (Development of *Pecten yessoensis* through Artificial Fertilization). *Nissuikaishi (Bulletin of the Japanese Society of the Scientific Fisheries)*. **11** (5-6), 219.

24. ——, 1964. Mutsu-Wan ni okeru hotategai zoshoku (Cultivation of *Pecten yessoensis* of Mutsu Bay). Association for the Conservation of Aquatic Resources in Japan. *Suisan zoyoshoku Sosho*, 6.

25. ——, 1950. Mutsu Wan san hotategai no zoshoku ni kansuru kenkyu (Cultivation of *Pecten yessoensis* of Mutsu Bay). Report on the Survey of Fisheries Resources of Aomori Prefecture (10), 75-148.

26. ——, 1950. Ecological Note of the Spawning Cycle of the Scallop, *Pecten yessoensis* Jay, in Mutsu Bay, *Sci. Rep. Tohoku Univ. Ser. IV (Biol)*, **18**, 477-481.

27. ——, 1951. Induction of Spawning in the Scallop, *Pecten yessoensis* Jay. *Ibid.*, **19**, 7-10.

CHAPTER III

Artificial Seeding of Abalones

1. Development and metamorphosis

Abalones are one of the important marine food resources in Japan. Their cultivation was considered for a very long time and in Meiji 15 (1882) Matsubara studied their spawning by examining the eggs and sperms. Isahaya (1882) also tried to describe the spawning of abalones in simple terms. However, it was Kishigami (1894-1895) who described the structure of the reproductive organs of abalones, tried artificial fertilization, and carried out large-scale experiments at Shirahama in the Chiba Prefecture and Kawana in the Shizuoka Prefecture. Despite his best efforts to rear this animal, the results were not good. In 1935 Murayama collected fertilized eggs of *H. gigantea* and succeeded in rearing the larvae for six weeks to a shell length of 1 mm. Iino (1952) succeeded in the artificial fertilization of *H. discus,* and was able to rear the larvae up to a 2 cm shell length in a period of 13 months. During this time, the details of the life history of this animal became clearer, and Iino was encouraged to try seedling production on a large scale.

Abalones are usually bisexual. The gonads are found on the right posterior part of the body in the area covering the stomach and liver. The spawning period differs according to regions; Ono (1932) reported that in the Hokkaido abalone (*Haliotis discus hannai*) it occurs between late July and late November, with a slight difference in the range in the various areas of Hokkaido. Kinoshita (1950) reported the water temperature for spawning to be 20°C, and that spawning occurred between late July and early November in the Iwate Prefecture, and between early August and late October in the Miyagi Prefecture. While the abalones found in warm currents are said to delay spawning until November-December (Kishigami, 1894), according to the observations carried out by Iino (1952), it occurs between mid-October and late December, with the peak period in November. The spawning during this time begins when the mean temperature (for five days) of the water drops below 20°C. Hence the spawning conditions for these abalones resemble those of *H. discus hannai.*

Spawning in *H. discus* was successfully induced during mid-April to late April, which is actually the post-spawning period, at the Chiba Fisheries Experiment Station (1963-1964). The workers in Tokushima Fisheries Experiment Station (1966) also succeeded in inducing spawning during the period between spring and early summer, which is the period of luxuriant growth for diatoms. However, when a large-scale production of seedlings is planned, a number of factors have to be controlled to obtain the viable fertilized eggs soon after collecting the mother abalones from the sea during the spawning period. Since these factors are difficult to control, large-scale production of seedlings is often unsuccessful. Hence, the optimum conditions for obtaining mother abalones capable of quick spawning, and the rearing conditions, including acceleration of gonadal maturation, have to be studied carefully.

According to Hamada (1964), it is very important to provide sufficient feed to the mother abalones held during the spawning period. While it is possible to obtain good specimens by taking the eggs from the gonads of the mother which fed well before spawning it is not possible to obtain good eggs from mothers which did not feed as their protoplasm breaks up during treatment. In view of this fact, it has been pointed out that normal mature eggs cannot be obtained unless the mother abalone has taken sufficient food. Sakai (1964) has also indicated that gonadal maturation does not occur unless the mother is fed on algae such as *Laminaria japonica, Undaria pinnatifida,* etc. during storage. Although the amount of feed taken during the spawning period decreases slightly, Sakai considers it abnormal if the mother abalone does not take any food during the spawning period.

Shiraishi (1964) claims that the maturation of the gonads can be accelerated and spawning induced about two weeks earlier, when the mother abalones are reared under suitable conditions. Just before spawning, the mature eggs separate and accumulate in the renal cavity; the renal area gradually turns green in color from the collection of eggs in the cavity, and from this coloration the spawning period can be predicted (Hamada, 1964). Sperms and eggs are released through the inhalant siphon, which opens on the dorsal part of the branchial chamber. When the gonads are completely mature, the sperms or ova can be liberated by physical or chemical stimuli. Murayama (1935) was able to induce spawning by adding sperms from mature abalones to sea water in tanks in which mature females were kept. Iino (1952), on the other hand, succeeded in inducing spawning by either increasing the pH or by heat stimulation (Table 6.14). Since the inducement of spawning through heat stimulation is now commonly practiced, a number of experiments are mentioned here.

Table 6.14. Spawning "inducing experiment by thermal stimulus" in abalones (Iino, 1952)

Species	Date	Water temperature, (°C)		Time lapse up to spawning		Fertilization
		Natural	Increased	Female	Male	
H. discus	Nov. 19, 1942	18.0	22.7-23.5	110	80	+
,,	Nov. 11, 1943	18.6	27.0-27.5	90	65	+
,,	Nov. 5, 1944	20.2	27.8-29.0	120	150	—
,,	Nov. 6, 1944	19.8	27.4-28.2	135	120	—
H. sieboldii	Oct. 27, 1949	18.8	20.4-25.9	12	77	—
,,	Oct. 29, 1949	21.4	22.9-24.6	6	20	—

Fig. 6.23 is an example of thermal stimulus method practiced at the Tohoku Fisheries Experiment Station (Kikuchi, 1963). Uno (1963) was able to induce spawning on October 10 in *H. discus* and *H. sieboldii* by thermal stimulation. The water temperature at that time was about 22°C and was changed only by 2-4°C. Several successful cases have been reported where the stimulation was in the range of 0.06 to 0.72°C/min. In fact the results would be better if the variation is minimal in this range (Iwate

Fisheries Station, 1966). At the Oyster Research Center, absorbed solar heat was used, and the temperature of the sea water running through this pipe gradually in-

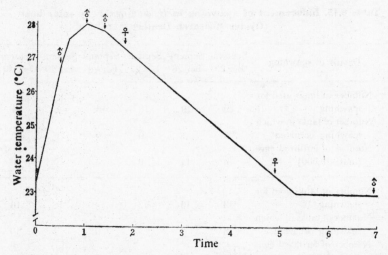

Figure 6.23. Inducement of spawning by thermal stimulus. Arrows in the figure indicate the time when the release of sperms and ova began. Each abalone released the gametes continuously for 10-20 minutes. Eight specimens each of male and female abalone were used in the experiment.

creased by 1.5 to 2.0°C. Since the temperature of the water decreased at night by 1.5 to 2.0°C, changes were made according to the rhythm of the daily changes in the surface water temperature under natural conditions. By increasing the range of

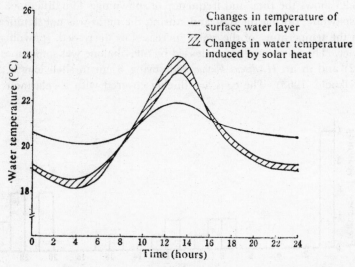

Figure 6.24. Changes in water temperature during the inducement of spawning when solar heat was used as the stimulus (Oyster Research Center).

changes in the water temperature by 3 to 4°C, it was possible to induce spawning (Figure 6.24, Table 6.15).

Table 6.15. Inducement of spawning in *H. discus hannai* **by solar heat**
(Oyster Research Center)

Method	Details of spawning	September 25	September 26	September 27	September 28	September 29	September 30	October 7
Control	Number of tanks used for spawning	19	19	18	18	6	·4	0
	Number of tanks in which spawning occurred	0	1	0	0	0	0	0
	Number of fertilized eggs (unit, 10,000)	0	44	0	0	0	0	0
Solar heat treatment	Number of tanks used for spawning	13	13	13	13	16	18	1
	Number of tanks in which spawning occurred	3	2	7	0	1	10	1
	Number of fertilized eggs (unit, 10,000)	244	493	836	0	0	1,446	200

There are reports claiming successful inducement of spawning by exposing the mother abalones to air; however, the data on the conditions of the exposure method are not yet adequate. Usually the spermatozoa are liberated earlier than the ova, in which case the rate of fertilization is high. It seems that the earlier release of spermatozoa, together with heat stimulation, induces spawning quicker than heat stimulation alone.

Figure 6.25 shows the time and frequency of spawning. The difference in the frequency of spawning during the day and during the night was not significant. However, when the temperature of the water increased or decreased, spawning was relatively heavier. The number of eggs released by one abalone was sometimes 1,400,000 (Iino, 1952) and in an *H. discus hannai* measuring 8 cm in shell length, 200,000 to 300,000 (Kikuchi, 1963). The eggs are thickly covered with a gelatinous substance

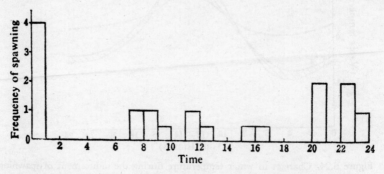

Figure 6.25. Relationship between the frequency of spawning and time.

and are perfectly round in shape. They have a tendency to sink but remain separate, i.e. they do not adhere to each other. When the eggs are forcibly ejected, they appear in the form of oval-shaped lumps. The size of the yolk is 0.18 mm for *H. discus, H. sieboldii*, and *H. gigantea*. The egg measures about 0.22 mm with membrane. The gelatinous substance which coats the membrane is 0.4 to 0.5 mm thick. The head of the spermatozoan is narrow and conical in shape and measures 8 μ in length; its tail is about 50 μ in length and helps the organism to move freely in the water.

Fertilization occurs in water and the size of the fertilized eggs is given in Table 6.16. According to Iino (1952) the fertilized eggs of *H. discus* produce the first polar

Table 6.16. Size of fertilized eggs of abalones (Iino, 1952)

Type	Fertilized egg size	Reference
H. discus	0.23 mm	Iino
H. sieboldii	0.28 mm	Iino
H. gigantea	0.27 mm	Murayama, 1935

body in 7 to 8 minutes after fertilization if the water temperature is 16 to 17°C. Soon after the second polar body is released. About 1 hour and 40 minutes later, the eggs begin to split. In 6 to 7 hours they become morula, and after 13 hours develop cilia and become trochophores. A trochophore larva engages in rotating movements inside the egg membrane and hatches after 20 hours. The size of a hatched trochophore is 0.17 mm in diameter and 0.23 mm in length. After hatching, the larvae move slowly toward the substratum. As the larvae require light during this phase, the unfertilized eggs should be removed and light rays used during water changes. In 27 to 28 hours,

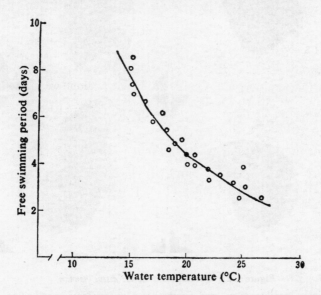

Figure 6.26. Relation between the free swimming period and the water temperature for the larvae of *H. discus hannai* (Oyster Research Center).

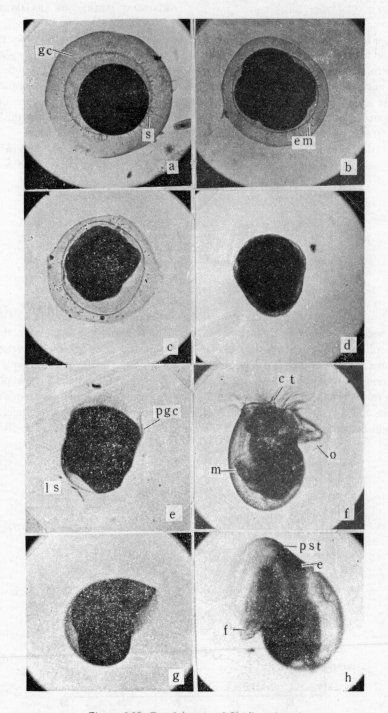

Figure 6.27. Development of *H. discus hannai*.

gc—gelatinous membrane; m—egg membrane; s—spermatozoa; pgc—ciliary band;
ls—larval shell; m—mantle; ct—cephalic tentacle; o—osphradium;
pst—peristomial shell; e—eye; f—foot.

the larvae transform into veliger and move in the water by means of cilia. From the 6th day onward, they begin to crawl either on the bottom or on the sides of the rearing tanks. When the veliger enter the sedentary phase, the cilia slough off and shells develop. Shell length increases to 2.3 to 2.5 mm in approximately 130 days, and the first inhalant siphon forms. At this stage the characteristics specific to abalones can be seen.

The course of embryonic development differs both according to animal type and rearing conditions. Kikuchi (1965) reported that cilia form in *H. discus hannai* 8 hours after fertilization, the larvae hatch after 10 hours, and the veliger phase is reached in 24. The veliger grow rapidly and enter a sessile life in 2 to 3 days. The number of days taken by the embryo from the time of fertilization to hatching, is closely related to the water temperature. Results of a study of this relationship are given in Figure 6.26. The first inhalant siphon forms in *H. discus hannai* in 6 weeks. The course of the larval development is shown in Figure 6.27.

2. Rearing and seedlings

The relation between the number of planktonic larvae reared and the survival rate, plays an important role in obtaining the maximum number of spat and, as shown in Figure 6.28, the number of larvae for *H. discus hannai* should be 100 to 200 larvae/liter. With reference to the suitable temperature range, Iino carried out elaborate investigations and found that in the larvae of *H. sieboldii*, which formed 35 hours after spawning, mass mortality occurred within 8 hours when the mean water temperature was 32.4°C, and within 24 hours when the temperature was 5.1°C. The larvae displayed considerable and uniform resistance when the mean water tem-

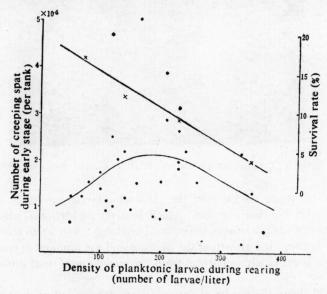

Figure 6.28. Optimum density and survival rate of planktonic larvae (Oyster Research Center).

perature was within the range of 7.8 to 28.0°C, particularly at 13.3°C or below (Iino, 1952).

With reference to salinity and the development of the larvae of *H. sieboldii*, Iino (1952) reported that 13.34 to 20.10°/$_{oo}$ was suitable. The development of the larvae of *H. discus hannai* was normal with a salinity of 19.16°/$_{oo}$, moderate at 16.15°/$_{oo}$, and differed in the latter through a slight abnormality in the foot. At 23.80°/$_{oo}$ larval development was retarded and the viscera showed degeneration. On the other hand, at 11.91°/$_{oo}$, too many abnormalities occurred; the head had separated from the shell in most specimens (Iwate Fisheries Experiment Station, 1966).

Feed during the planktonic larval stage has been discussed from various angles and the latest view is that the larvae should not be fed. However, since the larvae which grow rapidly enter a sessile life in 3 to 4 days, it would seem desirable to feed these at an early stage. Small *Navicula* sp., *Cocconeis* sp. (Kikuchi, 1964), *Amphora* sp. (Iwate Fisheries Experiment Station, 1966), *Platymonas* sp. (Shiraishi, 1964), etc. are said to be suitable as feed for larvae during the early creeping stage. Nonetheless, much research still needs to be done with regard to feed organisms.

When the spat have grown to 3 to 4 mm in shell length, large *Navicula, Licmophora*, etc. are given as feed. To accelerate growth in creeping spat, a special type of seedling collector must be used, on which diatoms can multiply easily. For this purpose, transparent or semitransparent, synthetic resin plates are used as seedling collectors (Figure 6.29). After allowing diatoms to grow on them, these collectors are introduced

Figure 6.29. Collectors for abalone seedlings (Oyster Research Center).

into the rearing water the moment the planktonic larvae are ready to adhere. The collectors attract the creeping spat, which feed on the diatoms adhering to them, as well as the walls of the rearing tanks, until attaining a 4 to 5 mm shell length. Thus the balance between the growth of the diatoms and the amount of feed consumed by the abalones, is extremely important from the viewpoint of spat growth and that of final yield.

Figure 6.30 shows the results of investigations carried out at the Oyster Research Center where the seedlings were grown in polyethylene tanks of about 1-ton capacity and an adhering surface of about 50,000 cm². According to the data, the survival rate is high when the number of spat during the initial stage of adherence is low.

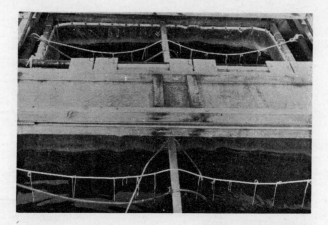

Figure 6.30. Polyethylene seedling tank (Oyster Research Center).

Nevertheless, the results were also good when 20,000 to 30,000 creeping spat were permitted to adhere in one tank, in order to obtain a larger number of spat of 5 mm shell length (Figure 6.31). The mortality rate of spat measuring 2 to 3 mm shell length

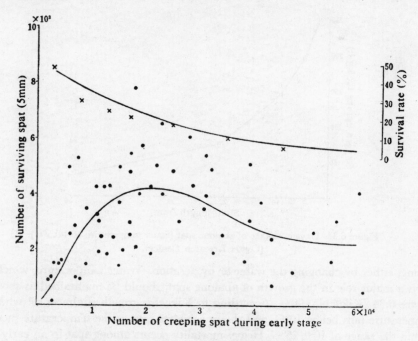

Figure 6.31. Number of creeping spat in early stage per seedling tank, number of spat surviving up to a growth of 5 mm shell length, and the final survival rate (Oyster Research Center).

was high, but mortalities rarely occurred among spat of more than 4 to 5 mm shell length when these are fed on seaweed (Figure 6.32).

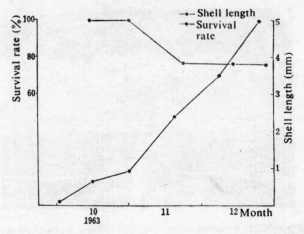

Figure 6.32. Growth and survival rate of spat (Oyster Research Center).

If the sea water is allowed to stagnate in the rearing tanks, mortalities are apt to occur. Since spat consume a much higher quantity of oxygen than adults (Figure 6.33), the rearing water must be maintained very close to the conditions of a natural

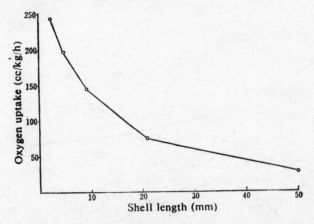

Figure 6.33. Oxygen uptake of abalone spat (water temperature $16\pm0.5°C$) (Oyster Research Center).

setting, either by changing the water or by aeration. Water temperature, which also plays a major role in the growth of abalone spat, should be maintained as shown in Figure 6.34 (Kikuchi, 1963). According to Kikuchi, growth is almost nil when the temperature falls below 10°C, and shows a higher rate as the temperature increases within the range of 10 to 25°C. Heavy mortality occurs among spat in an early stage of growth when the temperature falls below 2°C. When the spat attain a shell length of 4 to 5 mm, they are able to feed upon seaweed such as *Ulva pertusa, Endarachne binghamie,* etc.

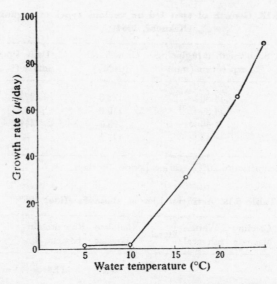

Figure 6.34. Relation between growth rate of *H. discus hannai* spat
(1-2 mm) and water temperature (Kikuchi, 1963).

Experiments in which the spat of *H. discus* were reared to a 1 cm shell length on a
diet of *Endarachne binghamie,* yielded the results shown in Table 6.17 (Iino, 1966).
Seaweed was given at the rate of 0.08 to 0.14 g per spat per day. Kikuchi (1964) also
carried out feeding experiments on *H. discus hannai* and found that spat growth differs

Table 6.17. Intake of *Endarachne binghamie* **by** *H. discus* **spat (Iino, 1966)**

Period	Mean shell length (cm)		Mean weight (g)		Gross weight (g)		Number of spat		Amount of diet (g)	Meat content
	Beginning	End	Beginning	End	Beginning	End	Beginning	End		
May 20	1.04	1.89	0.17	0.77	3.77	14.8	21	19	162.9	14.76
↓	0.86	1.25	0.09	0.21	4.05	8.1	41	37	118.4	29.24
July 20	1.16	1.61	0.21	0.51	1.30	2.58	6	5	34.2	26.71
July 20 ↓ September 20	1.96	2.13	0.83	1.18	15.93	21.27	19	18	168.2	31.49
	1.05	1.61	0.13	0.50	2.30	7.50	17	15	82.5	15.86

according to the diet (Table 6.18). Iino (1963, 1964) tried an artificial diet with the
same spat and obtained the results shown in Table 6.20. The composition of this
artificial diet is given in Table 6.19. Iino satisfactorily proved that spat can be grown
on an artificial diet prepared from white fish meal, starch, cellulose powder, and a
mixture of vitamins in the form of a gel of calcium algenate. His data also established

**Table 6.18. Growth of spat fed on various types of organisms
(Kikuchi, 1964)**

Diet	Shell length at beginning of experiment (mm)	Growth rate (μ/day)	Daily food intake (%)	Conversion rate (%)
Undaria pinnatifida	14.50	123.0	27.6	8.94
Enteromorpha instestir	14.55	110.0	16.6	14.18
Eisenia bicyclis diatoms	14.40	98.7	41.2	13.25
Adhering diatoms	14.45	118.3	—	—

Remarks: Water temperature 20°C, rearing period, 33 days.

Table 6.19. Artificial diet of abalones (Iino, 1963)

Group	Dry *Undaria pinnatifida*	Calcium algenate	White fish meal	Starch	Cellulose powder	Raw protein content, %	Comments
A	100					11.4	Various vitamins were included in groups B,
B	10	45	40	5		27.1	C, and D
C		60	20	15	5	11.5	Fish meal contained 7.7%
D		20	60	15	5	32.1	ash and 4.7% fat.

Table 6.20. Results of rearing spat on an artificial diet (Iino, 1963)

Group	Mean weight (number of larvae) Before W_1 (mg)	Mean weight (number of larvae) After W_2 (mg)	Number of dead spat during course of experiment	Period of experiment	W_2/W_1	Color of shell
A	210 (5)	523 (4)	1	(August–January) 153 days	2.5	green
B	177 (5)	722 (3)	2	(August–January) 153 days	4.1	green
C	258 (7)	308 (5)	2	(November–January) 78 days	1.2	green
D	247 (7)	584 (6)	1	(November–January) 78 days	2.4	green

that the minimum requirement of protein for normal growth in abalone spat is about 20%.

The final production rate of spat per unit area should be carefully considered when rearing seedlings. In other words, the extent of increasing spat density should be determined by controlling the normal growth. According to Kikuchi (1964), the relationship between the density of *H. discus hannai* spat (shell length, 1.5 cm) reared in rearing baskets of $53 \times 35 \times 40$ cm, and the growth of the spat when they were fed on seaweed such as *Eisenia bicyclis, Undaria pinnatifida, Ulva pertusa, Enteromorpha intestir,*